ELECTRONIC MEDIA

Electronic Media: Then, Now, and Later provides a synopsis of the beginnings of electronic media in broadcasting and the subsequent advancements into digital media. The Then, Now, and Later approach focuses on how past innovations laid the groundwork for changing trends in technology, providing the opportunity and demand for evolution in both broadcasting and digital media. An updated companion Web site provides links to additional resources, chapter summaries, study guides and practice quizzes, instructor materials, and more. This new edition features two new chapters: one on social media and one on personalizing the entertainment and information experience.

- The Then, Now, and Later thematic structure of the book helps instructors draw parallels (and contrasts) between media history and current events, which helps get students more engaged with the material.
- The book is known for its clear, concise, readable, and engaging writing style, which students and instructors alike appreciate.
- The companion Web site is updated and offers materials for instructors (an IM, PowerPoint slides, and test bank).

Norman J. Medoff, Ph.D., is Director of the School of Communication at Northern Arizona University. He has taught and served as an administrator at three different universities, produced numerous television and corporate video projects, and overseen the productions of many students. Dr. Medoff has authored articles in scholarly journals as well as trade and consumer magazines. He has also written textbooks on the Internet and mass media, television production, and electronic media. He has been the recipient of a Fulbright Specialist Grant. In addition, he is a past president of the Broadcast Education Association.

Barbara K. Kaye (Ph.D., Florida State University) is Professor in the School of Journalism & Electronic Media at the University of Tennessee–Knoxville. Her research interests are in the areas of media effects and consumer uses of new communication technologies. She examines how the emergence of the Internet, blogs, social media, and new television program delivery platforms influences political attitudes and how they have changed media use behavior. She also studies the uses and effects of profanity on broadcast and cable television programs. She has co-authored five textbooks, published more than 60 journal articles and book chapters, and taught in Italy and Austria.

ELECTRONIC MEDIA
Then, Now, and Later
THIRD EDITION

Norman J. Medoff, Ph.D.
and Barbara K. Kaye, Ph.D.

Routledge
Taylor & Francis Group

NEW YORK AND LONDON

This Edition published 2017
by Routledge
711 Third Avenue, New York, NY 10017

and by Routledge
2 Park Square, Milton Park, Abingdon, Oxon, OX14 4RN

Routledge is an imprint of the Taylor & Francis Group, an informa business

First edition published by Focal Press 2004
Second edition published by Focal Press 2010

Library of Congress Cataloging-in-Publication Data
Names: Medoff, Norman J., author. | Kaye, Barbara K., author.
Title: Electronic media : then, now and later / Norman J. Medoff, Ph.D. and Barbara K. Kaye, Ph.D.
Description: Third edition. | New York, NY : Routledge, 2016.
Identifiers: LCCN 2016016859 | ISBN 9781138903210 (hardback) | ISBN 9781138903203 (pbk.)
Subjects: LCSH: Broadcasting—History. | Mass media—History. | Digital media—History. |
  Telecommunication—History.
Classification: LCC HE8689.4 .M44 2016 | DDC 384—dc23
LC record available at https://lccn.loc.gov/2016016859

ISBN: [978-1-138-90321-0] (hbk)
ISBN: [978-1-138-90320-3] (pbk)
ISBN: [978-1-315-69703-1] (ebk)

Typeset in Giovanni
by Apex CoVantage, LLC
Printed and bound by CPI Group (UK) Ltd, Croydon, CR0 4YY

Dedication and Acknowledgments

This book is dedicated to B. Leigh Browning, Ph.D., 1964–2015, West Texas A & M University. Dr. Browning's help during the early stages of this third edition is much appreciated, as was the draft of the new chapter on social media she was working on at the time of her death. Thank you, Leigh.

We send a big thank you to Mary Beadle, Ph.D., John Carroll University, for all of her helpful suggestions and for writing Chapter 6, "Digital Devices: Up Close, Personal, and Customizable"; to Ross Helford for penning Chapter 12, "Feature Films: 'The Movies'"; and to Rebecca Lind, Ph.D., University of Illinois at Chicago, for creating the Instructor's Manual and Test Bank.

We would like to thank the following individuals with Routledge and Focal Press for initiating this edition and for moving it from inception to publication:

Linda Bathgate, Publisher
Ross Wagenhofer, Editor
Nicole Salazar, Editorial Assistant
Kathryn Morrissey, Editor
David Bevans, Editor
Brianna Bemel, Editor

We would also like to thank those individuals who reviewed the second edition of this book and made invaluable suggestions for this third edition: Mr. Roland Wong, Pima Community College; R. Glenn Cummins, Ph.D., Texas Tech University; and Mr. James Nadler, Ryerson University, Canada.

Our thanks also goes to the following individuals for providing personal information for the 'Career Tracks' feature: Trey Fabacher, Dan Hellie, John Montuori, Glenn Reynolds, Jay Renfroe, Ryan Kloberdanz, John Dille, Doug Drew, Norm Pattiz, Maria Hechanova, and Mark Casey.

Special thanks to our colleagues and friends for their thoughtful comments and help with various parts of the book: Gerald S. Adler, Alan Albarran, Paul Helford, Patrick Parsons, and David Pearson.

Thanks for outstanding research assistance from Samantha Boyd.

We also gratefully acknowledge the assistance of Carolyn Stewart at the MZTV Museum of Television in Toronto. Thanks for the wonderful images.

As always, thanks to my wife, Lynn Medoff, for her patience with me while I paced around trying to find the best way of writing phrases without repeating myself. Thanks also to Natalie Medoff, and Sarah, Jeremy, and Fox Clover, who are always mildly amused at my willingness to write books.

Norman J. Medoff

Hugs and kisses to my husband, Jim McOmber, for keeping me supplied with gallons of Starbucks' iced tea whenever I needed a caffeine jolt to keep writing. Hugs and pets to my funny cat, Jacky Paper, who kept me laughing and wrote parts of this book by simply walking across the keyboard. And, of course, love to my mom.

Barb K. Kaye

Contents

Preface

It was about 11 p.m. on a December evening in Flagstaff, Arizona. I was sitting in bed reading a novel on my iPad. Suddenly my bed began to shake and even roll a bit. At first I assumed that my small dog, who was sleeping under the bed, was having a nightmare and was moving around to get comfortable. The shaking continued, and my dog is really too small to cause the bed to move at all. I looked out the window but didn't see a large truck outside or any disturbance of any kind. It finally dawned on me that we were experiencing an earthquake.

In years past I would have quickly turned on my radio to hear the latest news about what happened and the magnitude of the quake. But now, instead I conveniently switched from the Kindle app on my iPad to my Facebook app. Within 5 minutes of the earthquake there were five posts on my news feed from locals who described what they experienced during the quake. One of the posts had a link to the National Oceanic and Atmospheric Administration (NOAA) Web site with exact information about the epicenter of the tremor and its strength.

My earthquake experience epitomizes how our information-seeking behavior has changed. Radio and television stations are no longer our 'go-to' sources for up-to-the-minute information. Instead, our first instinct is to access breaking news through our computers, tablets, or laptops. Our world of electronic media has changed dramatically since the Internet and World Wide Web first gave us rapid access to text, graphics, sound, and even video. And the tools we use are getting smaller—we no longer need bulky desktops computers—and laptops are lighter in weight than ever before, plus tablets and smartphones fit into purses and pockets. Although smartphones are only used by about 70% of the population, that number will increase to the point where the term 'smart' is no longer needed. All phones will have easy access to the Internet; all phones will be 'smart.'

Until about 20 years ago, teaching an introductory course about electronic media meant teaching the history, structure, economics, content, and regulation of *broadcasting*. Broadcasting and broadcasters were at the epicenter of all that was electronic media. In fact, the concept of a world of electronic media that didn't revolve around broadcasting and that wasn't based on the traditional mass communication model seemed far away and abstract. But much has changed dramatically in the past two or so decades. Today, students live in a nonlinear, digital world in which traditional broadcasting plays a greatly diminished role. For example, students no longer wait for over-the-air radio to play new music or favorite tunes. The Internet provides multiple streams of music, much of which can be shared and downloaded for future playback on computers or mobile devices. College classroom buildings and dorm rooms provide broadband Internet, which facilitates social and professional networking, and education, music, and movie streaming.

Furthermore, entertainment and leisure activities have changed. Gone are the multicomponent stereo systems and small-screen television sets that have been popular since the 1960s. Instead, students are designing their own entertainment and information systems by selecting among a variety of compact, portable MP3 devices, digital televisions, handheld devices, computers, and smartphones. Students handle, edit, and store media content for their personal use. Finally, they're also becoming online content providers through social networks like Facebook, Instagram, and Twitter.

But despite all of these developments, traditional electronic media are still worth discussing and the predigital world is not irrelevant or obsolete, because what happened before made what is happening today possible.

Electronic Media: Then, Now, and Later is rooted in the notion that studying the past not only facilitates understanding of the present, but also helps predict the future. Just as we can show how broadcast television spawned the cable industry, we can trace how the cable industry led to the satellite industry and how both have led to a digital world—one in which convergence has blurred the lines separating media functions and in which old-style broadcasters have expanded, consolidated, and adapted to the multiplatform system of contemporary electronic media.

The study of electronic media should address more than just the delivery systems used to reach mass audiences. Personal electronic devices that deliver information and entertainment selected by individual consumers should be covered as well. Devices such as smartphones and tablets—which allow users to surf the Internet, record and send video images, play music, and interpersonally communicate with voice or text—have changed the modern lifestyle to the point that they must be included in any discussion of the digital electronic media revolution. Digital video recorders have changed how audiences schedule their television viewing time, making the television network concept of scheduled viewing somewhat anachronistic. Online connections open the world to on the go, anytime entertainment.

This book links the traditional world of broadcasting and the contemporary universe of digital electronic media, which offers increasingly greater control over listening, viewing, and electronic interaction. As both emerging electronic media professionals and discriminating electronic media consumers, today's students must know about these changes and understand technology's enormous cultural impact and how newer digital devices and functions will affect the future of the industry.

ORGANIZATION OF THE TEXT

With the knowledge that what comes next is based on what came before, we would like to acknowledge Edward R. Murrow and his programs *Hear It Now* (1950–1951) and *See It Now* (1951–1958) for suggesting the structure of this text. Each chapter of the book is organized chronologically into these sections:

- *See It Then* begins with the invention or inception of the topic (e.g., television) and traces its development up to the Telecommunications Act of 1996 and into the new millennium.
- *See It Now* discusses activities and developments from about the year 2000 to the present.
- *See It Later* starts with the present and makes general predictions about what will happen in the digital world of tomorrow.

Chapter 1 summarizes the history of electronic media, introduces industry terms, and discusses current trends in media, such as digitization, convergence, and consolidation. Chapter 2 covers the history of radio from electrical telegraphy to satellite and Internet delivery. The chapter includes biographies of early radio pioneers and delves into laws and regulations regarding the transmission of radio content. Chapter 3 focuses on the development of television from the early days of broadcasting to cable, satellite, and online delivery systems. Chapter 4 concentrates on radio and television programs and programing. This chapter explains radio station formats and the various types of television program genres, such as dramas and situation comedies, and includes program origination and how shows make it to the 'airwaves.' Chapter 5 moves away from radio and television and focuses on the Internet. The chapter provides a history of the Internet—how it developed, how it works, and how it is used. The chapter specifically looks at the Internet from a media perspective and how it is used for communication and information delivery, whether in text, graphic, audio, or visual forms.

Chapter 6 is new to the third edition of this textbook. The chapter is titled "Up Close and Personal" to describe newer mobile and customizable technologies and platforms, such as cell phones, tablets, Netflix, Hulu, Sling TV, satellite radio, Pandora, and DVRs, through which users design their own communication and viewing experience. Media is a business, and so Chapter 7 is about advertising—the primary way media earn revenue. The chapter begins with an overview of advertising and the need to advertise. It then explains the development of radio and television commercials and moves into contemporary and online advertising, including social media, and discusses consumer targeting and tracking. Chapter 8 is an extension of the advertising chapter in the sense that by selling advertising time and online space the media are actually selling their viewers, listeners, and users to advertisers. But for the media to know how much to charge advertisers, they must know who is watching. The chapter, then, covers how diaries, meters, website cookies, and monitoring software are used to count the number of viewers, listeners, and online users. The chapter also considers the complex relationships among a medium, its audience, and its advertisers. Chapter 9, "Social Media," is another new chapter for this third edition. From 2011 when the second edition of this text was published until now, the number of social media sites and the number of users has soared. Social media are in the forefront of news dissemination and personal communication and thus deserve a chapter of their own. This chapter starts with a history of social media, then focuses on the 10 most-used social media sites. It concludes with discussion of the personal and social benefits and consequences of social media. Chapters 10 and 11 are both about the business of media. The chapters cover business models, ownership structures, and operations of the various types of media and delivery systems. Chapter 10 is more broadly focused, whereas Chapter 11 concentrates on radio and television stations, cable and satellite companies, and content distribution. Chapter 12 is about the film industry, which has strong ties to other forms of electronic media and to the Internet. The chapter is all about 'Hollywood'—the studio system, film creation, origination, distribution, and movie stars. Chapter 13 is the last chapter, but in no way less important than the others, and some could argue more important because it examines the social, cultural, and personal effects of media. The chapter discusses various effects theories and describes how mediated violence and positive and negative depictions affect audience behavior, emotions, and thought processes.

Opting Into Today's Media   1

Contents

Can you think of a recent day when you did not use some form of electronic media? It would have been a day when you were not at home, not in your car, not on campus, not in a hotel, and not in a fitness club, grocery store, coffee shop, or shopping mall. It would also have been a day when you did not watch television, listen to the radio, or use your cell phone or Internet-connected computer—a day that you did not use Google or Twitter or Facebook.

Chances are good that you cannot remember a day like that. It is almost impossible to totally escape electronic media in today's world. Signals from broadcast stations, satellites, and wireless Internet connections are pervasive yet invisible.

Electronic media is an integral part of everyday life. We use the media to learn about the world. We tune in to radio or television or click on links for world news, local traffic and weather reports, stock market analyses, and many other bits of information. We also use the media simply for entertainment. The *Super Bowl*, *The Simpsons*, *The Walking Dead*, *American Idol*, *The Big Bang Theory*, a Top 40 radio countdown show, and streaming audio and video programs entertain us for hours on end. Electronic media provide us with messages that influence and affect us in many ways. Our media world has changed from one of limited choices to a world of information and entertainment overload. It is a world of media and information

abundance that will not revert to a world of few media choices.

This book 'opts in' and examines the many aspects of electronic media. It investigates the history, structure, delivery systems, economics, content, operations, regulation, and ethics of electronic media from the perspectives of what happened in the past (See It Then), what is happening now (See It Now), and what might happen in the future (See It Later).

SEE IT THEN

ORIGINS OF ELECTRONIC MEDIA

The desire to communicate is a part of being human. We have always needed to express ourselves, but it took a long time before we could do so successfully. About 100,000 years ago, humans developed the ability to communicate using speech. About 40,000 years ago, humans drew pictures on the walls of caves.

Writing came into use about 5,000 to 6,000 years ago. With written language, we no longer had to rely solely on memory.

As early as 4000 BCE, people were writing on clay tablets using pictograms, a simple picture to symbolize an

FIG. 1.1 A 1,500-year-old cave painting from South Africa
Photo courtesy of iStockphoto. © Skilpad, image #10277331

FIG. 1.2 Hieroglyphics from inside a temple in Egypt
Photo courtesy of iStockphoto. © Tjanze, image #10353358

object or word, to communicate some basic information about everyday events. These tablets were portable and durable records. One thousand years later, about 3000 BCE, the Egyptians used the fibrous plant papyrus as a type of primitive paper. At the time, a form of picture writing called **hieroglyphics** evolved. About 2000 BCE, the Egyptians developed an alphabet of 24 characters. In the western United States, early Native Americans carved pictographs in rocks to show others what they saw and how they lived their lives.

Through the ages, various systems, such as smoke signals, semaphores (flags), and pigeons, have been used to send messages quickly and effectively. Each system worked when the conditions were just right, but each had its limitations. For instance, smoke signals and semaphore

FIG. 1.3 Petroglyphs from cultures dating back to 1500
Photo courtesy of Getty image #172711400

FIG. 1.4 The first printing press was built in the fifteenth century

systems did not work at night because they depended on sunlight for the receiver to see the signal. Pigeons could carry messages but were susceptible to natural predators and severe weather. Messengers were slow and could be captured during times of conflict or war.

The ability to create a permanent written record was held by persons with special talents who were known as scribes. Scribes learned to read and write and serve the government and businesses by recording the information using written language. They were present at many transactions both public and private to record what happened and what was involved, such as the exchange of goods and services for money. Scribes were an important part of business and government until the advent of printing.

In the middle of the fifteenth century, Johannes Gutenberg, a metalworker in Europe, developed a system to print multiple copies of an original page using a system of movable type. Using a modified wine press, Gutenberg printed pages by coating hand-carved three-dimensional alphabetic letters with ink and pressing them onto paper. The result was a printed page that could be duplicated many times. For the first time, *one* individual with a printing press could reach *many* people with print instead of handwritten copies of a book or manuscript, but more important, the printing press could make many more copies and did so much more quickly than was possible for even the most prolific scribe.

In 1844, Samuel F. B. Morse invented the telegraph, which used electricity to send messages through wires over long distances almost instantaneously. The telegraph sent messages from one source point to other points using a system of dots and dashes—short on/offs and long on/offs to spell out words one letter at a time—known as Morse code. The telegraph worked well as long as the distant point had the equipment and a skilled operator to receive and translate the coded message into words. One-to-one distance communication changed radically in 1876 when Alexander Graham Bell invented the telephone, a device that could be used by the ordinary person speaking into the mouthpiece. Both of these inventions were designed to facilitate **person-to-person** (or **one-to-one**) communication over distances.

As books and newspapers became popular, the practice of communicating to many people at once became common. This **one-to-many model** of communicating was not a balanced two-way model, however. The audience (the **many**) could possibly communicate back to the sender, but this communication, known as **feedback**, was limited. As such, the one-to-many model became known as **mass communication**. The mass media constitute the channel that uses a mechanical device (e.g., a printing press) or electronic device (e.g., broadcast transmitter) to deliver messages to a mass audience.

FYI: Communication Models

Shannon and Weaver Mathematical Model

Models are created to help us understand processes and concepts. Shannon and Weaver (1949) developed a model based on message transmission that helps explain the process of communicating. That model, also known as a *linear model*, works well to explain telephone communication.

The elements of that model are the *information source*, the *transmitter*, the *channel*, and the *destination*. The information source (a person) uses a transmitter (a telephone) to send a signal through a channel (telephone wires) that is received by a receiver (another telephone) and then heard at the destination (a person). In mass communication, the information source (say, a weathercaster at a television station) uses a broadcast television transmitter to send a signal using broadcast waves through the air (channel) that is received by a television receiver and then seen and heard by the viewer (destination). Additional concepts, such as noise that can interfere with the process, were added to the model to make it more generalizable.

Schramm–Osgood Communication Model

Schramm and Osgood (Schramm, 1954) used a simplified model to explain communication. Using only three basic elements—a *message*, an *encoder*, and a *decoder*—this model demonstrates the reciprocal nature of communication between two people or entities. It shows how communication is a two-way process in which the participants act as both senders and receivers of messages.

Schramm Mass Communication Model

In an attempt to create a model to explain mass communication, Schramm (1954) used one source to represent an *organization* that sends out *many identical messages* to the *audience* composed of many individual receivers, who are connected to groups of others and pass along information about the messages from the initial receiver. The dotted lines in the model represent **feedback** from the receivers, which is delayed and not explicit. The organization must then infer the meaning of the feedback (such as ratings for a program) and act accordingly.

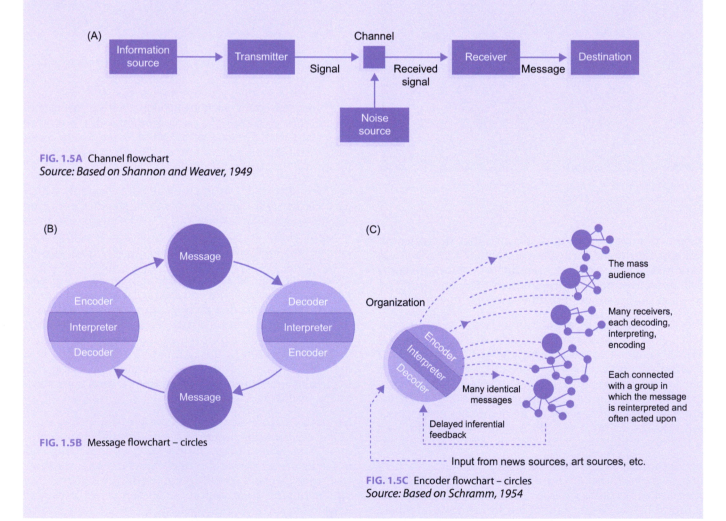

FIG. 1.5A Channel flowchart
Source: Based on Shannon and Weaver, 1949

FIG. 1.5B Message flowchart – circles

FIG. 1.5C Encoder flowchart – circles
Source: Based on Schramm, 1954

FYI: Communication Models (Continued)

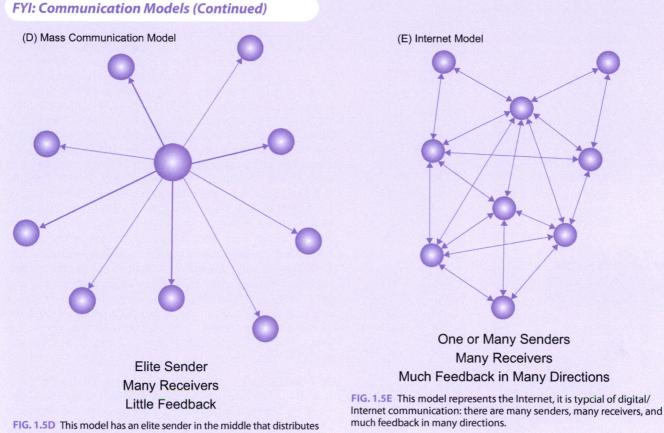

(D) Mass Communication Model

**Elite Sender
Many Receivers
Little Feedback**

FIG. 1.5D This model has an elite sender in the middle that distributes communication to many at once with little feedback from the audience to the sender. The model represents broadcasting or newspapers.

(E) Internet Model

**One or Many Senders
Many Receivers
Much Feedback in Many Directions**

FIG. 1.5E This model represents the Internet, it is typcial of digital/Internet communication: there are many senders, many receivers, and much feedback in many directions.

Although the concept of mediated mass communication was born in the 1400s with Gutenberg's printing press, it was not until 1690 that the notion was put into reality with the publication of *Publick Occurrences, Both Foreign and Domestick,* which was the first general-information pamphlet printed for mass consumption by the U.S. public. The pamphlet is thought of as the precursor to the modern-day newspaper. Its publisher stated, "the country shall be furnished once a month with an account of such considerable things as have arrived unto our notice" (National Humanities Center, 2004). Although the publisher had big plans, the government did not allow a second issue because the first was published without legal authority and because it contained salacious content—an item about incest in the French royal family. Other pamphlets/newspapers came and went, but finally in 1833, the *New York Sun* established itself as the first daily newspaper created for the mass audience. Publisher Benjamin Day set the price of his paper, the *New York Sun,* at one penny, hence the term the 'penny press.'

From 1833 on, the public was reached by print but not yet by sound. In the early 20th century, Guglielmo Marconi developed **radio telegraphy,** which sent a signal from a radio transmitter to a radio receiver, a system known as point-to-point delivery. Radio telegraphy worked in a similar fashion to Morse's telegraph but without the wires. Soon after the advent of radio telegraphy, other inventors produced a system for transmitting the human voice and other sounds, such as music. Radio signaled the beginning of broadcasting and eventually the start of commercial electronic media. Newspapers, magazines, civic clubs, and even schools promoted radio and stimulated interest in the new medium. In the late 1920s, the fascination with radio grew as sophisticated music and other entertainment programs hit the airwaves.

Radio enjoyed its place as the only instantaneous and electronic medium for more than 30 years. During this time, it developed most of the programming formats (some of which were later used for television), enjoyed financial success, and was a mainstay in American culture. Radio's stature changed after World War II, when television broadcasting got off to a roaring start. Many of the popular shows on network radio shifted over to

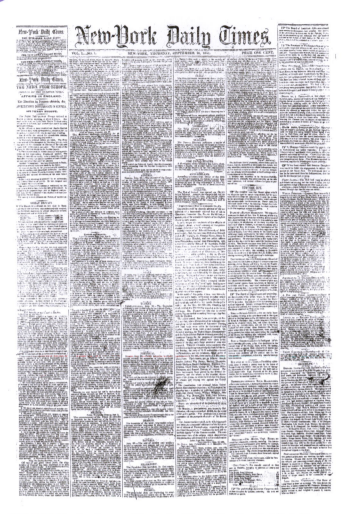

FIG. 1.6A The front page of the first edition of the *New York Daily Times*, which sold for a penny in 1851. The *New York Daily Times* eventually became the *New York Times*.

television, providing the new medium with an audience already familiar with the program.

Since the early 1950s, television has been a media powerhouse, dominating the national audience. Television is still in high demand by advertisers, who often have to wait in line to buy time on desirable network shows. Beginning in 1946 and lasting into the 1980s, the three major networks—ABC, CBS, and NBC—claimed almost 90% of the national prime-time viewing audience. Since then, the viewing audience has moved to cable, satellite, and the Internet. Even so, the broadcast television industry remains a dominant force in the media world. The reason for the dominance is not because the broadcast networks have huge audience shares as they did in the past, but rather that the companies that own the networks are diversified corporate media powerhouses that own a variety of media properties, including the legacy media (newspapers, radio, magazines, television stations, movie studios) as well as new media (cable systems, Web sites that stream media, and other digital properties). As they have been doing since the late 1920s, the networks produce program content. The content generates revenue because it is shown on broadcast networks, cable channels, satellite systems, and the Web.

Television, in its many forms—broadcast, cable, satellite, DVD, and now online streaming sites like Hulu.com and Netflix—has been the center point of media for almost 70 years. But television is now being nudged aside by the information and entertainment-rich Internet. Although everyday use of the Internet is still quite young relative to other media, the Internet is a **paradigm breaker**—a medium that defies previous models of electronic media. It has grown more rapidly than any other medium in history.

FIG. 1.6B The *New York Herald* sold for two cents in 1861.

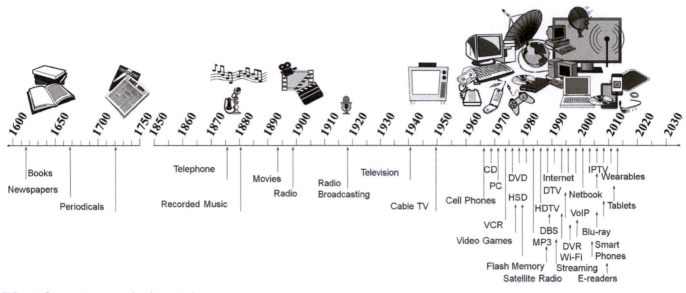

FIG. 1.7 Communication technology timeline
Source: Grant and Meadows, 2015

THE BEGINNINGS OF MEDIA REGULATION AND FREE SPEECH

The development of mass media in this country has been shaped by many factors, but to a large extent, law and regulation have set the parameters for how the media industries have grown. The first and perhaps most important regulatory guideline for media in the U.S. comes from our constitution.

The First Amendment to the Constitution was adopted in 1791 as one of 10 amendments that make up the Bill of Rights. This amendment states, "Congress shall make no law respecting an establishment of religion, or prohibiting the free exercise thereof; or abridging the freedom of speech, or of the press; or the right of the people peaceably to assemble, and to petition the Government for a redress of grievances." This provision for free speech accompanies the belief that in a democracy, access to information leads to making good decisions.

In fact, the argument that a free flow of information is vital to citizens' ability to make good decisions actually goes back further than the writing of the Constitution. In his 1644 book *Areopagitica*, English writer and poet John Milton expressed strong sentiment against any form of censorship. According to Milton, people should be exposed to a wide variety of opinions and ideas so they can apply reason and select the very best ideas. Even bad ideas and inaccurate information should be available because they present opinions and materials worth considering. Milton stated, "Let [Truth] and Falsehood grapple; who ever knew Truth put to the worse in a free and open encounter?" This concept, known as the **self-righting principle**, states that society can make adjustments and put itself on the right path if its citizens have many ideas and opinions from which to choose. In other words, 'truth' will win out over 'untruth' in a free and responsible society.

Closely related to Milton's self-righting principle is the concept of a **marketplace of ideas**. This term refers to a society in which ideas appear uncensored for consideration by the public, just as goods are presented in a marketplace. In the marketplace of ideas, people sample from the various sources of information and opinions, and because people are both rational and inherently good, they eventually find the truth. These concepts underlie the First Amendment to the U.S. Constitution.

The founders of this country, who put the First Amendment into law more than 200 years ago, certainly could not have envisioned the amazing technological innovations in mass media that have occurred since that time. Because the electronic media are technologically more complex than the print media, it should not be surprising that the federal government has always taken an interest in issues like preventing broadcast stations from blocking and interfering with their competitors' signals. The government also has shown interest in using electronic media for national security and in thwarting large corporations from monopolizing electronic media to prevent competition and avoid serving the public.

The right to speak freely is expected and often taken for granted by people who live in the United States. The First Amendment guarantees the right to free speech; however, this right is not always extended to the electronic media. The **scarcity theory** or **scarcity principle**, born in the early days of radio, states that because broadcasting uses parts of the electromagnetic spectrum and because the space on the spectrum is a limited natural resource, Congress retains the right to protect it on behalf of all of the people. In the 1920s, when the first broadcasting regulation was being debated, Congress determined that the 'airwaves' belong to everyone and should therefore be used in the best interest of all citizens, including children, who should be protected from exposure to inappropriate materials. Both Congress and the FCC, however, have conceded that it is nearly impossible to prevent children from hearing and seeing such materials over the airwaves.

The congressional right to regulate broadcasting was upheld in the U.S. Supreme Court in 1969 in *Red Lion v. FCC*. In this case, the Court held that Congress had the right to regulate the airwaves because of the scarcity principle (that the airwaves are a scarce natural resource that must be protected). In addition, the Court upheld the fairness doctrine, which required stations to make available opportunities for on-air rebuttal time to individuals with competing points of view. The FCC eventually abandoned the fairness doctrine in the late 1980s because of the government's general deregulatory climate, the perception that many electronic media options were available at the time (in other words, a 'lack of scarcity'), and the difficulty in trying to enforce it.

The downfall of the fairness doctrine has been something of a mixed victory for free-speech proponents. On the one hand, Congress can no longer force broadcasters to tell both sides of a story or force stations to find a spokesperson to give an opposing viewpoint to station commentary. On the other hand, audience members' opposing viewpoints will simply not be heard over the air, unless stations are motivated to provide time for them. Recently some attempts have been made to reinstate the fairness doctrine, but they have not been successful.

On the issue of free speech, the FCC treats subscription-based electronic media differently than broadcast media. Specifically, multichannel services, direct broadcast satellite (DBS), and cable and Internet-based subscription services are given stronger First Amendment protection than the broadcast media, because broadcasting is free and available to all, but subscription-based services are available only to those individuals who choose to pay for them. Thus, subscribers have control over what they watch.

The Internet also has the full protection of the First Amendment. Because almost anyone can publish on the Internet (anyone can theoretically publish a newspaper, magazine, or blog), Congress and the FCC have essentially allowed Internet content to be unregulated. However, Congress has tried to regulate some kinds of content for some kinds of audiences. For instance, the Telecommunications Act of 1996 originally included the Communications Decency Act (CDA), which attempted to limit the use of indecent material on the Internet. This act was ruled unconstitutional and was removed from the Telecommunication Act of 1996.

LISTENING AND VIEWING BEHAVIOR

Electronic media are powerful and very influential. Over the last hundred or so years since the origin of radio, they have affected the way we think, how we feel, and how we behave.

- **Cognitive effects:** Electronic media expose us to a flood of information. News reports and stories teach us about our world and open our minds to new ideas. The media are our eyes in places that we may never

go. Through the media, we vicariously experience the thrill of skydiving, the horrors of war, the inside of a prison, the views from the top of Mt. Everest, and the serenity of an isolated white-sand beach on the shores of the Caribbean Sea.
- **Emotional effects:** Electronic media influence attitudes. For instance, watching a show about an under-funded animal shelter that is forced to euthanize animals might make viewers more sensitive to the idea of spaying or neutering their pets. When we hear a sentimental or raucous song on the radio, it might cause a change in our mood.
- **Behavioral effects:** The electronic media persuade us to change our behavior or induce us to action. After watching a show about people who lost their homes to a forest fire, audience members might donate money to help provide emergency food and shelter. Hearing the pledge drive on a local National Public Radio station might prompt listeners to phone in their pledges for money.

How we experience the electronic media also influences how we live our lives. Starting in the late 1920s, people gathered around the living room radio in the evening to listen to popular network programs. When television finally became a reality after World War II, people sat around their television sets watching comedians Milton Berle and Lucille Ball, variety hosts Arthur Godfrey and Ed Sullivan, and band director Lawrence Welk, along with baseball games and boxing matches. Gone were the days when our entertainment needs were met by just relaxing on the porch, taking a stroll in the neighborhood, and singing and playing the piano in the front parlor.

After the introduction of television, radio narrowed its focus to attract specialized audiences—for instance, rock 'n roll music programming to attract teenagers. Stations that did not program rock 'n roll attempted to reach an adult or family audience by using a different type of music and less repetition. Radio listening became a popular activity outside the home with the advent of the small, lightweight, and portable transistor radio in the 1950s. Transistors found their way into the design of television sets as well. Audiences moved away from their large console television sets in the living room to small bedroom televisions. As television sets became smaller and prices dropped, multi-television households became the norm, and viewing shifted from being a family event to an individual activity.

SEE IT NOW

CHARACTERISTICS OF MASS MEDIA

Each mass medium has specific benefits and is best suited for specific types of communication, for example, television for broadcasting messages to a large, geographically and demographically diverse audience; radio for broadcasting local information to a local audience and delivering specialized programming to specialized audiences; and so on. Each medium can be differentiated from the others by considering the following characteristics: audience, time, display and distribution, distance, and storage.

AUDIENCE

Traditional media differ in the audiences they reach. Radio and television are single-source media that reach large audiences simultaneously, while other media, such as the telephone, reach only one person at a time. The Internet is unique in that it is a mass medium as well as an individual medium.

TIME

Media differ in terms of whether they transmit and receive information in an asynchronous or synchronous manner. With **asynchronous** media, there is a delay between when a message is sent and when it is finally received. Newspapers, books, and magazines—which are printed well in advance of delivery—are all asynchronous media, as are CDs, DVDs, and films. Conversely, with **synchronous** media, there is no perceptible delay between the time the message goes out and the time it is received. Synchronous messages from television, radio, and telephone are received almost instantaneously after transmission. In the past, electronic media were mostly synchronous because the programs were sent out at a given time as the audience listened or watched. Inventions like the audio tape recorder and the videocassette recorder made it possible for the audience to time shift, watch recorded programs at a more convenient time. Today, radio and television programs are commonly downloaded on a digital music device, recorded on a digital video recorder, or simply accessed through online services such as Netflix or Amazon Prime Video. Electronic media are now both asynchronous and synchronous.

DISPLAY AND DISTRIBUTION

Media also differ in how they display and distribute information. Display refers to the technological means (e.g., video, audio, text) used to present information to audiences or individual receivers. Distribution refers to the method used to carry information to receivers. Television's audio and visual images are distributed by broadcasting, cable, microwave, or direct broadcast satellite. Radio is generally transmitted over the air, although direct broadcast satellites send radio signals to subscribers across the country.

Digital audio and video information and programs that are available over the Internet are streamed live or stored on a server (*host*) and distributed on demand to a computer, tablet, iPod, or smartphone (*client*). Individuals connect to the Internet via **cable modem** or **digital subscriber line** (*DSL*), a 4G cell phone connection, or public or private *Wi-Fi* (wireless local area network). Mobile devices like tablets and smartphones play music from radio stations (generally through apps like Tune In or iHeart Radio) and video from program suppliers like Netflix simply by using the app to access a library of movies or television programs.

INTERACTIVITY

The Internet is highly interactive, but some parts of it are not as interactive as others. Many Web sites are informational and do not contain interactive components other than a contact email address. Other sites are filled with interactive games and puzzles, and some, like social media sites and blogs, are highly interactive.

Radio and television broadcasts are generally not considered interactive per se. Listeners might call radio request lines or television viewers might fill out a survey, but this is **feedback** (an individual audience message sent back to the source of the mass communication) rather than true interactivity.

This aspect of feedback has changed primarily because of digital technology. Audience members can send texts and emails to a radio DJ, who can respond immediately. Pandora listeners can fast forward on a song they do not like or give it a thumbs down and delete it instantaneously from the playlist. Viewers are often able to cast their votes online for their favorite performer on talent shows like *The Voice*.

DISTANCE

Mediated messages are transmitted over both short and long distances. Some media are better suited for long-distance delivery and others for short or local transmission. Print media need to be physically delivered to their destinations, a process that can be cumbersome and expensive over long distances. Electronic media deliver messages through the airwaves, telephone lines, cable wires, satellites, and fiber optics, giving them a time and cost advantage over print.

The Internet has changed the definitions of local, regional, or national media. The Internet is a global medium that can be used just as easily to send a message to someone across the room as to someone across the globe. Since the message travels at the speed of light, there is almost no signal delay across distances.

STORAGE

Message storage is limited to media that have the means of housing large amounts of information. For instance, CDs, DVDs, and computer flash and hard drives have the capacity to store millions of bits of data, whereas newspaper publishing offices typically have limited space for storing back issues. Until recently, television stations had to rely on videotape libraries, but most have now changed to digital storage of all programs.

Digitization obviates the need for paper, videotapes, and DVDs. Small memory devices have the capability of storing large amounts of digital data from television, radio, newspapers, and film. The **cloud** is also used to store programs and other data without taking up physical space in production studios, businesses, and private homes. Although the term 'cloud' implies that the data is stored in some vault in the sky, all it means is that huge amounts of data are stored on multiple powerful computers, often in multiple locations. Cloud users can access their information from anywhere, using any computer. Cloud storage can be bought or leased from a **hosting** company like Google that provides the cloud-based servers.

WHY DIGITAL?

Digitization of mass mediated content is probably the most revolutionary innovation since the printing press. Digitization is transforming the media and ways consumers use the media. Translating analog signals (continuous waves) into binary or discontinuous signals compresses (reduces) data so they are more easily stored and sent. In binary format, large amounts of information are archived and retrieved. Users no longer have to search through torn pages or garbled video- and audiotapes to find the information they are seeking. Plus, digitized material fits onto miniature but powerful portable devices, such as laptops, smartphones, iPods, and electronic books. A digitized dictionary, encyclopedia, and 10 years of *The New York Times* can all be stored on a small device and slipped into a back pocket for use anytime, anywhere.

The Federal Communications Commission (FCC) regulates electronic media and controls the use of portions of the electromagnetic spectrum in this country. It was the FCC that mandated that all broadcast television stations change their signals from analog to digital on June 12, 2009, and return the frequencies used for analog television broadcasting to the FCC to allot to other services, such as cell phones.

ZOOM IN 1.1

For more information about the electromagnetic spectrum, go to http://imagine.gsfc.nasa.gov/science/toolbox/emspectrum1.html

Digitization has also been very kind to our ears. Satellite direct digital radio service began in 2002 with satellite services XM Satellite Radio and Sirius Satellite Radio. These services sent out audio signals to a satellite, which then transmitted them to a satellite receiver radio. The best part is that the signals can reach the listener almost anywhere in the country. Someone on a cross-country trip can listen to the same station the whole way—there is no such thing as losing the signal. Of course, satellite radio requires a special receiver and a paid subscription, but it can certainly be worth it to have more than 200 channels of music, talk, sports, and news. Satellite radio was a bit slow to catch on, and in 2008 XM and Sirius pooled their resources and became one company, Sirius XM Radio.

Digital terrestrial radio, also known has **HD radio (high-definition radio)**, is also a recent digital audio option. Digital broadcasting (technically called **in-band on-channel, IBOC**) lets a radio station use the same frequency to broadcast its analog and digital signals, which translates to clearer, static-free radio, and it is compatible with both analog and digital sets. Better yet, stations can offer more than one 'channel' in their allotted frequency. In other words, an HD FM radio station that is broadcasting at 103.1 MHz can program several subchannels with a variety of formats.

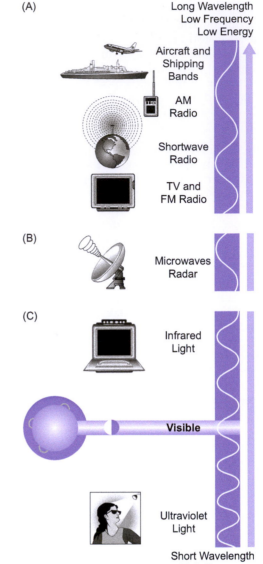

FIG. 1.8 The electromagnetic spectrum is the range of electromagnetic energy that radiates in waves from a source. At different frequencies the waves have different properties and can be used for a variety of purposes.

TRENDS AND TERMINOLOGY

The electronic media industry has changed dramatically over the past 20 years, most notably since 1996, when the Telecommunications Act was signed into law. As the technology and rules regarding ownership change, it will be important to understand a number of issues in the field of electronic media.

NEW MEDIA AND OLD MEDIA

New media often refers to programming that is easily accessible via the various forms of digital media. However if broadcast television can be accessed digitally on the Web is it new media or old media? The lines between new and old media have become blurred and it is best to consider distribution of media rather than whether the medium is analog or digital. New media delivery is digital and on demand rather than being only available at a particular time and via a particular receiver (e.g., tuning it to an FM radio program on an FM receiver). The content of much of a broadcast networks can be viewed not only through old media but through new media as well. A network show like *Elementary* can be viewed at a certain time on a television station or viewed via CBS All Access (or Netflix or Amazon Video). New media would certainly be the appropriate term for viewing programs within social media. In a sense, the terms new and old media are much more relevant to the delivery system than the content.

Is radio that is streamed to a smartphone old or new media? If the content is streamed from a new, nonlicensed audio service like Pandora or Spotify, it is considered new media. If the content is streamed from the Web site of a licensed radio station, it is considered old or legacy media but delivered on a new media platform.

AUDIENCE SELECTION OF PROGRAMMING: LINEAR LISTING VERSUS RECOMMENDATIONS AND RANDOM ACCESS

Beginning in the 1920 when radio became a source of free entertainment for the audience, daily newspapers listed the shows on that night or the Sunday paper would have program listings for the week. This was how radio audiences kept track of the programs and learned when to gather in the living room to hear their favorite network or local program. The linear-designed program order was created to attract and keep an audience listening by stacking shows that might appeal to the audience members who had an inclination to watch particular programs, often of the same genre, such as situation comedy. Keeping an audience listening for more than one show was both a science and an art. Programmers had to decide which programs attracted a large audience that would stay tuned for the next program and the next. Program listings in the magazine *TV Guide*, or a local paper, gave the times, channels, and program descriptions.

Television programs are now available on Hulu, the network Web site, or other delivery systems that allow video on demand (VOD). Audience members can select any program available and view it at any time. Consideration for keeping the audience from one show to another is no longer only tied to the network program schedule. The newer process involves recommendations from friends on social media and recommendations from the video provider. Watch a program on Netflix and Netflix will recommend many other shows to try. As viewers scroll through the Netflix menu they will find a group of shows with the heading, "Because you watched _____." This list offers similar shows in content or style . . . perhaps even country of origin (e.g., British comedies) to the program they watched earlier.

The important change in program selection is that recommendations are becoming a strong predictor of audience

program selections rather than relying on linear program schedules supplied by a network, satellite, or cable channel.

INTEGRATION: VERTICAL AND HORIZONTAL

Vertical Integration

When a company expands its business into areas that are at different points on the same production/distribution path, such as when a broadcast network owns its production studios (e.g., NBC owns Universal Studios) to create programs for its network, cable channels, and Web site. Vertical integration helps companies reduce costs and improve efficiency by decreasing distribution expenses and simplifying negotiation, among other advantages. All four of the major broadcast networks are part of large media companies that own companies (or divisions of the main company) that produce programming and distribute it as well.

Horizontal Integration

Horizontal integration in electronic media is the process that a company follows to increase its audience share and revenue by increasing the number of outlets. A company may do this via internal expansion, acquisition, or merger. All of the major broadcast networks and many of the large media companies own many media outlets. The broadcast networks not only have bought numerous television stations, but some actually own more than one network. CBS is also part owner of the CW Television Network and owner of the Showtime cable channels.

CONVERGENCE

One of the dominant trends in electronic media in the past 20 years is **convergence**. Convergence refers to the blurring of boundaries between the different types of electronic communication media. The media and other telecommunication services, like voice telephony and online services, have traditionally been distinct, using different methods of connecting with their audiences or users as well as different **platforms**, such as television sets, telephones, and computers. Moreover, these services have been regulated with separate laws.

With digital technology, various media are used at the same time over one device. In other words, when connected to the Internet via a broadband connection, listeners can access an online radio station, retrieve email, listen to iTunes music, download a book, or use instant messaging to have a real-time conversation (including both audio and video) with others. Smartphone users receive calls, send and receive text messages, digital pictures, and video, store and play MP3 music files, and surf the Web. Smartphones have the capability to remotely program household devices like a thermostat or light fixtures.

Digitization has also changed how we read books. Buy an iPad, Kindle Fire, or another type of tablet or e-reader device and download hundreds of pages of favorite reading. Instead of lugging a backpack full of books, one small, thin tablet is all that is needed. So, are books still considered a print medium if they can be downloaded onto a computer or tablet? Clearly, digitization and convergence have blurred the lines that distinguish one medium from another, such that the traditional definitions of these media need to be reevaluated.

DECONVERGENCE

The trend toward media convergence was energized by digitalization. Large media companies combined their print and media operations to facilitate a sharing of stories, information, personnel, and infrastructure. However, in the last few years, the trend has been for companies to **deconverge**, which has resulted in several large media companies 'spinning off' their print business from their electronic media business. In 2014, Gannett Company announced a plan to separate its broadcast operation from its newspaper business, yielding two separate companies, Gannett (print) and Tegna (broadcast and digital). Other recent deconvergers include the Tribune Co. (now rebranded as tronc). that split into Tribune Media and Tribune Publishing and the E.W. Scripps Company that merged with the broadcast properties of Journal Communications, and the two companies spun off their newspapers to form a new company, Journal Media Group. Although technological convergence is a reality in mass media, convergence between print culture and electronic media culture has often been difficult. The style of journalism between print and electronic versions varies greatly, and the personalities who create content and the goals of the platforms differ as well. In addition, business practices and investment strategies have encouraged media companies to allow both print and electronic media to operate in their own style and with their own goals.

CONSOLIDATION

Deconvergence has not stopped companies from buying more broadcast stations. In 2013, Gannett bought the Belo Corporation for $2.2 billion, which resulted in Gannett owning a total of more than 40 television stations. Consolidation concentrates media ownership, giving a smaller number of owners more media outlets and a larger share of the audience. Media companies are spinning off parts of the conglomerates to better concentrate their business efforts in specific media platforms.

When one media company merges with or acquires another media company, the overall ownership of media outlets is consolidated, which is what the Gannett Company did when it acquired the Belo Corporation. The acquisition gave Gannett an additional reach of 14% of the national television audience. This type of business **consolidation** was facilitated by the Telecommunications Act of 1996, which relaxed many of the limits on media ownership. Consolidation is also occurring in the radio business. The top 10 radio station groups have almost two-thirds of the radio audience. Radio station ownership by local businesses local to the community of the radio license has dropped by more than one-third since 1975.

Big electronic media companies are not stopping with consolidation within electronic media; they are also acquiring other media such as newspapers. If a local broadcast station parent company acquires the local daily newspaper in the same market, it creates a situation referred to as **cross-media ownership**. This strategy is not a new concept; it actually can be traced to the 1920s, but the government has mostly discouraged it, because it reduces the diversity and number of different media 'voices,' and therefore opinions, in a given market.

Localism

Broadcast radio and television stations have been licensed as local media. They are licensed to local communities, and the FCC has long required broadcasters to serve the needs and interests of the communities to which they are licensed. Congress has also required that the FCC assign broadcast stations strategically located to serve communities around the country. Broadcast stations are meant to be local and serve their communities, but broadcast networks have no such requirement because they serve the entire country. Cable channels like ESPN or HBO also have no such requirement; they are national. In television, the concept of localism gets a bit lost because television station signals are often carried to distant markets via cable television systems. For example, Phoenix television stations are carried in many smaller communities in Arizona, including some that are 150 miles from Phoenix. Serving the community becomes somewhat nebulous when the community is almost statewide.

Localism is still an important concept in radio because radio has no equivalent to television cable systems. Radio station signals remain local because of the nature of how they are broadcast and received. Localism is an advantage for radio and may be a factor in keeping local radio stations in business. A radio station serves its community with local-interest entertainment programming, news, weather, sports, and community affairs and information. Small advertisers use radio to reach potential customers. If a small business (e.g., a used-car dealer) had to pay for advertising on a station that reached audiences in a 250-mile radius of the community of license, the advertising cost might be prohibitive, and most of the audience might not consider driving that far to buy a vehicle.

IT'S NOT CALLED SHOW ART, IT'S CALLED SHOW BUSINESS

Although most media consumers like to think that the sole purpose of the media is entertainment, the harsh reality is that stations are on the air to make money. Broadcast stations, cable and satellite channels, and media-related Web sites cost money to operate and often have investors who demand a return on their investment. The 'media are here to entertain us' fantasy may have been somewhat true when many broadcasting stations were small, privately held businesses that took pride in having a dedicated audience that they bent over backward to please with interesting programming. With profit margins at about 10% to 25%, independent broadcasters were thriving, at least until the 1990s, when media became much more competitive and the business model emphasized profits. Nowadays consolidation has led to fewer companies owning more stations. Plus, many ownership groups are publicly held and as a result look to the bottom line to please their stockholders.

Buying a successful station is an expensive endeavor, and the purchasers more often than not incur a huge debt that takes many years to pay off. Station owners entice the largest possible audience to their programming so they can sell advertising for the highest possible price to bring in enough profit to support the enterprise and pay back the investors and debt holders. In a weakened economy, most recently in 2008, advertising is cut back, often leaving stations in dire straits and ripe for takeover or purchase by a media conglomerate.

MONETIZATION

Electronic media companies are scrambling to find ways to make up for lost advertising revenue by attempting to monetize content or generate revenue from the newer delivery systems (e.g., Web sites, podcasts). Media companies are faced with the challenge of reaching their target audience that is distracted by other entertainment outlets. When the audience moved to the Internet, media created their own Web sites. As the audience moved to smartphones and Twitter, companies followed with promotional tweets. As the audience moved to social network sites, media companies started their own social network groups. Radio and television stations find themselves chasing their listeners and viewers from one media landscape to another. But if they do not, they could lose them, and when the audience leaves, so does advertising dollars and the ability to finance new electronic and digital endeavors to lure the audience back.

BIG DATA

Big data is the term for sets of data that are so large that traditional data processing procedures are inadequate. Big data are generated by numerous sources such as a popular Web site. Server logs from the Web site can show information like who connected to the Web site, when they connected, how long that person stayed on the Web site, if the person went deeper into the Web site by clicking on a story or banner ad, what Web site the person went to after the first site, and so on. In other words, as compared with typical audience measurement as supplied by a company like Nielsen, digital media provide much more information about the viewers' behavior. This type of information is a result of being connected to the Internet and can involve millions of people and require much work to capture, analyze, share, visualize, and interpret. Big data yields reams of information but requires talented and educated analysts and complex computer programs to interpret the data. As the Internet of Things (connected devices in homes, work, autos) becomes more commonplace, big data will become more commonplace as well.

AUDIENCE DEMOGRAPHIC CHANGES

The size of the population of the U.S. is expected to grow by 100 million people by the year 2050. The minority population, currently about 30% of the total, is expected to reach 50% by 2050. Latin and Asian populations are expected to increased dramatically, perhaps tripling within the next 30 years. How will changes like these affect media and media programming? There are no simple answers to this question, but it is apparent that media programming will change as the population changes.

SEE IT LATER

This book provides an overview of electronic media, which have traditionally been defined as mass media delivered electronically. The book also presents an expanded view of electronic media including newer personal media. When attempting to predict the future of media, it is important to consider the concept of **disruptive technologies**. In his book *The Innovator's Dilemma*, author Clayton Christensen coined the term 'disruptive technologies' to describe a technology that displaces an existing technology or a product that creates a new industry. Personal computers changed the way people 'type,' eliminating the market for typewriters and paving the way for changes in how people communicate. Email has changed the way we communicate with others, for personal and business reasons, supplanting letter writing and disrupting the flow of mail via the postal service. Related industries that produced personal stationery and greeting cards have been changed forever.

Innovations in personal media have prompted noticeable changes in the way people listen to music and watch television. Specifically, listeners are moving away from broadcast radio and toward the Web. This movement has caused a shakeup in the radio and music industry. The radio industry is losing listeners, and the music industry is losing money due to decreased CD sales. Renting media by streaming on demand is displacing buying media. Vast libraries of media are now available from a variety of providers, which has created a need for 'media now.' People no longer want to wait for a program to air, nor do they desire to travel to a store to purchase a CD or DVD.

New ways of accessing television programming have dramatically changed the thinking of television producers. In addition to pitching a program idea to one of the big four television networks, many producers are pitching to other companies. Cable networks like USA and TNT, HBO, and Showtime welcome program ideas and have increased audience loyalty and critical success. Companies that stream video like Netflix or Amazon Prime Video are succeeding with programs like *House of Cards* and *Bosch* that are shown only through streaming. Now the traditional television networks are just players in the mixture of many companies that produce and distribute television shows. In addition, program distributors are going 'over the top' by providing programs via streaming devices that bypass traditional cable and satellite technologies.

Individuals are now generating their own content and displaying it on the Internet to potentially millions of users. Since 2005, people from all over the world have uploaded millions of videos onto YouTube. Almost any kind of video can be quickly uploaded using FFmpeg or Adobe Flash encoding. YouTube is a phenomenon and a favorite among young people who enjoy short clips of video about anything and everything. Purchased by Google in 2006, the site attracts millions of viewers each day who spend at least a few minutes every day watching user-generated video content.

Amateurs produce most YouTube videos, attesting to the changes in the video production process. In the last 10 years, more and more of the process of creating and producing programs has been accomplished using high-quality but inexpensive video cameras and laptop computers. Although quality differences are present, small computers now perform many of the production tasks that were once performed by expensive stand-alone equipment. People edit video and audio on their personal laptops and tablets and thus produce television or radio programming in their own homes. GoPro cameras placed on a surfboard, mountain bike handlebars, or helmet provide amazing live-action video in HD quality. These high-quality, miniature cameras that cost less than $400 are changing adventure videography forever.

The ease of producing and displaying homemade video is causing the networks and cable channels to rethink their scheduling and programming strategies. Networks and cable channels are looking for ways to make money without having to produce 30- or 60-minute programs, which are costly and financially risky.

The ability to produce programs with personal computers and inexpensive high-quality video cameras and audio equipment may lead to greater audience **media literacy**. Media literacy is the ability to access, analyze, evaluate, and create media in a variety of forms, and the understanding not only of the meaning of the content of the media but also of the power of the media, the intent of the media, and the influence of the media.

Not only do individuals produce media content that in the past could only be produced by large companies on expensive equipment, but the problem of distribution of the content is no longer a technological problem. Distribution on the Internet is accomplished by posting video on YouTube or any Web site. The challenge is not distribution, but instead the challenge is *being found*. Since millions of people produce content, how would any individual's content be located on the Web? The answer to this question is now evolving. Posting video on Facebook, YouTube, or Vimeo enables millions of people to see it, but is the audience looking for it? How would the audience search for it? These are the challenges that accompany attempting to reach large audiences like the broadcasters, cable channels, and movie studios have been doing for many years.

ACCELERATING CHANGE

A final consideration in looking forward to the future of media is the concept of **accelerating change**. Accelerating change is the increase in the rate of technological progress.

Technological change is occurring at an ever-increasing pace, as are the resulting social and cultural changes. Two forces are driving this acceleration. One is explained through Moore's Law, an observation made in 1965 by Gordon Moore, cofounder of the Intel Corporation, which first posited that the number of transistors in a dense integrated circuit doubled approximately every 2 years. In everyday applications of the law, the quality and computing ability of microprocessors in computers improves dramatically in short periods of time. Another obvious application of the law is seen in the capacity of digital memory devices. As computing speeds and storage capabilities improve, the ability of media outlets to send programming out to many people, either synchronously or asynchronously, is greatly enhanced. A practical example of the benefit of enhanced storage capability is the ability of a company like Netflix to make its enormous library of television programs and feature films available to all of its subscribers anywhere and at any time.

A second driving force behind accelerating change is the power of the billions of people and billions of 'things' that are now 'connected' to the Internet. Ideas and information are being shared constantly among people who can change and improve almost anything. In his book *Digital Wisdom*, futurist author Shelly Palmer suggests that inventors and innovators can be assisted by billions of people who share their abilities and experiences. Apps are created daily to help us perform many daily tasks. The saying "There is an app for that" epitomizes the reality that technological advances alter our lives by changing the way we behave, entertain ourselves, and communicate with others. Our media world is changing rapidly, and according to futurists like Shelly Palmer, the rate of change is getting faster and faster.

10 QUESTIONS YOU ABSOLUTELY SHOULD ASK ABOUT STUDYING ELECTRONIC MEDIA. THE ANSWERS WILL SURPRISE YOU!

The questions that follow might just be the questions you need to ask before you continue to read this book.

1. How much time do you spend each week with various forms of electronic media?
   Few activities command as much time and attention in our lives as our interactions with electronic media. In each household that has a television set, it is on for longer than the time spent working, going to school, shopping, or exercising. Sleep is the only activity that is more time consuming, and then not always for all people.
2. Has your electronic media viewing/listening changed now that you are in college?
   Today's video program content is viewed on more than just television sets. Consumers are watching via the Internet and on cell phones, in-home and out-of-home, live and time-shifted, free and paid, and rebroadcasts and original program streams. Radio, traditionally listened to at home, at work, in autos, on the beach, and while working out, is now listened to on cell phones, tablets, MP3 players, and all Internet-enabled devices.
3. How familiar are you with the icons of pop culture? And where do you get most of the information about your favorite celebrities? Most likely the images and information come from electronic media.

4. What social benefit do you get from watching and listening to electronic media?

   Media users often talk about what they see on television and hear on the radio: *American Idol, Orange Is the New Black, Survivor, Dancing With the Stars, Wheel of Fortune, Star Trek, Adult Swim,* and the weather forecast. They talk about television programs and movies they have downloaded, streamed, and seen on DVDs. Music and musical performers are common discussion topics at home, work, and school.

5. Do you think that the electronic media are ambassadors of our culture?

   American electronic media content is pervasive in many parts of the world. That means that the perceptions that people in other countries have formed about us are often based on what they have seen in the movies and on television.

6. Is watching/listening to the electronic media enough. Why do we know so much about media personalities?

   Celebrities get quite a bit of news coverage. Some shows are dedicated to news about the electronic media and movie personalities. Many people are fascinated with the lives of prominent and famous people. Shows like *Entertainment Tonight, TMZ,* and *Access Hollywood* get more viewers when famous people commit foolish or scandalous acts that generate entertainment news stories.

7. How do electronic media actually influence you?
   - Speech—We learn new phrases and meanings for words and slang.
   - Customs and traditions—The portrayal of holiday festivities, like the dropping of the 'apple' on New Year's Eve, shapes how we observe these holidays.
   - Styles of clothing, cars, and technology—We see and hear about these products through electronic media, and we are tempted to try them out.
   - Sense of ethics and justice—We view many stories of good and evil and even experience real courtroom dramas by viewing the many courtroom shows on television.
   - Perceptions of others in our society and distant countries—*National Geographic* programs show us how people in South America or remote Alaska live.
   - Lifestyles—We learn about other people's lives and our own by watching talk shows, self-help shows, and advice shows.

8. Are you different than college students were years ago because of electronic media?

   By the time you finish high school, you have been subjected to many thousands of hours of electronic media. What effect does that have on you? Are you different than your parents or grandparents because you have used so much electronic media? Do electronic media have a quick and direct effect on you or a slow, subtle, cumulative effect? Many people who study media, including psychologists and sociologists, believe that contemporary digital media equipment sets young people apart from older people (e.g., those over 30 years old). Young people do not remember a time when they weren't constantly connected and available to their peers and to the world. Entertainment can be customized and even individualized. People born since 1990 were introduced to technology early in life. Young people are comfortable at multitasking and can seek out and even create their own entertainment content. They expect change and innovation at a much faster pace than people who grew up with traditional analog media.

9. Can studying electronic media help you in your everyday life?

   Studying electronic media and becoming media literate will help you be a discriminating consumer who can make good media choices. By knowing more about the history, structure, economics, and regulation of electronic media, you can better understand and even predict the future of media. Understanding the storytellers that create media content helps us realize the effects that the stories have upon audiences. It also helps you understand how the constant connectivity of today will influence us in the future.

10. Will studying electronic media help you in your career?

    Few industries have undergone and continue to undergo the dramatic technological and business changes that we have seen in electronic media in the past 10 years. The electronic media are always changing, and as they change, so do you. For college students interested in a career in electronic media, knowing about these changes will present appropriate strategies for job seeking.

FIG. 1.9 This toddler will grow up in an age in which all news and entertainment is digital and available anywhere and at anytime. A small tablet with an Internet connection can become a window to the world.
Photo courtesy of Sarah Clover

SUMMARY

Until recently, the number of people we could communicate with was limited to those we could see face-to-face or contact by letter. Since the mid-19th century, electricity has enhanced various forms of communication and allowed us to communicate to one person or to many people over long distances with one message. Through the use of electronic media—radio, television, cell phones, and the Internet—we now communicate with a huge number of people almost instantaneously.

Traditional mass media share characteristics such as audience, time, display, distribution, distance, and storage. Electronic media are not constrained by time and distance. Electronic media can have cognitive, emotional, and behavioral effects on the audience that can influence us and can change our lives.

The Internet has emerged as a new mass medium at an unprecedented speed. It was adopted rapidly and represents a combination or convergence of various mass media. The Internet enables communication with a large audience for low cost and short turnaround time. The process of digitization has simplified the format through which information is transmitted. Numerous trends are changing the media industry and how we relate to and use electronic media. Convergence is the combining of media and thus the blurring of the distinctiveness among them. Deconvergence is the process of splitting media companies to better focus on individual media platforms and conform with regulation. Consolidation involves fewer companies owning more electronic media stations and businesses. Some trends have resulted directly from changes in technology. For example, desktop production has been fueled by digital technology and faster computers, which allow individuals to create content for electronic media on a single computer.

Technological change is closely related to changes in the media. As computers get faster and memory storage becomes less expensive and easier, our ability to access media transforms. We can rent or use media without having to buy it. We can get media at any time, practically from anywhere. These changes, both technological and cultural, alter our ways of creating and using media.

The study of electronic media is important not only as a field of intellectual pursuit but also as a means of preparing for a successful career in a media-related field. In addition, because electronic media are so pervasive, we need to be critical consumers of both the content and the activity that consumes so much of our time. Electronic media provide a window for the world to view many different cultures. Finally, electronic media are dynamic forces in society that are constantly changing. We need to study the changes and understand that they affect us deeply.

BIBLIOGRAPHY

Bangemann, M., & Oreja, M. (1997). Green paper on the convergence of the telecommunications, media and information technology sectors, and the implications for regulation towards an information society approach. Retrieved from: http://ec.europa.eu/avpolicy/docs/library/legal/com/greenp_97_623_en.pdf

Bonchek, M. (1997). From broadcast to netcast: The Internet and the flow of political information. Doctoral dissertation, Harvard University, Cambridge, MA. Retrieved from: www.engagingexecutives.com/from-broadcast-to-netcast

Chester, J., & Larson, G. (2002, January 7). Something old, something new. *The Nation*. Retrieved from: www.thenation.com/docPrint.mhtml?i520020107&s5chester

Corporation for Public Broadcasting. (2003). *Time spent with media*. Retrieved from: www.cpb.org

Dizard, W. (2000). *Old media, new media*. New York: Longman.

Dominick, J. (2013). *The dynamics of mass communication: Media in translation* (12th ed.). New York: McGraw Hill.

Eggerton, J. (2010, January 20). Kaiser: First drop for real-time TV viewing among youth. *Broadcasting and Cable*.

Ga, B. (2004, September 19). How people spend their time. *Google*. Retrieved from: http://answers.google.com/answers/threadview?id=403376

Grant, A., & Meadows, J. (2008). *Communication technology update and fundamentals* (11th ed.). Boston: Focal Press.

Grant, A., & Meadows, J. (2012). *Communication technology update and fundamentals* (13th ed.). New York and London: Technology Futures, Inc., and Focal Press Taylor & Francis.

Jayson, S. (2010, February 23). The net set. *The Arizona Republic*.

How many online? (2001). *NUA surveys*. Retrieved from: www.nua.ie/surveys [January 14, 2002]

Kaye, B. K., & Johnson, T. J. (2003). From here to obscurity: The Internet and media substitution theory. *Journal of the American Society for Information Science and Technology*, 54(3), 260–273.

Kaye, B., & Medoff, N. (2001). *The world wide web: A mass communication perspective*. Mountain View, CA: Mayfield.

Longley, R. On an 'average' American day. *About*. Retrieved from: http://usgovinfo.about.com/od/censusandstatistics/a/averageday.htm

Markus, M. (1987). Toward a "critical mass" theory of interactive media: Universal access, interdependence and diffusion. *Communication Research*, 14(5), 491–511.

Medoff, N., Tanquary, T., & Helford, P. (1994). *Creating TV projects*. White Plains, NY: Knowledge Industry.

National Cable Television Association. (2003). *Industry overview*. Retrieved from: www.ncta.com

National Telecommunication Information Agency. (1999). *Falling through the net: Defining the digital divide*. Retrieved from: www.ntia.doc.gov/ntiahome/digitaldivide

Neufeld, E. (1997). Where are the audiences going? *Media Week*, 7(18), S22–S29.

Palmer, S. (2012). *Digital Wisdom: Thought Leadership for a Connected World*. Stamford, CT: York House Press.

Pavlik, J. (1996). *New media technologies*. Boston: Allyn & Bacon.

Potter, W. (1998). *Media literacy*. Thousand Oaks, CA: Sage.

Pruitt, S. (2002, September 25). Online piracy blamed for drop in CD sales. *PCWorld.com*. Retrieved from: www.pcworld.com/news/article/0,aid,86326,00.asp

Roper Starch. (1997). America's watching: Public attitudes toward television. Retrieved from: www.fcc.gov/Bureaus/Mass_Media/Notices/1999/fcc99353.pdf

Schramm, W. (1954). *The process and effects of mass communication*. Urbana: University of Illinois Press.

Schramm, W. (1988). *The story of human communication: Cavepainting to microchip*. New York: HarperCollins.

Shane, E. (1999). *Selling electronic media*. Boston: Focal Press.

Shannon, C., & Weaver, W. (1949). *The mathematical theory of mass communication*. Urbana: University of Illinois Press.

Silverblatt, A. (1995). *Media literacy*. Westport, CT: Praeger.

Sterling, C., & Kittross, J. (2002). *Stay tuned: A history of American broadcasting*. Mahwah, NJ: Erlbaum.

Straubhaar, J., & LaRose, R. (2004). *Media now* (4th ed.). Belmont, CA: Wadsworth.

Turrow, J. (2014). *Communication in a converging world* (5th ed.). New York and London: Taylor & Francis.

Walker, J., & Ferguson, D. (1998). *The broadcast television industry*. Boston: Allyn & Bacon.

Why Internet advertising? (1997). *Media Week*, 7(18), S8–S13.

Wigfield, M. (2009, May 13). FCC increases consumer choice by speeding number portability. *Federal Communication Commission*. Retrieved from: www.fcc.gov

From Marconi to Mobile Listening 2

Contents

The authors would like to acknowledge Dale Hoskins and Gregory Newton for their contributions to earlier versions of this chapter.

SEE IT THEN

EARLY INVENTORS AND INVENTIONS

The 19th century was a time of tremendous technological growth around the world. The Industrial Revolution, which began in England, took off in the United States just after the Civil War and continued into the early 1900s.

Printed books, newspapers and magazines were the main ways of communicating to the public at large, but much of the population was semiliterate, and printed publications were expensive to purchase; thus they were a communication luxury for the elite. Inventors and progressive thinkers were experimenting with new ways of transmitting messages. They were not sure of where it would all lead but were hoping to somehow connect people from across the globe.

This chapter focuses on the development of radio, the first electronic medium, from the discovery of airwaves to how it became a mass medium. The technological advances that led to modern-day radio and that are paving the way to the radio of tomorrow are also covered.

ELECTRICAL TELEGRAPHY

Samuel F. B. Morse, a well-known American painter and inventor, was the first to send an electromagnetic signal through wires. He invented what he named the 'electric telegraph' in 1835. The instrument used pulses of current to deflect an electromagnet, which moved a marker to produce a written code on a strip of paper. The next year, he changed the system so that it embossed paper with a system of dots and dashes, which later became known as Morse code.

Each letter of the alphabet was assigned a series of dots and dashes. For instance, the letter H is represented by a series of four dots, an E is one dot, an L is a dot, a dash, and two dots, and an O is three dashes. To transmit the dots and dashes, an operator would push down with one

FIG. 2.1A The first telegraph, circa 1840
Photo courtesy of iStockphoto © Jonnysek, image #3785353

FIG. 2.1B Portrait of American inventor and painter Samuel Morse with his telegraph.
Photo courtesy of Shutterstock 244389145

finger on a lever connected to the telegraph. To send the message "hello," the operator would tap out the representative dots and dashes for each letter, with the dots requiring a short tap and the dashes a long tap. The code was universally accepted and simple enough that telegraph operators could quickly tap out and decode messages. Telegraph messages were sent across wires strung from pole-to-pole. The system was not very secure, and messages did not always get to the receiver. Thunderstorms would disrupt the electrical current, and high winds would blow over the poles or break the wires. Also, outlaws would sometimes deliberately cut the wires to block transmission.

Morse patented the telegraph in 1840, and the U.S. government provided funds for a public demonstration. An electrical line was set up between Washington, DC, and Baltimore. After dealing with some technological problems (some of which were remedied by Ezra Cornell, for whom Cornell University is named), Morse sent the first official message, "What hath God wrought," on May 24, 1844. The message was taken from the Bible to suggest that the telegraph would change not only communication but also the world forever—and maybe not for the best.

The government permitted private businesses to develop an electrical communication industry, a policy that would be repeated in later years with other communication technologies. By 1861, a transcontinental line was in place, and telegraph messages buzzed across the United States.

Even though the electrical telegraph conquered the problems of distance and speed, it still presented some challenges:

1. It required building a costly system of wires.
2. It worked only as long as the wires were in place.
3. The telegraph was a restricted system that required trained telegraph operators who knew how to send and receive in Morse code. Western Union was the dominant company in the business, and it controlled all messages.
4. Once a message was decoded, delivering it to the appropriate receiver sometimes proved problematic. The local address of the receiver had to be found, and a courier had to physically travel from the telegraph office to the receiver's home or place of business. Ironically, physically delivering a message to a nearby receiver took a great deal more time than sending the message across the country.
5. The cost to send a telegraph message was determined by the length of the message; therefore messages were often short and somewhat cryptic. Because of the need for brevity, messages often lacked specific meaning or emotion.

ELECTRICAL TELEPHONY

Despite its shortcomings, the telegraph was the fastest way to communicate across distance and was quickly adopted by the public. Alexander Graham Bell, however, came up with a better method of two-way communication that could be used by individuals without special training. He invented the electrical telephone, which combined two scientific principles to achieve electrical conduction of sound wave patterns that were converted to electrical patterns and sent through a wire. So instead of tapping out a signal that needed decoding, a message could be spoken and transmitted by sound waves. On March 10, 1876, Bell made the first 'telephone call' to his assistant—"Mr. Watson, come here; I want to see you"—through his experimental system. One year later, the first telephone line was constructed between Boston and Somerville, Massachusetts. Like the telegraph, telephone lines were strung pole-to-pole.

Electrical telephony was an astounding new invention. For the first time ever, people could speak to one another across distances without a special coding/decoding system. The costs and logistics of building a telephone system inhibited widespread use at first. As in the case of the telegraph, telephone calls could only be made between places that were wired to receive and transmit them. Even areas that had been wired sometimes had problems, as when bad weather caused telephone lines to break and created lapses in service. Despite these drawbacks, however, the growth of distance communication was assured, and the telephone soon became an integral part of life.

POINT-TO-POINT ELECTRICAL COMMUNICATION

Both electrical telegraphy and electrical telephony were designed and used as systems to facilitate point-to-point communication. Using either of these systems, one person could send a message to another person at a distant location. The speed at which the signal traveled through the wires was the same as the speed of light (186,000 miles per second), which meant the message reached its destination almost instantaneously.

Electrical point-to-point communication proved its value in many situations, such as announcing the arrival of incoming stagecoaches, trains, and ships into port. By the late 1800s, telegraph messages and telephone calls were commonly used to alert manufacturers and merchants about when materials, supplies, and products would be arriving, and telegraphed and telephoned information about incoming storms and weather forecasts were invaluable to farmers.

The U.S. economy depended somewhat on maritime shipments of goods and products, but there was no way to contact ships when they were at sea. A way of sending electrical messages without wires was needed. Given the importance of the shipping business and the potential for profit, possible solutions were sought from scientists, innovators, and inventors to develop a wireless system of communication.

WIRELESS TRANSMISSION

JAMES CLERK MAXWELL

As early as 1864, James Clerk Maxwell, a Scottish physicist, predicted that signals could be carried through space without the use of wires. In 1873, he published a paper that described invisible radiant waves. These waves, which were later called radio waves, were part of **electromagnetic theory** that suggested that wireless signals could travel over distances and carry information. The theory used mathematical equations to demonstrate that electricity and light are very similar and both radiate at a constant speed across space.

HEINRICH HERTZ

Many scientists from around the world contributed to the development of radio. Heinrich Hertz, a physicist from Germany, expanded Maxwell's electromagnetic theory to build a crude detector of radiated waves in 1886. Hertz set up a device that generated high-voltage sparks between two metal balls. A short distance away, he placed two smaller electrodes. When the large electric spark jumped across the gap between the two large balls, Hertz could see that a smaller spark appeared at the second set of metal balls. It was proof that electromagnetic energy had traveled through the air, causing the second spark. Hertz never pursued the idea of using the waves to transmit information, but his work is considered crucial to the use of electromagnetic waves for communicating. In fact, the basic unit of electromagnetic frequency, the **hertz**, was named after him.

In the late 1890s, English physicist Sir Oliver Lodge devised a way to tune both a transmitter and a receiver to the same frequency to vastly improve signal strength and reception. Another scientist noted for contributing improvements in the wave detector and antenna

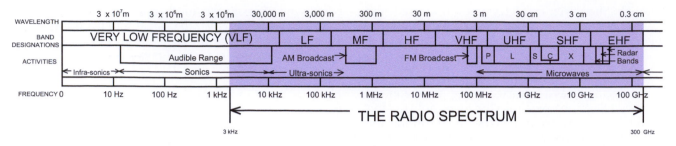

FIG. 2.2 The radio bands represent a portion of the electromagnetic spectrum.

is Alexander Popoff, a Russian who experimented in the 1890s. Interestingly, Popoff's work was dedicated to finding a better way to detect and predict thunderstorms.

GUGLIELMO MARCONI

Italian inventor, Guglielmo Marconi, is generally credited with the first practical demonstration of the wireless transmission of signals. After reading about Hertz's experiments, Marconi used electromagnetic waves to transmit Morse code signals. But he also noted that having an antenna above ground improved signal transmission. Marconi also improved Hertz's wave detector and discovered that an antenna could strengthen signal transmission.

When Marconi approached the Italian government for a patent and financial support, they expressed no interest. Fortunately for Marconi, his mother came from an Irish family with connections in Great Britain, which was a strong maritime power and thus very interested in developing a system to contact ships at sea. Marconi contacted the head of the telegraph office of the British Post Office, William Preece, who had also done some wireless experimentation. Preece was very receptive to the idea of sending wireless signals and Britain eventually provided Marconi with a patent and the financial support he needed to further develop his wireless system.

Towards the end of the 1890s, Marconi carried out a series of demonstrations showing that his wireless radio system could work over short ranges. In 1899 he successfully sent a signal across the English Channel, the longest distance yet. Just two years later, he sent a signal (the letter *S* in Morse code) across the Atlantic Ocean from Great Britain to North America, convincing many that wireless communication across great distances was possible.

Marconi's system could carry only dots and dashes, or Morse code, and not the human voice. As the telephone had been invented more than 25 years earlier, the public believed that radio would not be very useful unless it too transmitted sound.

American inventor Nathan Stubblefield is often credited as the first person to successfully transmit the human voice using radio waves, although he used ground conduction rather than transmitting through the air. Regardless, Stubblefield supposedly communicated the words "Hello Rainey" to his assistant during an experiment near Murray, Kentucky in 1892.

FIG. 2.3 Guglielmo Marconi at work with his wireless radio
Photo courtesy of MZTV Museum

REGINALD FESSENDEN

A Canadian electrical engineer, Reginald Fessenden, and an engineer from General Electric, Ernst Alexanderson, constructed a high-speed alternator (a device that generates radio energy) to carry voice signals. On Christmas Eve, 1906, Reginald Fessenden transmitted the first radio broadcast using modulated continuous electromagnetic waves carrying sound wave patterns instead of Morse code.

From his home at Brant Rock, Massachusetts, Fessenden sent out sounds of violin music, vocal Bible readings, and general season's greetings that were picked up by ships at sea along the East Coast of the United States.

LEE DE FOREST

In 1899, American inventor Lee de Forest earned a Ph.D. from Yale University. His dissertation investigated wireless transmissions. One year later, he developed a wireless system to compete with Marconi's by constructing the *audion tube* that could amplify sound transmission.

The audion expanded earlier work by English engineer John Fleming, who developed a radio wave detector in a sealed glass tube that came to be known as the *Fleming valve* or *diode tube*. This device, which resembled a household light bulb, detected voice waves, but it could not amplify the sound. De Forest added a third element to the diode tube to make a triode tube, which became the audion. The audion amplified sound so it could be heard easily using radio waves.. De Forest filed for a patent for his invention in 1906.

| 0.03 cm | 3 x 10⁵Å | 3 x 10⁴Å | 3 x 10³Å | 3 x 10²Å | 3 x 10Å | 3Å | 3 x 10⁻¹Å | 3 x 10⁻²Å | 3 x 10⁻³Å | 3 x 10⁻⁴Å | 3 x 10⁻⁵Å | 3 x 10⁻⁶Å | 3 x 10⁻⁷Å |

The figures in the chart row represent: 0.03 cm, 3×10^{5}Å, 3×10^{4}Å, 3×10^{3}Å, 3×10^{2}Å, 3×10Å, 3Å, 3×10^{-1}Å, 3×10^{-2}Å, 3×10^{-3}Å, 3×10^{-4}Å, 3×10^{-5}Å, 3×10^{-6}Å, 3×10^{-7}Å

| INFRARED | VISIBLE | ULTRAVIOLET | X-RAY | GAMMA-RAY | COSMIC-RAY |

Sub-Millimeter | Visible | ←—— Ultraviolet ——→ | ←—— Gamma-ray ——→ | ←—— Cosmic-ray ——→

←———— Infrared ————→ | ←———— X-ray ————→

Frequency row: 1 THz, 10^{13}Hz, 10^{14}Hz, 10^{15}Hz, 10^{16}Hz, 10^{17}Hz, 10^{18}Hz, 10^{19}Hz, 10^{20}Hz, 10^{21}Hz, 10^{22}Hz, 10^{23}Hz, 10^{24}Hz, 10^{25}Hz

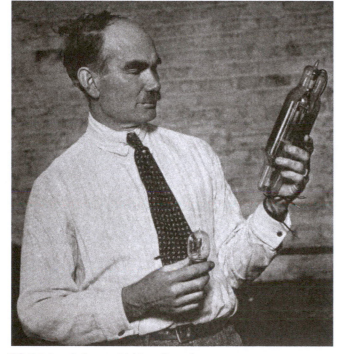

FIG. 2.4 Lee de Forest with his audion tube
Photo courtesy of MZTV Museum.

De Forest was very good at generating publicity with events like playing a phonograph record from the top of the Eiffel Tower in Paris and transmitting the sound across radio waves for 500 miles, but he was not a skillful businessman. He spent much of his time trying to create a place for himself in the burgeoning radio industry, but he suffered many financial setbacks and legal problems with patents. His biggest legal battle regarded the 1914 invention of the regenerative or feedback circuit, which achieves higher amplification than the audion. Edwin H. Armstrong, another radio inventor, was credited for inventing the circuit while he was an undergraduate at Columbia University, but De Forest claimed that he was the real inventor. De Forest prevailed even though during his testimony in court, he could not explain exactly how the system worked. Regardless, historians often see it the other way and give Armstrong, who later invented FM radio transmission, credit for the circuit.

THE BASIS FOR REGULATORY POWER OVER BROADCASTING

Congress is the law-making body of the U.S. government and has the right to regulate broadcasting, because the airwaves belong to the people. Unlike newspaper companies that print and distribute newspapers wherever and whenever they like to as many people as possible, broadcasters use the electromagnetic spectrum to transmit their information. In other words, newspapers create their own medium, and broadcasters must use an existing natural resource and limited medium—the electromagnetic spectrum—to reach their audience. The important point to remember about the electromagnetic spectrum is that different portions of it are used to transmit different types of signals. Only a small portion of the spectrum is usable for broadcasting. The government considers it a scarce resource, much like it might consider water a scarce resource in some areas of the country. Use of the electromagnetic spectrum is therefore subject to regulation by the federal government on behalf of the people. The government wants to prevent monopolistic use and squandering of the electromagnetic spectrum in particular. Hence, in the early days of radio, the government stepped in and began regulating how broadcasting could use the spectrum.

A TRAGIC LESSON

In the early 1900s, radio was somewhat of an oddity. Although many hobbyists were experimenting with radio transmissions and radio demonstrations were used for publicity purposes at fairs and department stores, only two companies were making parts for radio receivers. Beginning as early as 1903, government representatives of the leading industrial countries began holding annual conferences (called Radio Conferences) to discuss humanitarian and international uses of wireless radio. By 1910, many of these nations had established regulations to guide the use of radio, particularly in terms of maritime uses. The U.S. Congress passed the Wireless Ship Act of 1910, which required a ship with more than 50 passengers to carry a radio that could send signals to another radio within 100 miles. A ship also needed to have at least one person aboard who is capable of operating the radio.

Unfortunately, the requirements set forth by the Wireless Ship Act were not enough to save the victims of *Titanic*. In mid-April 1912, the 'unsinkable' luxury liner set out on its maiden voyage from England, bound for the United States. Its passengers, mostly wealthy and well-known people, expected a luxurious trip across the Atlantic Ocean on the newest and most sophisticated ocean liner ever built. *Titanic* was equipped with the most modern technology available at the time, including a wireless radio and trained operators.

Late at night on April 15, *Titanic* collided with a huge iceberg in the North Atlantic Ocean, ripping open the hull and causing the ship to rapidly take on water. Supposedly, the ship's radio operator had received warnings about icebergs being dangerously close to the ship's path, but he did not heed them. Instead, the operator requested that other ship radio operators clear the airwaves to allow *Titanic* to send personal messages from the ship's famous passengers to friends and family in Europe and the United States.

After *Titanic* collided with the iceberg and began to sink, the radio operator on board frantically sent out SOS signals. Unfortunately, the collision occurred late at night, and most of the wireless operators on other ships in the area had already gone off duty. One ship, the *U.S.S. California*, was less than ten miles away and could have rescued survivors from the icy waters, if only its wireless operator had been on duty ten minutes longer. Because the Wireless Ship Act of 1910 failed to require 24-hour staffing of wireless systems (as agreed to in the 1906 international agreements), *California* had no one on duty to receive *Titanic's* SOS. Only one operator, on the ship *Carpathia*, picked up the SOS signal and sped to *Titanic*. Although *Carpathia* was able to rescue more than 700 passengers, over 1,300 perished when *Titanic* went down.

At the time *Titanic* was sinking, a young wireless operator named David Sarnoff was stationed inside Wanamaker's Department Store in New York City. He claims to have picked up *Titanic's* distress signals and the responses from *Carpathia*. Sarnoff says he stayed at his wireless station for the next 72 hours, receiving information about survivors. Sarnoff relayed the events of the sinking to other wireless

operators and newspapers. As news of *Titanic* spread, the government ordered that the airwaves be cleared of other wireless operators, leaving Sarnoff as the only one to send messages to coordinate rescue traffic and to pass along exclusive information to *New York American* daily newspaper.

FIG. 2.5 A young David Sarnoff at his wireless station in Wanamaker's Department Store
Photo courtesy of MZTV Museum

FIG. 2.6A The news reported about the sinking of the *Titanic* in April 1912 came from wireless radio transmission.
Photo courtesy of Getty Images #80680699

Some historians, however, dispute Sarnoff's claims that he was one who picked up the signals, because there was no way to identify the person who was operating the transmitter. Whatever the case, Sarnoff went on to become president of Radio Corporation of America (RCA), and the story marks an interesting and important milestone in the development of the wireless. The sinking led to government scrutiny of the role of wireless radio and to the provisions of the Wireless Ship Act of 1910. Governments worldwide sought to increase wireless conformity and compatibility. Congress amended the Wireless Ship Act of 1910 to require two trained radio operators, an auxiliary power supply, and the ability of the radio operator to communicate with the bridge of the ship. Also, the act was extended to include not only ships at sea but also those on the Great Lakes.

Congress also passed the Radio Act of 1912, which required the licensing of radio operations used for the purpose of interstate commerce. It also required licensed operators to be citizens of the United States and stated that licenses must be obtained from the U.S. Secretary of Commerce and Labor, who had jurisdiction over commercial radio use in this country. The act also required that ships had to be equipped to send out a distress signal at any time and that all ships must monitor distress frequencies continuously.

RADIO BECOMES A MASS MEDIUM

While Fessenden and de Forest were working mainly on the mechanics of radio, Charles D. 'Doc' Herrold was thinking about what could be transmitted over the airwaves. Three years after Fessenden's Christmas Eve broadcast, Herrold came up with the idea of regularly scheduling transmissions. Herrold operated a technical college in San Jose, California, and, with the help of some students, began transmitting music, speeches and other types broadcasts at scheduled times. In a sense, Herrold created the first radio station, specifically the first college station. Herrold's station was one of the first to operate regularly after the Radio Act of 1912, and it continued operating until World War I. It came back on the air in early 1922 with the call letters KQW. The station was later sold and then moved to San Francisco, where it became KCBS, a station that still broadcasts today.

Station KDKA in Pittsburg, PA, however, is officially considered the first commercial broadcast station because it was the first one formally licensed by the Secretary of Commerce. Dr. Frank Conrad, an engineer at Westinghouse, formed KDKA in 1916. Conrad sent both voice and music programs from his home in Pittsburgh to the Westinghouse plant located about 5 miles away. Conrad devised a system for broadcasting music by placing a microphone next to a phonograph record as it played. Conrad's broadcasts were so popular that he regularly scheduled them on Wednesday and Sunday evenings so listeners would know when to tune in.

Another station that claims to be the first licensed station on the air is 8MK-WWJ in Detroit. It started out as 8MK, an amateur station that first went on the air from a 'radio phone room' in the *Detroit News* building. The license for this station was eventually issued by the Department of Commerce to the *Detroit News* for station WWJ on March 3, 1922.

BROADCASTING

Early transmissions reached a small segment of the general population. Few people had a radio receiving set, and there were not enough programs on the air to induce people to buy one. Moreover, there was not a universally accepted word for what to call these transmissions. Eventually, the word 'broadcasting' was used to describe the mass transmission of radio programs to the public. Broadcasting is actually an agricultural term that describes a way to plant seeds by 'broadcasting'—casting them in all directions using a circular hand motion rather than planting them in rows (Lewis, 1991).

Radio was still in its infancy, and no one was really sure of the direction it should take. David Sarnoff, the person who supposedly received and transmitted news about the *Titanic*, had by 1916 become commercial manager of the American Marconi (a subsidiary of British Marconi).

FIG. 2.6B An artist's impression of a newspaper seller bringing news of the *Titanic* disaster
Photo courtesy of Getty Images #79024197

Sarnoff thought that money could be made from radio, but he just had to figure out how to do it. He came up with an idea that he penned in a memo addressed to the manager of American Marconi. In it, Sarnoff outlined the essence of what he thought radio broadcasting should become.

A plan of development, which would make radio a 'household utility' in the same sense as the piano or phonograph. The idea is to bring music into the house by wireless. . . . The problem of transmitting music has already been solved in principle and therefore all the receivers attuned to the transmitting wavelength should be capable of receiving such music. The receiver can be designed in the form of a simple 'Radio Music Box' and arranged for several different wavelengths, which should be changeable with the throwing of a single switch or pressing of a single button. . . . The box can be placed on a table in the parlor or living room, the switch set accordingly and the transmitted music received. There should be no difficulty in receiving music perfectly when transmitted within a radius of 25 to 50 miles. . . . The same principle can be extended to numerous other fields as, for example, receiving of lectures at home, which can be made perfectly audible; also events of national importance can be simultaneously announced and received.

(Benjamin, 1993)

In addition, Sarnoff's memo stated specific ways to generate revenue from radio. Sarnoff suggested that large profits could be gained from the sale of radio receivers to the general public. In retrospect, it seems that the idea of a 'radio music box' would have been adopted immediately. But American Marconi and other companies ignored it. The idea of Sarnoff's 'radio music box' simply did not catch on for many reasons, including the need for listeners to wear earphones to hear mostly static-filled broadcasts. Additionally, the radio receiving sets were large, heavy, and expensive, and the public was generally uninterested. Also, engineers and executives who held the power in the radio industry were people who were not particularly interested in entertaining the masses.

WORLD WAR I

World War I started in Europe in 1914, but the United States did not formally get involved until 1917. After April 6, 1917, when the U.S. entered the war, the U.S. Navy took over operations of all commercial radio enterprises and began recruiting amateur radio operators and experimenters to provide the military with knowledgeable and experienced radio operators.

The federal government used radio to communication within its armed forces and among the armed forces of its allies. In addition, for security reasons, the government prevented foreign radio operators from transmitting in the United States. Therefore, the federal government took over the operation of all high-power radio stations in the country, including the point-to-point sending and receiving stations owned by American Marconi (the American subsidiary of Marconi's Wireless Telegraph and Signal Co.). And in 1917, it even closed down the amateur and experimental stations, which stopped the progress of radio as an entertainment medium.

The government's action led to two important developments in radio: First, operating a station during the war required the government to train many people to work as radio operators and technicians, and second, the government took control of all patents related to wireless communication and placed them in a 'patent pool' for all scientists and engineers during the war. This pooling of patents helped the war effort by stimulating the technological development of radio for military purposes. In turn, these developments helped stimulate the growth of the radio industry after the end of the war.

When the war ended in November of 1918 and recruits became civilians again, the Navy realized that it lacked the experience and expertise to maintain its control over the radio industry, but the federal government was in the position to maintain control of radio. Considering the vicious competition between telephone and telegraph companies, the monopolistic leanings of radio companies, and examples in Europe of government control, governmental control seemed like the best move to guide and regulate radio in the U.S. Nonetheless, experimenters, hobbyists, and commercial radio companies like AT&T and Marconi pressured Congress and President Woodrow Wilson to return radio to citizens and private industry. On July 11, 1919, the president acquiesced, and the military takeover of radio ended 8 months later on March 1, 1920.

RADIO CORPORATION OF AMERICA (RCA)

After World War I, the British-owned Marconi Company sought to strengthen its position as the leader in long-distance radio communication. It tried to buy a large number of the powerful Alexanderson alternators produced by General Electric (GE) for its American subsidiary, American Marconi, which essentially would have given Marconi a near monopoly in transatlantic radio. Because the U.S. government had just taken steps to avoid foreign control of radio in this country, it was opposed to selling equipment to British-owned Marconi. The U.S. government considered continuing its control of the radio industry, but it lacked the skilled operators needed to do so. Moreover, opposition from American Telephone and Telegraph (AT&T), the Marconi Company, General Electric, and other companies that contributed patents to the government patent pool during the war, along with amateur radio operators, was strong enough to convince the government to back off, allowing the radio industry to be guided by private enterprise.

GE bought a controlling interest in American Marconi, but soon thereafter GE decided that its strength and expertise were in manufacturing, not radio communication. It then formed RCA in 1919 to conduct its radio business. RCA took over the radio stations that were formerly owned by American Marconi. In the next few years, much legal wrangling occurred among AT&T, RCA, and GE over which company controlled broadcast equipment patents. From 1919 through 1921, GE, RCA, Westinghouse, and AT&T signed agreements to pool their patents, leading to a consortium of companies that would move the technology and business of broadcasting to a national level. By 1922, GE, Westinghouse, AT&T, and United Fruit Company (a small company that held desirable patents for radio equipment) had become the corporate owners of

RCA. Approximately 2,000 patents for radio equipment were pooled, and an effective manufacturing and marketing plan was enacted in which radio receivers would be manufactured by GE and Westinghouse but sold exclusively by RCA; in turn, AT&T would be allowed to charge for messages that were sent over its wires.

THE 1920s

Broadcasting as it is known today began during the 1920s. Some of the groundwork for commercial broadcast radio was laid as early as 1909 with Doc Herrold's and Frank Conrad's broadcasts. By mid-1920, Conrad had convinced his superiors at Westinghouse that the company could make money by selling premanufactured radio sets that could pick up programming from a radio station operated by Westinghouse. The inaugural broadcast for station KDKA was the presidential election of 1920. KDKA's programs consisted mostly of music, much of which came from live bands that performed in a tent on the roof of the building that housed the station. After high winds destroyed the tent, a bona fide studio was built so bands could play indoors with much better sound reproduction.

In 1920, Westinghouse was manufacturing and selling radio sets and was looking for ways to increase radio set sales. The company promoted and broadcast a program each evening in the hope that people who were listening at their neighbor's house would consider buying their own radios and get into the habit of listening nightly. The real goal for Westinghouse was to sell receivers and promote the name of the company. At this point in time, Westinghouse was aggressively marketing radio sets to the masses. It encouraged local appliance and department stores to sell its radios. For example, a store in Pittsburgh ran an ad in *The Pittsburgh Press* for amateur wireless sets selling for $10, a high-priced unit in 1920 dollars, but the sets could yield many hours of free entertainment.

FYI: 1920 Presidential Election

Until 1920, the sounds of presidential campaigns had been heard only by phonograph record. That changed on election night, November 2, 1920, when returns from the presidential election were broadcast live. With this, radio became a force in the political process, bringing the live events and real sounds of political campaigns directly to the voters.

Another innovation that came out of the 1920 presidential election was that election returns within a radius of 300 miles of Pittsburgh were received and transmitted by wireless telephone. This process was created by the Westinghouse Electric & Manufacturing Company and its subsidiary the International Radio Telegraph Company. The returns were received directly from an authoritative source and sent by a wireless telephone stationed in East Pittsburgh. Receiving stations of almost any size or type could catch the messages within the radius by using a crystal detector, a tuning coil, a pair of telephone receivers, and a small antenna. Using a two-stage amplifier, the operator would attach the receivers to the speaker on a phonograph so that messages could be heard anywhere in a medium-sized room.

At first the number of stations that went on the air after KDKA grew slowly. In fact, only 30 stations had been granted licenses to broadcast by January 1, 1922. But that changed, and by May 1, the number of stations increased to 218 and by March 1923, there were 556 licensed stations on the air. During 1923, many thousands of radio receivers were sold, which increased the demand for programming and thus stations.

Department stores and hotels set up radio stations on their properties to draw customers. Large stores and hotels were especially active in radio because they often owned tall buildings that provided good locations for antennas.

Live music was the most popular form of radio entertainment, and some sporting events, like heavyweight prize fights and World Series baseball games, drew huge audiences, consisting of as many as half a million listeners. Other people were drawn to radio for political programming, like President Warren G. Harding's Veterans Day speech from Arlington Cemetery near Washington, DC.

Factory-built receivers became common in American homes in 1921. A sophisticated receiver cost $60 and a simple one only $10. As the daily pay for the average worker at that time was about $1, few Americans could afford a sophisticated receiver, but many bought the less-expensive model. Westinghouse promoted sales of the sets in the towns where it had manufacturing plants by broadcasting radio programming. In 1922 inventor Edwin Armstrong boosted radio broadcasting with his new superheterodyne receiver. The superheterodyne allowed radio to be received at long distances with the help of a superheterodyne receiver from the transmitting station. Later that same year, a broadcast originating from London was received at station WOR in New York City.

By 1923, more than 600 licensed stations were broadcasting, although many of them went off the air after only a short time. Most of the owners of one or more stations were radio receiver manufacturers and dealers and businesses involved with electrical device repair. Many of the radio stations were put on the air as a sideline to the main business of the company that held the license.

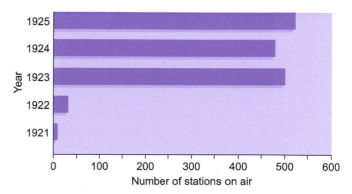

FIG. 2.7 Radio stations on the air, 1921–1925. The increase in radio stations on the air began in 1921 and exploded in 1922–1923, when the number went from 30 to more than 550. A slight decrease occurred after that boom, followed by another increase in 1924–1925.
Source: Sterling and Kittross, 2002, p. 827

In 1923, the federal government adopted the four-letter call sign rule, such that stations west of the Mississippi River were assigned K as the first letter and stations east of the Mississippi were assigned W as the first letter, with very few exceptions. KDKA in Pittsburgh is an exception because it was licensed shortly before this rule went into effect. Many colleges and universities put stations on the air during 1923 in the hope that doing so would help supplement the education of their students.

FYI: A Radio in Every Home, the Internet in Every Hut

The rapid technological advancement of early radio led H. P. Davis, a Westinghouse vice president (in 1922), to state, "A receiving set in every home, in every hotel room, in every schoolroom, in every hospital room. . . . It is not so much a question of possibility, it is rather a question of how soon" (Hilliard & Keith, 2001, p. 33).

Interestingly, President Bill Clinton made a similar statement regarding access to the so-called Information Superhighway, as presented by the Internet. In October 1996, Clinton stated, "Let us reach a goal in the twenty-first century of every home connected to the Internet and let us be brought closer together as a community through that connection" ("Clinton Unveils Plan," 1996). He also later stated, "Our big goal should be to make connection to the Internet as common as the connection to telephone today" (Internet in Every Hut, 2000). In 2009, President Barack Obama stated that a major component of putting the American Dream within reach of the American people is by expanding broadband lines across America to give everyone the chance to get online.

THE RISE OF COMMERCIAL RADIO

In the early 1920s, radio stations were often started for the purpose of supporting or promoting a product or service offered by the station owner. This strategy generally did not work, however. The number of radio receivers in the United States grew dramatically during the 1920s. In just 1 year, from 1923 to 1924, the number of sets went from 0.5 million to more than 1.25 million. This phenomenal growth was due in part to the fact that manufacturers were marketing inexpensive sets, which were affordable to most people. At this point, radio had truly become a mass medium. Even though the demand for radio receivers exceeded the supply, having a large audience did not guarantee success. The radio industry had yet to come up with a way to make radio pay for itself, let alone make a profit.

Each radio station aired its own programming. Lectures, news, political information, weather announcements, and religious sermons were the most common. Sports broadcasting came along later when transmitting equipment became portable.

Politicians were among the first to seize the opportunity to reach the public. Politicians preferred that voters hear their messages directly, not as interpreted (or edited) by newspaper writers. Going live on the radio was the best way for local, state, and federal politicians to speak directly to their constituents with an immediacy never before achieved.

WEAF

In 1922, WEAF—the AT&T-owned station in New York—came up with a novel way to generate revenue. Using part of the economic model for making money from the telephone system, WEAF acted as a **common carrier**, a company that transports goods or services for the public, and sold time to advertisers. This method of generating revenue, known as **toll broadcasting**, was similar to that used when a long-distance call was made and charged to the caller. Like the caller, the advertiser would pay a toll for the time used on the air. This concept was a critical part of the new economic model for supporting radio. It provided funds to the broadcaster to pay expenses, helped the advertiser reach an audience to sell a product or service, and kept broadcast programming free to the audience.

In retrospect it would seem that toll broadcasting should have been an instant sensation among broadcasters, but it did not catch on immediately, partly because of cross-licensing agreements among the companies that had shared in the patent pooling of World War I, which gave AT&T the sole right to 'charge' for messages. At a radio conference in 1922, U.S. Secretary of State Herbert Hoover disparaged the idea of toll broadcasting, stating that a service with as much promise as radio should not "be drowned in advertising chatter" (Hilliard & Keith, 2001, p. 30).

STATION INTERCONNECTION OR CHAIN BROADCASTING

WEAF also pioneered interconnecting its stations. Just after the beginning of 1923, WEAF sent a musical performance over the telephone lines (owned by AT&T, its parent company) to a station in Boston, and both stations broadcast the program simultaneously. This interconnection was called **chain broadcasting**, and though this term is not commonly heard today, it still appears in legal documents. The more common term used now is 'network,' but until 1926 it was referred to as chain broadcasting. Both terms refer to stations that are interconnected for the purpose of broadcasting identical programs, simultaneously.

THE COPYRIGHT ACT OF 1907

For years artists, writers, and composers had envied the legal protection given to inventors for their patents. Similar protections were extended to inventors, engineers, and scientists for their 'intellectual product' when Congress passed the Copyright Act of 1907. The American Society of Composers, Authors, and Publishers (ASCAP) was established in 1914 to collect 'royalty fees,' payment for the right to use a product, such as a song, on behalf of the composers and authors of songs and other owners of copyrighted material.

As programming included more and more recorded (phonograph) music, musicians, composers, and lyricists began to complain that radio stations were broadcasting their work without permission (and, more importantly, without the artists receiving any compensation). The stations felt, however, that broadcasting copyrighted phonograph music actually benefited the artists by promoting their work. This logic is still used today by stations and audio services when negotiating with copyright holders.

As a result of this conflict, ASCAP negotiated a fee with WEAF in 1923 to use copyrighted material. Specifically, WEAF agreed to pay $500 for the year. After a court case upheld ASCAP's right to negotiate these fees, the organization made broadcasting agreements with other stations. The broadcasters responded by forming the National Association of Broadcasters (NAB), a trade group that, among other actions, negotiated rates charged by ASCAP.

THE NETWORK SYSTEM

By 1926, individual stations were having a difficult time filling airtime. There were few phonograph recordings to play, so performers had to come to the station, which was burdensome. It became apparent that a system of shared programming broader than AT&T's existing chain broadcasting was needed. RCA, GE, and Westinghouse together formed the National Broadcasting Company (NBC), which was established as a programming network. NBC's main purpose was to provide programs to stations. NBC then bought AT&T–owned WEAF, which essentially took AT&T out of the ownership and chain broadcasting business and gave NBC a programming monopoly. Instead of just supplying stations with programs they could air at any time, NBC came up with the idea of establishing station affiliation. Stations would affiliate with NBC and agree to broadcast programs simultaneously. NBC initially affiliated with 19 stations. NBC's first live network program featured live orchestras and singers. The broadcast was carried by the affiliated stations and reached millions of listeners. By the end of 1926, NBC was successfully operating two major networks. The original NBC network was renamed NBC Blue, and the newly acquired AT&T WEAF chain broadcasting system was named NBC Red. NBC Blue and NBC Red—along with their later-formed lone network competitor, CBS—dominated broadcasting for the next 15 years.

The financial success of NBC led to other program networks. In 1927 the United Independent Broadcasters (UIB) started a network but it had limited success, primarily because it was not well funded. In fact, AT&T would not lease interconnecting lines to UIB because of the fear of nonpayment. Columbia Phonograph Company rescued UIB. The Columbia Phonograph Company and the Victor Phonograph Company were in direct competition. Victor was about to merge with RCA (the parent company of NBC), a move that worried Columbia because of RCA's name recognition and business power and that Victor might gain a competitive edge by having its records played on NBC stations. To hold off Victor, Columbia merged with UIB to form the Columbia Phonograph Broadcasting System (CPBS) so that it could play its records on its own network.

At first CPBS was financially successful mostly because it attracted big dollar advertisers, but the network quickly encountered financial difficulties. Realizing that CPBS could become a network powerhouse, William S. Paley, a wealthy cigar company magnate, bought a controlling share of the network. Paley became president of the network and changed its name to Columbia Broadcast System (CBS). Paley controlled CBS until 1983, becoming one of the most well-known electronic media moguls in the United States.

THE RADIO ACT OF 1927

The Radio Act of 1912 was adequate for regulating point-to-point radio as it existed from 1912 to about 1922. But when radio went beyond point-to-point communication and also provided point-to-multipoint communication, or **broadcasting**, this early legislation was no longer adequate to regulate commercial radio. The government was unprepared to deal with the proliferation of radio stations in the early 1920s. Instead of assigning each station to its own frequency on the dial, all stations were put on the same wavelength: 360 meters, about 830 KHz on the AM dial. With all of the stations in a given reception area operating on the same frequency, listeners could not hear any one station clearly; instead they heard a jumble of several stations whose signals were interfering with each other.

The government initially encouraged stations in the same geographical area to take turns broadcasting. For instance, Station A would broadcast from 6:00 to 8:00 p.m. and Station B would broadcast from 8:00 to 10:00 p.m. Some stations voluntarily ceded airtime to more popular stations by going off the air at night. This type of cooperation did not last long, especially when it cut into a station's advertising time.

In early 1927, Congress passed the Radio Act of 1927, which formed the Federal Radio Commission (FRC). The FRC was charged with organizing and administrating radio in the United States. Its specific responsibilities included issuing licenses, redesigning the use of the electromagnetic spectrum, and providing 107 channels for radio stations. Each station was then assigned its own frequency and power level, which significantly reduced interference among stations.

The FRC established licensing criteria. A station had to prove that it could provide enough funding to operate, and to be able to design and create its programming, a station also had to prove it was not attempting to form a monopoly and that it was not owned by a telephone company or trying to control a telephone company.

The FRC developed regulations for stations and networks of stations. According to the 1927 act, the U.S. Secretary of Commerce was authorized to inspect radio stations, license operators of stations, and assign call letters. The act stated that the FRC would be composed of five members, each representing one of the five geographical areas of the country, and each member was appointed to a 6-year term.

In addition, the act addressed the equality of transmission facilities and issues of reception and service. The act restated the concept that the public owns the airwaves, but individuals and corporate entities could be licensed to operate stations. Criteria for ownership also were established because the number of applicants competing for frequencies exceeded the number of frequencies available on the electromagnetic spectrum. The statement reflecting these criteria—of operating in the 'public interest, convenience, and necessity'—set the tone for case law in subsequent legal disputes over who controlled radio and why. In addition, it was made clear that government censorship of radio was not allowed.

Another feature of the 1927 act was that all existing radio licenses became null and void 60 days after its passage. This requirement forced all stations operating at the time to reapply for licenses, which allowed the FRC to assign frequencies to stations with the intent of minimizing interference and bringing some order to the chaos of the radio band. The result was that the powerful stations were treated well and given desirable frequencies, while the less powerful stations were given less desirable frequencies. Other stations, such as college stations, which had little power in a business or political sense, were simply forced off the air or bought out by commercial stations.

THE COMMUNICATIONS ACT OF 1934

By 1933, the importance of broadcasting had become very clear both to the federal government and to the population in general, but the government did not have a clear vision for the future. Although the Radio Act of 1927 had helped to organize the electromagnetic spectrum, assign channel allocations, and regulate radio, there were still some problems that Congress felt it needed to address. Several government agencies—including the Interstate Commerce Commission, the Federal Radio Commission, and the Department of Commerce—were regulating various aspects of electronic communication, but they were not working closely together. Realizing that radio needed better government oversight, President Franklin D. Roosevelt established a committee to study the nine different government agencies involved in public, private, and government radio. At the end of that year, the committee recommended that one agency should regulate all radio and related services. The result of that recommendation and subsequent bills sponsored in Congress was the Communications Act of 1934.

The Communications Act of 1934 incorporated most of the Radio Act of 1927 but made several updates. A new governing body, the *Federal Communications Commission* (*FCC*), replaced the FRC. The act established that the president of the United States would appoint seven FCC commissioners, no more than 4 could be from any one political party, and each would serve a 7-year term.

Substandard Programming

The FCC first set about tackling substandard radio programming, which ranged from fortune telling, huckstering medicinal cure-alls, and excessive advertising to issues of obscenity and religious intolerance. Although it could not censor, the FCC had the right to evaluate stations' policies and programs to make sure they were operating in the public interest. Between 1934 and 1941, the FCC examined many stations, but only two licenses were revoked, and only eight others were not renewed. Perhaps more important, the FCC made radio stations aware that it was watching them. A letter of inquiry from the FCC, often referred to as a 'raised eyebrow' letter, would usually get the station's attention and initiate station action to correct problems.

In 1934, the U.S. had three nationally operating radio networks: NBC Red, NBC Blue, and CBS. Radio receivers were highly desirable, and set sales continued to climb rapidly. Approximately 15 million American homes had radio receivers. Moreover, automobile manufacturers began putting a radio receiver in vehicle dashboards. In-vehicle radio began a love affair between drivers and stations and boosted radio listenership.

The Communications Act of 1934 included a section devoted to regulating political programming on the radio. *Section 315* of the act stated that any radio station that allowed a candidate for an elected office to use the station's time for political purposes had to allow all bona fide candidates for the same office an equal opportunity for airtime. It meant that opposing candidates must have the opportunity to buy an equal amount of time. News and public-affairs programs were excluded from this provision.

Sixty days before an election, candidates could purchase time at the lowest unit rate (cost-per-spot) available to any station advertiser. For example, if one candidate bought 25 60-second announcements in prime time, then all the other candidates also must be given the opportunity to buy 25 60-second announcements in prime time at the same price. *Section 312(a)(7)* stated that stations had to make a reasonable amount of time available to candidates for federal office.

The Communications Act of 1934 simplified government regulation of radio in the United States. But because it was enacted before the introduction of television, cable, satellite, microwave transmissions, and the Internet, the act required numerous additions and revisions over the next 60 years. Regardless, it was the single most important piece of legislation in terms of how it shaped the development of electronic media until the 1996 Telecommunications Act rewrite.

A NEW NETWORK

In 1934, another radio network entered the picture. The Mutual Broadcasting System (MBS) was a cooperative programming network, which, unlike the other networks, did not own any stations. The programming was created by member stations and sent out over the network. *The Lone Ranger* was Mutual's best-known radio show. The program originated at member station WXYZ in Detroit, MI. Mutual did not have the big-name stars or entertainment that was typical on NBC or CBS and thus could not attain comparable audience size or advertising desirability.

PRESS/RADIO WAR

As radio gained in popularity, newspaper readership and the number of daily newspapers in the United States began to decline. Although many factors likely brought about newspapers' hardship, the availability of news over the radio was a major contributor. Radio newscasts began in the 1920s, and NBC started a regular network nightly newscast with Lowell Thomas, a well-known newscaster, on its Blue network in 1930. NBC's initiation of nightly news signaled the beginning of a serious news effort to expand radio's influence and use.

Newspapers, already wary of radio stealing their readers, tried to force radio stations and networks to limit their newscasts. Both newspapers and radio stations employed staff reporters, but they were also highly reliant on the wire services for the bulk of their news. Wire services are news operations that supply stations and newspapers with information and stories. The wire services had more money than did the stations or networks and thus could send many more reporters out into the field. The newspaper industry, however, did not like that the wire services were sending stories to radio stations and thus in 1933 began pressuring the news wire services to release news reports only to the papers and to cease delivering stories to the stations. The newspapers started the press/radio war by getting the wire services to align with them and against the radio stations. The newspapers also refused to print radio program schedules without charge. The radio industry retaliated by hiring more freelance reporters to help its small news staff gather and report the news.

ZOOM IN 2.1

Radio and newspapers settled the war in 1933 by signing the Biltmore Agreement (named after the hotel in New York where the agreement was signed). This agreement limited network radio in the ways listed below:

1. Two newscasts per day: one before 9:30 a.m. and one after 9:00 p.m. (to protect the morning and evening editions of newspapers)
2. Commentary and soft (features and stories that are not necessarily time sensitive) news rather than hard (e.g., crime and time-sensitive) news reporting
3. Use of the Press-Radio Bureau, which would supply the networks and stations with news through subscriptions
4. No radio news-gathering operations
5. No sponsorship of news shows
6. A required statement at the end of each radio newscast: "You can read more about it in your local newspaper."

The Biltmore Agreement did not last long. Because the agreement was between the wire services and the radio networks, newspapers, and independent and local stations—especially those in big cities—the radio industry felt that it should continue gathering and reporting news for its demanding audience. Moreover, newscasters, who had previously simply read the news reports, also became **commentators** who therefore skirted the issue of reporting news. Essentially commentators did not just read the news; they made editorial comments as well. Eventually the wire associations ignored the restrictions of the Biltmore Agreement. Soon after the agreement was signed, the two major wire associations (International News Service and United Press) broke the agreement and continued tailoring their news feeds for broadcast. Even though the Biltmore Agreement was short lived, it is a classic example of efforts by an existing medium to slow the growth of and competition from a newer medium.

FYI: Disney v. Sony

The type of action that the newspaper industry took against the radio industry in the 1930s has been seen periodically since then, as in the case of *Disney v. Sony* in 1976. At the time, Disney Studios was trying to prevent the copying of Disney movies and other products by individuals with home videocassette recorders, the earliest of which were made by Sony. Disney was unsuccessful in this action. As in the case of the Biltmore Agreement, actions that have tried to stop new media technologies from developing have been mostly unsuccessful.

After the press/radio war, the radio industry became free to develop its news operations and cultivate broadcast journalists. Edward R. Murrow became one of the best reporters of his time. In the late 1930s, he broadcast dramatic and descriptive live reports of prewar events in Europe. After the war broke out, he took to the rooftops and reported on the live bombings of London. Murrow later hosted the radio show *Hear It Now*, which later moved to television and was renamed *See It Now*. Murrow set the tone for radio news reporting, which was emulated by other successful newscasters like Walter Cronkite and Dan Rather, who started in radio but became famous on television.

WORLD WAR II

After the Japanese attack on Pearl Harbor on December 7, 1941, the radio networks interrupted their regular programming to announce the attack. And the next day, President Franklin D. Roosevelt gave a speech that was listened to by 62 million people. In his speech, one of the most influential and often replayed speeches from that era, he described the attacks as being part of "a date which will live in infamy."

Upon entering World War II, the U.S. government immediately took steps to support the overall war effort. Amateur radio transmitters were shut down to prevent the possibility that military information would be sent to the enemy. Regular broadcasting at some stations in the western U.S. was curtailed to prevent enemy aircraft

116

THE SATURDAY EVENING POST

April 9, 1927

. . .   . . .

ATWATER KENT RADIO

Model 35, six-tube
receiver illustrated
with ONE Dial.
Speaker Model H.

Calming down

AN UPSTAIRS SET

Many are finding invaluable
a second radio installed upstairs. A bit of gentle, soothing music puts youngsters
quietly to sleep. It is price-
less in a sick room. When
the family set downstairs is
playing jazz, it is delightful
sometimes to slip away by
yourself and listen to music
more suited to your mood
and cultivated taste.

What a day! Adventure, thrills for the
youngster; for you—housework, irrita-
tions, fretting cares.

But now, before supper, music comes to
soothe nerves, to smooth the mind; bringing pleasant, quiet thoughts—sleepy
dreams.

"Music before meals," say the doctors.

You've no idea how comforting it is,
after you have selected a radio, to find
that in every way it is superior to sets of
your neighbors. To come home and appreciate how much purer, sweeter and
more *natural* is the tone of yours, how
much richer its volume.

And after watching others fiddling with
their sets, what a joy of simplicity is

your *single dial*, that filters from the air any
kind of music you want with just one turn!

The rather startling compactness of your
Atwater Kent Radio was achieved, not by
leaving anything out, but by engineering
skill, great precision and fine assemblage.

It has been the experience of more than
one million owners that Atwater Kent
Radio is peculiarly free from trouble—
never goes back on you.

Its big popularity explains the smallness
of its cost.

EVERY SUNDAY EVENING: The Atwater Kent Radio Hour
brings you the stars of opera and concert, in Radio's finest program. Hear
it at 9:15 Eastern Time, 8:15 Central Time, through:

WEAF	New York	WOC	Davenport
WEEI	Boston	KSD	St. Louis
WRC	Washington	WWJ	Detroit
WSAI	Cincinnati	WCCO	Minneapolis-St. Paul
WTAM	Cleveland	WGY	Schenectady
WGN	Chicago	WSB	Atlanta
WFI	Philadelphia	WSM	Nashville
WCAE	Pittsburgh	WMC	Memphis
WGR	Buffalo	WHAS	Louisville

Write for illustrated booklet telling the complete story of Atwater Kent Radio

ATWATER KENT MANUFACTURING COMPANY   *A. Atwater Kent, President*   4703 Wissahickon Avenue, Philadelphia, Pa.

FIG. 2.8 Advertisement for Atwater Kent Radio: 1927

ZOOM IN 2.2

Hear a clip from the audio of President Roosevelt's speech on December 7, 1941, by going to www.archives.gov/education/lessons/day-of-infamy

from using the broadcast signal to locate west coast cities. All shortwave stations capable of sending a signal overseas were brought under government control, and manufacturers of radio parts and equipment were required to convert from manufacturing consumer equipment to producing equipment that would directly aid the military. The building of new radio stations stopped. Materials that had been used to construct stations were deemed 'scarce resources' by the government. A small number of new stations got on the air between 1942 and 1945, but the government curtailed most of the growth of the industry.

Despite the hardships faced by Americans during these years, radio continued to be popular. It was free entertainment to those who owned radios, and it kept listeners aware of what was happening in the world. Radio provided entertainment and relaxation in a time of tension and, hence, drew large and devoted audiences.

Radio's prosperity during the war years was furthered by tax breaks. Fearing that some companies would gain huge profits from government contracts during World War II, lawmakers imposed a 90% excess-profit tax on American industry. Basically, the tax meant that for every dollar of profit a company made, it had to give back 90 cents in the form of taxes. But there was a loophole in the tax law that allowed companies to get a bit more for their money. That is, they could use their profits to pay for advertising and be taxed at the rate that existed before the war, or about 10%. Companies now had a huge incentive to ramp up their advertising, even during the hard times of the war. Business could either spend the money advertising or pay the money in taxes. Radio profited from the increased advertising. Even when the war effort minimized the number and variety of products a company had available to sell, they kept advertising for the purpose of keeping their name or product in front of the audience.

Additionally, because paper was mostly relegated to the government's war effort, newsprint was scarce. Hence, the daily editions were limited to fewer pages than before the war. Advertisers thus flocked to radio, which would fill its many hours of available airtime with as many commercial announcements and program sponsorships as advertisers were willing to buy. Newspapers suffered as they reluctantly turned away advertisers, who then bought radio advertising instead.

THE BLUE BOOK

After World War II, the U.S. economy was very prosperous, and restrictions on radio were lifted. The FCC again turned its attention to establishing guidelines for the radio industry. Although both broadcasters and the FCC were familiar with the phrase "public interest, convenience, and necessity," dating back to the Radio Act of 1927, opinions differed as to what it meant in practical terms. Additionally, many broadcasters had questions about the FCC's authority to enforce this vague standard for programming.

In 1946, the FCC attempted to clear the air about programming standards by issuing a report known as the *Blue Book* (because it had a blue cover), which outlined its philosophy of broadcast programming. The purpose of the report, officially named the Public Service Responsibility for Broadcast Licensees, dealt with four types of radio programming that were of interest to the FCC in relation to stations' responsible programming: (1) sustaining programs (not sponsored), (2) local live shows, (3) public-issue discussions/programs, and (4) advertising. Although stations complained that the *Blue Book* was restrictive, the FCC contended that it merely explained its policy as it related to programming.

The *Blue Book* was used as a benchmark for FCC policy over the next three decades despite changes in the industry with the advent of network television. Moreover, the *Blue Book* received generally favorable reviews from both the government and citizens' groups. One policy derived from the Blue Book that dealt with broadcasters' handling of controversial issues eventually led to implementation of the fairness doctrine that mandated the stations to seek balance when discussing controversial issues.

AM RADIO EVOLVES

Until the 1960s, most radio stations aired on the AM band from 540 KHz to 1610 KHz. All stations that broadcast in this band use amplitude modulation (AM) to carry voice communication. AM is a method of combining audio information with the basic carrier wave that is sent from the broadcast antenna to a receiving antenna. Amplitude modulation combines the audio with the carrier wave by varying the size or height of the wave (its amplitude).

AM radio's popularity diminished as television's popularity grew. Although the networks continued to supply programming to radio stations into the early 1950s, most radio comedy and dramas were being refashioned for television broadcasting. If a show was good enough to hear, it was probably good enough to be seen as well. Audiences eventually preferred the television versions, which made the radio versions unprofitable and left radio to come up with a way to fill the many hours of daily airtime. By 1955, AM radio had reinvented itself by becoming more music oriented, with an in-station announcer spinning records. Radio of the 1950s was no longer the daytime companion and nighttime focus of attention. Radio stations began to have their in-station announcers take air shifts in time blocks—for example, from 6:00 to 10:00 a.m. each morning—and play music, announce song titles and artists, and read weather or brief news reports. These announcers became known as disk (or disc) jockeys, or DJs, because most of the time they were playing phonograph records, or *disks*, on the air. Many stations, trying to differentiate themselves in a competitive market, selected a specific style of music and played it most of the time. The result was that stations specialized in musical genres like country-and-western music, African American–influenced music (known as rhythm and blues), classical music, popular music, and so on. Other stations like KFAX in San Francisco adopted an all-news format, and KABC, an ABC–owned station in Los Angeles, adopted an all-talk format.

In 1993, the AM band was expanded from the range of 550 to 1600 to include 1610 to 1700 kHz. This addition allowed local low power stations (e.g., college, religious, non-English, and government owned stations in locations like airports and national parks) to operate on the AM band without interfering with existing stations.

FM RADIO CAPTURES THE AUDIENCE

AM is susceptible to the static caused by thunderstorms and electrical equipment, which creates noise distortion on the receiving end. The fidelity (or sound reproduction) is limited such that AM cannot reproduce very-high-frequency sounds (such as the high notes from a violin or piccolo) or very-low-frequency sounds (such as the low notes from a bass drum or bass violin).

Edwin H. Armstrong, the inventor of the superheterodyne radio receiver, sought to eliminate the static and improve the fidelity of the radio signal. After many years of experimentation, Armstrong's patents were finally granted in 1933. In 1935, Armstrong gave a public demonstration of FM (frequency modulation) radio. He explained that

FM's audio quality was superior to AM radio because the **frequency** of each wave was modulated by sound rather than the **amplitude** of each wave.

FYI: The Armstrong/Sarnoff Conflict

Edwin H. Armstrong's FM radio invention seemed like a natural for the radio networks: less static, better sound, and a receiver that picked the strongest signal on the frequency without interference. Despite those technological advancements, David Sarnoff, the head of RCA and a friend of Armstrong's, decided against supporting FM. Rather, he wanted to spend more time and energy on the development of television and to avoid having to pay Armstrong for his invention. Later, Sarnoff testified in court that "RCA and NBC have done more to develop FM than anybody in this country, including Armstrong" (Lewis, 1991, p. 317). Armstrong fought Sarnoff and his company for patent infringement, vowing to continue "until I'm dead or broke" (Lewis, 1991, p. 327).

This conflict began with a lawsuit by Armstrong against RCA and NBC in 1948 and continued through 1953. By then, Armstrong had run out of money to pay his lawyers, and the prospect of receiving damages from RCA in the near future (lawyers estimated it would take until 1961) seemed remote. On January 31, 1954, despondent over a dispute with his wife and the continuing battle with Sarnoff and RCA, Armstrong jumped to his death from his 10th-story bedroom window, "the last defiant act of the lone inventor and a lonely man" (Lewis, 1991, p. 327).

How AM and FM Differ

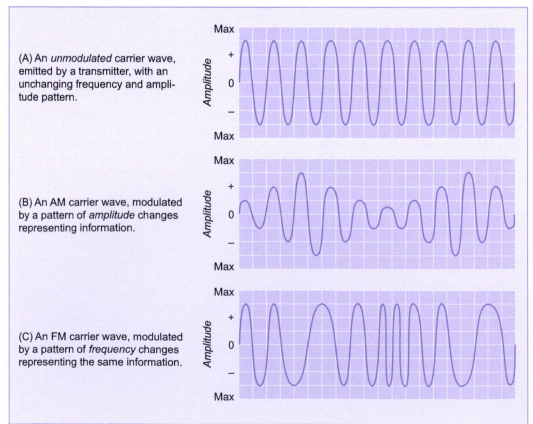

(A) An *unmodulated* carrier wave, emitted by a transmitter, with an unchanging frequency and amplitude pattern.

(B) An AM carrier wave, modulated by a pattern of *amplitude* changes representing information.

(C) An FM carrier wave, modulated by a pattern of *frequency* changes representing the same information.

DIAGRAM 2.1

Despite providing better sound and no static, FM broadcasting started out slowly, faltered, and then got a new life beginning in the 1960s. One important discovery that Armstrong made was that frequency modulation required more bandwidth. Instead of the 10 KHz channel used by AM broadcasting, FM required 20 times more space for each channel, or 200 KHz. The government set aside the 42 to 50 MHz band for FM radio beginning January 1, 1941. By the end of 1941, there were about 40 FM stations on the air, but many of the stations were not fully powered, and some were experimental. Further, the FM audience was limited because by 1941 only about 400,000 receivers had been sold that could pick up FM signals.

During World War II, interest in FM waned, and only a few stations remained on the air. The government decided that the original band reserved for FM radio was needed for government services, so it moved FM back to the original space in the electromagnetic spectrum, where channel 1 in the VHF television band was located. The FM band was later reassigned to the 88 to 108 MHz band in 1945, and FM radio broadcasting in the 42 to 50 MHz band ceased in 1948. As a result, many listeners owned FM radios that would no longer receive FM radio signals. FM stations did not operate profitably for some time, and total national FM revenues did not pass $1 million until 1948.

Despite providing better sound and little static, FM broadcasting started out slowly, faltered, and then got a new life beginning in 1961 when the FCC authorized FM **stereo broadcasting**. When listeners started to notice the superior sound quality of FM, and inexpensive receivers from Germany and Japan became readily available, the FM audience grew. The demand for new radio stations in the 1960s prompted the FCC to push new licensees into the FM band. However, car manufacturers were slow to include radios with FM receivers in their new models. Therefore, in the 1960s (with FCC urging), Congress passed legislation requiring all car radios to have FM receiving capability. At that same time, the FCC turned down a proposal for AM stereo, signaling the long, slow decline of AM as a music broadcaster and the rise of FM music. AM radio was better suited for talk than for music and FM radio for music rather than talk.

The counterculture of the 1960s, the rise of rock 'n roll, and high-fidelity stereo systems boosted FM listenership. The rise in audience popularity did not immediately translate into profitability for FM stations, however. Although revenues climbed from the early 1960s into the mid 1970s—from $10 million in 1962 to $308.6 million in 1975—more FM stations were losing money than making money.

It took until the late 1970s and early 1980s before FM radio gained an equal footing with AM radio. In 1978, the FM audience surpassed the AM audience for the first time. By the late 1980s, the FM audience was much larger than the AM audience, commanding almost 75% of radio listeners. Almost all car and portable radios then had an easily tunable FM receiver, which made FM as easy to find and listen to as AM.

SEE IT NOW

SATELLITE RADIO

The biggest disadvantage to over-the-air radio (also referred to as terrestrial radio) is the limited signal range. Some stations are more powerful than others, and AM typically covers a larger area than FM, but when driving long distances, signals are often lost between cities. Long-distance drivers have to frequently retune to local stations. Satellite-delivered radio, however, solves the range problem because satellite signals cover huge distances.

In 2001, XM Satellite Radio began sending a satellite signal to cars and homes equipped with special satellite receivers. XM was formed with the help of investors from automobile manufacturing, broadcast radio, and satellite broadcasting, namely, General Motors, American Honda, Clear Channel (the nation's largest radio group), and DirecTV. XM's only competitor, Sirius Satellite Radio, also entered the market in 2001. Both services provide hundreds of channels of audio service, including commercial-free music and news, sports, talk, and children's programming. Slow subscriber rates led to the merger of SiriusXM in 2008. The combined company is now known as Sirius XM Radio.

SiriusXM beams radio signals to two satellites. The signal is then downlinked to satellite receivers, both stationary, as in a home, or mobile, as in a vehicle. In addition to satellites, small terrestrial transmitters called *repeaters* augment the signal to ensure that listeners in big cities can pick up a satellite signal even if it is blocked by tall buildings.

Satellite radio got off to a slow start. Auto manufacturers did not begin to install satellite radio–capable receivers in new cars until 2003. All other cars had to be retrofitted with a new receiver or adaptor of some sort to receive the signal.

Although it was predicted that satellite radio would lead to the demise of over-the-air broadcasting, it has not become the major competitor to local broadcast radio that many feared. Radio stations often program with localism in mind and that has a strong appeal. Perhaps listeners rely on broadcast stations for local information more so than for selection of music. The absence of local radio, as a weakness of satellite radio, demonstrates the strength of local broadcast radio.

RADIO GOES DIGITAL

Radio broadcasting has traditionally been an analog medium and, compared to analog television broadcasting, uses far less of the electromagnetic spectrum for each channel. AM radio channels are 10 KHz wide and FM radio channels are 200 KHz wide. Compared to the size of a television channel (6 MHz or 600 times the width of an AM channel), radio is a more efficient use of the spectrum. Both broadcasters and the FCC have been seeking a more efficient use of the spectrum through digital rather than analog broadcasting.

Digital broadcasting allows several separate signals to use the same space as one analog signal. In addition, digital broadcasting provides better signal reproduction and essentially better sound.

Although the FCC has not mandated that radio stations change to digital, many stations have initiated the change on their own. Both the FCC and the radio industry wanted a switch to digital while keeping all existing radios from becoming obsolete (which is what happened when the FCC mandated the FM band to change frequencies in 1945). A system called **IBOC (in-band, on-channel)**, enables broadcasters to use their existing frequencies to broadcast in digital and analog at the same time. The digital signal is of higher quality than the existing analog service. Audience members who are happy with analog can stay with analog. iBiquity Digital, which licenses the software to radio stations (and consumer electronics manufacturers) under the brand name HD Radio, owns IBOC technology.

HD radio has been slow to make inroads, as the lack of a mandated change has presented broadcasters and listeners with a 'chicken or egg' quandary. In 2009, less than 15% of all the radio stations in the United States were broadcasting a digital signal. By 2016, more than 2,300 stations were broadcasting in HD. Although 78% of all radio listening is on stations broadcasting with HD radio, many listeners are unaware of the difference between analog radio and

ZOOM IN 2.3

You can go to www.ibiquity.com for more information about terrestrial broadcast HD radio. For examples of other radio/audio services that are Web based, go to www.pandora.com or www.accuradio.com

HD radio because all of the main channels for radio can be received on any radio receiver, HD capable or not.

Digital or HD radio provides additional signals to each station. More signals (or channels) gives broadcasters more to sell and thus additional revenue streams for local stations. The size of the bandwidth available to licensed broadcasters and the ability of digital broadcasting to make very efficient use of the band through compression, additional signals, known as digital subchannels, are available (although the more subchannels that are created, the smaller the available bandwidth, which can affect sound quality).

For example, one classic rock station can offer a narrowly targeted classic 1970s channel, a 'deep cuts' album track channel, and a 1990s 'classic alternative' station, all on the same frequency that previously allowed only one analog signal. The IBOC receiver is capable of receiving and separating these signals so they will be easily tunable by the digital radio audience. The primary channel is typically designated HD1 (e.g., WCBS-HD1), with digital

subchannels designated HD2 and HD3. Other ancillary services are being tested. For instance, breaking news, sports, weather, and traffic information can be delivered separately from the main audio program on a subchannel. Another available feature is iTunes tagging, which allows audiences to capture and store artist and song information, which can then be synced with their MP3 player. Users can also pause, store, fast forward, index, and replay audio programming, giving audiences much more control over their radio listening experience.

Consumers in many markets have very little incentive to purchase digital radio receivers because their old radios receive the main signal of all radio stations. The relative lack of consumer interest slowed the rollout of both home and car radios capable of receiving digital signals, and auto manufacturers were at first slow to make the radios available in new cars. By 2016, all major car manufacturers supported HD radio technology with optional HD radios. Since most of the HD radio stations are located in larger markets, many consumers in medium, small, or rural markets do not gain any radio signals by spending the extra money for an HD-capable radio.

Elsewhere in the world, many broadcasters began digital radio broadcasting in 2003. However, they operate in a different portion of the spectrum and use a different standard than the one adopted by the United States.

INTERNET RADIO

Internet radio began in the late 1990s, when streaming via the Internet became a technologically feasible method of transmitting music. Broadcast stations added an online component to their over-the-air signal to provide their listeners additional ways of connecting to their stations. Internet radio can be heard via computers, tablets, some MP3 players, and smartphones.

In addition to licensed broadcast stations streaming their signals over the Internet, many nonlicensed 'radio' stations have been streaming music, talk, news, and entertainment. Notable among these 'netcasters' are music services like Pandora and Spotify. These services have a strategy of enabling the listeners to interact with the service to create individual playlists and personal 'radio stations.' SiriusXM has an Internet component for its subscribers to provide its music, talk, sports, and entertainment programming anywhere.

In a study done in 2016, it was found that two-thirds of all smartphone owners stream music daily (Washenko, 2016, March 11). This finding points to a dilemma faced

ZOOM IN 2.4

Not all 'radio' stations are broadcasters. Many stations are Internet only.

Finding radio stations on the Internet is easy. You can try these sites for up-to-date lists of stations: www.radio-locator.com and www.radiotower.com

by traditional radio stations. Unless people are in their cars, a large portion of the listening audience is getting audio service through their smartphones.

NEW RADIO TECHNOLOGY AND COPYRIGHT ISSUES

Until the late 1990s, radio was an over-the-air service that was programmed by radio station programmers and listened to by a passive audience. Listeners could call in to request music or respond to talk show hosts, but for the most part radio was a one-way, linear programmed service. Audience members could record shows on tape and play them back later, but that was not a very common practice. The passive radio began to change when digital technology spawned online radio stations, audio services, and peer-to-peer music file sharing.

Beginning in the late 1990s, online file sharing of copyrighted music became a serious problem for the music industry. Internet users could go to various **peer-to-peer sharing** sites, such as Napster, and download music by copying files from other users who had connected to the site. The music industry claimed that it lost substantial revenue because so many people were getting free music online instead of buying CDs. Despite the threat of legal action, individual users have continued to download music and even feature-length movies without paying for them. The music and movie industries continue to pursue copyright violators.

Copyright law originated in Article I, Section 8(8) of the U.S. Constitution, which allowed authors rights to use their 'writings and discoveries' for their own benefit. Copyright law has since been revisited and revised, and it has been upheld in the courts, generally protecting the creators of original works from unauthorized use.

Specifically, someone who creates an original work can receive copyright protection for his or her lifetime plus 70 years for work created after January 1978 and 95 years for work created before that time. For example, a song written in 1985 and copyrighted by the composer will receive copyright protection until 70 years after the songwriter dies. If anyone wants to use the words or music to that song for commercial purposes, he or she must obtain written permission from the songwriter or an agency designated to grant permission on the songwriter's behalf. Usually, the user of the material pays a fee to the creator for the privilege of using it.

Sometimes authors and others allow a licensing agency to collect usage fees for them. In the case of musicians who create original music, certain organizations negotiate and collect fees from others who wish to use their music. In the United States, there are two large music-licensing agencies that perform this function. ASCAP was started in 1914, and Broadcast Music, Inc. (BMI) was started in 1940. BMI was formed during the height of network radio popularity to compete with ASCAP and be friendlier to the radio stations and networks.

These organizations negotiate **blanket fees** with users like broadcasters and production companies. A blanket fee is determined by using a formula to calculate the yearly amount that a radio station will be charged to use all of the music licensed by the organization. That amount is based on factors such as the percentage of that organization's music the station plays per week, the size of the station's market, and the station's overall revenue. Large stations in large markets pay more than small stations in small markets.

If a copyright expires, then anyone can use the material without asking permission or paying a fee. Once a copyright has expired or if a work was never copyrighted, the material is considered to be in the **public domain**. Advertisers, performers, and writers like to use material that is in the public domain for their projects, because no permission or payment is necessary. Public-domain material is particularly attractive to those producing low-budget projects.

Educators and others can use copyrighted material without getting permission or paying a fee if their use of the material is noncommercial and limited. This allowance falls under Section 107 of the 1976 Copyright Act and is referred to as **fair use**. Using copyrighted material in this way must be carefully done, however, to be legal. To determine fair use, four issues should be considered: (1) the purpose or use of the material, (2) the characteristics of the original work, (3) the amount of the original work used, and (4) the possible impact that the use might have on the market for the original work.

The *Recording Industry Association of America* (*RIAA*) is the trade organization that represents the people and companies that produce 90% of the recorded music in this country. It aggressively attempts to identify people who illegally share files containing copyrighted music, and when it does, it often prosecutes them for copyright violation. In late 2003, the RIAA filed hundreds of lawsuits against individual music file sharers.

The *Digital Millennium Copyright Act* (*DMCA*), passed in 1998, was designed to protect creative works in this digital era. It prohibits the manufacture and distribution of devices or procedures that are designed to violate copyright law in the digital environment. In addition, this law requires Internet service providers (ISPs) to identify their customers who violate copyright law by using file-sharing services, and the RIAA uses this information to take action against these people.

Webcasters interested in playing copyrighted music can work with SoundExchange, an organization that represents record labels similarly to how ASCAP represents composers, authors, and publishers for music licensing. SoundExchange was originally a division of the RIAA that was formed to collect royalties resulting from the DMCA. It was spun off and became an independent nonprofit organization in late 2003.

An 'intellectual property' right closely connected to copyright is performance right. Performance rights provide the performer with the legal protection of their product, such as the performance of a musical piece or an announcer's voice in a commercial. Traditionally, performance rights were negotiated in a performer's contract with a show's producer. In some instances, such as commercials, actors as performers were granted residuals (similar to royalties) if the airing of the commercial exceeded a set number of airings. However, in the early part of this century, commercial announcers demanded extra payment for commercials produced for broadcast that aired on the Internet. With the widening distribution of performers' works on the Internet (and other digital media), performers are now demanding intellectual property right protection of their work, similar to copyright and patents. This extension of intellectual property rights protection into the area of performance rights, especially on the Internet, appears to be the next 'copyright' issue.

SEE IT LATER

INTERNET RADIO

Competing services now supply much of what radio has supplied for the last 90 years. Music is heard over the Internet 24 hours a day, 7 days a week, from thousands of sources all over the world. Internet and satellite providers mimic licensed radio stations free from the content restrictions imposed on broadcast stations by the FCC.

Until several years ago, online radio could not compete with broadcast radio because it was not portable. But now with wireless Internet signals (Wi-Fi) and mobile devices (smartphones, tablets, iPods), Internet radio can be listened to almost anywhere. Broadcast radio stations need a strategy for maximizing their potential via the Internet. Even the automobile, once the exclusive domain of broadcast radio, now offers online radio stations and audio services.

Broadcast radio is going through the same challenge that it did in the 1950s, when television took over its audience. Radio once again must reinvent itself to ensure its viability. Digital technologies are important parts of that process, but the key is developing new content. Radio stations need to offer audiences programming and other services that they cannot get anywhere else. A return to more local content is one option, as is a general emphasis on talent rather than narrow and formulaic music offerings.

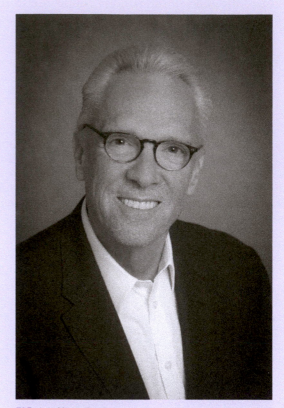

FIG. 2.9 Norm Pattiz

Pattiz's new venture is called Podcast One. Taking advantage of an audience that is comfortable with talk entertainment about a large variety of topics, Podcast One distributes talk shows that program to very specific audiences as well as general audiences. This business model avoids using radio stations as the middleman, essentially going 'over the top' in audio by bringing podcast programs directly to the listening audience. Podcasts about sports, comedy, politics, news, society and culture, TV and film, and health topics are all available to audiences through one Web site: podcastone.com

ZOOM IN 2.5

Changing Audio Business

The business of audio has been changing, due at least in part to technology, but also in part to audience behavioral changes. One industry leader, Norm Pattiz, has been aware of these changes and has created companies to take advantage of them.

Norm Pattiz began Westwood One in the 1970s to take advantage of distributing syndicated audio programs to radio stations via satellite rather than tape or LP. At first Westwood One distributed its programming to its Mutual Broadcast System affiliates. The business grew dramatically to eventually serve more than 5,000 radio stations, providing more than 150 news, sports, music, talk, and entertainment programs, features, and live events. Westwood One also provided local news, sports, and weather to over 2,300 stations.

OTHER TECHNOLOGICAL CONSIDERATIONS

Although broadcast radio has always been portable, it has been virtually blocked from reaching mobile users on their cell phones. Most smartphones have the technological capability to receive radio broadcast signals, but as of 2016, very few phone users have activated the chip in smartphones necessary for FM reception. In 2014, one company, Blackberry, activated the chip that provides in its phones the capability to receive radio, but Blackberry phones make up a very small percentage of total cell phone ownership, and it appears that Blackberry phones will not regain their former market strength.

Smartphone apps, however, may be the answer broadcasters are looking for to reach the mobile audience. One such app, TuneIn Radio, lets users access local stations or distant stations with a few taps on their phones. Some industry observers predict that broadcast companies may choose to start a business that competes directly with Pandora. This represents a paradigm shift for radio, from a linear programming model and a passive audience to a random-access model and an interactive audience. One study found that 68% of the respondents who own a smartphone streamed music daily and spent almost 45 minutes each day listening (Washenko, 2016, March 11).

In-car radio listening has long been the domain of terrestrial radio. Although radio listening in the car is still strong despite competition over the years from 8-track tapes, cassette tapes, satellite radio, CDs, and MP3s, the connected car (with a mobile Internet connection) threatens that dominance. If listening to a digital Internet-based system becomes as easy as pushing a button on the car radio, broadcast radio audience size may suffer additional setbacks.

Radio will probably experience a continued decline in the amount of time listeners spend on broadcast delivery because of competition from the many competing audio services and other entertainment and information options. However, mobile delivery through smartphones, tablets, and other devices could actually boost radio listening.

THE ROLE OF GOVERNMENT

The FCC has long been interested in preserving radio's localism. Over the years, the agency has encouraged radio stations to serve their communities. Local broadcast television and radio stations enjoy a competitive advantage over satellite services, because they provide the programming, news, and talk important to people in their community.

The FCC has also granted licenses to several hundred low-power FM (LPFM) stations across the United States. These stations offer nonprofit organizations, such as religious groups, academic institutions and schools, and community organizations the opportunity to reach local audiences with just 100 watts of power, enough power to reach listeners within a few miles of the station. Localism might be rejuvenated, at least somewhat, through stations like these. Hyperlocalism, localism within a small neighborhood, has become possible through broadcast radio.

So far, SiriusXM has not been required to provide local stations to subscribers, although it does provide local traffic and weather. The FCC may decide to allow SiriusXM to provide local programming, commercials, and even news. If the FCC does make this decision, it could change the revenue stream for all electronic media and have serious implications for local broadcasters. Local broadcasters would be forced to compete for local advertising dollars with SiriusXM, and some siphoning off of local dollars would certainly occur. Thus, local broadcasters would have to fight even harder to keep their revenues from shrinking.

Local radio faces additional regulatory battles. Traditional radio stations enjoy one continued advantage over newer digital services: satellite radio, Internet music services, and digital cable music providers all pay two separate performance royalties for the music they play, the first to the songwriter and music publisher and the second to the record labels and recording artists. Because of a quirk in U.S. copyright law, traditional radio stations only pay the songwriter and music publisher. As music sales dropped in the early years of this century, the music industry scrambled for new revenue streams and, in an echo of the 1920s copyright battles, the RIAA began actively lobbying Congress to revise the law and introduce an additional royalty for broadcast radio that would be paid to the labels and artists. The rule for broadcasters requiring payment only to the songwriter and music publisher was maintained because Congress believed that playing a song on the air encourages people to buy the music; the promotional value was thus considered to be adequate compensation to the performers because the record labels and artists made money from the sales of their albums.

FINANCIAL OUTLOOK

Online services like Pandora and Spotify continue to gain strength in subscriber revenues and advertising revenues. Satellite radio continues to aggressively seek program content to attract subscribers and advertisers. As audience targeting becomes more sophisticated, advertisers seek listeners regardless of the audio content or the audio delivery system. Radio station owners continue to pursue listeners to try to convince sponsors that they can deliver the desired target audiences.

Recently broadcast radio advertising revenues have declined slightly. There was a 1% decline from 2014 to 2015, but digital radio (including the streaming of broadcast radio) rose 5%, reaching more than $1 billion for the first time. This increase indicates that advertisers are recognizing radio's digital reach.

SUMMARY

Electronic media communication has changed over the years in response to the human desire to go beyond face-to-face contact. Since the beginning of the 20th century, humankind has developed the technology to reach people over long distances in a matter of seconds. First using the wired telegraph and telephone and then using radio telegraphy and telephony, people have been able to communicate both one-to-one and one-to-many. The ability to communicate one to many using radio signaled the beginning of electronic mass media. The excitement generated by broadcasting lured many people to experiment with radio, both transmitting and receiving. Many hobbyists built their own radio receivers, and a number of them also dabbled in radio transmitting.

Entrepreneurs and inventors like Guglielmo Marconi, Lee de Forest, Edwin H. Armstrong, Frank Conrad, and David Sarnoff propelled radio from an experimental system to an industry and storehouse of American culture. From its modest audience size in 1920 to its peak in 1950, radio was the dominant mass medium.

The U.S. government has played a role in the development of the radio industry by ensuring that control stayed in the hands of American companies, as evidenced by its seizure of all powerful radio transmitters during World War I. Rather than keep control after the war, the radio industry became a commercial enterprise, guided by market factors more so than government intervention. At first, radio stations experienced numerous problems with technology, some of which stemmed from all stations broadcasting on the same frequency. The government corrected that problem by establishing separate frequencies for stations in the same market and region with the Radio Act of 1927 and the establishment of the Federal Radio Commission. The government also established the philosophy that the airwaves belong to the people and that broadcast stations must operate in the "public interest, convenience, and necessity." The print media embraced radio to a certain extent. For example, many newspapers added a section for radio programming schedules, discussions of programs, and even technical tips for better reception. There was a definite ambivalence displayed by newspapers when radio began broadcasting live news reports, a practice that newspapers tried unsuccessfully to impede during the press/radio wars in the early 1930s.

Radio exposed the American audience to the concept of free entertainment and information programming (once the initial price for the receiver was paid). Although newspapers were very inexpensive, radio programming was free and could be enjoyed in unlimited amounts by the audience. Moreover, it did not require literacy. Radio also encouraged people to stay home and listen to free programs rather than go to vaudeville shows at their local theaters. 'Talking' motion pictures, a product of the late 1920s, drew large audiences but did not seem to slow down radio's growth. The phonograph record industry was forced to cope with their customers who now could receive free music on the radio, rather than pay for phonograph records that were expensive and had lower-quality sound.

Radio also exposed listeners to the voices of politicians, celebrities, sports heroes, and even common people. Audiences heard different regional dialects and accents. Politicians embraced radio as a means of reaching their constituency with messages that were tailored to their audience.

Networks NBC Blue and NBC Red and CBS provided radio programming from the late 1920s through the 1940s. These networks' programming innovations set the stage for many years of audience loyalty and appreciation. In fact, many of the program types developed during these years made the transition to television and continue to the present. Radio strongly influenced American society by providing free entertainment and information and exposing listeners to voices of celebrities and government officials. Radio also siphoned some of the interest away from newspapers.

AM radio lost its network entertainment programming and prominence in the minds of the audience when television was introduced after World War II. But AM radio reinvented itself by developing music formats hosted by disc jockeys. FM radio, which rebounded from a serious setback when the FCC changed its band location, gained dominance in musical programming after the introduction of stereo broadcasting in the 1960s. By the 1980s, the FM audience was larger than the AM audience. Once again, AM had to reinvent itself, which it did by concentrating programming more on talk, news, and religion instead of music.

The Telecommunications Act of 1996 triggered a dramatic increase in broadcast station owner consolidation, because it relaxed ownership rules, and this consolidation has led some industry watchers to criticize radio for losing its localism. Alternative delivery systems have further fragmented the radio audience and, along with various economic and regulatory issues, created challenges for the radio industry. However, many of these services cannot compete with traditional radio when it comes to providing locally oriented news and entertainment using a technology that is both portable and without direct cost to the audience. Changes in audience behavior signal a change in how the industry will continue to operate. Podcasting may change the audience for talk radio. A recent study found that more than two-thirds of smartphone owners spend time streaming music daily. Most people acquire radios when they buy a car. Few radios are bought for homes or dorm rooms. The future points to a world where mobile listening is done on a smartphone via streaming or in an automobile, where the dashboard is changing to include Internet connectivity and therefore easy access to non-broadcast audio services.

BIBLIOGRAPHY

Abbot, W. (1941). *Handbook of broadcasting* (2nd ed.). New York: McGraw-Hill.

Banning, W. (1946). *Commercial broadcasting pioneer: WEAF experiment, 1922–1926*. Cambridge, MA: Harvard University Press.

Beauchamp, K. (2001). *History of telegraphy*. London: Institute of Electrical Engineers.

Benjamin, L. (1993, Summer). In search of the "radio music box" memo. *Journal of Broadcasting and Electronic Media*, 325–335.

Broadcasters start digital radio service. (2003, June 17). *USA Today*. Retrieved from: www.usatoday.com/tech/news/2003-06–17-digital-radio_x.htm [July 1, 2003]

Clinton: "Internet in every hut." (2000). *Reuters Wired News*. Retrieved from: www.wired.com/news/print/0,1294,34065,00.html [July 15, 2002]

Clinton unveils plan for "next generation of Internet." (1996). *CNN*. Retrieved from: www.cnn.com/US/9610/10/clinton.internet [July 15, 2002]

Faruk, I. (2014, April 23). 2 headwinds for Sirius XM radio you shouldn't ignore. *The Motley Fool*. Retrieved from: www.fool.com/investing/general/2014/04/23/2-headwinds-for-sirius-xm-radio-you-shouldnt-ignor.aspx

Ferguson, D., & Greer, C. (2011). *Local radio and microblogging: How radio stations in the U.S. are using Twitter*. UK and Europe: Broadcast Education Association, Taylor & Francis.

Floherty, J. (1937). *On the air: The story of radio*. New York: Doubleday, Doran & Co.

Goldman, D. (2010, February 2). Music's lost decade: Sales cut in half. *CNN.money.com*. Retrieved from: www.money.cnn.com/2010/02/02/news/companies/napster_music_industry/

Goodman, M., & Gring, M. (2003). The radio act of 1927: Progressive ideology, epistemology, and praxis. In M. Hilmes (Ed.), *Connections: A broadcast history reader* (pp. 19–39). Belmont, CA: Wadsworth.

Gordon McLendon and KLIF. (2002). *Encyclopædia Britannica*. Retrieved from: www.britannica.com/%20/%20facts/5/71004/KLIF-as-discussed-in-Gordon-McLendon-and-KLIF

Gross, L. (2003). *Telecommunications: Radio, television, and movies in the digital age*. New York: McGraw-Hill.

Heine, P. (2009, July 19). Stream it like you mean it. *Media Week.com*. Retrieved from: www.adweek.com/as/content_display/special-reports/other-reports/e3i4d0b1b4303c8399766735f5b52963ebe

Hendricks, J., & Mims, B. (2015). Keith's *radio station: Broadcast, internet, and satellite* (9th ed.). Burlington: Focal Press.

Hewitt, L., Krause, A., & North, A. (2014). *Music selection behaviors in everyday listening*. USA and North America: Broadcast Education Association, Taylor & Francis.

Hilliard., R., & Keith, M. (2001). *The broadcast century and beyond* (3rd ed.). Boston: Focal Press.

Lessing, L. (1956). *Man of high fidelity: Edwin Howard Armstrong*. Philadelphia: Lippincott.

Lewis, T. (1991). *Empire of the air: The men who made radio*. New York: HarperCollins.

Pandora, You Tube, AM/FM, Spotify compete for ears as American listenership evolves: Infinite Dial. (2016). *Rain News*. Retrieved from: http://rainnews.com/pandora-youtube-amfm-spotify-compete-for-ears-as-Americanlistenership-evolves

Public interest, convenience, and necessity. (1929, November). *Radio News*. Retrieved from: www.antiqueradios.com/features/frc.shtml

Radio is officially America's number one mass research medium. (2015, June 24). New York: Radio Advertising Bureau. Retrieved from: www.rab.com

Settel, I. (1960). *A pictorial history of radio*. New York: Citadel Press.

Siepmann, C. (1946). *Radio's second chance*. New York: Little, Brown and Company.

Sterling, C., & Kittross, J. (2002). *Stay tuned: A history of American broadcasting*. Mahwah, NJ: Erlbaum.

Washenko, A. (2016, March 4). Radio's digital revenue breaks $1 billion in RAB 2015 survey. *Rain News*. Retrieved from: rainnews.com/radios-digital-revenue-breaks-1-billion-in-rab-2015-survey

Washenko, A. (2016, March 11). Two-thirds of smartphone owners stream music daily (study). Rain News: Chicago. Retrieved from: rainnews.com/two-thirds-of-smartphone-owners-stream-music-daily-study/

Webster, G. (1998). *The Roman imperial armies of the first and second centuries* (3rd ed.). Norman: University of Oklahoma Press, p. 255.

Whitmore, S. (2004, January 23). Satellite radio static. *Forbes.com*. Retrieved from: www.forbes.com/2004/01/23/0123Whitmore.html

Television: From Radio With Pictures to Video Screens Everywhere

3

Contents

SEE IT THEN

THE EXPERIMENTAL YEARS

EARLY INNOVATIONS

While some experimenters and inventors worked with radio waves to send audio across distances, others were more interested in transmitting live pictures. Experiments in television began in the 1880s. Early thinking about how to send pictures using electricity was divided between two methods: mechanical scanning and electronic scanning. Mechanical scanning employed a spinning disc system that used one disc to record and send the visual image and another disc for viewing it. Paul Nipkow developed a mechanical scanning system in 1884 in Germany. In 1926, Scottish engineer John Logie Baird developed a workable mechanical scanning system to send live television images. The British Broadcasting Corporation (BBC) adopted his system and began broadcasting television programming in 1936.

By today's standards, the John Logie Baird system was primitive, using only 30 horizontal lines of information. While the UK was standing behind mechanical scanning, in the U.S. attention was on electronic scanning. Eventually, the UK dropped Baird's mechanical system in favor of electronic scanning because electronic scanning yielded more lines of resolution. The more lines, the clearer the image. Until 2009, analog NTSC (National Television Standards Committee) television used 525 lines; now digital ATSC (Advanced Television Systems Committee) television can use up to 1,080 lines per frame. The biggest problem with the mechanical system was the blurry images, and there did not seem to be a way to improve the picture. In 1922, inventor Philo T. Farnsworth, an Idaho high school student, sketched out a system for electronic television. Farnsworth joined the Navy a few years later, but he learned that if he pursued electronic television, any patents he developed would become government property. So he left the Navy and enrolled at the University of Utah. While there, he made many connections and eventually lined up backers to fund his research. Farnsworth set up a lab in Los Angeles, where he continued his work developing an electronic scanning system.

With more experience than Farnsworth and with Westinghouse to back him, researcher Vladimir K. Zworykin is historically credited as the primary force behind electronic scanning. In 1923, he developed a working electronic television scanning system that produced a better

41

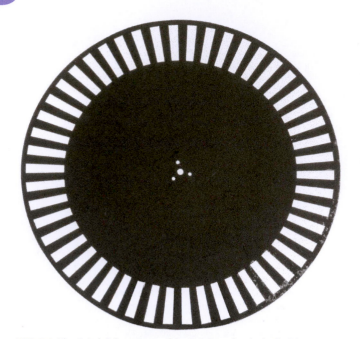

FIG. 3.1 The Baird disk was part of the Baird mechanical television system.
Photo courtesy of MZTV Museum

FIG. 3.2 An actor performs in an experimental TV studio in 1928.
Photo courtesy of MZTV Museum

FIG. 3.3 British family watching a 1930 Baird Televisor, a mechanical scanning TV.
Photo courtesy of MZTV Museum

picture without using the spinning disks of mechanical television. Instead, his system relied on the **iconoscope**, a special photosensitive camera tube that converted light into electrical energy.

ZOOM IN 3.1

View a video clip about the origins of television at www.farno vision.com/media/origins.html

By the early 1930s, Zworykin had been granted a number of patents that improved the electronic system, but he could not get his superiors at Westinghouse to pay much attention to his work. But RCA managers were interested and recruited Zworykin to work for them. RCA was also eyeing Farnsworth's work and tried to recruit him as well, but Farnsworth instead joined the Philco company and moved to Philadelphia. Farnsworth's time at Philco lasted only a few years, and he eventually returned to the laboratory in Los Angeles. In the meantime, Zworykin and Farnsworth took several patent battles to the courtroom. Although he gained some smaller victories, Zworykin lost out to Farnsworth when RCA agreed to pay to license Farnsworth's patents for a workable electronic scanning system. But Zworykin was still researching television. He developed the iconoscope, a cathode ray that substantially improved picture brightness.

In 1930, the leaders in radio technology—RCA, GE, and Westinghouse—joined forces to develop electronic television. Zworykin worked with engineers from RCA and GE, and by 1936, an experimental television station—W2XF in New York—began transmitting television pictures.

Development of electronic television continued throughout the 1930s, and television made its debut at the New York World's Fair in 1939. RCA president David Sarnoff introduced television to the public, and President Franklin D. Roosevelt gave the first presidential television address from the World's Fair.

By 1939, a 441-line electronic picture had been developed, and stations were transmitting programming on a regular schedule. By 1941, the television picture had improved to a relatively sharp-looking 525-horizontal-line picture

In 1941, the National Television System Committee (known as the NTSC) advised the Federal Communications Commission (FCC) about the broadcast technical standards for operation. The result was that commercial television broadcasting began by FCC approval on July 1, 1941.

FIG. 3.4 Vladimir K. Zworykin with the iconoscope, the cathode ray television tube he invented
Photo courtesy of MZTV Museum

Compared to radio, television required much more space (i.e., bandwidth) on the electromagnetic spectrum. The technical standard for broadcast television, according to the NTSC, required 6 MHz of spectrum. By comparison, AM radio requires only 10 KHz and FM requires just 200 KHz. Television required 30 times as much space as FM and 600 times as much as AM radio.

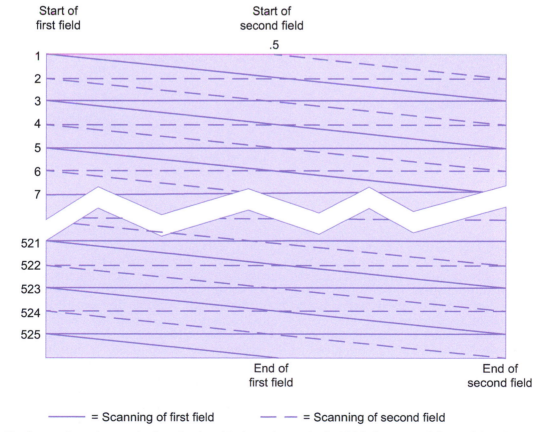

DIAGRAM 3.1 The diagram shows the scanning lines developed in electronic scanning for NTSC television. Each frame of the video was composed of two fields, or half frames with 262.5 lines. The first field began in the upper left and ended in the middle of the screen at the bottom. The second field began in the middle of the screen at the top and continued to the bottom of the screen at the lower right. The frames were shown at 30 per second, fast enough to not be noticeable to the viewer.

FIG. 3.5 Philo T. Farnsworth with an early television pickup tube
Photo courtesy of MZTV Museum

FIG. 3.6 David Sarnoff, president of NBC, makes an introductory speech for live TV at the 1939 World's Fair.
Photo courtesy of MZTV Museum

FIG. 3.7 A 1939 RCA television
Photo courtesy of MZTV Museum

FIG. 3.8 This image of Felix the Cat was the result of early electronic scanning experiments.
Photo courtesy of MZTV Museum

BROADCAST TELEVISION AND WORLD WAR II

Commercial television broadcasting was ready to begin business in 1941, but U.S. involvement in World War II essentially halted its development. In early 1942, the federal government noticed that the manufacturing of television stations and receivers used materials and equipment that could be used for the war effort, especially in the production of radar equipment, and so it stopped television manufacturing and thus broadcasting almost entirely.

As World War II drew to a close, resources became plentiful, and restrictions on television set manufacturing were gradually removed. Television, which had been talked about by many but seen by few, was about to get a real test in the marketplace. Yet even after the war ended, it took almost 2 years to resume television station construction, set manufacturing, and broadcasting.

OFF TO A SLOW START

In 1945, there were only 6 television stations on the air, and 3 years later, on January 1, 1948, there were only 16. There were many reasons for the initial slow growth of television, but perhaps the most important was that building a television station was difficult. Television added pictures to the sounds, making the construction of a broadcast facility much more complicated. Additional technological complications resulted in additional expenses. Television required more space, more equipment, and more personnel than radio. Also, investors were concerned that not many people owned television sets. It was not until about 1948, when the U.S. economy began to boom and prices of television sets became more affordable, that the number of new programs increased substantially, which set the stage for massive growth in the television industry.

By late 1948, there were only 34 stations on the air, but numerous applications were being submitted to the FCC for new licenses. Many of these applications came from

AM radio broadcasters who wanted to start television stations. Some newspaper companies were also interested in adding a television station to their holdings. Several newspaper companies built powerful stations that have stayed on the air for many years. For instance, WGN (whose call letters are also an acronym for the **W**orld's **G**reatest **N**ewspaper) in Chicago was built by the Tribune company, then owner of the *Chicago Tribune* and other papers, WTMJ was built by the owner of the *Milwaukee Journal*, and WBAP in Ft. Worth, Texas, was built by the Ft. Worth *Star Telegram*.

Similar to what happened in the early days of the feature film industry, the new television industry offered opportunities for many people. Veterans returning from the war who had radar experience often became television engineers. Others moved from camera operation to director to producer in a matter of months. The race was on to provide a lot of television programs, but the talent pool of people qualified to work in television was still quite small.

The broadcast system for television was similar to that of radio: a signal was sent out from a single antenna to many receivers in a given geographic area, the one-to-many model of mass communication. Television stations were interconnected or networked from a single source, with the television network headquarters in New York. The network sent out the programming signal to the affiliated stations via telephone wire. Individual stations received the signal and broadcast it via a transmitter and antenna. Within a given time zone, the signal was sent out simultaneously to many stations, and the program was aired at the same time on all stations. Television stations in the western part of the U.S. received a later version of the show. From 1946 until the late 1960s, a show that was broadcast live in the eastern half of the U.S. had to be reperformed for broadcast in western states. For example, a live variety show on the air at 8:00 p.m. EST was reshot for live broadcast at 8:00 p.m. PST. This practice changed when videotape was used to record a program and hold it for broadcast later in the western time zones.

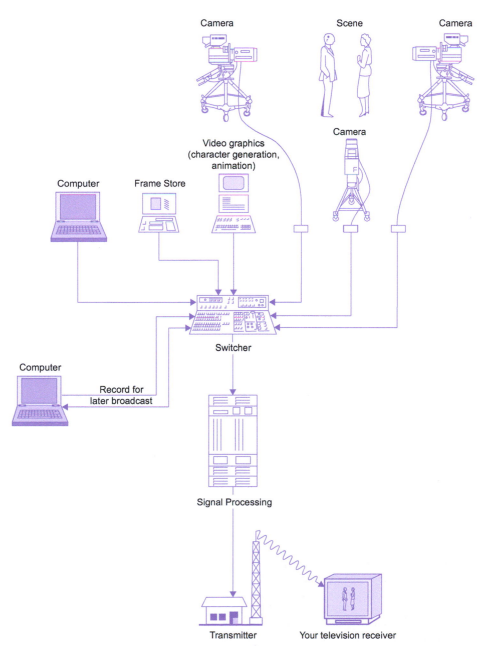

DIAGRAM 3.2 TV studio to transmitter to home service. This is the signal path for broadcast television.

THE BIG FREEZE (1948–1952)

The post–World War II audience demand for television sets and programming was the catalyst for the increased number of new applications for television licenses, especially in the large markets, where available channels were scarce because there were only 12 channels located in the VHF band for television. The number of new station applicants overwhelmed the FCC and though it had a set procedure for allocating radio stations to markets, the FCC simply was not prepared to deal with licensing television stations. And so in September 1948, the FCC essentially threw up its hands and yelled, "Freeze!"

The FCC put a 6-month freeze on the application process to reconsider the licensing process and other issues associated with the industry. Owners whose applications had been approved before the freeze, however, were allowed to continue constructing their stations and to begin broadcasting. The FCC expected that it would take 6 months to resolve the issues, but it actually took about 3.5 years. The *Sixth Report and Order* was signed on April 15, 1952, officially ending the freeze. Several of the major issues were settled as follows.

ADDITIONAL UHF CHANNELS

Before the freeze, all stations were licensed to the very-high-frequency band, or VHF channels, which were channels 2 through 13. The *Sixth Report and Order* made the ultra-high-frequency band (UHF) available, consisting of channels 14 through 83, to television stations across the United States. The high end of the UHF band, channels 70 through 83, is reserved for **translators**, which receive television signals from stations and retransmits them on different frequencies, and **repeaters** that receive a television signal then retransmit it to improve reception in cities distant from the city of license.

Broadcasting television signals on UHF band seemed like a great idea, except that none of the television sets manufactured up to that time could receive those channels without using a UHF converter. VHF tuners clicked into place on channels with each turn of the dial, but UHF converter dials were often not easy to operate, and they required time-consuming and sometimes frustrating channel adjusting. Stations that were given licenses to broadcast on the UHF band were disadvantaged in comparison to their VHF rivals simply because most television sets could only receive VHF programs, and viewers were reluctant to buy clumsy-to-use converters. Moreover, UHF signals do not travel as far as VHF signals. Even though UHF stations pumped out more power than VHF stations just to reach the same geographical area, in many cases, viewers also had to buy a special UHF antenna to pick up the signal. Exasperated viewers found themselves up on their roofs adjusting their VHF antenna in one direction and their UHF antenna in another, and even combo VHF and UHF antennas required multidirectional adjusting.

To make it easier to watch both VHF and UHF programs and to give UHF stations a more competitive foothold in the market, Congress passed the All-Channel Receiver Act in 1962. The 1962 act authorized the FCC to mandate that all television sets manufactured in 1964 and beyond have the capacity to receive both VHF and UHF stations without a UHF converter. Despite this legislation, UHF stations were still thought of as inferior to VHF. In fact, a common saying was that getting a VHF license was like getting a "license to print money," while UHF stations often lost money.

FIG. 3.9A In some cities, multiple antennas were needed to receive the broadcast stations. In this picture, there is a VHF antenna to receive channels 2–13 and several directional UHF antennas to receive channels 14–83.

Photo courtesy of shutterstock.com/XL1200

FIG. 3.9B These rooftops show an unsightly array of antennas and satellite dishes for receiving television and radio signals. This type of view was typical in major cities in the United States and developed countries throughout the world from the late 1940s until the 1990s. Cable and satellite systems greatly reduced the need for antennas, but the 2009 switch to digital broadcasting has actually increased the number of rooftop and indoor antennas.
Photo courtesy of iStockphoto. © dubassy, image #7418420.

EDUCATIONAL CHANNELS

As part of *Sixth Report and Order*, the FCC reserved 242 channels spaces for educational television, which made up about 12% of the 2,053 channels allocated for television station use across the United States. Unlike the allocation set aside for FM noncommercial radio (i.e., 88.1 to 91.9 MHz), these channels were not located within one part of the television band but rather were spread throughout the UHF and VHF bands. Although commercial entities opposed this generous allotment of educational channels on the grounds that it made commercial channels less available in many cities, the ruling was upheld.

COLOR TELEVISION

In the 1950s, while the public was just beginning to get used to the idea of black-and-white television, the networks were experimenting with full-color broadcasting. The FCC's oversight of spectrum space helped determine the development of color television, because at the time, the existing color system (called the **field sequential** system) required more channel space (18 MHz) than black-and-white television (6 MHz).

RCA and CBS had both developed their own color systems, which they each presented to the FCC. The two competing systems both fit into the existing 6 MHz of channel space that was allotted for every television station. The CBS system used a mechanical color wheel that transmitted a color signal. There were several drawbacks to the CBS color system. The signal produced 'noise' that distorted both the electronic picture and sound, but more critical was that existing black-and-white sets could not

receive the CBS color signal at all. Any program that was broadcast in color could not be seen on a black-and-white set, even in black and white. Despite these disadvantages, in 1950 the FCC announced support of the CBS color system. The public, however, was not quite ready to buy color sets, had CBS or its manufacturing partners even produced them in the months following the FCC's decision. Very few programs had been produced for color broadcasting, and very few audience members could afford color television sets, which were very expensive.

FIG. 3.10 The first color television was made by RCA in 1954 and sold for around $1,000—or about $6,000 in today's money.
Photo courtesy of MZTV Museum

Even the FCC's support of the CBS system did not last long. A little over 2 years later, the FCC reversed its decision in favor of RCA's electronic color system. The reversal came about in part because the NTSC preferred the RCA system and pointed out that CBS was not developing its system as quickly and efficiently as needed to meet viewers' demands. No doubt David Sarnoff, head of RCA, also lobbied the FCC for acceptance of his company's system. After adoption of the RCA color system, there was not a huge demand either to manufacture color sets or to broadcast color programs. Because almost all television cameras were capable of producing only black-and-white images, very few programs were made for broadcasting in color. Also, both CBS and ABC were not strongly motivated to support the RCA system because RCA owned their rival network, NBC. The strong point of the RCA system was that programs broadcast in color could be picked up by black-and-white sets, though the picture was still in black and white. Because of this compatibility, viewers did not have to run out and buy color sets, and most did not. The viewing public was generally satisfied with receiving color-produced programs in black and white on their black-and-white sets. It took more than 20 years before color sets became standard in most homes.

Although other better-performing systems were adopted worldwide, the NTSC (RCA) system remained in use until the United States replaced it with digital television in 2009. The 525-line color broadcasting system was criticized repeatedly over the years, not only for its lower resolution as compared to other systems but also because engineers often found it unreliable. Some engineers jokingly referred to the NTSC acronym as meaning 'never twice the same color' because viewers had to adjust the color each time they changed the channel. In other words, the hues and tones varied from channel to channel such that flesh-colored tones on one channel sometimes looked green on another. Colors also changed with the brightness of a scene.

DOMINATION OF THE NETWORKS

Beginning in the early days of radio and continuing into the era of television, the networks exerted quite a bit of control over the affiliates. The reason behind this was simple: the networks provided high-quality entertainment. Big-name stars from Hollywood and New York could be heard on local stations in small towns across the country. Without the big-name stars and high-quality programs, a local station was nothing special—often a so-called mom-and-pop operation owned by a group of small businesspeople. The networks' ability to bring stars and desirable programs to the affiliates continued into the television era.

As had been the case in the 1930s and 1940s when the networks dominated radio programming, they also controlled television programming from its inception. The fortune the networks made from radio was used to bankroll television, which ironically cannibalized radio's audience and its advertising dollars and revenue.

RELATIONSHIPS WITH AFFILIATES

The freeze helped solidify the television networks' dominance. During the 4 freeze years, existing stations scrambled to affiliate with the two powerful networks, CBS and NBC. Thus, 2 years after the freeze ended, CBS and NBC had more than three-quarters of all stations that were affiliates. ABC, which was formed after NBC divested its Blue radio network in 1943, was always a distant third in number of stations and audience size. ABC was so financially strapped that in 1951 it merged with United Paramount Theaters to receive a cash infusion and stay in business. A fourth network, the DuMont network (owned by a TV set manufacturing company), had many affiliates in medium- and smaller-sized markets, but it experienced problems similar to ABC with audience size and ceased operation in 1955. Television stations with network affiliations did well, while stations without a network affiliation often struggled for audiences, programming, and money.

Independent stations were forced to either produce shows on their own or seek programming material from independent producers or syndicators. In the early 1950s, quality programming came almost entirely from the television networks. It was easier and less expensive to get programs from a network than to produce them at the station or to get them from other sources. More important, network programs were usually of higher quality than locally produced or independently produced programs. Some local television stations produced news, public affairs, children's, and sports programming, but drama, situation comedies, and even variety shows were too expensive for most local stations and thus were produced by the networks.

These harsh facts of life in the television industry meant that network affiliation was highly valued. The networks had their pick of stations in a given market and were in a very strong bargaining position with their affiliates. In other words, the networks could often dictate financial terms and the availability of airtime to the local stations. The top three stations in a market affiliated with CBS, NBC, and ABC. In most cases, the independent stations were newcomers to a market or broadcast on a less desirable UHF channel, and in television markets that had three stations or fewer, there were no independent stations.

The networks also had quite a bit of freedom from regulation. Although local stations were regulated directly by the FCC, the networks were regulated only through the stations they affiliated with because the stations, not the networks, used the publicly owned airwaves.

The affiliations between local stations and networks were renewable every year. However, from the post–World War II years until recently, an affiliation with a network usually lasted for many years. Often, the relationship between a network and an affiliate began in the very early days of the station's existence and remained unchanged. The affiliate relationship has several components. The local station provides its airtime (known as **clearance**) and its audiences to the network, and the network provides

FARNSWORTH TELEVISION

In those cities where television programs are broadcast, a limited number of Farnsworth table model television sets will soon be available. Like the Farnsworth portable radio, table model, and phonograph-radio, the new television receiver combines modest price with the quality you expect from the home of television. The inventor, Philo T. Farnsworth, developed the first practical system of electronic television, and it is the company which bears his name that today offers for your entertainment and pleasure the fruits of electronic research. Prices: Farnsworth radios and phonograph-radios, **$25 to $350**

A color-action photograph from "Song of Norway," Broadway musical triumph based on the life and music of the Norwegian composer, Edvard Grieg

Capehart and Farnsworth television will bring the greatest stage shows to your home—in sparkling, detailed black-and-white action pictures

CAPEHART TELEVISION

In the field of musical reproduction, one phonograph-radio stands supreme, and that is the Capehart. Only by comparison with the human voice, or with the original musical instrument, can its clarity and purity of tone be appreciated. That standard of excellence will be inherent, also, in the new Capehart television receivers for the home. Just as Capehart now brings you the finest instruments for musical reproduction, so will Capehart bring you the finest instruments for your visual entertainment. Phonograph-radio prices: The Panamuse by Capehart, **$300 to $700**. The Capehart, **$925 to $1500**

N. W. AYER & SON

FARNSWORTH TELEVISION & RADIO CORPORATION, FORT WAYNE 1, INDIANA

FIG. 3.11 Advertisements for early television sets

a dependable schedule of high-quality shows to the local affiliate for most of the broadcast day. In addition, the affiliate is paid for its airtime. This practice, known as **station compensation**, is based on the size and the demographic makeup of the audience delivered by the station to the network, its advertisers, and the competition for the station's affiliation. However this practice is no longer common.

In the early days of television, NBC and CBS (ABC was not considered a strong competitor) were both financially strong, had highly-rated VHF affiliates, and an inventory of programs that were set to make the transition from radio to television. The shows that moved from radio to television often took their sponsors and listeners with them. An established radio show with a loyal following could create an instant audience of loyal television viewers.

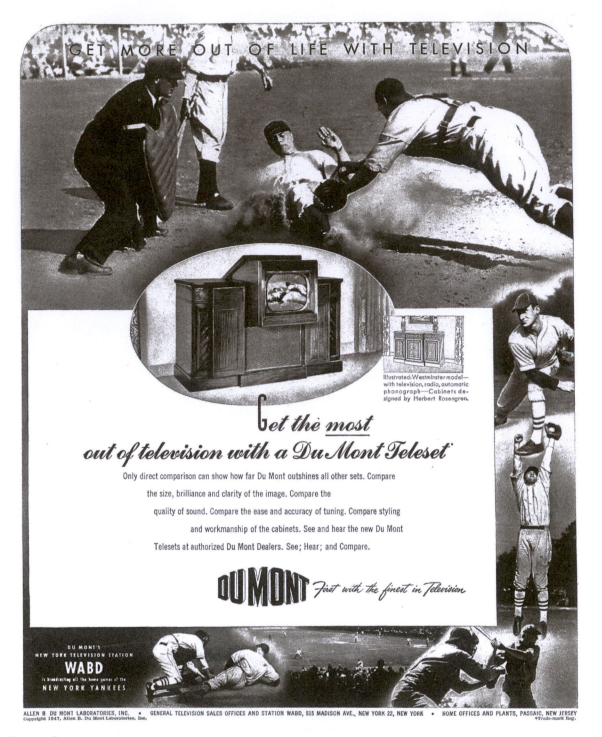

FIG. 3.11 (Continued)

Kingsize pictures are 2½ times bigger—
126 square inches!

There's MORE to see today in television
and RCA Victor shows it to you better!

Television sound is static-free FM
through the famous
"Golden Throat" tone system!

You follow every play—
pictures are <u>locked</u> <u>in</u> <u>tune</u>
with sending stations!

Big console-size
speaker!

RCA Victor 8T270

Kingsize pictures—126 square inches
big. But that's not all! These are
Eye Witness pictures—brilliant,
clear, steady, actually *locked in tune*
by RCA Victor's Eye Witness Picture Synchronizer. Powerful circuits
adjust *automatically* to television
signals of varying strength. New
Multi-Channel Automatic Station
Selector, improved controls, make
tuning simpler,
easier than ever.
Mahogany or wal-   **$495.00†**
nut finish. Blond   *Installation extra*
slightly higher. AC.   *plus Fed. excise tax.*
                        Buy the Optional
                        RCA Victor Owner
                        Contract.

Tuning
is simpler,
easier than
ever!

HERE IS 16-INCH RCA VICTOR

RCA Victor 8TR29

*New Eye Witness television table
model has AM-FM radio and lovely
matching stand . . .* all for the one
low price. Pictures, 52 square inches
big, are *locked in tune* by RCA
Victor's Eye Witness Picture Synchronizer. Tuning is simple with the
new Multi-Channel Automatic Station Selector. Television sound and
powerful AM-FM
radio reception   **$375.00†**
are heard through   *Including matching*
the famous   *stand (not shown).*
                    Installation extra
"Golden Throat"   *plus Fed. excise tax.*
tone system. AC.   *Buy the Optional*
                    RCA Victor Television
                    Owner Contract.

EYE WITNESS
TELEVISION

Installation by experts of RCA's own service organization, the RCA Service Company, is yours when
you purchase the RCA Victor Television Owner
Contract. This optional Contract also covers all
charges for antenna and it guarantees you a full year
of fine Eye Witness performance. No other company
offers such extensive television service facilities.

†All prices subject to change without notice. Zone 2
prices slightly higher.

**SEE YOUR RCA VICTOR TELEVISION DEALER
FOR A DEMONSTRATION AND FREE BOOKLET**

RCA VICTOR

DIVISION OF RADIO CORPORATION OF AMERICA
WORLD LEADER IN RADIO...FIRST IN TELEVISION

FIG. 3.11 (Continued)

If there was a problem with network programming in the early days of television, it was that television was really radio with pictures. Many of the same programs, with the same stars, switched from radio to television. Although this transition was comfortable for the audience, it did not encourage much experimentation or the development of new program types and styles. Despite this, the years after World War II were the Golden Age of television, when audiences and advertisers flocked to the tube. In the 1950s, television was severely restricted by technological factors. Cameras were large and heavy, and strong lighting (which generated quite a bit of heat) was required to get a good video image. Portable video cameras did not exist, and neither did videotape (until after 1956). Although about one-quarter of the prime-time programs were recorded on film, most shows had to be produced live.

FYI: Kinescopes

In the early days of television, kinescope recording was the most common way to preserve live television programs for airing across time zones. Developed by Du Mont, NBC, and Kodak, kinescopes stored visual images by aiming a film camera at a television monitor as the show was airing live. The recorded broadcast could then be re-shown in a different time zone. Kinescope recordings were of poor quality and did not hold up as well as programs recorded on film.

After videotape became available in 1956, kinescopes ceased to be a viable medium for storing video programs. Kinescopes are now a collectors' item.

For more information, go to http://www.museum.tv/eotv/kinescope.htm

BLACKLISTING AND BROADCASTING

After World War II, the U.S. government and the public in general became very aware of the growing power and nuclear capabilities of the communist-controlled Soviet Union. In addition, Americans feared that communism was spreading in many parts of the world.

The general attitude toward communism was not just that it was different from the American system but that it was a political ideology that would be used to take over the world. Politicians seized on these fears and used them for political gain. In the early 1950s, a small group of former Federal Bureau of Investigation (FBI) agents published a newsletter called *Counterattack*. Its purpose was to encourage Americans to identify and even shun people who demonstrated sympathy or ideological agreement with communism. Another publication, *Red Channels: The Report of Communist Influence in Radio and Television*, described the communist influence in broadcasting and named 151 people in the industry who supposedly had communist ties. The names of these people were put on a blacklist, and they were essentially no longer permitted to

work in broadcasting or related industries. *Red Channels* was published just as North Korea, a communist country, invaded South Korea. The United States got involved in the conflict, and hence the nation's role in stopping communism began, as American troops were sent to fight in a foreign land.

The Korean War, which lasted from 1950 to 1953, reinforced many Americans' feelings that communism had to be stopped, and the idea that communism had infiltrated broadcasting triggered activities to stop it. Although no celebrity or individual employed by the television industry could be linked to the Communist Party, the mere listing of a person's name in *Red Channels* prevented him or her from continuing a career in broadcasting. The networks and advertising agencies even employed individuals to check the backgrounds of people working in the industry. If a person's name showed up in any list of communist sympathizers, he or she would not be hired to work for a network, an advertising agency, or any project or program. New employees were expected to take a loyalty oath to the U.S. before working. People who were suspected of communist activities were often forced to confess and name their communist associates.

The most prominent of the politicians who used Americans' fear of communism to strengthen his own political power was Joseph McCarthy, a U.S. senator from Wisconsin. He used televised congressional hearings about communism in the U.S. Army to further his notoriety. Eventually, McCarthy's tactics caught up with him. Edward R. Murrow challenged him on a personal interview show called *See It Now* that aired on March 9, 1954. In the show, excerpts of speeches given by McCarthy were replayed to uncover his inconsistencies. In the end, McCarthy was shown to be a bully who ignored fact and used innuendo to level accusations at his adversaries and innocent people who worked in the media.

ZOOM IN 3.2

Learn more about blacklisting by going to www.museum.tv and click on the Education tab, then select Encyclopedia of Television. Select B then click on Blacklisting.

The *Red Scare* created a chilling atmosphere for broadcasting and its employees. The cloud of blacklisting continued until 1962, when radio comedian John Henry Faulk, who had been blacklisted in 1956, won a multimillion-dollar lawsuit against AWARE, a group that had named him as a communist sympathizer. The effect of the lawsuit was that the blacklisting practices went from up front and public to secretive and private. The **blacklist** became a **graylist**, which was used in broadcasting to identify people who might have subversive ideas, especially those who embraced communism. The practice of graylisting lasted into the 1960s.

FIG. 3.12 Advertisement for the 1958 Philco Predicta

CABLE TELEVISION: TELEVISION BY WIRE

The development of cable is best understood by going back to the late 1940s, when the television industry was just beginning. Television found an eager audience for its programs, many of which were taken directly from radio and modified for the screen. Viewers wanted to **see** their favorite radio stars and **watch** the new shows, such as the *Texaco Star Theatre*, starring Milton Berle, and *Your Show of Shows*, with Sid Caesar, Imogene Coca, and Carl Reiner.

The public heard all about television and viewed some programs in their neighbors' homes, appliance stores, or bars. They were eager to buy sets and begin watching in their own homes. The problem was that television stations were located primarily in large cities like New York, Philadelphia, and Chicago. Each of these markets had several stations and brought big-name talent to the audience. People were rushing to buy television sets, and television stations were rushing to get licensed to deliver programs.

Broadcast television signals were (and still are) sent from a station's antenna to a receiving antenna that is connected to a television set. Television antennas located in rural areas and other smaller markets outside major metropolitan areas sometimes received a signal from one or more major market stations with the help of small transmitters located nearby. These retransmitters, called translators, received the television signal from a station and then retransmitted it to antennas in the rural area. Translators would send the signal out on a special frequency to avoid interfering with the signal of the originating station.

Translators were not used in the very early days of television, however, and many of the people who lived in areas far from big cities or where the television signal was blocked by hills or mountains could not receive television signals. Whether television signals were blocked or there simply were not any nearby stations, many viewers throughout the country went without television altogether. The desire for television led to the birth of an alternative delivery system: community antenna television, commonly known as cable television.

COMMUNITY ANTENNAS

As the saying goes, "Necessity is the mother of invention," and in this case, it was the lack of television that led to invention. Appliance store owners in smaller cities and towns had a hard time selling television sets because the areas did not have a local station and could not receive signals from distant stations. These store owners needed to find a way to bring a television signal to their town.

There are competing stories about who first came up with a specific solution to the problem of reception of distant television signals. Some contend that John Walson, an appliance storeowner in rural Pennsylvania, was the first to bring **community antenna television** (CATV) to his community via cable, yet others claim that L. E. Parsons of Astoria, Oregon, was the first to do so. Essentially, both of these men came up with the idea of placing a television antenna on top of a hill or mountain to bring in television signals from distant stations. But bringing in television was just part of the plan; making money was the other incentive. The Music Box memo written by David Sarnoff years earlier was a plan to sell radios by providing interesting programming. Applying Sarnoff's idea to television, the plan was for appliance store owners to come up with a way to bring television programs to their area and then sell the television sets so customers could watch the programs.

The idea of sharing or distributing television signals was not new. It first started in New York City when apartment dwellers found that other buildings blocked the signals and they could only get television by placing an antenna on the roof of the building. Landlords soon became wary of granting permission for rooftop placement because of potential disputes over the best locations, problems over maintaining the antennas and their masts, and the basic ugliness of the proliferating metal antennas. Sharing an antenna system solved these problems. The **master antenna system** was used to receive the television signal and to distribute it to the apartments through a system of wires. This type of strong antenna was the forerunner of cable television delivery (CATV).

One of the first hook-ups of cable television took place in 1948 when L. E. Parsons in Oregon placed a television antenna on top of a hotel in Astoria to receive the signal of a Seattle station. He then connected a long wire from the antenna to his apartment. Local interest grew considerably when word got out that Parsons had the only

television set within 100 miles that could receive a signal. He also placed a line from the antenna to the television set in the lobby of the hotel.

Eventually, Parsons obtained permission to run wire through the underground conduits to businesses in downtown Astoria. Consumer demand led to Parsons wiring many homes, although he did not initially have permission to string his wires on utility poles. The result was a somewhat primitive delivery system that nonetheless became a viable business—the Radio and Electronics Company of Astoria. Parsons also began consulting with others who wanted to set up similar systems elsewhere in the country.

Bob Tarlton, another early community antenna entrepreneur, connected his appliance store in Lansford, Pennsylvania, to an antenna on top of a nearby hill, presumably to demonstrate television so he could sell more sets. He was the first to conceptualize CATV as a local business and used modified equipment purchased from Milton Shapp, owner of Jerrold Electronics and future governor of Pennsylvania. Eventually, Jerrold Electronics became the leader in manufacturing equipment specifically for the community antenna television industry.

MICROWAVE TRANSMISSION

CATV systems eager to expand their reach came up with yet another way of doing so—microwave transmission. By using microwave signals, a CATV system imported television signals from distant large markets rather than just from the closest market. Microwave signals were especially important in the western part of the United States where cities are far apart. For example, a CATV system in Flagstaff, AZ, could offer its subscribers programming from a television station in Phoenix (144 miles away). Bringing in distant signals gave the audience more programming choices and thus a reason to subscribe and pay for CATV.

COMMUNITY ANTENNA BECOMES CABLE TV

The advent of the cable industry was not a result of an earth-shaking technological breakthrough. Rather, it was an idea that grew out of strong demand for television from an audience hungry for entertainment. In fact, few patents for cable television were issued during the beginnings of cable television, because the technology it used was based on the simple distribution of existing broadcast signals. In some ways, broadcast television was simply 'hacked.' Instead of a simple community antenna system that picked up the broadcast signals from a nearby town, community antenna systems became complex cable television systems that added signals from distant markets and eventually added non-broadcast television channels to their lineup of channels.

Community antenna television became **cable** television. A cable system picks up a variety of television signals (from its broadcast antenna, microwave antenna, and satellite dish antenna) at its head end and sends them through wires to homes in the community.

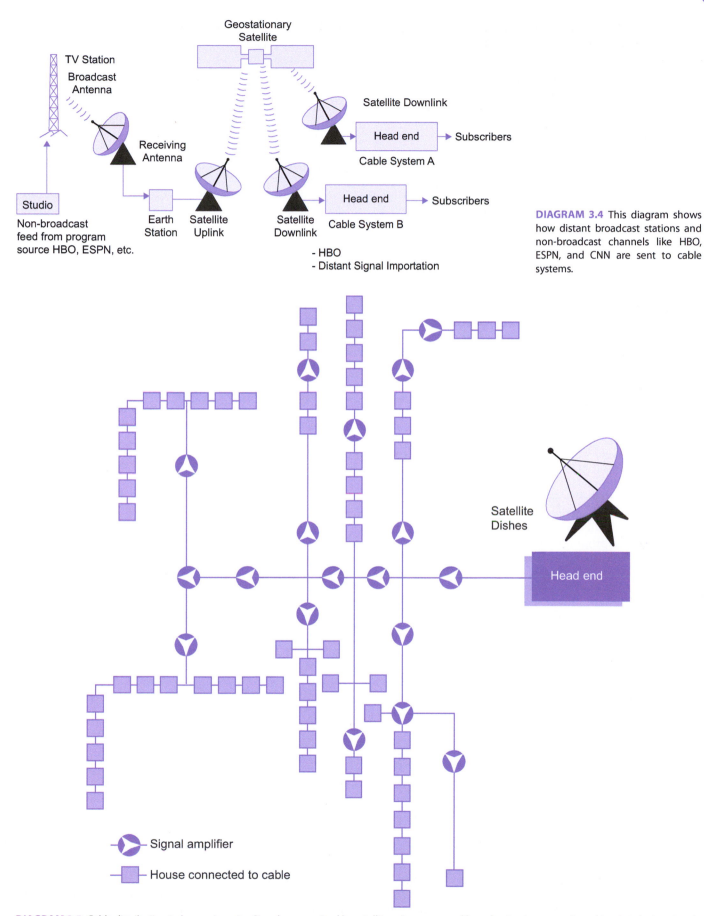

Geostationary
Satellite

TV Station
Broadcast
Antenna

Receiving
Antenna

Satellite Downlink

Head end → Subscribers

Cable System A

Studio

Non-broadcast
feed from program
source HBO, ESPN, etc.

Earth
Station

Satellite
Uplink

Satellite
Downlink

Head end → Subscribers

Cable System B

- HBO
- Distant Signal Importation

DIAGRAM 3.4 This diagram shows how distant broadcast stations and non-broadcast channels like HBO, ESPN, and CNN are sent to cable systems.

Satellite
Dishes

Head end

⊙— Signal amplifier

▢— House connected to cable

DIAGRAM 3.5 Cable distribution to homes in a city. Signals are received by satellite, microwave, and broadcast antennas and combined at the head end. From the head end, the signal is sent through a system of wires that have amplifiers (i.e., the round symbols in the drawing) to boost the signal strength and continue the signal to houses connected to cable (i.e., the square symbols).
Source: Franklin, 2004

REINING IN CABLE

Carrying signals via cable or microwave from distant stations did not go over well with local stations that sometimes were excluded from the cable system's channel lineup. Whether or not a local station was carried by a local cable system was a decision made on the whim of the system's owner. As more and more viewers subscribed to cable, local stations not carried by the system were losing their audience and, thus, advertising revenue. Local stations lobbied the FCC to compel cable systems to transmit local station programming. The FCC agreed, and in 1965 it issued the **must-carry rule**, which stipulated that cable systems must carry signals from all 'significantly viewed' stations in their market. In other words, cable systems had to carry local broadcast stations on their systems.

Cable argued that FCC restrictions hampered its development. The FCC capitulated to cable and eased its regulations and dropped most **carriage rules** (i.e., what channels the systems carried) for small systems and allowed unrestricted importation of foreign-language and religious programs and rescinded the minimum-capacity rule that required systems of 3,500 or more subscribers to carry at least 20 channels and to furnish equipment and facilities for access by the public. In 1972, the FCC adopted new legislation that was a result of the FCC's continuing desire to preserve local broadcast service and to create an equitable distribution of broadcast services among the various regions of the country. The legislation created the **nonduplication** (or **syndicated exclusivity**) rules that prevented cable systems from showing a **syndicated** program (reruns of a show that formerly aired in prime time on a network) from a distant station if a local station in the market was airing the same program. In addition, the FCC ruled that new cable systems had to have at least 20 channels and two-way capacity, allowing signals to travel both to and from the audience. The added technical requirements made building cable systems more complicated and thus more expensive, which slowed down the growth of cable systems in all but the largest markets. The syndicated exclusivity rules were dropped in 1980 after causing 8 years of problems for cable operators.

FYI: 'Renting a Citizen' for Cable

Some cable companies tried to influence city officials with attractive stock deals. For instance, in 1982, a Denver councilwoman received a terrific stock deal from one of the cable franchise applicants and observed that the only person she knew who did not own cable stock was the coach of her son's little league football team. (She later discovered that he did not live in the city of Denver.) And in Milwaukee in 1982, an applicant for the cable franchise retained two local political consulting firms: one to help pitch its case and the other to prevent the first firm from helping any of the competitors (Hazlett, 1989).

HBO

In 1972 the television industry changed forever with the advent of *Home Box Office (HBO)*. Initiated by Time, Inc.'s Sterling Manhattan Cable in New York City, HBO was the first service to deliver television programming via satellite. HBO placed a transponder on an existing satellite, Westar 1. HBO then uplinked a program signal to the transponder. The signal was then downlinked to cable services' satellite dish receivers. Not only did HBO bring subscribers live sporting events and full-length movies, but it also specialized in programming that was not offered, or offered minimally, by the networks. Rather than just delivering broadcast network programming via cable, HBO supplied subscribers with original programming.

But it took some legal wrangling (*HBO v. FCC*, 1977) before HBO could be delivered via satellite rather than through microwave only. Microwave signals are similar to point-to-point transmissions and require a system of towers to relay the signal. Rules regulating cable systems were relaxed regarding the types of programs HBO could present and the size (and therefore the cost) of the receiving equipment necessary to receive satellite signals. Once these technological and regulatory issues were settled, the door was open for cable companies to offer satellite-delivered (premium) channels to their customers, for an extra fee. HBO, thus, became the first company to deliver non-broadcast, original programming via satellite.

Even though HBO used satellites, it became known as the first **pay-cable channel**. HBO delivered its programs via satellite to a cable service that in turn delivered it to its cable subscribers. HBO was also called a **premium channel** because cable subscribers had to pay extra for it. Other cable-only networks soon followed such as ESPN (1979) and CNN (1980).

HBO filled the void for new and varied programming and ushered in a new world of television viewing. When satellite distribution of video programs to cable companies became both technically and economically feasible, the cable industry grew rapidly. It is somewhat ironic that direct broadcast satellite delivery (DBS, which began in the mid-1990s) eventually posed competition to the cable industry it helped to launch with the advent of satellite-to-cable program (HBO, etc.) delivery.

UPHEAVAL AND EDUCATION

THE TUMULTUOUS 1960s

In 1961, recently appointed FCC Chairman Newton Minow stated at the National Association of Broadcasters convention that television programming was a "vast wasteland." Critics latched onto that remark as accurately depicting the quality of the programming offered by most television stations.

Although few would argue that television has traditionally offered many hours of intellectually light programs, it has and still does serve a function beyond pure entertainment. For example, television allowed Americans to witness history during the tumultuous decade of the 1960s. This journalistic function both solidified the importance of television in American society and gave it real credibility as a provider of valuable information.

Some of the many events covered by television during the 1960s included the presidential debates between candidates John F. Kennedy and Richard M. Nixon; the 1962 Cuban missile crisis; the 1963 assassination of President John F. Kennedy and the subsequent killing of his accused assassin, Lee Harvey Oswald; the 1968 assassinations of Martin Luther King Jr., and Robert Kennedy; and humanity's first walk on the moon in 1969. In addition, continuing coverage of the escalating war in Vietnam, domestic unrest regarding racial issues, and violent demonstrations during the 1968 Democratic Convention in Chicago gave the public a view of historical events in an up-close-and-personal way never before possible.

Viewers were deeply affected by the events and issues presented on television—for example, realistic footage of rioting in the streets and the horrors of war. In addition, the proliferation of violent action shows on television led many to wonder whether the media were somehow encouraging people to behave in violent ways.

The U.S. government responded to the violence on television and in real life in 1968 by creating a research commission, the Commission on the Causes and Effects of Violence. In 1969, the Senate asked the U.S. Surgeon General to investigate the relationship between television and violent behavior. The results, published in early 1972, stated that violence on television might lead some individuals to violent behavior. Although the findings fell short of pointing to television as the *cause* of increased violence in society, they fueled the efforts of citizen action groups like the Action for Children's Television (ACT), which sought to focus congressional attention on the content of television and its effect on children.

EDUCATIONAL TELEVISION GOES PUBLIC

During the television freeze that followed World War II, the FCC was lobbied both by commercial broadcasters and the Joint Committee on Educational Television (JCET) regarding noncommercial television stations. The commercial broadcasters tried to prevent the FCC from reserving television channels for noncommercial television stations, while the JCET lobbied for noncommercial television use. As part of the *Sixth Report and Order*, which ended the station-licensing freeze, the FCC increased the number of channels reserved for noncommercial stations from 10% to 35% of the available station allocations (242 channels—80 VHF and 162 UHF) for noncommercial use. Since then, the FCC has increased the number of channels dedicated for noncommercial stations to a total of 600.

In 1959, noncommercial television producers formed the National Educational Television (NET) network to operate as a cooperative, sharing venture among stations that sent prerecorded programs by mail. After one station aired a program, it sent it to the next station, and so on. This inexpensive and low-tech network, which became known as a **bicycle network**, did not allow stations in different locations to air the same program at the same time.

In 1967, the Carnegie Commission on Educational Television (CET)—a group composed of leaders in politics, business corporations, the arts, and education—published a report about noncommercial television that recommended that the government establish a corporation for public television. Until that point, noncommercial television had been strongly associated with educational television. The CET wanted to change the direction of noncommercial television so as to provide a broader cultural view.

The eventual result of the report and discussion that followed was the Public Broadcasting Act of 1967. The term 'broadcasting' was used instead of 'television' because Congress included radio as well as television in the legislation. The act provided that a corporation would be set up, with the board of directors appointed by the president of the United States, and that financial support would come from Congress. Having a corporation oversee public television was intended to provide some distance between the government and the noncommercial network.

The corporation that was formed in 1968, the Corporation for Public Broadcasting (CPB), was meant to support both the producers who created programs and the stations that aired them. However, CPB was not allowed to own or operate any stations. Rather, the Public Broadcasting Service (PBS), the television network arm of CPB, would operate the network that connected the participating (but not owned) stations. PBS received funding to support public stations and producers. The stations could then select and produce programs without direct government pressure. PBS went on the air in 1969 and began distributing programming to member stations 5 nights a week, including a children's daytime show called *Sesame Street*.

ZOOM IN 3.3

Action for Children's Television (ACT) was a national grassroots organization founded by Peggy Charren in 1968. The goal of the organization was to raise the quality and diversity in television programming for children and adolescents. In addition, the group sought to eliminate commercial abuses directed at young audiences. The organization had thousands of members across the United States and had a major impact on the content and scheduling of children's television programs and advertising. After The Children's Television Act of 1990 was passed, much of what the organization lobbied for was achieved, and ACT disbanded in 1992.

PBS, like CPB, is a private, nonprofit corporation whose members are public television stations. Its mission has been direct involvement in program acquisition, distribution, and promotion for its stations. Although PBS does not produce programs, it does support programs produced by PBS stations and helps acquire programs from independent producers around the world. PBS has also been involved in developing engineering and technology for the creation and distribution of its programs and in marketing video products (e.g., DVDs of programs) to the public. In addition, PBS administered the PBS Adult Learning Service, which provided televised educational courses for credit to as many as 450,000 students each year. This service was discontinued in 2005.

PBS is unlike the commercial networks because it does not sell advertising time. Instead, PBS gets its funding from a variety of national, regional, and local sources. Audience members provide almost 25% of the funding through direct donations. State governments provide about 18%, and CPB (along with federal grants and contracts) adds about 16%. Businesses add another 16%, state universities and colleges add more than 6%, and foundations provide an additional 5%. PBS's programming philosophy is oriented to providing programs of cultural and educational interest. Although PBS and public television stations do not sell advertising, they do seek **underwriting**, which is similar to advertising, but without a call to action. In other words, an underwriting announcement can have a message very similar to an advertisement, but without a call-to-action phrase like, "call 888 555-1212 now for a 20% discount."

When the Carnegie Commission first recommended establishing the new, noncommercial network (PBS), the commercial television networks supported the concept because it would provide programming to that segment of the audience that was critical of the commercial network's entertainment programming. The audience that had opposed the move from 'serious' cultural and information programming of the 'live' era of television would now have an alternative to the entertainment program content of the commercial networks. The commercial stations and networks could now concentrate on entertainment that drew a less sophisticated audience that was more attractive to advertisers. Also, the commercial networks would be under less pressure from the FCC to program to the public interest if the public broadcasters were fulfilling the role of providing information and cultural entertainment.

Because the President of the United States appoints the leaders of CPB, it sometimes gets caught up in politics. For example, President Richard Nixon vetoed a funding bill for CPB because he did not like that it allowed PBS to air information programs that showed his administration in an unfavorable light. Nixon's action resulted in the forced resignation of some CPB officials, who were replaced by people who favored Nixon's view of the role of CPB and PBS. Although this event was somewhat unusual in the history of CPB, it shows that public television is influenced by the presidential administration in power.

INCREASED CHOICE AND COMPETITION

The television broadcast industry was affected by major changes that began in the 1970s. Services like Home Box Office (HBO) used satellite distribution of television programming with great success. Cable channels proliferated and videocassette recorders (VCRs) sales exploded. Viewers were no longer tied to television viewing schedules but could now record programs for viewing at their convenience, an action that was called 'time-shifting.'

By the 1970s television sets, even color sets, had become affordable for the low and middle classes. Television set manufacturers lowered prices as their costs decreased because of the economy of scale: the more sets that were produced, the less expensive it became to manufacture each one. Additionally, Sony, Panasonic, and Toshiba began importing high-quality, competitively priced sets from Japan. Along with the increased number of set purchases, sometimes two or more for one home, came a demand for high-quality, innovative programs and an increase in the amount of time spent viewing. And best of all for the networks, advertising revenue jumped.

The networks' high times started diminishing in the 1980s. As more and more cities were wired for cable, it started cutting into the broadcast networks' 90% share of the viewing audience. Interesting programs on new cable channels like CNN, MTV, and ESPN and pay channels like HBO and Showtime lured viewers away from broadcast television. Video stores popped up all over the country, giving viewers yet another viewing option that further eroded the network audience. Cable television, video rentals, and VCR time-shifting all contributed to broadcast losing about one-third of its viewers. Along with the smaller share of viewers came a loss in advertising revenue.

As viewers slipped away, so did the networks' power over their affiliates and advertisers. Affiliated stations took revenue matters into their own hands by pre-empting network shows for ones that they produced or obtained from syndicators on which they could sell advertising and keep revenue at the station level. Affiliates even questioned whether payments for clearance were high enough and whether long-term affiliate agreements were financially wise.

Despite these challenges, many executives of the big three networks decided that their strategies were working fine and did not require overhauling. New ideas clashed with old, and the direction of the television industry became a source of tension. Generally, the big three networks managed to change very little with the times, leading some industry critics to refer to them as the "three blind mice" (Auletta, 1991).

SATELLITE DELIVERY

CABLE SYSTEMS AND SATELLITES

Satellite delivery changed the use and purpose of cable television. Television signals delivered by a satellite can reach many cable company receivers throughout the country at once.

Three different types of satellites are used for communication purposes:

1. Geosynchronous satellites, like the one used by HBO, are parked in an orbit 22,300 miles above the earth's equator. These stationary satellites provide services to cable companies and television stations along with direct broadcast satellite video and audio to homes.
2. Middle-earth orbiters are satellites that travel in a lower orbit—beneath 22,300 miles but more than 1,000 miles above the earth. These nonstationary satellites are used for voice and data transmission and also for global positioning system (GPS) devices.
3. Low-orbit satellites are nonstationary and travel from 100 to 1,000 miles above the earth. They are used for personal communication services, such as mobile phones, Internet access, and video conferencing.

ZOOM IN 3.4

The idea of using satellites for communication purposes was first publicized in a 1945 article appearing in Wireless World, written by novelist and scientist Arthur C. Clarke. He theorized that three satellites in geostationary orbit (22,300 miles above the equator) could be used to relay information to the entire globe (Clarke, 1945).

The popularity of HBO demonstrated viewers' eagerness for a wider range of programs than were provided by the broadcast networks. The cable industry quickly realized that it could make a fortune by expanding its offerings from simple retransmission of broadcast shows to creating new cable programs. Premium channels, such as HBO rival Showtime, and the first advertiser-supported networks like USA Network, ESPN, CNN, MTV, and BET began popping up in the late 1970s to early 1980s. Different from premium channels, advertising-supported cable channels were not billed separately but were offered as part of a package of programs that included the broadcast networks. But all of these cable channels worked the same way—they were uplinked to satellite and then downlinked to a cable service that distributed them to subscribers' homes.

Once viewers took their first look at the new cable channels, they wanted more of them. And viewers who lived in areas that did not have a cable system began rallying their local governments to get them in their communities. Cable systems were expensive to set up and miles and miles of cable wire had to be laid to connect each home to the system. In addition, existing cable systems had to add expensive satellite receiving dishes to downlink the new cable-only channels. But demand fostered supply. In 1978, there were only 829 **satellite-enabled** cable systems in the U.S., but within 2 years, that number had spiked to 2,500 such systems.

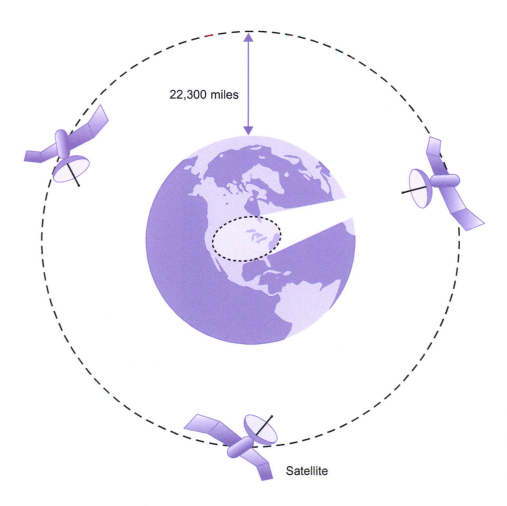

22,300 miles

Satellite

DIAGRAM 3.6 Geosynchronous satellites orbit the earth 22,300 miles above the equator. At that distance from the earth, the satellite stays in the same location relative to the earth. A narrow-beam signal from the satellite creates a large signal pattern or 'footprint' in which the television signal can be received with a small receiving dish.

The first satellite-receiving dishes measured 8 to 15 feet across. The dish was costly, and a large space was needed to install it, and thus use was restricted to cable companies. The receiving dish was large because its size was directly related to the power of the transponders and the frequency at which the signal was sent. The C band—the space in the electromagnetic spectrum used by communication satellites—required a large dish to best receive the signals.

SATELLITES AND SUPERSTATIONS

Satellite delivery and cable networks were turning viewers' attention away from broadcast television. In response, the broadcast industry started its own hybrid satellite/broadcasting system. Independent stations WGN (Chicago), WTBS (Atlanta), and WOR (New York) started uplinking their signals to satellite. Their signals were then downlinked by cable systems for delivery to their subscribers. Programming from these 'superstations' could be seen in distant cities across the country. Viewers in San Diego could watch a Chicago Cubs game or see a local New York weathercast on the superstations.

However, superstations only came about because of earlier FCC rulings. In 1972, the FCC authorized satellite distribution of signals to cable systems in any part of the country. Cable systems were required to carry signals from the big three networks (ABC, CBS, and NBC) and to obtain those signals from the station in the market closest to the cable system. This requirement became known as the **antileapfrogging rule**, because cable was not allowed to skip over closer stations in favor of stations in distant markets. Importing signals from independent stations was similarly limited. The tide turned from the regulatory attitude of creating more cable regulation to deregulation sentiment in 1976. When the leapfrogging rule was

eliminated, cable systems could import a wider variety of signals, including signals from the superstations.

SATELLITE MASTER ANTENNA TELEVISION (SMATV)

SMATV is an acronym for **satellite master antenna television**, which is also known as **private cable**. Essentially, an SMATV system is similar to a very small cable system. SMATV serves one or several adjacent buildings in a part of a city. Apartment buildings and housing complexes, large hotels and resorts, and hospitals commonly use SMATV systems as an alternative to each unit having to subscribe to cable separately. An SMATV head end is usually placed on the roof of a building. The head end receives the satellite signals. The signals are then delivered to each unit within the building.

MULTICHANNEL MULTIPOINT DISTRIBUTION SYSTEMS (MMDS)

Multichannel multipoint distribution systems provided channels similar to cable but used a microwave signal to send the channels to individual homes. Some cities were better suited to MMDS than others. Factors such as housing density and terrain led to MMDS systems becoming successful, at least for some time. In cities where the cable franchise was limited to specific parts of the city (or metropolitan area), some parts of the city could not get cable service. Those parts were sometimes served by MMDS, which did not require the stringing of cable because MMDS provided more than 30 television channels via a microwave signal. Subscribers needed only a small antenna on their rooftop and a receiver near their television set to receive MMDS.

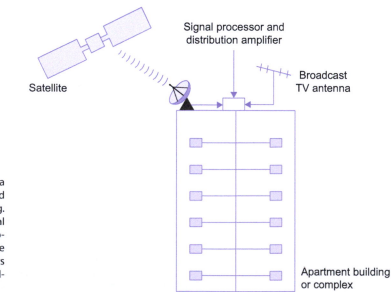

DIAGRAM 3.7 This diagram shows how a satellite master antenna system works. A satellite sends program information and is received by a satellite receiving dish on the top of an apartment building. A broadcast antenna receives broadcast signals from the local broadcasters. These signals, channels of television, are then programmed to channels that are sent to individual apartments. The system is designed to give uniformly good signals to all subscribers in the building and avoid having individual apartments install multiple antennas or receiving dishes on the rooftop.

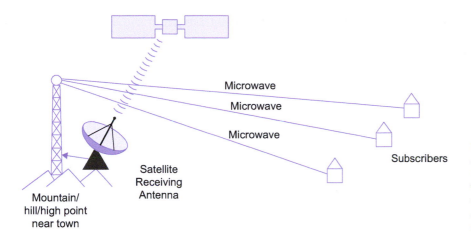

Microwave

Microwave

Microwave

Subscribers

Satellite Receiving Antenna

Mountain/ hill/high point near town

DIAGRAM 3.8 MMDS program content is sent from a geostationary satellite to a receiving dish. The signal is then sent from the dish to an MMDS transmitter that uses microwaves to send the signal to subscribers in town. Each subscriber needed a microwave antenna and signal processor to receive the television channels.

CONSUMERS GET THEIR OWN SATELLITE TV DISHES

In the 1970s and 1980s satellite receiving antennas were available to the public. These antennas were not connected as part of a multichannel delivery system business but were used to 'intercept' the signals sent by program providers to the cable systems that distributed them to homes. The systems were referred to as *TVROs*, which stood for 'television receive only.'

TVROs appealed to those households that were not served by cable and could afford the expensive installation and equipment. At first the biggest downside to these first generation home satellite systems was the complex installation and the size of the receiving dishes. At about 8 feet by 15 feet, the dishes were cumbersome and unsightly. Neighbors complained and even some homeowners' associations (HOAs) enacted rules banning the dishes because they affected the appearance of the homes in their neighborhoods.

At first, satellite subscribers could receive premium channels (HBO, Showtime, and some adult entertainment) for free. The cable networks were not happy about this type of pilfering, so they came up with a way to block premium reception by scrambling the signal. After that, when a premium channel was tuned in, subscribers saw a picture that was nothing more than zig-zaggy lines of color. The scrambling of the premium channels led to a decrease in demand for the big dishes, and growth of this method of receiving multiple television channels slowed considerably.

The satellite television industry changed in 1994 with DirecTV, a subsidiary of General Motors owned by Hughes Electronics. DirecTV was the first multichannel delivery service that provided television directly to subscribers. The service became known as *DBS*, or direct broadcast satellite. Subscribers could now downlink satellite television programs to a small dish provided by DirecTV instead of subscribing to a cable company. DirecTV was without competition for 2 years, when the DISH Network hit the marketplace. Although consumers paid more for DBS than for cable, it offered about 150 channels of high-quality television, often many more than the number of channels provided by cable.

Another disadvantage of early DBS was that until 1999, it was not allowed to provide local broadcast stations to individual subscribers' homes. This limitation forced subscribers who wanted local programming, such as evening newscasts, either to subscribe to cable or to capture over-the-air signals with an antenna in addition to their DBS service. For many consumers, this was not acceptable. Paying for both cable and DBS was expensive and involved paying for quite a bit of program duplication. Also, very few people wanted to switch to an antenna while viewing just to see the local news or other programs. This problem was rectified in 1999, when the Satellite Home Viewers Act (SHVA) mandated that local broadcast signals be carried on DBS.

As DBS technology improved, so did satellite dish receivers. Not only did dishes become smaller (18 inches in diameter), but they became capable of receiving high-quality digital transmission. Local neighborhood associations and neighbors themselves no longer had a reason to complain, as dishes could be discreetly placed on a rooftop, on a windowsill, or in a backyard.

FIG. 3.13 This picture shows both an older C-band satellite receiving dish and a newer DBS dish mounted below it on the same mast. The DBS dish is about one tenth the size of the C-band dish above it.
Photo courtesy of iStock #2687905

To date, DirecTV and the DISH Network are the two major players in the DBS business. These two companies have been fighting each other and cable for subscribers. In March 2002, DirecTV and DISH Network proposed to the FCC that they merge and create one national provider of DBS service. The government rejected the merger idea, however, preferring to maintain two providers that could compete with each other and give the audience a choice of services.

SEE IT NOW

DIGITAL TELEVISION

In 1996, the FCC appointed a task force to study the conversion of all television stations to digital broadcasting. On the task force's recommendation, the FCC ordered all television stations to upgrade to digital broadcasting by 2006. Station owners were up in arms, not so much because they opposed digital but because of the millions of dollars it would cost them to comply. Consumers were also concerned because conversion meant they had to buy new sets that could pick up digital signals. After much brouhaha from the industry and viewers, the FCC capitulated by postponing the conversion until February 17, 2009. But continued pressure led to the Obama administration and Congress persuading the FCC to postpone the digital conversion one more time, to June 12, 2009.

Digital television (DTV) transmits television programs in high-definition television (HDTV), which is a widescreen, high-resolution format. It also uses standard definition (SDTV), similar to an analog television picture, but with better color reproduction and less interference. An analog picture is made up of 525 lines of information (top to bottom) of resolution, but the HDTV picture has up to 1,080 lines, more than twice the picture information. In addition, the HDTV picture has a wider *aspect ratio* (the relationship between screen width and height), yielding a 16:9 picture (16 units wide, 9 units high) that more closely resembles a widescreen movie picture than the 4:3 picture of analog television. The HDTV standard adopted by the Advanced Television Systems Committee is known as ATSC A/53 or simply ATSC.

In making the move from analog to digital, the FCC forced all television stations to make large investments in new digital equipment. This switch to digital did not offer immediate financial rewards, as stations do not have a way to generate any more income with a digital picture than they did with analog-only broadcasting. One advantage to the digital system, however, is that stations have enough room in the 6 MHz channel to send out more than one program at a time. For example, a local station might have its main programming on its first channel, that is, a network feed, such as NBC. A second channel could carry weather and news, and a third could show foreign-language programs. Digital conversion also meant that broadcasters had to give up the analog frequencies that they had been using since the 1940s. The FCC has auctioned these frequencies to telecommunications companies and, as a result, the government has received billions of dollars in revenue.

> ### FYI: Analog Versus Digital
>
> Duplicating an analog signal is like pouring water from one jar to another. After you finish pouring, the new jar is almost full, and a few drops will be left in the old jar. When you duplicate in analog, the entire signal is not duplicated. Some of the signal gets lost in transit. Duplicating a digital signal is like pouring marbles from one jar to another. In this case, the new jar will be full of marbles, and the other jar will be completely empty. Digital transfer allows the entire sampled signal to be duplicated, and the copies are identical to the original. Although some say that the process of digital sampling misses some of the original (like the content represented by the spaces between the marbles), the process is very accurate.

Before the official switch to digital, the audience was slow to embrace the new digital broadcast technology, although it was often available, but on different channels than the original station channel. For example, a station

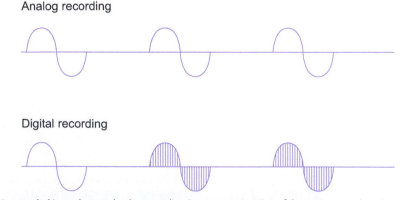

Analog recording

Digital recording

DIAGRAM 3.9 When content is recorded in analog mode, the recording is an approximation of the entire sound or picture (or both). The recording process is slightly distorted, and some information is lost. When the analog recording is played back, more distortion occurs, and more information is lost. When content is digitally recorded, the recording is sampled and encoded digitally. The process of sampling is recording (or measuring) the content at regular, very rapid intervals. The more samples of the content, the more accurate the recording and playback.

FIG. 3.14 2009 coupon for digital TV converter box

normally broadcasting an analog signal on channel 7 would be given another channel, perhaps 47, to broadcast the digital signal. Before the mandated switch to digital, viewers were skeptical about the benefits, and they were reluctant to buy a new digital television set. Preconversion, only about 10 to 20% of viewers watched 'over-the-air' broadcasting. About 80 to 90% of the audience received television stations via either cable or DBS. The remainder of the audience received over-the-air TV signals via indoor or rooftop antennas. It was only this small segment of the total audience that was affected by the digital conversion, as both cable systems and DBS continued to provide analog signals for those subscribers with analog sets. Households that did not subscribe to one of these services were given the opportunity to purchase set-top converter boxes that received the digital signal over the air and converted it to an analog signal that was compatible with older television sets. The government offered free coupons worth $40 toward the purchase of these set-top boxes. About 34 million of these coupons were sent to audiences (up to two per household) prior to the conversion, and the audience experienced few problems. But once digital television was up and running and cable and satellite systems added HDTV delivery, viewers were thrilled with the picture quality. Analog sets are no longer sold as new products in this country but may often still be found at garage sales or thrift stores.

TV SETS AND VIEWING HABITS CHANGE

Television sets have become much more similar to computers than they have been in the past. Computers often serve as centers for home entertainment. They play any digitally recorded medium and receive any streaming service, both audio and video. The missing link has been the ability to receive live television from broadcast network, cable, or satellite television on a computer monitor without special hardware or software. Digital broadcasting (and digital cable and satellite delivery) blur the lines between using television sets for entertainment only and using television monitors for computer work.

This blurring of distinction has encouraged some of the big names in the computer industry—Apple, Microsoft, Dell, and Samsung—to market equipment for both computing and entertainment needs and it also changes how audiences use their computers and where they place them in their homes.

TV-on-DVD has changed television viewing. The sales of television programs on DVD generated $1.5 billion in 2003. After sales peaked in 2004, DVDs have seen a decline in sales over the past decade. Many viewers became willing to buy a whole season of a television series on DVD to avoid annoying commercials, to get a better-quality picture, and to set their own viewing times. This trend slowed somewhat because download sites—both legal and illegal—made buying the physical DVD less necessary. Using devices like the iPad and Ultrabooks as replacements for laptops and desktop computers is significant, because they do not have internal DVD drives, which encourages viewers to seek entertainment that is stored or transported in other forms, such as flash memory or 'jump' drives or streaming. The movie industry has encouraged an increase in streaming movies by releasing digital versions of new movies online (pay-per-view) 1 to 3 weeks before they become available on DVD.

> ### ZOOM IN 3.5
>
> The Advanced Television Systems Committee (ATSC) is an international nonprofit organization that helps develop voluntary standards for digital television. More information about this group is available at www.atsc.org, and more information about digital television is available at www.dtv.gov

INTERNET DELIVERY OF TELEVISION PROGRAMS

Television stations do not currently have a viable model for delivering live television programs directly online or to cell phones, although some attempts have been made. Some devices, such as the Google Chromecast, allow a mobile device with a Chromecast app to stream video to a television with a Chromecast device attached to it. This is not the same as delivering live television; this service is limited to subscription streaming services such as Netflix or free streaming services such as YouTube. Until broadcasters find a technological solution to reaching audiences on mobile devices by using their broadcast signal, they are limited to streaming solutions from their Web sites via the Internet.

The broadcast networks are now providing program streaming to all devices. Full episodes and clips from shows are available free at the network Web sites. This service has made the network libraries of programs available to the audience 24/7.

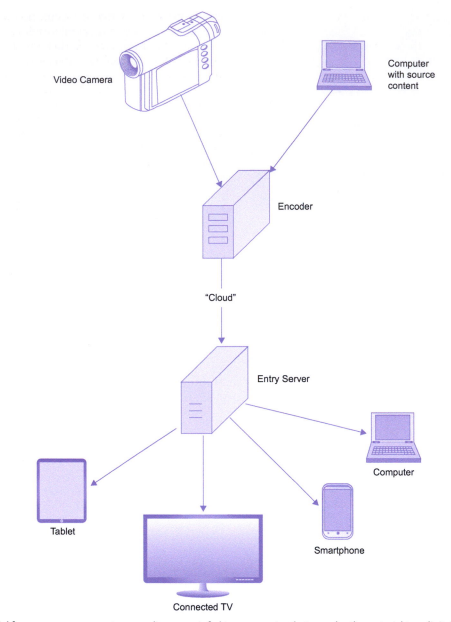

DIAGRAM 3.10 Material from a camera, computer, or audio source is fed to a computer that encodes the material to a digital signal stream. The stream is sent to the 'cloud' and is retrieved by an entry server. The entry server decodes the signal and sends it to receiving devices (tablets, smartphones, TV sets, and computers) that play the audio and video signals.

FIG. 3.15 This small USB device, the Google Chromecast, allows a video signal retrieved on a mobile device to be sent to a large television via the Internet.

ZOOM IN 3.6

Go to CBS.com/all-access to learn about All Access that enables subscribers to access a giant library of programs from the CBS archives, new episodes of current shows the next day after airing live, and, in many markets, streaming live TV from its network-owned or network-affiliated stations. Subscribers pay about $6 per month for the service. Although full episode streaming from the broadcast networks has been around for years, the difference in this service is that streaming live TV is available. Until recently, the audience could not stream live television for sports events and news.

Stations are now using the Web to connect with viewers through the use of online video, social networking, and other features. Stations host blogs about local news issues, up-to-the-minute traffic and weather coverage, community chat rooms, in-depth coverage of stories, and—most importantly for station revenues—advertising space. News personalities offer email contact with the audience as well as a presence on Facebook and Twitter. Some stations post information and content from audience emails directly into local news shows.

Cable and telecommunication companies are reaching out to their users by encouraging them to subscribe to 'wired' programming services that are viewable on most mobile devices. For example, Comcast, the largest cable company in the country, uses a system named XFINITY, which allows subscribers to access television shows and movies from computers, tablets, and smartphones. Subscribers can also download many television shows and movies to mobile devices. The downloaded shows can be viewed when the subscribers are offline. Verizon Wireless has expanded its telephone service and now also offers television shows via its FiOS cable system as well as its FiOS Mobile app that lets subscribers stream television programs and movies to their smartphone or tablet.

CONTEMPORARY CABLE

In 2013, cable and satellite delivery systems lost 250,000 subscribers, the first decline ever. This decrease in subscriptions indicates that 2012 may have been the peak for subscribers to multichannel program providers. Cable subscriptions peaked in 2001, when there were almost 67 million subscribers, but by 2010 the number of subscribers dropped to 63 million and by 2014 to just under 50 million. Viewers are watching television programs via DBS or other delivery methods. A study conducted in 2014 showed that more than half of cable subscribers would terminate their cable television subscriptions if they had a choice of another cable service. In addition, 73% of the respondents feel that "cable companies are predatory in their practices and take advantage of the consumers' lack of choice" (Sherman, 2014). This attitude is reflected in consumers 'cutting the cord' and removing cable and satellite television from their homes. Despite the decline in the number of subscribers, cable is still a profitable business. Although there are fewer cable television subscribers, the financial loss subscriptions is made up for by increased prices for television and Internet service. Cable television is not coming to an end, but many feel as if the cable model is broken. Between 1995 and 2005, cable companies doubled the size of the 'bundles' that were marketed to subscribers, and the charges to the customers rose at a rate three times the rate of inflation. According to a former FCC chairman (Kevin Martin), "the average cable subscriber was paying for more than 85 channels that she didn't watch in order to obtain the approximately 16 channels that she does" (Popper, 2015, April 22).

Except for VOD (video on demand) and the possibility of a more 'a la carte' approach allowing the subscribers to select only the channels they want, the cable industry is consolidating, raising prices, and holding on to its ISP (Internet service provider) business, the cable industry seems to be technologically stagnant. At present, cable is the only or the best way to receive a broadband Internet connection. Some cable companies offer a companion streaming service through the Internet for its subscribers, but these efforts are overshadowed by the growth in services Netflix and Hulu+.

SEE IT LATER

The television industry has changed immensely since 2010. Subscriptions to cable and satellite systems have declined, while subscriptions to streaming services like Netflix and Hulu continue to increase. Television sets have changed dramatically. Analog sets are no longer manufactured. Flat-panel smart television sets that come with built-in connections to the Internet have decreased in price, with a typical 47-in. diagonal HD set selling for about $500 or less. Increasingly, television sets are used as monitors that receive inputs from a variety of sources: DVD players, streaming devices like Apple TV and Roku, tablets like the iPad, and laptop computers. In addition, sets are all capable of receiving signals from broadcast television, cable, and satellite delivery systems, also known as 'legacy' delivery systems. With all of these inputs, viewers are becoming somewhat 'agnostic' about how they get their television. In other words, the audience does not care who or what delivers television as long as they get it where and when they want it.

TELEVISION AND THE INTERNET

Essentially, the television industry is in flux. Both broadcast networks and premium cable services will launch streaming services that do not require a cable or satellite subscription. Instead of paying the average of $85 per month for cable or satellite services, the audience will be moving to an a la carte subscription model. Netflix and Hulu now offer an enormous selection of programs, while each service charges only about $10 per month. Some subscription services, including both Netflix and Amazon Prime, are now producers of content. It appears that young people are early adopters of this model. These streaming services, known as 'over the top' or OTT, can collectively cost more than a single subscription to cable or satellite, but they will provide mobility to subscribers. In other words, the television viewing audience does not have to stay at home to watch their favorite programs. The new subscriber model will be available anywhere the viewers can get an Internet connection. Simply put, streaming will disrupt the traditional television business in a way that is similar to how the Internet disrupted the music and print industries. 'Cord cutting,' referring to viewers dropping their subscriptions to cable and satellite services, is the wave of the future.

NEW TELEVISION SETS

Television sets are changing as well. A few years ago, 3D sets were the big fad, but audiences did not like wearing 3D glasses to watch a show. Now 3D sets are scarce in electronics showrooms and are replaced by new sets with curved instead of flat screens. Purportedly, a curved screen creates a viewing experience that draws the viewer into the program, but one that comes at a significantly higher cost than flat-panel sets. The second innovation, UHD television, holds more promise for the consumer market.

It is expected that in the near future, HD sets will give way to UHD (ultra-high-definition) sets. HD sets are capable of reproducing a television picture with 1920 (wide) × 1080 (high) pixels (picture elements). UHD sets are capable of reproducing a television picture with at least 3840 × 2160 pixels, yielding about four times as much picture information per frame. The term 4K refers to the 3,840 or roughly 4,000 pixels wide. The UHD picture yields a clearer, more realistic picture. The change in resolution is technically similar to the change from SD (standard definition 480 lines of resolution) to HD (1,080 lines of resolution). Although UHD sets are sold at premium prices now, in the near future the prices per set will drop as they did for HD sets. Consumer adoption of UHD sets depends in part upon the availability of UHD programs. Currently the production companies that produce television programming are slowly changing from HD production to UHD production. Netflix is providing 4K streaming for some of its shows, but to receive the shows, a 25 Mbps or faster Internet connection is needed, which is not a typical download speed . . . yet.

Further boosting the potential for 4K sets is the availability of 4K content on Blu-ray disks. Blu-ray expects to have 4K content programs available on disc in the near future, but the transformation means viewers will have to buy a special 4K player. DVDs and Blu-ray discs have been enormously popular since the mid 2000s, but younger audiences are moving away from discs to online delivery. In 2014, DVD sales dropped 28% from the previous year for a total loss of $12.2 billion. A further decline to about

$8 billion is expected by 2018. If 4K catches on, it will be through online streaming rather than DVD.

Even before 4K becomes mainstream, there is already buzz about 8K television sets. These sets double a 4K picture's pixel width and height and yield an amazingly realistic television image. Although these sets are demonstrated at technology and industry conventions, they will most likely be used only in movie theaters because they require a screen size that is too large for most homes. In addition, there is no commercially produced 8K content currently available to general audiences.

The change in resolution does not just change television sets but also affects cameras, smartphones, tablets, and all monitors. In other words, anything that records or plays back a television image will have to change to accommodate 4K or 8K content.

ORGANIC SETS

Organic is not just a type of food; it also refers to a type of television set. *OLED* sets are constructed with **organic light emitting diodes** that produce an amazingly vivid image with very strong contrast levels. Although smaller OLED sets are affordable, larger sized ones (40+ inches) are very expensive and thus they have not made a dent in the consumer market.

BROADCAST TELEVISION

The broadcast network audience share will continue to decrease in the years to come, because the audience will have more choices. The networks may find that their profitability depends increasingly on delivering programs that other services cannot provide or that they, the networks, can best provide. For example, the networks can deliver live programs, like news and sports, and can operate profitably by offering reality shows that do not require large payments to stars, writers, and independent producers. The networks will continue to be challenged by technological changes and other delivery systems. As a result, they will continue to vertically integrate by buying program-production houses and program-syndication companies to gain control over programming sources and outlets. They will also utilize new programming services on other delivery systems (e.g., cable, satellite, and the Internet) and even new technologies to keep their audience share large enough to attract advertisers. CBS, the network that carried the 2011 Super Bowl, offered a live stream of the broadcast for the first time.

Although it does not now pose a threat to large television producers, user-generated television content looms on the horizon. Digital tools for the production of high-quality television, once the exclusive domain of 'big media,' are now becoming available to ordinary people. Personal technology of this type may change the future production of television programs.

Local broadcast stations are now producing video content not only for their broadcast newscasts but also for online streaming. A video package produced for a nighttime newscast is often archived as is or edited to a different

FIG. 3.16 Curved TV
Photo courtesy of Shutterstock

length and repurposed for later use on the Web site. Stations owned by broadcast groups often share content in this way.

Television stations compete not only with traditional media, other television stations in their market, cable, satellite, and commercial streaming services but also with the streaming service YouTube that shows content created by the audience. YouTube (purchased by Google in 2006 for $1.65 billion) had close to 100 million viewers per month in 2010. Just 5 years later, YouTube had more than 1 billion unique users each month, and half of these users watch YouTube on a mobile device.

Each of the broadcast networks and most of the cable networks have their own Web site that streams video. NBC, ABC, and Fox created Hulu to attract viewers to clips and full episodes of shows. Hulu also offers two subscription rates, one with some commercials inserted into the programs and another slightly more costly option that eliminates commercials within programs. It is not yet clear if online viewing will cause local television stations serious problems in the near future. The challenge to traditional television is to provide compelling content that audiences will watch and advertisers will support.

The future of broadcast television might depend upon its ability to provide live local news and big sporting events from the professional sports world. Since people do not want to watch time-shifted sports, broadcast television's (and the cable and satellite providers') ability to show live sports, at least for the time being, sets it apart from competing systems like Netflix and Amazon Prime. Although some have predicted the end of broadcast television, its total demise is not in sight, although an overhaul of its structure is overdue.

CABLE TELEVISION

Cable channels and networks are considering streaming, DVD, and videogame use when attempting to reach young audiences. They are experimenting with putting prime-time shows on later at night or perhaps repeating them at a later time to reach those that use other entertainment options during prime time.

Cable systems will struggle in the future. They are fighting the 'cutting the cord' trend and dealing with a new generation of 'cable nevers,' young people who, after leaving their parents' homes, have never subscribed to cable television because their viewing needs are met by Internet delivery.

SATELLITE TELEVISION (DBS)

Satellite television delivery faces the challenges of providing a unique service and making itself user friendly. Satellite delivery to homes had a marketing advantage by offering many more channels of high-quality video than cable, but it does not offer high-speed broadband at a price and speed competitive with cable. Although satellite receivers provide high-quality reception and recording, the user interface is, at best, a bit clumsy. Additionally,

the remotes are tricky to use and sometimes difficult to read without proper lighting. Searching for specific programs is slow and often frustrating. Improvements are being made but might still fall short of computer interfaces that are easily controlled by a keyboard and mouse, which are familiar to most users. The computer keyboard-and-mouse interface is better suited for Internet television than the complicated remotes offered by cable and satellite companies.

Although it is common for people to pay $100 or more per month for a satellite television subscription, many subscribers are saving money by dropping the middleman and using their computers or Internet-capable television to receive programs directly. Dropping a subscription to satellite might become as common as dropping the daily delivery of the newspaper has been for the past 10 years. People do not want to pay for a costly subscription to a newspaper when they could get news from reliable sources on the Internet for free. The same changes might occur in the satellite television program industry.

Consumers would like to get their television in an a la carte style, meaning that consumers would rather pay per channel than pay for a bundle of television channels, many of which they do not watch. In this way, the satellite television industry is at least as vulnerable as the cable industry to cord cutting. The difference between the cable and satellite industry relationship to consumers is that the cable industry has become the main provider of Internet connections. While any home connected to cable television can also be easily connected to cable's ISP service, this is not true for satellite subscribers. Internet service through a satellite service is primarily provided in remote areas, where cable television cannot provide service. Until 2015, the satellite industry's lack of interactivity put it at a disadvantage relative to cable as a multichannel provider, because interactivity will continue to be an essential part of the television viewing experience. In July of 2015, AT&T acquired DirecTV, which opens up possibilities for DirecTV to offer its customers broadband access through its parent company.

SUMMARY

Early experimenters in television tried two methods to obtain pictures: a mechanical scanning system and an electronic scanning system. The electronic system was eventually adopted as the standard. Although many inventors were involved in the development of electronic television, two of the most important were Vladimir K. Zworykin and Philo T. Farnsworth.

The FCC authorized commercial television broadcasting in 1941, but the industry did not catch on until after World War II, when the materials needed for television equipment manufacturing became available. After the war, television grew so rapidly that the FCC could not keep up with license applications or technical issues. In 1948, the FCC put a freeze on all television license applications that lasted until 1952. During the freeze, the FCC considered the allocation of spectrum space to stations, designed the UHF band, and dealt with the issues of VHF and UHF stations in the same markets, color television, and educational channels.

At first, television was a live medium. In the 1950s, many high-quality dramatic programs were written for live theater-type performances to attract educated audience members who could afford the cost of a television set. Program production changed when videotape became available and programs were no longer produced live in the studio. The television audience changed when the price of a television set dropped and programs were adapted to appeal to a mass audience. The early days of television had some challenges, including blacklisting people working in the industry who supposedly had communist sympathies.

Stations with network affiliations did well because of network programming. Independent stations had to resort to older programs from syndication as well as sports and locally produced programs, especially children's programs and cooking shows.

The importance of television became more evident in the 1960s, when events such as the assassinations of John F. Kennedy, Martin Luther King Jr., and Robert Kennedy were covered extensively by the networks. That decade also saw a growth in local news coverage and national news coverage of the Vietnam War, violence in the streets, racial tension, and the first man on the moon. In 1967, Congress approved the Public Broadcasting Act, signaling the birth of a network dedicated to noncommercial broadcasting.

In the 1970s, cable television grew rapidly, and many cities, large and small, became wired for cable. Cable television started out in the late 1940s as a small-time system to provide rural audiences with television programming. Strong growth began in the mid-1970s, when cable systems began to use communication satellites to receive programming from all over the country and also began to develop their own programming through national cable channels, pay channels, local origination, and pay-per-view.

Throughout the 1980s, VCRs gave the audience the ability to rent movies and record programs off the air. The addition of cable networks like CNN, MTV, and ESPN began to change the dominance of the broadcast networks. Broadcast deregulation was a guiding principle for the FCC, and many rules were modified or removed entirely. Television technology improved, and electronic news gathering using portable video equipment became common in stations across the country. During this decade, a new major commercial network, Fox, began operating. Overall, the audience size grew, but the networks' share of the audience declined because viewers had more choices, primarily because of cable television.

The 1990s began with extensive coverage of the Gulf War, giving television an opportunity to show live video from the other side of the world. In the mid-1990s a new technology, direct broadcast satellite, began service to audiences, providing many high-quality channels and direct competition to local cable companies. The passage of the Telecommunications Act of 1996 changed many of the rules regarding ownership of electronic media stations and allowed television group owners to acquire as many stations as they wanted, up to a cap of 39% of the national audience. This relaxation of ownership rules resulted in fewer groups owning more stations and thus consolidating the television station business. In the 1990s, three new television networks emerged: the WB, UPN, and PAX, which have since evolved into the CW (UPN and WB) and ION. In the last half of the 1990s, the Internet became popular and signaled huge changes in audience media behavior. People began to spend more time online and less time in front of their television sets.

At the beginning of the new century, consolidation of station ownership began to raise issues about diversity and localism. The role of the networks continues to change from one of a delivery system for independently produced programs to one that delivers network-owned and -produced programs. The convergence of computers and television will continue drawing corporations like Apple, Microsoft, Dell, and Samsung into the television business. Digital broadcasting replaced analog broadcasting in 2009 and ushered in a new era of high-quality video and sound and different delivery methods, including multiple signals and Web delivery of content to the audience. Television stations are now using new media to provide additional viewing opportunities and social media to connect with the audience.

An important trend in television delivery systems is vertical integration. Large media companies now own networks, television stations, newspapers, magazines, cable channels, book publishing companies, movie studios, syndication companies, and television program–producing companies. Television networks are producing content for other services to distribute. This practice contributes to the profitability of the parent companies and signals a strategy change by the networks. The television networks are no longer just in the business of exhibiting their content. They are owners of content regardless of the manner of distribution.

The audience of the 2010s consists of many who get television through streaming video from the Internet. This audience has increasingly cut the cord by dropping cable and satellite TV subscriptions and relies upon subscription service like Netflix and Amazon for television viewing. Many content providers (e.g., HBO and ESPN) will continue to seek direct relationships with the audience and will skip the multichannel program services. This model of television delivery will be a major force in shaping the future of television.

Television sets have changed dramatically in the past 10 years. The old, bulky cathode ray television sets are now replaced by slim, light, large television sets that have HD or UHD resolution that yields amazingly clear pictures. These digital sets, once very expensive, have now become affordable and dependable entertainment centers in homes. A variety of inputs brings television programs, movies, social media, and audio music services to the audience with the touch of the remote control.

Technology continues to present different viewing opportunities to the audience that will force the networks and multichannel television program services (cable companies, satellite companies, and telecommunication companies) to continually adjust their delivery and revenue models to stay financially viable in the competition for viewers and advertisers.

BIBLIOGRAPHY

Auletta, K. (1991). *Three blind mice: How the TV networks lost their way.* New York: Random House.

Bellamy, R., & Walker, J. (1996). *Television and the remote control: Grazing on a vast wasteland.* New York: Guilford Press.

Bray, H. (2015, Mach 15). Boston globe. *Despite cord cutting, cable TV's future looks very familiar.* Retrieved from: www.bostonglobe.com/business/2014/10/29/despite-cord-cutting-cable-future-looks-very-familiar/L0seHkhxonu646PQhml/story.hmtl

Clarke, A. (1945, October). Extraterrestrial relays: Can rocket stations give worldwide radio coverage? *Wireless World*, 305–308.

Cohen, D. (2009, October 27). Quality content drives 3D home. *Variety.* Retrieved from: www.variety.com/article?VR1118010455.html

Davidson, J. (2014, September 24). 5 packages that will replace pay TV as we know it. *Everyday Money TV.* Retrieved from: http://time.com/money/3421921/cable-tv-cord-cutters/

Duncan, J. (2009, August 7). Americans viewing more TV and movies via internet streaming. *Digital trends.* Retrieved from: http://news.digitaltrends.com/lifestyle/americans-viewing-more-tv-and-movies-viainternet-streaming/

Faulk, J. (1964). *Fear on trial.* New York: Simon & Schuster.

Federal Communications Commission (FCC). (1998, November). *Digital television consumer information.* Retrieved from: www.dtv.gov/

Foley, K. (1979). *The political blacklist in the broadcast industry: The decade of the 1950s.* New York: Arno.

Fotrell, Q. (2014, September 3). Cable companies should be afraid of this trend. California: Market Watch, Inc.

Franklin, C. (2004). How cable television works. Retrieved from: http://electronics.howstuffworks.com/cable-tv.htm [April 1, 2016].

The future of television: The weak staff. (2014, December 9). *The Week.* Retrieved from: http://theweek.com/articles/441909/future-television

Gillmor, D. (2004, January 11). Putting ads in the hands of the public. *San Jose Mercury News.* Retrieved from: www.siliconvalley.com/mld/siliconvalley/business/columnists/dan_gillmor/7684606.htm [January 20, 2004]

Halberstam, D. (1993). *The fifties.* New York: Ballantine Books.

Hazlett, T. (1989). Wiring the constitution for cable. *Regulation: The Cato Review of business and government, 12*(1). Retrieved from: www.cato.org/pubs/regulation/regv12n1-hazlett.html

Hendrik, G. (1987). *The selected letters of Mark Van Doren.* London, LA: Louisiana State University Press.

Higgins, J. (2004, January 5). The shape of things to come: TiVo's got a gun. *Broadcasting & Cable*, p. 33.

Hilliard., R., & Keith, M. (2001). *The broadcast century and beyond: A biography of American broadcasting.* Boston: Allyn & Bacon.

Home video index. (2004). New York: Nielsen Media Research.

Hutchinson, T. (1950). *Here is television: Your window to the world.* New York: Hastings House.

Jarman, M. (2008, August 30). Intel bets on web TV concept. *The Arizona Republic.*

Jayson, S. (2010, February 23). The net set. *The Arizona Republic.*

Jessell, H. (2002, August 5). Sink or swim: With a set-top box of their own, TV broadcasters can take their place in the digital future. *Broadcasting & Cable*, pp. 14–16.

Kumar, V., and Schechner, S. (2009, June 30). High court boosts remote DVR. *The Wall Street Journal.* Retrieved from: http:online.wsj.com/article/SB124640368173.html

Labaton, S. (2003, September 4). U.S. court blocks plan to ease rule on media owners. *New York Times.* Retrieved from: www.nytimes.com/2003/09/04/business/media/04FCC.html?hp5&pagewanted5print&position5 [October 8, 2003]

Local television in 2018: Evolving with shifting audience demos. *Nielsen.* Retrieved from: www.nielsen.com/us/en/newswire/2013/local-television-in-2018-evolving-with-shifting-audience-demos/

Maynard, J. (2004, February 4). TV-on-DVD popularity a threat to networks. *Arizona Republic*, p. D1.

Mitchell, C. (1970). *Cavalcade of broadcasting.* Chicago: Follett Books.

Perez, S. (2009, August 6). Livestation brings TV to the iPhone. *New York Times.* Retrieved from: www.readwriteweb.com/archives/livestation_brings_live_tv_to_the_ihone.php

Popper, B. (2015, April 22). The great unbundling: Cable TV as we know it is dying. *The Verge.* Retrieved from: www.theverge.com/2015/4/22/8466845/cable-tv-unbundling

Ritchie, M. (1994). *Please stand by: A prehistory of television.* Woodstock, NY: Overlook Press.

Romano, A. (2004, January 5). The shape of things to come: Keeping up with the nerds. *Broadcasting & Cable*, p. 32.

Schatzkin, P. (2003). *The boy who invented television.* Burtonsville, MD: Team Com Books.

Schneider, M. (2009, August 27). TV audiences are growing older. *Variety.* Retrieved from: www.variety.com/article/VR1118007846.html

Schwartz, E. (2000, September/October). Who really invented television? *Technology Review.* Retrieved from: www.technologyreview.com/Biztech/12186/?a5f

Sherman, E. (2014, June 9). Most consumers ready to dump their cable companies. *Money Watch.* Retrieved from: www.cbsnews/news/most-consumers-ready-to-dump-their-cable-companies/

Smith, F., Wright, J., & Ostroff, D. (1998). *Perspectives on radio and television* (4th ed.). Mahwah, NJ: Erlbaum.

Sterling, C., & Kittross, J. (2002). *Stay tuned: A history of American broadcasting.* Mahwah, NJ: Erlbaum.

Television viewing becomes increasingly fragmented as overall consumption grows, Accenture global survey finds (2009, April 20). Retrieved from: www.newsroom.accenture.com/article_display.cfm?article_id54822

Vaughn, R. (1972). *Only victims: A study of show business blacklisting.* New York: Putnam.

Walker, J., & Ferguson, D. (1998). *The broadcast television industry.* Boston: Allyn & Bacon.

Television and Radio Programs and Programming

4

Contents

In the early days of radio, amateur operators just put something on the air—often on the spur of the moment. Maybe someone would drop by a station to sing or play an instrument or to talk about some issue to whoever was listening. Some years later, radio station licensing required the transmission of *scheduled* programming. Listeners crowded around their radios to hear sports, news, music, dramas, and church services. By today's standards, the level of static and poor audio quality would make radio unlistenable, but to yesterday's audience, radio was magic.

This chapter covers the development of radio programming, starting with station formatting, how stations obtain music, how songs get on the air, and who decides how often they are played. The chapter then moves on to television programming. It describes the different types of television programs and explains how program ideas are developed, how programs make it to the air, and how program-scheduling strategies are used to keep the audience tuned to a channel.

SEE IT THEN: RADIO

TYPES OF PROGRAMS

MUSIC

In one of the first live radio performances, opera singer Enrico Caruso sang "O Sole Mio" from the Metropolitan Opera in New York in 1910. Lee de Forest, the disputed inventor of the audion tube, which amplifies radio voice, masterminded the performance to promote radio.

ZOOM IN 4.1

Hear Caruso as he sang live from the Met almost 100 years ago at www.youtube.com/watch?v=u1QJwHWvgP8

By the late 1920s, music was the main source of radio programming, For example, NBC Blue and NBC Red network-affiliated stations in New York devoted about three-quarters of programming to music. Because the quality of phonograph records was very poor, most of the music was broadcast live. Musicians performed in studios and often in front of a live audience. The studios were often decorated with potted plants, such as palm trees, to make them seem like real concert halls or ballrooms; thus, radio programming of that era is often referred to as **potted palm music**.

Radio-transmitting equipment was bulky, heavy, and hard to move around, and thus on-location broadcasts were rare and troublesome. Station KDKA broadcast the 1921 Dempsey–Carpentier boxing match with a temporary transmitter set up in New Jersey. It was fortunate for KDKA that Carpentier was knocked out in the fourth round, because shortly after the fight ended, the station's transmitter melted into one big heap of metal.

Developments in the early 1930s set the standard for radio programming for the next 20 years. Portable transmitting equipment made possible coverage of live sports and other types of on-location events, and creative writers developed new and sophisticated types of radio programs. By the late 1940s, only about 40% of the programming was music oriented. Comedies, soap operas, dramas, quiz shows, and children's programs filled the airwaves and were a prominent form of in-home entertainment until the late 1950s.

Early programs were not always scheduled at regular times; the audience rarely knew what was going to be playing when they turned on the radio. Station executives realized that to capture and maintain a steady listenership, programs needed to air on a regular schedule. Eventually, newspapers ran radio program listings, and special radio program guides were printed so listeners knew when it was time to run inside to hear their favorite shows, such as *The Green Hornet*, *The Jack Benny Program*, and *The Lone Ranger*.

DRAMAS

Among the most popular programs were daytime serial dramas, which came to be known as *soap operas* because laundry soap manufacturers were the primary sponsors. However, the shows could have just as easily been called *cereal operas*, because the first serial drama, *Betty and Bob*, was sponsored by General Mills. These 15-minute continuing-storyline dramas appealed largely to females and dominated the daytime airwaves. In the evening, a more diverse audience tuned in to a different type of dramatic programming, episodic dramas, which resolved the story within a single episode.

ZOOM IN 4.2

Every Sunday from 1930 to 1954, the Mutual network aired one of the most popular radio programs ever, *The Shadow*. The program enthralled listeners with its famous opening line, "Who knows what evil lurks in the hearts of men; the Shadow knows," which was followed by a sinister laugh ("Famous Weekly Shows," 1994–2002). Listen to audio clips of *The Shadow* on YouTube. Just enter the search term: The Shadow radio show.

COMEDIES

Comedians such as Jack Benny, George Burns, Gracie Allen, and Bob Hope all got their start in radio programs between the late 1920s and mid-1940s and later made a successful transition to television.

The comedy *Amos 'n' Andy* made its radio debut in 1928 and was the first nationwide hit on American radio. Avid fans would stop what they were doing and crowd around the radio to laugh at the latest antics of their two favorite characters. In addition to its popularity, *Amos 'n' Andy* was also one of the most controversial programs on the air. The title characters were "derived largely from the stereotypic caricatures of African-Americans" and played by White actors who "mimicked so-called Negro dialect" (*Amos 'n' Andy Show*, 2003). Freeman Gosden and Charles Correll, the White creators and voices of the program, "chose black characters because blackface comics could tell funnier stories than whiteface comics." Even so, the program quickly came under fire from the Black community and the National Association for the Advancement of Colored People (NAACP). As the show progressed, the characters grew beyond being caricatures. In fact, NBC claimed the program was just as popular among Black listeners as among White listeners. The program moved to CBS television in 1951 but only remained on the air for 2 more years and had its final radio broadcast in 1960.

It was not until 1947 with the debut of *Beulah* that an African American starred in a radio sitcom. Well-known actress Hattie McDaniel took over the part of Beulah from a White male actor. Beulah was a stereotypical simple-minded but warm, caring, funny maid who always outwitted the family for whom she worked. When McDaniel took the lead, the program was renamed *The Beulah Show* and remained on radio until 1954. The show also aired on television from 1950 to 1953 with the role of Beulah played by various African American actresses, including Ethel Waters.

FIG. 4.1 Charles J. Correll and Freeman F. Gosden wore blackface as characters Amos and Andy.
Photo courtesy of RKO Pictures/Photofest. © RKO Radio Pictures.

ZOOM IN 4.3

Listen to clips of popular old radio shows on Old Time Radio https://archive.org/details/oldtimeradio

Watch television programs such as *Amos 'n' Andy*, *Bob Hope*, *Fibber McGee and Molly*, *The Red Skelton Show*, *Abbott and Costello*, and *The Adventures of Ozzie and Harriet* on YouTube and enter in the name of the program. Or click on http://oldmovietime.com/tv/

The Burns and Allen Show hit the radio airwaves in 1932. The show featured George Burns as a curmudgeonly husband and straight man to his wife, the very funny Gracie Allen (they were also married in real life). After a successful 18-year run, the program moved to television, where it aired for eight seasons and 291 episodes.

The Jack Benny Program aired from 1932 to 1955. The episodes were usually based around a comedy sketch that featured Jack Benny and his regular cast but also included guest stars who sang or played a musical instrument.

The Bob Hope Radio Show (1938–1955) was a variety-type program that featured guest stars joking around with Hope and singing and performing audio skits. During WWII, many of Hope's programs were recorded live from military bases around the world.

BACK TO THE MUSIC

Radio thrived throughout the 1930s and 1940s and became the primary source of entertainment. People could not imagine anything better than having entertainment delivered right to their living rooms. But then came television—a new-fangled device that combined sound with images. Television quickly won the hearts and eyes of the public, who proudly displayed their new television sets and pushed their radio receivers into a dark corner of their living room. In 1946, only about 8,000 U.S. households had a television set, but just 5 years later, some 10 million households were enjoying the small screen.

Television programs were in huge demand and were desperately needed to fill airtime. Industry executives convinced radio show producers to move their programs over to television. As radio soaps, comedies, quiz shows, and dramas were adapted for television, radio found itself scrambling to replace these programs or face its own empty airtime. Considering that the loss of network radio programs also meant the loss of advertising revenue, radio executives knew they had to do whatever they could to save radio from dying off altogether.

The easiest way to fill radio airtime was with music. After all, radio is an audio-only medium, so music programming is the perfect fit. Several technological innovations helped remake radio into a music-dominated medium. Most importantly, music was now recordable on vinyl records and reel-to-reel tape, which vastly improved the overall sound quality. Further, thanks to the invention of the transistor, radio sets could be made much smaller. Pocket-sized transistor radios hit the consumer market in the mid-1950s

FIG. 4.2 Radio's portability, as well as its ability to provide soothing background noise, lends it to listening while doing other things. *Photo courtesy of Arnoud R. Beem*

and forever changed music listening habits, as listeners could now hear their favorite tunes outside of their homes. Radio listenership rose even higher as automobile manufacturers offered optional in-dash transistor radios. By turning back to music, radio distinguished itself from television and reemerged as a medium in its own right.

But radio station managers knew music itself was not enough to draw and maintain an audience. It had to be the right kind of music, the kind that appealed to the local broadcast market. Stations saw value in developing their own identities and attracting particular listeners through **music formatting**. Different stations in the same broadcast area began playing different types of music. Maybe one station played classical, another blues and jazz, another big band, and so on. Realizing that music fans were spending a great deal of money on records, enterprising radio programmers soon got into the habit of checking record sales to predict what songs and what performers would go over well on their station and keep within the format. Station formatting caught on and is still the mainstay of radio programming today.

Rock 'n' Roll

Despite the success of music formatting, stations needed something new to draw listeners back to radio—and rock 'n' roll saved the day. Young people went crazy over this new sound, and parents went crazy trying to keep their teens away from this wild new music. The more teens were told they could not listen, the more they wanted to listen, and rock 'n' roll stations flourished.

About the same time that rock 'n' roll steamrolled its way onto the music scene, radio stations were experimenting with DJs whose job it was to play the tunes and establish rapport with listeners. Cleveland's Alan Freed was the first DJ to introduce teenagers to this exciting new sound and to play rhythm-and-blues and Black versions of early rock to his mostly White audience. He is probably the most influential DJ of all time and is best known for dubbing the new music 'rock 'n' roll,' a term that had been used in many blues songs as a euphemism for having sex.

The new sound combined the rhythm-and-blues sound of Memphis with the country beat of Nashville and thus became the first "integrationist music" (Campbell, 2000, p. 72). Rock 'n' roll was embraced by many different types of listeners and thus provided a way to break away from the "racial, sexual, regional, and class taboos" of the 1950s (Campbell, 2000, p. 76). The music united Blacks and Whites, rich and poor, men and women, and Northerners and Southerners.

Rock 'n' roll fans eagerly tuned to stations that dared to play the new beat, while anti–rock 'n' rollers listened to stations that refused to broadcast the controversial sound. Some parents, teachers, and legislators called for banning rock 'n' roll, and some communities even held record-burning bonfires. Teenagers and rock 'n' roll enthusiasts were fiercely pitted against those who tried to keep the new tunes off the airwaves and out of the record stores. Newspapers ran front-page stories featuring the conflict between antirock crusaders and pro-rock activists. The resulting

publicity sparked a renewed interest in radio, and as ratings shot up, protests against rock 'n' roll subsided. Rock 'n' roll became the music that defined the baby-boom generation and continues as a popular station format today.

Capitalizing on the appeal of rock 'n' roll, radio programmers Todd Storz and Bill Stewart created the first **Top 40 format** for a station in Omaha, Nebraska. Storz and Stewart listened to the music that other stations were playing and discovered that at any one time, the number of different hit songs was about 40—hence the name Top 40. Storz and Stewart's new format featured up-and-coming hit songs and current hits.

The Top 40 format caught on with many stations across the country and appealed especially to young people, who wanted to keep up with the latest and most popular songs. But Top 40 music playlists were largely based on local or regional preferences and decency standards. Before cable music entertainment programs, pop culture magazines, the Internet, and other music sources, there were few ways for teens on the East Coast to know what West Coasters were listening to—that is, until Casey Kasem aired *Billboard* magazine's Top 40 national records of the week on his *American Top 40* program. At its peak, the program aired on more than 500 radio stations across the country, reaching millions of young music fans. After 39 years on the air, Kasem recorded his last show in July 2009. Kasem, who was also the voice of 'Shaggy' Rogers of *Scooby-Doo*, passed away in 2014 at age 82.

Rock 'n' roll was boosted, ironically, by television. American teenagers were wild for *American Bandstand* (1957–1989), a Saturday-afternoon dance show that featured

FIG. 4.3 Philadelphia—Circa 1957: Television host Dick Clark presides over the set of his show *American Bandstand* as teenagers dance to Top 40 popular music circa 1957 in Philadelphia.

Photo courtesy of Michael Ochs Archives/ Getty Images

(A)

(B)

FIG. 4.4A & 4.4B Fats Domino and Chuck Berry were among the Black performers who helped usher in rock 'n' roll.
Photo courtesy of Warner Bros./Photofest. © Warner Bros. Courtesy Photofest.

artists playing the latest hits. Originally broadcast from a studio in Philadelphia, operations moved to Los Angeles in 1963 at the height of the show's popularity. As many as 20 million viewers learned the latest dance moves and were sometimes the first to see new bands and singers. Host Dick Clark helped launch classic artists like Jerry Lee Lewis, the Beach Boys, Madonna, Steely Dan, The Jackson 5, and KISS. *American Bandstand* largely promoted mainstream rock 'n' roll to White, middle-class teens. There was not a dance show counterpart for African Americans and the rhythm and blues (R&B) scene until Don Cornelius introduced *Soul Train* in 1971. For the next 12 years, *Soul Train* popularized R&B, soul, and hip-hop artists such as Paula Abdul, Mary J. Blige, James Brown, the Supremes, Sly and the Family Stone, and Smokey Robinson. *Soul Train* had such a broad appeal that even White musicians like Elton John and David Bowie made appearances. Both dance programs were highly influential in determining radio station playlists. *American Bandstand* and *Soul Train* fans set the demand for certain songs and artists. Rock 'n' roll was initially thought of as a flash-in-the-pan musical style that would be popular only among a small number of teenagers. However, it proved to be the magic formula that rekindled radio listenership.

SEE IT NOW: RADIO

TYPES OF PROGRAMS

MUSIC

Since the late 1950s, it has been standard for stations to establish an identifying format. A station chooses a format based on budget, local audience characteristics and size, the number and strength of competing stations, and potential advertising revenue. Stations often change their format in accordance with these factors.

ZOOM IN 4.4

Learn more about different types of music and station formats at these Web sites:

- Guide to Radio Station Formats: www.newsgeneration.com/broadcast-resources/guide-to-radio-station-formats/
- Radio Station World: radiostationworld.com/directory/radio_formats/

FYI: Top Radio Station Formats by Share of Total Listening 2013

1. Country (14.8%)
2. News/Talk (11.3%)
3. Pop Contemporary Hit Radio (8.0%)

(Continued)

NEWS AND INFORMATION

About two-thirds of 'all-news' and 'news/talk' stations in the United States are on the AM dial. In 1961, the Federal Communications Commission (FCC) opened up spectrum space for FM, which has a sound quality superior to AM, and authorized stereo broadcast FM. Almost all stations remained on the AM spectrum, largely because most radios were manufactured with only an AM tuner and could not receive stereo FM. As listener demand for stereo FM increased, more stations began moving to the FM dial, and manufacturers included an FM dial on most radio sets. Once FM radios were available in cars, stations were lured away from tinny-sounding AM to the richer-sounding FM. Where the best and hippest stations went, so did listeners.

By the early 1970s, only about 30% of listeners tuned exclusively to FM stations, but 20 years later, three-quarters of radio listeners preferred FM, and now about 85% of radio listeners tune to the FM band. Music-formatted AM stations could not come close to the sound quality emanating from their FM competitors, so they either migrated to the FM band or stayed on the AM band but changed to a talk or news format for which audio quality was not so important.

The news/information format falls into three basic but overlapping categories:

1. **All-news** stations primarily air national, regional, local news, weather, traffic, and special-interest feature stories. The reports are usually scheduled throughout the day. The news cycle may occasionally be interrupted for a special in-depth report or talk show or another program that is not part of the usual schedule. It is rare to find a local station that employs beat reporters. Instead, stations air news that they obtain from wire services or other news sources.

2. The **sports/talk** format is similar to the news/talk format but focuses mainly on sports issues and news and live sporting events. Sports talk usually includes call-in programs and may emphasize local sports, especially if a professional or big-name college team is located within the broadcast area.

3. The **news/talk** format consists of a combination of call-in talk shows and short newscasts that usually come on once an hour or between segments. Most talk programs air during regularly scheduled times and usually last from 1 to 4 hours. News/talk shows are often politically oriented and are usually hosted by nationally syndicated conservative hosts. To counteract the conservative voice, some local radio stations pick up syndicated shows that are moderate or lean liberal, or they produce their own progressive programs. But in fact, there are few big-name national hosts to spread the left (liberal) philosophy. When Air America, a progressive radio network, began broadcasting in early spring 2004, listeners with forward viewpoints were happy to have a forum in which they could express their views without "being called dopes, morons, traitors, Feminazis, evildoers and communists" by conservative hosts (Atkins, 2003). Air America was never profitable, and it ceased live programming in January 2010.

NONCOMMERCIAL RADIO

Noncommercial stations are usually owned and operated by colleges and universities, religious institutions, and municipalities. The stations usually broadcast at the lower end of the FM dial, between 88 and 92 Mhz. Because these stations are noncommercial, they do not bring in revenue from advertising dollars. Instead, they rely on other forms of revenue, such as monetary donations, government grants, and underwriting, which is typically an on-air promotional announcement or program sponsorship. Unlike the advertisements that run on commercial stations, underwriting spots must abide by FCC regulations that limit promotional content.

Most noncommercial stations rely on their own music library of donated CDs, but they also produce their own community-affairs and news shows. Currently, about 3,500 noncommercial stations broadcast in the United States, many of which are housed on college campuses.

Community and college stations are usually not as rigidly formatted as commercial stations. Listeners are often treated to a variety of music, such as classical, jazz, alternative rock, and blues. Additionally, news, talk, and sports programs may be thrown into the mix. The eclectic nature of noncommercial stations draws listeners who are tired of the same old rotated music offered by most commercial stations. College stations are especially known for introducing new artists. Well-known bands and artists such as Nirvana, U2, The Cure, Elvis Costello, Nine Inch Nails, and R.E.M. owe much of their success to college radio.

Congress established the **Corporation for Public Broadcasting (CPB)** in 1968. The CPB, in turn, set up the **National Public Radio (NPR)** network in 1970. NPR's mission "is to work in partnership with its member stations to create a more informed public—one challenged and invigorated by a deeper understanding and appreciation of events, ideas and cultures" (What Is NPR? 2010).

National Public Radio distributes both its own and independently produced programs to its member stations. NPR programs air on slightly more than 1,000 public stations, attracting about 34 million listeners each week. NPR has won many programming awards and is considered one of the most trusted news sources. NPR delivers such favorites as *Morning Edition*, *All Things Considered*, and *The Motley Fool Radio Show*. *Car Talk*, one of NPR's most well-known shows, first aired on a Boston radio station in 1977. The call-in program about auto mechanics was so popular that NPR picked it up for national broadcast in 1986. Callers explained what was wrong with their car, and brothers Ray and Tom Magliozzi took it from there with humorous banter and spot-on diagnosis and recommendations for repair. Tom passed away in late 2014 but recorded episodes are still enjoyed by 3.2 million listeners each week.

ZOOM IN 4.6

- Learn more about NPR at www.npr.org
- Tune in to your local NPR member station. Listen to at least one news program, one music program, and one other program of your choosing. How do NPR programs differ from those on commercial radio stations?
- Listen to your college radio station. How is it different from the commercial stations in your area?

WHERE RADIO PROGRAMS COME FROM

LOCAL PROGRAMS

Some radio stations produce their own newscasts, weather and traffic reports, and sports shows featuring university and high school coaches, local sporting events, and other types of community programs. Radio stations generally foot the bill for their own local programming. News, talk, and sports programs are generally very expensive to produce, so it is usually only the big-market stations that create their own programs. They pay for the talent, production equipment, physical space, and copyright fees (which go to music licensing agencies), but most of the music itself is provided free of charge by recording companies. It is very common for recording companies to send demo discs of new releases to stations across the country. This arrangement works well for stations that play current music but not so well for stations that play classical music or other old tunes. There is little financial incentive for recording companies to send out free CDs of 200-year-old classical pieces, which will draw small audiences, or 30-year-old classic rock 'n' roll that listeners are not going to rush out to buy, so stations that specialize in older or more specialized music often have to purchase their own music libraries.

NETWORK/SYNDICATED PROGRAMS

Given the cost of producing enough programming to fill airtime 24/7, some radio stations subscribe to network and syndicated programs. Network programs are usually all inclusive, meaning that music, news, commercials, on-air talent, and other program elements are supplied to a station. The station merely puts on the air 24 hours of packaged programming.

Syndicated programs, on the other hand, are individual shows that a station airs in between its own programs. For instance, a local station may broadcast its own music until noon, air a syndicated talk show until 3 p.m., and then return to its own programming for the rest of the day.

ZOOM IN 4.7

Learn about some different radio networks at these sites:

- Premiere Radio Networks: www.premrad.com
- Cumulus Media: www.cumulus.com/
- Westwood One: www.westwoodone.com

Network and syndicated programs are commonly delivered via satellite from radio networks and syndicators. Stations either broadcast the live satellite feed or tape it for later airing. Programs hosted by Delilah, Sean Hannity, and Steve Harvey are among the most popular programs syndicated by Premiere Radio Networks, which provides about 90 programs to 5,500 radio affiliates.

Stations pay cash and/or negotiate for commercial time in exchange for network and syndicated programming. Some stations might cut a deal in which they get programming free of charge, but the network or syndicator reaps revenue by selling commercial time within a show to national advertisers.

HOW RADIO PROGRAMS ARE SCHEDULED

Airtime is a valuable commodity, and so every second needs to be scheduled. Station management sets up program clocks (also known as **hot clocks** and **program**

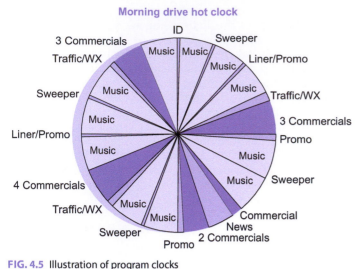

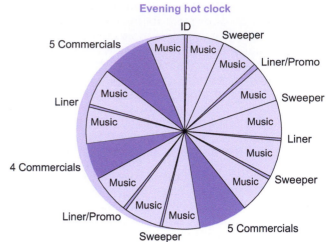

FIG. 4.5 Illustration of program clocks
Source: Eastman and Ferguson, 2002

wheels), which are basically minute-by-minute schedules of music, news, weather, commercials, and other on-air offerings for each hour of the day. For example, the first 5 minutes of the noon-to-1:00 p.m. hour may be programmed with news, followed by 2 minutes of commercials, followed by a 15-second station promotion, followed by a 10-minute music sweep (music that is uninterrupted by commercials). Each hour may vary, depending on audience needs and time of day. For instance, morning drive-time music sweeps may be cut by 30 seconds for traffic updates, and an extra 5 minutes may be set aside for news during the 6:00 p.m. hour.

WHO SCHEDULES RADIO PROGRAMS

In the early days of music formatting, songs that made it to the airwaves and how often they were played was largely left to the DJ's discretion. After the **payola scandal** in the late 1950s, when DJs were caught taking money from record companies in exchange for playing and promoting new releases, stations shifted the programming function from the DJs to a **program director**.

The program director is responsible for everything that goes out over the air. The director comes up with a playlist of music, which is basically a roster of songs and artists featured on the station. The playlist reflects the station's format and is usually determined by audience demand and album sales.

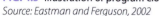

ZOOM IN 4.8

Try to find a radio station's playlist or rotation from a station or its Web site.

The program director also sets the **rotation**, which is the frequency and times of day the songs will be played. Favorite songs and current hits are usually rotated

frequently, and older and less popular numbers may have a less frequent rotation. Most stations try to avoid playing different songs by the same artist too closely together or playing too many slow-tempo or similar-sounding selections in a row.

Program directors also make a point of varying nonmusic programs. For example, a station may air two family- and relationship-oriented/advice call-in programs consecutively in the early afternoon to keep the 25- to 50-year-old females listening but then switch to a political talk program during drive time to attract commuters and males.

SEE IT THEN, SEE IT NOW: TELEVISION PROGRAMS

TYPES OF TELEVISION PROGRAMS

In the early days of television, executives were faced with the challenge of what to put on the air. After all, people were not going to buy TV sets if there was nothing to watch. Program executives smartly adapted existing radio programs for television. Instead of just having sound stages, where radio actors spoke into microphones, sets were built and scenes were acted out in front of a camera. Many radio programs soon had televised counterparts, and in the process loyal fans followed the programs from one medium to the other.

There are many **genres** (or types) of television programs, each of which is distinguished by its structure and content. Television programs can be categorized as either narrative or non-narrative. **Narrative programs** weave a story around the lives of fictional characters played by actors. Dramatic programs and situation comedies, such as *The Big Bang Theory*, *Secrets and Lies*, *Nashville*, and *The Mentalist*, present fictional stories that are scripted and acted out.

Even though it can be argued that every television program tells a story, **non-narrative programs** tell stories that are real and do not come from fictional scripts. Game shows, talk shows, reality shows, sports, and news are all types of non-narrative programs. Non-narrative programs present real situations and real people (e.g., game show contestants, news anchors, sports stars), not actors.

NARRATIVE PROGRAMS

Anthologies

In the early days of television, New York City was the central location for network headquarters and television studios, and thus programs were influenced by Broadway plays. Live productions of serious Broadway dramas became common fare on television in the late 1940s and early 1950s. These anthologies were hard-hitting plays and other works of literature that were adapted for television. They were often cast with young talent from radio and local theater, some of whom went on to become major theatrical and television stars. Robert Redford, Joanne Woodward, Angela Lansbury, Chuck Connors, Paul Newman, and Vincent Price all got their starts acting in televised anthologies. In its 11-year run on television, *Kraft Television Theater* produced 650 plays, featuring almost 4,000 actors and actresses.

> **ZOOM IN 4.9**
>
> Read more about anthologies on the Museum of Broadcast Communications Web site: www.museum.tv/eotv/anthologydra.htm

Anthologies helped boost the fledgling television industry by attracting deep-pocket sponsors and well-educated and affluent viewers who were most likely to buy expensive television sets. Anthologies were enormously popular into the late 1950s. They lost their appeal toward the end of the decade when expanded production capabilities led to on-location, action-oriented shows that drew viewers away from anthologies, and where the viewers went, so did the sponsors.

Dramas

A dramatic series presents viewers with a narrative that is usually resolved at the end of each episode; in other words, the story does not continue from one episode to the next. A drama typically features a recurring cast of primary characters that find themselves involved in some sort of situation, often facing a dilemma that gets worked out as the action peaks and the episode comes to a climax and resolution.

Dramas are often subcategorized by subject matter. For example, police and courtroom dramas are popular today, as demonstrated by the ratings of such programs as *NCIS*, *CSI*, and *Law & Order SVU*. These shows give a look into the lives of cops on the streets and lawyers in the courtroom. Medical dramas fade in and out of popularity. Nonetheless, ABC's *Grey's Anatomy*, which premiered in 2005, is still going strong after more than a decade,

FIG. 4.6 Ryan O'Neal, Mia Farrow, and Barbara Parkins in *Peyton Place*
Photo courtesy of ABC/Photofest. © ABC

and *ER*, which catapulted George Clooney, John Stamos, and Julianna Margulies to stardom, had a very successful 15-year run (1994–2009).

Serials

More commonly known as **soap operas**, these programs have an ongoing narrative from one episode to the next. Serials are different from other types of dramas in several ways. There is little physical action; instead, the action takes place within the dialogue. Also, there are many primary characters. Serials typically have many storylines going on at the same time, such that characters are involved in several plots simultaneously that may not be resolved for years, if at all. When it finally seems like a resolution is at hand (say, a marriage and a happy life), a twist in the story leads to more uncertainty (did she unknowingly marry her long-lost brother?) and to a new, continuing storyline. *General Hospital* (1963–present) and *The Young and the Restless* (1973–present) are popular soap operas of today. *Guiding Light*, however, was the king of soap operas and holds the distinction of being the longest-running scripted (nonanimated) program in radio and television history. The show made its radio debut on January 25, 1937, and began airing on television in 1952. It was still broadcast on radio until 1956, when it moved exclusively to television. Fans were heartbroken when the 72-year-old soap opera ended on September 18, 2009. Long-time favorites *One Life to Live* and

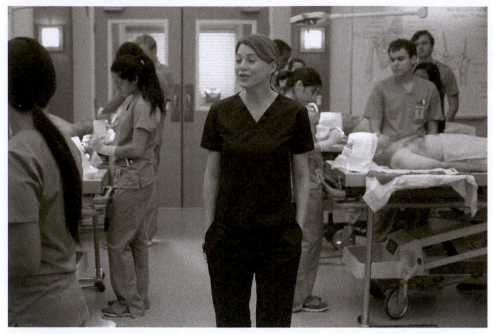

FIG. 4.7 *Grey's Anatomy*—"Walking Tall"
Photo courtesy of Richard Cartwright/ABC via Getty Images

All My Children were cancelled in 2011, but both were revived on the Internet. Picked up by the production company Prospect Park, both shows moved to online-only delivery. Although the shows each amassed a loyal following, breach of contract and licensing lawsuits between Prospect Park and ABC eventually led to the end of both shows.

Although most soap operas air during the daytime and are targeted primarily to women, the prime-time hours (8 p.m.–11 p.m. EST and PST; 7 p.m.–10 p.m. CST and MST) have also seen their share of soaps (however, to attract male viewers, the networks are careful not to call them 'soaps'). These programs were particularly popular on prime time in the 1980s, when viewers were treated to a peek into the fictional lives of the rich on such programs as *Dallas* (1978–1991), *Dynasty* (1981–1989), and *Falcon Crest* (1981–1990), and they seem to have resurged with *Desperate Housewives* (2004–2012), *Mad Men* (2007–2015), and *Nashville* (2012–present).

Telenovela ('tele' meaning television and 'novella' meaning a literary work, or in some languages meaning 'romance') is a type of short-run serial that originated in Latin American countries. Hugely popular telenovelas captivate viewers in Mexico, Central and South America, Spain, and even Russia and China. In the United States, the emotionally charged, convoluted, passionate telenovelas are typically aired on Spanish channels, such as Univision and Telemundo. A telenovela is usually shown 4 or 5 days per week for about 150 episodes, though they can vary in length. From the opening kiss through romantic rivals, break-ups, and tragedies, the telenovela captures viewers through its emotional intensity.

Through its MyNetworkTV, Fox Television brought telenovelas to the United States in 2006, but with little success. Ultimately, the channel aired six telenovelas (*Desire,*

Fashion House, Wicked Wicked Games, Watch Over Me, American Heiress, and *Saints & Sinners*), each showing 5 nights per week for 13 weeks. Despite the genre's popularity in other countries, MyNetworkTV telenovelas did not draw an enthusiastic audience in the U.S. and was dropped after only 1 year.

Situation Comedies

Situation comedies (sitcoms) are usually half-hour programs that present a humorous narrative that is resolved at the end of each episode. Sitcoms feature a cast of recurring characters who find themselves caught up in some situation. Situation comedies are perfect for television, because they can be shot in-studio on a typical three-sided stage that is decorated to look like a simple apartment or home. Most early sitcoms were family oriented, and the comic aspect was noted in the dialogue between the characters rather than in visual gags. However, Lucille Ball excelled at physical comedy and changed the face of television by insisting on using three cameras to film *I Love Lucy* (1951–1957). If a physical gag or antic failed, the scene could be reshot or edited for maximum effect. Multiple-camera filming paved the way for other comedies that featured more physical comedy than comedic dialogue.

ZOOM IN 4.10

- To learn more about *I Love Lucy*, go to www.museum.tv/encyclopedia.htm
- Listen to the *I Love Lucy* introduction music at: www.youtube.com/watch?v=-t4ql-r406Q

FIG. 4.8 Jackie Gleason, Art Carney, Audrey Meadows, and Joyce Randolph on the set of *The Honeymooners* (from left to right)
Photo courtesy of CBS/Photofest. © CBS

FIG. 4.9 The cast of *I Love Lucy* (from left to right): Lucille Ball, Vivian Vance, Desi Arnaz, and William Frawley
Photo courtesy of CBS/Photofest. © CBS

Even though they are dubbed 'situation comedies,' sometimes the situation itself is not funny, but it is handled in a humorous fashion. For example, on *Friends*, it was not very funny when Ross told girlfriend Mona that Rachel, mother of his baby, is moving in with him, but the plot took a humorous twist when Mona unexpectedly showed up on his doorstep. At other times, the situation itself creates the humor, like on *Frasier*, when class-conscious Roz found the perfect boyfriend but then discovered he was a garbage man. The television sitcom has been a prime-time staple that has kept viewers laughing for decades.

Situation comedies are often criticized for stretching the limits of what is considered funny. The airwaves of the 1970s and 1980s were filled with social-consciousness sitcoms. *All in the Family* (1971–1979) and *Maude* (1972–1978) lightheartedly poked fun at serious issues like the Vietnam War, abortion, feminism, and racism. These shows boldly took on controversial subjects. They artfully and humorously addressed sensitive topics, and they elevated the collective consciousness to greater acceptance and understanding of people of all types.

Missing from the networks' line-up since the cancellation of *Beulah* in 1953 were sitcoms featuring African Americans. When *Julia* premiered in 1968, it was hailed as the first starring television role that depicted an African American woman as an intelligent, educated, sensible person. The character Julia, played by Diahann Carroll, was a widowed single mother who worked as a nurse. The groundbreaking show's 3-year run, however, was fraught with controversy, as critics claimed that Julia's middle-class life as a single mother was unrealistic.

Julia led the way to other African American–centered sitcoms, such as *Sanford and Son* (1972–1977), *The Jeffersons* (1975–1985), *What's Happening* (1976–1979), *Diff'rent Strokes* (1978–1986), and *Benson* (1979–1986). The 1980s brought *Webster* (1983–1989) and *The Cosby Show* (1984–1992). Over the next three decades, about 60 African American sitcoms premiered, including the popular *Family Matters* (1989–1998), *Fresh Prince of Bel-Air* (1990–1996), *Moesha* (1996–2001), *The Bernie Mac Show* (2001–2006), *That's So Raven* (2003–2007), and *Everybody Hates Chris* (2005–2009). The number of such programs on the air at any given time fluctuates, but the concern that such shows are overly stereotypical continues, as does discomfort that some Black characters, such as on *The Cleveland Show* (2009–2013), were voiced by White actors.

Facing formidable competition from comedy cable channels and humorous online videos, contemporary network sitcoms, such as *Modern Family* (2009–present) and *The Big Bang Theory* (2007–present) continue toppling long-standing television taboos in the guise of humor. Modern sitcoms jab at drug addiction, drunkenness, casual sex, venereal disease, vomit, and teenage pregnancy, among other gross and sensitive topics. The Fox network refused to air a 2009 *Family Guy* episode that centered on abortion. Those who condemn television shows for tackling sensitive issues scream the loudest; those who are nonplussed just sit back and enjoy the show. Humor often makes it easier for viewers to confront what they dislike most, but laughing at ourselves and at our foibles is a healthy way to release tension.

> ### *FYI: 2015 67th Annual Emmy Award Winners*
>
> Outstanding Comedy Series: *VEEP*
>
> Outstanding Drama Series: *Game of Thrones*
>
> Outstanding Lead Actor in a Comedy Series: Jeffrey Tambor as Maura Pfefferman *(Transparent)*
>
> Outstanding Lead Actress in a Comedy Series: Julia Louis-Dreyfus as Vice President Selina Meyer *(VEEP)*
>
> Outstanding Lead Actor in a Drama Series: Jon Hamm as Don Draper *(Mad Men)*
>
> Outstanding Lead Actress in a Drama Series: Viola Davis as Annalise Keating *(How to Get Away with Murder)*
>
> ---
> *Source: Academy of Television Arts & Sciences*

FIG. 4.10: *Modern Family*
Photo by Eric McCandless/ABC via Getty Images

FIG. 4.11 The characters in *The Simpsons* (top left to right): Bart, Homer, and Marge; (bottom left to right) Lisa and Maggie
Photo courtesy of Fox/Photofest. © Fox Television

Movies and Miniseries

Theatrical movies, albeit mostly low-budget ones such as westerns, made their way onto television screens in the mid-1950s, but only after American movie studios took a cue from their British counterparts that profits could be gleaned by renting films to television stations. For example, eccentric millionaire and RKO studio owner Howard Hughes sold older films to General Tire & Rubber Company to air on its New York television station. NBC's *Saturday Night at the Movies* debuted in 1961 and was quickly followed by ABC's *Sunday Night at the Movies*. By the mid-1960s, television schedules across the country were filled with theatrical films, which drew very large audiences. When Alfred Hitchcock's *The Birds* hit the television airwaves in 1968, 5 years after its theatrical release, it reaped 40% of the viewing audience. Another movie classic, *Gone With the Wind*, captured half of all viewers for two nights in 1976.

In the 1970s, the overabundance of theatrical movies on television led to a new program format, **made-for-television movies**. Different from theatrical releases, which are produced for showing in movie theaters, made-for-television movies are written, produced, and edited to accommodate commercial breaks. Although many made-for-television movies are low-budget productions, some—such as *Brian's Song* (1971), *Women in Chains* (1972), *Burning Bed* (1984), and *The Waltons' Thanksgiving Story* (1973)—drew larger audiences than some hit theatrical films.

The made-for-television movie spawned the television **miniseries**, which is a multipart, made-for-television movie that airs as several episodes rather than in one installment. Miniseries often tackle controversial issues. Successful miniseries from the 1970s and 1980s included *Lonesome Dove, Holocaust, Shogun, The Winds of War*, and, most notably, the 12-hour *Roots* (1977), which won nine Emmys and one Golden Globe award. Just over half of U.S. viewers tuned in for the final episode of *Roots*, and 85% watched at least some part of the program.

Contemporary mini-series are usually scheduled to air on consecutive nights or over successive weeks. As such, they are known as short-form and long-form productions, respectively. After the riveting 30-hour *War and Remembrance* (1988–1989), which cost more than $100 million to produce, ratings for such long-form miniseries plummeted. The networks turned instead to shorter 4- to 6-hour miniseries, such as *Dune* (2000) and *Generation Kill* (2008).

These days, made-for-television movies and theatrical films abound on television. Films are so much in demand that the movie industry adapted its release schedule to accommodate later television viewing. A theatrical film is first released to the movie theaters and then pulled when its box office receipts drop. Sometimes a film will be held for a second release, but usually within about 6 months of the final theatrical run, the movie is distributed for DVD rental and then later made available for sale. Sometimes a film will be shown on pay-per-view and later licensed to pay-cable networks, such as HBO and The Movie Channel. After a film has made the rounds on the cable channels, it may be shown by a network-affiliated or independent station.

NON-NARRATIVE PROGRAMS

Variety Shows

Radio variety shows were largely limited to audio acts such as singing and comedy and thus were one of the first types of programs to transition to television. Visual acts such as juggling, dancing, and magic tricks are much better suited for television than for radio. The audience loved these programs that often showcased new talent. Programs such as *The Ed Sullivan Show* (1948–1971), *The Red Skelton Show* (1951–1971), *The Jackie Gleason Show* (1952–1971), and *The Carol Burnett Show* (1967–1978) showcased singing and dancing, stand-up comedy, and comedic skits, along with other lighthearted entertainment.

Variety shows often provided viewers the first glimpse at new talent such as Humphrey Bogart, Bob Hope, and Lena Horne. Elvis Presley's 1956 appearance on *The Ed Sullivan Show* made headlines when producers refused to show his gyrating hips and only aired close-ups from the waist up. Nevertheless, Elvis's appearance attracted 83% of the viewing audience, the largest share in television history. On February 9, 1964, *The Ed Sullivan Show* hosted the Beatles' American television debut in front of a screaming studio audience of mostly teenage girls.

Although many variety shows were quite successful and enjoyed long runs, others could not produce the big ratings to justify their production expenses. During its 23 years

FIG. 4.12 The Beatles made their U.S. debut on *The Ed Sullivan Show* in New York on February 9, 1964.
Photo courtesy of CBS/Photofest. © CBS

on air, *The Ed Sullivan Show* (1948–1971) attracted a large but mostly older audience. CBS finally pulled the show to make way for more modern, hip programming. *The Smothers Brothers Comedy Hour* (1967–1969) was probably the most controversial variety show of all time. It was wildly popular among young adults, who loved how it mocked politics, government, church, family, and almost everything and everyone. Network censors often butted heads with the show's stars over the irreverent nature of the material, and many viewers were offended by the content. The show was cancelled by CBS but later taken up by ABC and still later by NBC. The antiwar, left-wing gags ultimately ran their course, and the program just faded away after a brief reprise in 1975. The *Carol Burnett Show* (1967–1978) premiered the same year as the Smothers Brothers but had a longer and less controversial run. A parody scene of *Gone With the Wind*, in which Burnett as Scarlett O'Hara majestically walks down a staircase wearing a dress made from a window drape that includes the curtain rod, is known as one of the funniest moments in television.

Variety shows have generally fallen off the programming schedules. *All That* (1994–2005) and *The Amanda Show* (1999–2002) are examples of modern variety/comedy sketch programs that had somewhat successful runs. *The Maya Rudolph Show* aired in May of 2014 as a one-time special with the hopes of becoming a regularly scheduled program. The show's concept is a throwback to older skit comedy/song-and-dance shows. Although the show was well received, it has yet to be put on as a recurring show.

Saturday Night Live is the most well-known, longest-running, non–prime-time variety/comedy/skit program. On the air since 1975, it has become a cultural icon. Grizzled *SNL* viewers contend that the show reached its peak in the early years, while Millennials think it is more hilarious now than ever before.

Game and Quiz Shows

Studio-based game and quiz shows have always been very popular with viewers. Old shows like *What's My Line* (1950–1967), *I've Got a Secret* (1952–1967), and *To Tell the Truth* (1956–1968) were all variations on simple ideas that invited the audience to get involved and play along. Shows like *Beat the Clock* (1950–1961) and *Truth or Consequences* (1950–1988) centered on contestants performing outrageous stunts for money and prizes.

Television viewers love the competitive nature of game shows. It is fun to watch contestants compete for prizes and money. Many game shows are formatted so the television audience can play along. Most viewers have 'bought a vowel' while watching *Wheel of Fortune* or guessed the cost of a refrigerator on *The Price Is Right* and wondered how far they could go without using a lifeline on *Who Wants to Be a Millionaire?* Viewers root for their favorite contestants and develop a feeling of kinship with them, especially when they keep winning, like 2004 *Jeopardy!* champ Ken Jennings, who holds the record for the longest winning streak in television game show history. Viewers sat on the edge of their chairs as they rooted Jennings on through 74 consecutive wins that amassed $2.5 million in prize money.

At first, quiz show contestants competed for small prizes. Then big-money quiz shows, such as *The $64,000 Question*, became enormously popular after their introduction to prime-time television in the 1950s. Charles Van Doren, a young, handsome, witty faculty member at Columbia University, was selected as an ideal contestant for NBC's *Twenty-One*. Van Doren's winning streak lasted for 15 weeks. In early 1957, he became a media celebrity, was featured on the cover of *Time* magazine, and was offered a job on the *Today* show. All was well in Van Doren's life until defeated contestant Herb Stempel came forward, claiming that the show was rigged and that he had been forced to deliberately lose to Van Doren. Stempel divulged that the program's producers gave the answers to the favored contestants and coached them on how to create suspense by acting nervous and uncertain about their responses.

As it turned out, Stempel's allegations that the producers of *Twenty-One*, under pressure from program sponsors to get high ratings, had fixed the outcomes were true. The investigation kicked off television's first major scandal by revealing that cheating was prevalent on many of the quiz shows. After rounds of televised congressional hearings, Van Doren and many other contestants, producers, writers, and others who worked behind the scenes were indicted by a federal grand jury for complicity in the deception. Most defendants received suspended sentences, but Van Doren, who was probably the most beloved and well-known contestant, suffered greatly. He lost his teaching position at Columbia, as well as his job on the *Today* show, and he still lives in relative obscurity.

Although it took years for the public to regain its trust in game shows, the power of a good contest eventually overcame the skepticism. Quiz and game shows, such as *Joker's Wild* (1969–1974), *Concentration* (1958–1978, 1987–1991), and *$10,000 Pyramid*, which later changed its title to *$25,000 Pyramid*, continuing up to *$100,000 Pyramid* as the top prize increased, have been long-time mainstays of daytime television. Today, the most popular game shows, *Jeopardy!* (1964–present) and *Wheel of Fortune* (1975–present), air not during daytime hours but during the prime-time access hour. After the quiz show scandals, it took until the 1999 debut of *Who Wants to Be a Millionaire?* before viewers could again enjoy a quiz show during the prime-time hours.

FIG. 4.13 *Twenty-One* contestant Charles Van Doren answering questions from inside of the isolation booth.
Photo by Hulton Archive/Getty Images

FIG. 4.14 The set of *Jeopardy!*, which is hosted by Alex Trebeck
Photo by Ben Hider/Getty Images

ZOOM IN 4.12

- Listen to former game show host Sonny Fox talk about the quiz show scandal at www.pbs.org/wgbh/amex/quizshow/sfeature/interview.html
- Read what the Museum of Broadcast Communication has to say about the quiz show scandal at www.museum.tv/eotv/quizshowsca.htm

Reality Shows

Reality Situation

Candid Camera (1948–1979, 1991–1992, 1996–2004, 2014), which was the first reality-type program, premiered in 1948 under its original radio title, *Candid Microphone* (1947–1948, 1950). The show featured hidden-camera footage of unsuspecting people who were unwittingly involved in some sort of hoax or funny situation that was contrived as part of the program. After catching people's reactions to outlandish and bizarre situations, host Allen Funt would jump out and yell, "Smile, you're on *Candid Camera*!"

America's Funniest Home Videos (1989–present) breaks the typical reality format. Instead of professional camera operators taping antics, ordinary people tape each other doing silly things and send in their recordings with the hopes of having them shown on television and winning a cash prize.

ZOOM IN 4.11

- Listen to an NPR report about the history of quiz shows and hear short audio clips from quiz programs at www.npr.org/player/v2/mediaPlayer.html?action=1&t=1&islist=false&id=93749534&m=93749516
- See how well you would have done on the quiz show *Twenty-One*. To answer some of the original questions asked on the air, go to www.pbs.org/wgbh/amex/quizshow/sfeature/quiz.html

FYI: Surviving an Avalanche—How America's Funniest Home Videos Are Selected

"When *America's Funniest Home Videos* premiered in 1990, it hit with a bang. Within two weeks we were receiving about 5,000 tapes a week so we quickly had to develop a system to handle that avalanche. After each VHS cassette was assigned a unique number and logged into a database, it then went to the screeners. To find the 60 or more clips needed for each half-hour episode the 10 screeners had to sift through at least 4,000 per week. Their selections were then scored on a scale of 1 to 10 with all 5 and

Reality Family

On reality family shows, the camera literally follows family members through their daily lives, and most dialogue and action is unscripted. The first such program was *An American Family*, which aired on PBS in 1973. It followed a year in the real life of the Loud family from Santa Barbara, California. The program documented the parents' marital discord (which later led to divorce) and the lives of their five teenagers. The audience watched as son Lance struggled with his sexual identity. He became the first openly gay person to appear on television. *An American Family* reflected the changing lifestyles of the times and stood in stark contrast to unrealistic, idealized, fictional family programs, such as *The Brady Bunch* (1969–1974). In 2011 HBO aired its *Cinema Verite*, a drama about the making of *An American Family*.

A contemporary version of *An American Family* was MTV's *The Osbournes* (2002–2005), which portrayed the everyday family life (sans oldest daughter Aimee) of Ozzy Osbourne, former member of heavy-metal group Black

FIG. 4.15 The Osbournes
Photo courtesy of MTV/Photofest. © MTV

Sabbath. The cameras were even there when Ozzy's wife, Sharon, underwent treatment for colon cancer. Attracting an average of 6 million viewers per week, *The Osbournes* was one of the most-watched shows ever on MTV. Whether they were throwing a baked ham at the noisy neighbors, potty training their dogs, or turning on yard sprinklers to repel nosy fans, the Osbournes created a whole new program genre: the reality sitcom.

Reality Crime and Medicine

COPS (1989–present) is one of the longest-running reality programs. The basic premise is simple: A camera crew follows police on their beats and tapes their confrontations with suspects. Other reality programs, such as *America's Most Wanted* (1988–2012), depict real-life crime situations but were scripted and played by actors reenacting the real-life situation. In *Cold Justice* (2013–present), a former prosecutor and a former crime scene investigator travel around the country helping local law enforcement solve cold cases. The show has led to many arrests and convictions. Emergency medical shows like *Rescue 911* (1989–1996) and *Trauma: Life in the ER* (1997–2002), kept viewers on their seats wondering if a life would be saved or if a victim would succumb to his or her injuries. Each episode of *Dr. G. Medical Examiner* (2004–2012) showed real-life medical examiner, Dr. Jan Garavaglia, conducting an autopsy to piece together the chain of events that led to a person's death.

Reality Dating

The forerunner to reality dating programs was *The Dating Game*, which first appeared in 1965 and aired on and off for 35 years. A bachelor or bachelorette questioned three hopefuls who were hidden from view and then selected one for a date. Although most contestants were ordinary people, occasionally stardom's knowns and still unknowns, such as Farrah Fawcett, Steve Martin, Burt Reynolds, Arnold Schwarzenegger, Tom Selleck, Ron Howard, Sally Field, and Michael Jackson, would appear on the show hoping to win a date.

Other dating shows, such as *Elimidate* (2001–2006) and *The Fifth Wheel* (2001–2004) followed couples as they went on dates and got to know each other. Afterward, the couples talked about their dates, often trashing each other and calling each other derogatory names if the date did not go as well as hoped. Other dating programs have a different spin on love and relationships. On *Parental Control* (2005–2010), parents interviewed potential dates and chose the most likely person to steal their son or daughter away from their current girlfriend or boyfriend. *Dating in the Dark* (2009–2010) literally took the idea of blind dating and made it into a show. Three single men and three single women would live together, but they were not allowed to see each other. They could only talk and get to know each other in the dark. The point was to take personal appearance out of the dating equation. On *The Bachelor/Bachelorette*, the bachelor or bachelorette chooses from among 25 suitors for a spouse. Each episode ends with potential mates being eliminated. In the final episode of the season, two potential spouses remain—one is 'broken up with' and the other gets a red rose and

a marriage proposal. But perhaps the most risqué dating show is *Dating Naked* (2014–present). Billed as a social experiment, the show focuses on two daters who each go on three naked dates, including with each other. The contestants frolic around a tropical island, enjoying beach and poolside dates in the nude. However, when shown on television, the daters' genitals are blurred, as are the women's breasts—only the buttocks of both genders are shown.

Reality Game

The newest reality rage is the so-called reality game show, such as *Survivor* (2000–present), *The Amazing Race* (2001–present), *American Idol* (2002–2016), and *America's Next Top Model* (2003–2015). In these shows, contestants do seemingly impossible tasks to be declared the winner. On *Survivor*, a group is sent to a remote location and left to deal with nature, all while each contestant tries to position himself/herself as the most valuable person. Each week the castaways vote one person off the island until only the winner remains. *American Idol* (2002–2016) is a kind of an updated *Gong Show* (1976–1989) but with an interactive twist, in that viewers vote for the winners either by phone, text, app, Facebook, or Google. *Idol* centers on a panel of judges who verbally jab at the contestants after they have sung their hearts out and tried their best to win. In the program's first season, about 110 million votes were cast, and by the tenth season votes were coming in to the count of 750 million.

Reality Home Improvement

Another new reality format revolves around home and gardening. Such programs, which are mostly on cable networks, include *Trading Spaces* (2000–2008), in which neighbors redecorated each other's homes; *Design on a Dime* (2003–present), redecorating for under $1,000; *House Hunters* (1999–present), in which a realtor shows buyers who are relocating nationally or internationally a selection of homes that are for sale; and *Yard Crashers* (2008–present), in which a team of landscapers transforms homeowners' yards into amazing outdoor spaces.

Big ratings, audience demand, and low production costs make reality shows a network executive's dream. Such unscripted reality shows account for about 25% of all prime-time broadcast programming. Reality shows generally follow the same format week after week. Will they make the deadline? Which house will they buy? Will they love the remodeling job? Will they get married? How much weight will they lose? Will she be cured? Even though the answers are usually apparent and predictable, reality shows pull in the audience.

But all that glitters is not gold—at least not for the contestants. Reality show participants do not receive union protection, as do most television actors. Contestants complain of physical isolation, grueling work, sleep deprivation, and bad food. They often become ill tempered and act irrationally, fueled by an endless supply of alcohol. Additionally, participants may not be thoroughly vetted.

On VH1's *Megan Wants a Millionaire* (2009), one of the contestants went on the lam after being suspected of murdering his ex-wife. After the man was found dead of an apparent suicide, negative publicity forced the network to cancel the show even though it had only been on the air for 3 weeks.

Sports

Televising live sporting events in the early days presented several technological challenges. Directors had to figure out how to get heavy, bulky cameras and other equipment to the site of an event; then, once there, they had to strategize how to position the cameras to capture all the action. Boxing, wrestling, bowling, and roller-derby matches were easy to cover because they were played in relatively small arenas and required little camera movement. The first televised baseball game was between collegiate rivals Columbia and Princeton on May 17, 1939. The action was covered by only one camera that was positioned on the third baseline.

As production capabilities increased and cameras became more portable, television moved to covering more action-packed sports. Multiple cameras could be set up to follow basketball and football games and even to track golf balls as they flew long distances across fairways.

FIG. 4.16 The first-ever televised baseball game was played on May 17, 1939, between Princeton and Columbia.
Photo courtesy of Getty Images Sport Collection

The 1960 Winter Olympics in Squaw Valley, California, was the first Olympics televised in the United States (1936 Summer Games in Berlin was televised in Germany). Hosted by Walter Cronkite, viewers for the first time witnessed the outdoor excitement and challenges of the games. *Wide World of Sports* (1961–1998) was a groundbreaking Saturday-afternoon program that spanned "the globe to bring you the constant variety of sport, the thrill of victory, and the agony of defeat,

the human drama of athletic competition." Gymnastics, track and field, skating, rodeo, skiing, surfing, badminton, and demolition derbies were just a few of the sports featured on the show.

Live sports and sports shows are now found on many cable and broadcast networks 24/7. Sports fans rabidly devour shows like *College Gameday*, *Baseball Tonight*, *Sports Center*, *NFL Live*, *Sports Nation*, and *NBA Fastbreak*.

Career Tracks: Dan Hellie, NFL Network—Host Total Access (M-F football show) also host NFL Game Day Live (Sunday's during football season). Host live coverage from Super Bowl, NFL Combine & NFL Annual meetings

FIG. 4.17 Dan Hellie

What was your first job in electronic media?
My first paying job was as a news photographer in Knoxville, Tennessee. An internship at the local ABC affiliate let me take the position. As I knew that I would eventually have to shoot my own video in small markets when I got on the air, this was a perfect scenario for me. It was an opportunity to learn how to shoot and edit while getting paid. The job benefited me tremendously in my next two jobs in Alexandria, Minnesota, and Florence, South Carolina, where I was often 'one-man banding' my own stories. I truly believe this should be mandatory for every on-air person, because it makes you realize that shooting is really an art and forces you to realize the duties involved with shooting and editing a story. It really helped me become a better storyteller and helped me learn that it's the video telling the story.

What led you to your present job?
Other jobs! It's like minor-league baseball. I spent my rookie season as a cub reporter doing news/sports in a four-person newsroom in Alexandria, Minnesota, where I shot, wrote, produced, reported,

and anchored. After 9 months in Alexandria, I headed back down south to Florence, South Carolina, to an actual newsroom. In Florence, I was promoted from weekend sports anchor to main sports anchor. After roughly 2 years in South Carolina, I headed to West Palm Beach, Florida, which was Market 38. It was my first opportunity to cover professional sports and interact with pro athletes on a regular basis. I always thought of West Palm as AA or AAA stop in baseball terms. After progressing to the main sports anchor position in West Palm, I moved to the ABC affiliate in Orlando as the main sports anchor. Orlando is a top 20 market, #19 to be exact, and had its own professional franchise (the NBA's Orlando Magic). It was the first time that I had a producer and didn't have to worry about editing my own sportscast.

After 3 years in Orlando, I moved north again, this time to Washington, DC, and the NBC–owned station WRC, where I anchored the weekday 5:00, 6:00, and 11:00 p.m. sportscasts. I also hosted the Redskins coaches show during football season as well as cohosted a roundtable show called *Redskins Showtime*. When I wasn't anchoring, I was in the field reporting. Reporting may actually be the most rewarding part of my job, because we are actually in the field interacting with the athletes, coaches, and front-office people that we are covering.

Since July 2013 I've been working for NFL network hosting Total Access (M-F football show) and NFL Game Day (Sunday's during football season). I also host live coverage from the Super Bowl and NFL Scouting Combine.

The key for me was networking and really keeping an ear to the ground to hear about when and where jobs were open. An agent helps for on-air people like myself, but it always helps to keep in touch with people you have met along the way. Broadcasting is a very small business, and maintaining contact with former coworkers and managers is crucial.

What advice would you have for students who might want a job like yours?
The television biz has undergone a complete makeover in the last several years. Be open minded and be willing to do everything. Nowadays talented reporters don't have to make as many minor-league stops as I did. Big markets and even networks are willing to hire younger people because of tightening budgets. Get a resume tape together as soon as possible, be willing to move anywhere, and focus on being well rounded. Don't be afraid to take a job as a photographer, Web producer, assignment editor, or writer if it gets you in the door. The more knowledge you have, the more valuable you are.

ZOOM IN 4.13

Learn more about talk shows at the Web site of the Museum of Broadcast Communication: www.museum.tv/eotv/talkshows.htm

Late-Night Talk

Late-night talk shows have long been the rage since 1954, when *The Tonight Show Starring Steve Allen* hit the airwaves. Jack Parr, Johnny Carson, Jay Leno, Conan O'Brien, and Jimmy Fallon have all been at the helm of *The Tonight Show*. Whether *The Tonight Show*, *Jimmy Kimmel Live* (2003–present), *The Late Show With Stephen Colbert* (David Letterman retired in 2015 after hosting the show for 22 years), *Late Night with Seth Myers* (2014–present), or *Last Call with Carson Daly* (2002–present), these programs follow the same basic format. The shows open with a short comic routine, followed by a skit or funny bit about a current issue or event, and lighthearted banter or serious talk with the nightly celebrity guests. The shows often wrap up with a band promoting a new CD. Regardless of the comedic content and ironic nature, late-night talk shows have a strong influence on public opinion.

News and Public Affairs

After the press–radio war (1933), radio news was limited to special report bulletins and information programs, such as President Franklin D. Roosevelt's Depression-era 'fireside chats' that kept Americans abreast of the country's economic situation. But with the entry of the United States into World War II, there was renewed interest in and growth of radio news.

Edward R. Murrow was perhaps the best-known journalist of that era. He broadcast dramatic, vivid, live accounts from London as it was being bombed. After the war, Murrow hosted *Hear It Now* (1950–1951), a weekly news digest program, which led to *See It Now* (1951–1958), the first nationwide televised news show. *See It Now* was also the first investigative journalism program, a style that has since been imitated by contemporary programs such as *60 Minutes* (1968–present) and *Dateline NBC* (1992–present). Murrow focused not only on the major events of the day but also specifically on the everyday people involved in those events. Many Americans hailed Murrow for his heroic role in bringing an end in 1954 to Senator Joseph McCarthy's rampage about imagined communists. At a time when both legislators and the media were too timid to challenge McCarthy, Murrow stood up for the people who were being persecuted by claiming that McCarthy had gone too far with his accusations.

ZOOM IN 4.14

To learn more about Edward R. Murrow and hear audio clips of his news reports and read about this program, *See It Now*, click on these sites:

- statelibrary.dcr.state.nc.us/nc/bio/literary/murrow.htm
- www.otr.com/murrow.html
- www.museum.tv/eotv/murrowedwar.htm

Making its debut in 1947, *Meet the Press* was the first television interview show and is still the longest-running network program. Program guests discuss politics and current events and treat subjects in a serious manner. *Meet the Press* was the precursor to public-affairs programs such as PBS's *Washington Week* (1967–present) and *PBS NewsHour* (1975–present), which as gone through several title changes since its debut.

At a time when there were few opportunities for African Americans to appear on television, *Like It Is* (1968–2011) was a New York–based Sunday-morning public affairs program. The program focused on the Black experience, with the hosts discussing the issues of the day with such celebrities as Sammy Davis Jr., Andrew Young, Lena Horne, and Muhammad Ali. The show filled the void left by mainstream public affairs programs that focused on White, middle-class issues. The program was well regarded and won seven local Emmy awards.

Generally, a television news show is classified as a local newscast, a broadcast network newscast, or a cable network newscast:

- A local newscast is produced by a television station, which is usually a network affiliate. Local news typically

FIG. 4.18 Stephen Colbert

features community news and events. Depending on the size of the market and the community, most local stations have a small staff of reporters (who sometimes double as anchors) who gather the news and write the stories. It is economically unfeasible for each station to send reporters all over the world, so local stations also receive news stories from the wire services, such as the Associated Press (AP) and Reuters, and from their affiliated networks, which transmit written copy and video footage via satellite.

- Broadcast network news specializes in national and international rather than local news. Prime-time broadcast news programs, as known today, were slow to arrive on television. NBC and CBS expanded their 15-minute newscasts to 30 minutes in 1963, and ABC followed in 1967. In 1969, the networks aired newscasts 6 days per week and finally every day of the week by 1970. It was Walter Cronkite who brought respect and credibility to television news. On the air from 1962 to 1981, Cronkite's commanding presence, avuncular nature, and stoic and professional delivery earned him the title of the most trusted news anchor. Although he rarely expressed an opinion, when he did, it had a profound impact. For example, when Cronkite called the Vietnam War a stalemate, President Johnson fretted, "If I've lost Cronkite, I've lost Middle America." Cronkite, who died in 2009, set high standards for those who followed, such as Charles Gibson (ABC), Tom Brokaw (NBC), and Katie Couric (CBS).

Networks also produce morning and prime-time news programs. *Today* (NBC, 1952–present), *Good Morning America* (ABC, 1975–present), and *This Morning* (CBS, 1998–present) wake up viewers with light-hearted information and entertainment fare, sprinkled with a few minutes of hard news. The networks turn to more serious and concentrated reporting of important issues in their prime-time news programs. Shows such as *20/20* (ABC, 1978–present) and *48 Hours* (CBS, 1988–present) bring in-depth coverage of current issues and events.

- Competition for broadcast network news arrived in 1980, when Ted Turner started Cable News Network (CNN). His critics dubbed it the Chicken Noodle Network and mocked, "It'll never work. No one wants 24 hours of news." But Turner proved them all wrong. There was and still is an appetite for 24/7 news. CNN's success as a 24-hour cable news network demonstrates the need for an all-day newscast. Viewers are no longer content to wait for the 6:00 or 11:00 local or network newscast to learn about an event that happened hours earlier. They want to see the action when it happens—"All the news, all the time." Cable networks are perfect for filling that need. Today, CNN continues to be one of the most widely watched and relied-upon news sources around the world. Fox News Channel, MSNBC (a Microsoft and NBC partnership), ESPN (sports news), and CNBC (financial news) all specialize in around-the-clock international and national news, in-depth analysis, and special news programs.

Television news—whether local, broadcast network, or cable network—is certainly not without its critics. Even with 24-hour news programs, there is way too much going on in the world than can be shown on a single television newscast. So it is up to the news producers to be the gatekeepers and decide what news should go on the air and how it should be presented. Viewers complain that too much attention is given to stories that are of little importance but have flashy video, that the news is often sensationalized, that stories are too superficial, that not all stories are believable, that the news is politically biased, and that news organizations will not air reports that show certain politicians or companies in an unfavorable light.

FIG. 4.19 *Today Show*
Photo by Peter Kramer/NBC/NBC NewsWire via Getty Images

Many viewers are concerned about biased and unfair stories, especially when news sources such as Fox News Channel air reports that clearly reflect their political points of view. In such a heavily competitive environment, news networks are always trying to distinguish themselves. Fox's answer is, "Wave the flag, give 'em an attitude, and make it lively" (Johnson, 2002, p. 1A). It has positioned itself as an opinion page, sharing views to the right of center. As competition remains strong, the fear is that other networks may, too, be tempted to "claim political niches the way Fox has" (Johnson, 2002, p. 1A). Just as liberals deem Fox News low in credibility, conservatives judge progressive MSNBC as such. Moreover, many cable news interview segments have become nothing more than shouting matches between the host and the guests in which fingers point and tempers boil over. Feuds even occur across networks—the most notable between Fox's Bill O'Reilly and MSNBC's Keith Olbermann in 2009. The feud garnered so much press that the networks' top management finally stepped in and told both O'Reilly and Olbermann to drop it. Although the battle increased the ratings for both shows, the personal barbs flying back and forth between the two hosts had gone too far even for cable television.

The G. W. Bush administration was heavily criticized for blurring the lines that separate news and public relations. There are many examples of so-called video news reports that were shown to millions of homes under the guise of being 'news' but were actually prepackaged press releases created by the federal government. For example, a 'story' touting the benefits of the Medicare Prescription Drug Improvement and Modernization Act was actually a video press release created by Department of Health and Human Services, Center for Medicare & Medicaid Services. A video 'news' release that championed the Bush Administration's efforts to open markets for American farmers was actually masterminded by the Agriculture Department's Office of Communication.

The viewing public casts a critical eye toward television news. Only about one-half of viewers think MSNBC, Fox News, and CNN are very believable. CNN edges out its competitors, with 58% of viewers claiming that it can be trusted, compared to 50% for MSNBC and 49% for Fox News. Credibility ratings for cable networks are down from 60% in 2010. Local television news shows fare a bit better. About 65% of viewers state that local television news is somewhat or very credible. The lack of trust in cable news is moving many viewers from television to the Internet as their primary source of news.

The expansion of broadcast network news and the emergence of cable news have cut into newspaper readership. In a 2012 survey, about 3 of 10 respondents read "a newspaper yesterday." Newspaper readership has been declining for decades; in 2006 about 40% of people read a paper daily, in the late 1990s about 50% did so, and in 1965 slightly more than 7 of 10 people read a paper daily. The good news for the printed press is that decline in readership is slowing (except among 18- to 24-year-olds), and the online versions are taking up some of the slack.

About 45% of newspaper readers access a newspaper digitally either through the Web or mobile application or in a combination of both digital and print.

Newspapers and broadcast and cable television executives are concerned that they cannot seem to capture young adults. Only about 20% of those between the ages of 18 and 24 read a newspaper daily compared to 55% of those older than 65. Newspapers are not the only news medium shunned by young people. Despite the availability of around-the-clock television news, only about 4 out of 10 Millennials 'get' news 7 days a week. In fact, they know less and care less about news and public affairs than any other generation in the past 50 years. This disinterest among young people brings new challenges to the television news industry to lure this group of viewers to the screen.

CHILDREN'S SHOWS

Most children's shows in the early days were variety-type programs that featured clowns, puppets, and animals. *Kukla, Fran, and Ollie* (1948–1957) focused on real-life Fran and her puppet friends. Like many shows of its time, it aired live, but unlike other shows, the dialogue was unscripted. *The Howdy Doody Show* (1947–1960), hosted by Buffalo Bob Smith, featured Howdy Doody, an all-American boy puppet with 48 freckles (one for every state in the union). Narrative programs such as *Zorro* (1957–1959) and *The Lone Ranger* (1949–1957) also were very popular with children, as were science fiction/adventure shows like *Captain Video and His Video Rangers* (1949–1950).

> **ZOOM IN 4.15**
>
> Watch this 6-minute clip of the last episode of Howdy Doody and find out Clarabelle the Clown's secret: www.youtube.com/watch?v=lJ6ybvlsb4s

Captain Kangaroo (1955–1984), one of the longest-running children's programs of its time, prided itself on being slow paced and calming. Captain Kangaroo and his sidekick, Mr. Green Jeans, taught children about friendship, sharing, getting along with others, and being kind to animals. Segments of the program were interspersed with the cartoon adventures of Tom Terrific and Mighty Manfred the Wonder Dog.

The longest-running and probably the most popular and highly regarded children's show in television history is *Sesame Street* (PBS 1969–2015; HBO/PBS 2015–present; new episodes first air on HBO then are rebroadcast on PBS). The program set the standard for contemporary educational and entertainment children's shows. *Sesame Street* teaches children about words, spelling, math, social skills, logic skills, problem solving, hygiene, healthy eating, pet care, and many other subjects. *Sesame Street* shows that learning is fun by using puppets, like the beloved Muppets,

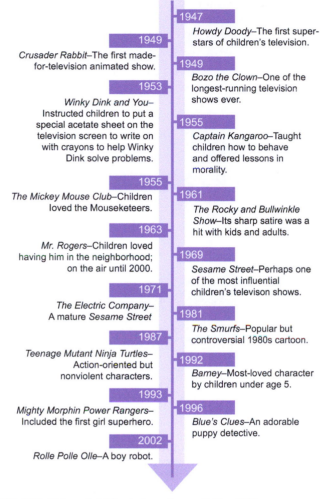

1947
Howdy Doody–The first superstars of children's television.

1949
Crusader Rabbit–The first made-for-television animated show.

1949
Bozo the Clown–One of the longest-running television shows ever.

1953
Winky Dink and You–Instructed children to put a special acetate sheet on the television screen to write on with crayons to help Winky Dink solve problems.

1955
Captain Kangaroo–Taught children how to behave and offered lessons in morality.

1955
The Mickey Mouse Club–Children loved the Mouseketeers.

1961
The Rocky and Bullwinkle Show–Its sharp satire was a hit with kids and adults.

1963
Mr. Rogers–Children loved having him in the neighborhood; on the air until 2000.

1969
Sesame Street–Perhaps one of the most influential children's televison shows.

1971
The Electric Company–A mature *Sesame Street*

1981
The Smurfs–Popular but controversial 1980s cartoon.

1987
Teenage Mutant Ninja Turtles–Action-oriented but nonviolent characters.

1992
Barney–Most-loved character by children under age 5.

1993
Mighty Morphin Power Rangers–Included the first girl superhero.

1996
Blue's Clues–An adorable puppy detective.

2002
Rolle Polle Olle–A boy robot.

FIG. 4.20 History of children's shows *Source: McGinn, 2002*

animation, and games. The show has almost 200 Emmy awards, more than any other television program.

Other PBS children's programs include *Curious George* (2006–present), *Arthur* (1996–present), and *Postcards From Buster* (2004–2008, 2012). Nickelodeon also airs many children's shows, such as *Sponge Bob Square Pants* (1999–present) and *Dora the Explorer* (2000–2015). The Disney Channel's Playhouse Disney features programs like *Mickey Mouse Clubhouse* (2006–present) and *Imagination Movers* (2008–2013).

NONCOMMERCIAL TELEVISION

In addition to establishing National Public Radio, the Corporation for Public Broadcasting also created **Public Broadcasting Service** (PBS) in 1969. PBS is devoted to airing "television's best children's, cultural, educational, history, nature, news, public affairs, science and skills programs" (About PBS, 2010). PBS has about 350 member stations that are watched by almost 100 million viewers every month. PBS is watched about 8 hours per month by about 73.2% of all U.S. television-owning families.

In addition to its children's shows, PBS is known for airing quality educational and entertainment programs. PBS program milestones include *Masterpiece Theater* (1971–present), *NOVA* (1974–present), *Great Performances* (1972–present), *American Playhouse* (1982–1993), and *Civil War* (1990), a Ken Burns epic documentary.

Some contemporary PBS programs include *American Family* (2002), about the everyday life of a Hispanic family living in East Los Angeles; *Antiques Roadshow* (1979–present), a traveling show that offers appraisals of antiques and collectibles; *Frontline* (1983–present), an

FIG. 4.21 Buffalo Bob Smith poses with Howdy Doody and Clarabell the Clown on the set of the *Howdy Doody Show* to celebrate its 10th anniversary, 1957.
Photo courtesy of NBC/Photofest. © NBC

FIG. 4.22 Several of the *Sesame Street* characters
Photo courtesy of PBS/Photofest.
© PBS

investigative documentary program that delves into the issues of today; *American Masters* (1983), biographies of the best U.S. artists, actors, and writers; *Nature* (1982–present), documentaries about various animals and ecosystems; and *Keeping Up Appearances* (1990–1995), one of several British comedies.

ZOOM IN 4.16

- Learn more about the Public Broadcasting System (PBS) at www.pbs.org
- Watch a PBS program and consider how it differs from a broadcast or cable program of the same genre.

FYI: Most-Watched Programs in U.S. Television History

1. Super Bowl XLIX (Feb. 2015) Patriots/ Seahawks	114.4 million
2. Super Bowl XLVIII (Feb. 2014) Seahawks/ Broncos	112.2 million
3. Super Bowl XLVI (Feb. 2012) Giants/Patriots	111.3 million
4. Super Bowl XLV (Feb. 2011) Packers/Steelers	111.0 million
5. Super Bowl XLVII (Feb. 2013) Ravens/49ers	108.7 million
6. Super Bowl XLIV (Feb. 2010) Saints/Colts	106.5 million
7. M*A*S*H finale (Feb. 1983)	106.0 million
8. *Roots* (miniseries) Part IV (Jan. 1977)	100.0 million
9. Super Bowl XLIII (Feb. 2009) Steelers/Cardinals	98.7 million
10. Super Bowl XLII (Feb 2008) Giants/Patriots	97.5 million

Sources: Bibel, 2015; Nielsen, 2014

WHERE TELEVISION PROGRAMS COME FROM

The process of getting a program on the air usually starts with a **treatment**, a description of the program and its characters. Next, the writer or producer pitches the treatment to a network or production company. Even though treatments are accepted from anyone, those submitted by established writers and producers usually get the most serious consideration. The treatment is basically a detailed description of the program and the characters as well as outlines of some episodes. If network executives think the project is worthwhile, they will go forward with a **pilot**, a sample episode that introduces the program and characters to viewers. Pilots are often spectacular productions that draw in viewers. Unfortunately, subsequent episodes are not always nearly as grand as the pilot, and disappointed viewers stop watching the show.

About 2,000 to 4,000 program proposals are submitted each year to networks and production companies. Of those, maybe 100 will be filmed as pilots, between 10 and 20 will actually make it to the air, and about 5 will run for more than a single season.

WHO PRODUCES AND DISTRIBUTES PROGRAMS

Producers

Networks and local stations acquire program broadcast rights from independent producers. Most major production companies—such as Columbia TriStar, Warner Bros., and 20th Century Fox—started as movie production houses and have been around for years. Some of the major production companies own or have financial interest in some television networks and may produce programs for them. Major production houses have brought such programs as *According to Jim* (ABC Studios) and *Cold Case* (CBS Paramount Networks). In contrast to the huge conglomerates with ties to the networks, there are small independent production houses. For example, Bochco Productions produced *NYPD Blue*, *Philly*, and *Hill Street Blues*, and Cherry Productions kick started *Desperate Housewives*.

Independent producers also distribute their programs on cable networks such as HBO and Showtime, and along with pay-per-view and video-on-demand channels, they offer movies, sports, and special events.

Broadcast and Cable Networks

The FCC in 1970 enacted **financial interest and syndication rules (fin/syn)**, which stated that a broadcast network could produce only 3.5 hours' worth of its weekly prime-time programs; the rest had to come from outside production companies. In mandating fin/syn, the FCC hoped that independent producers would bring more diversity to prime-time programming and break what it considered the networks' monopoly over the production and distribution of television programs. At the time, Fox network was airing fewer than 15 hours of prime-time programming, so it was exempt from fin/syn. The exemption allowed Fox to produce more hours of programming, which helped it develop into a full-fledged network.

Many cable networks produce their own programs or receive programs from independent producers. Shows on cable channels such as HGTV, the Golf Channel, and Animal Planet are produced especially for their viewers.

Even though PBS is a network, it is noncommercial and nonprofit. Thus, it is not financially feasible for PBS itself to produce programs. Instead, it obtains programs from independent producers, from PBS member stations that produce their own shows, and from other various sources. PBS acts only as a distributor in that it transmits programs to its member stations via satellite.

As new competition from cable networks and satellite-delivered programs eroded network audiences, the FCC relaxed and then rescinded fin/syn in 1995. As a result, the competition between the networks and independent producers has heated up. Today, about one-half of all network programs are produced by network-owned or -affiliated production companies. Independent producers complain that their programs are less likely to make it to the airwaves than network-produced shows. They also claim that because the networks stand to make quite a bit of money down the road from syndication, they are more likely to keep their own poorly rated programs on the air and instead cancel independently produced programs that have higher ratings.

The networks contend, however, that this ongoing competition has led to higher-quality programs. Moreover, they counter charges of keeping low-rated programs on the air with examples of networks canceling their own programs, such as when ABC canceled *Ellen* (1998). Nevertheless, networks are not as quick to cancel shows as they were in the past several decades. In earlier days of television, programs were given a chance to grab an audience. Hit shows like *M*A*S*H* (1972–1983), *Cheers* (1982–1993), *Seinfeld* (1989–1998), and more recently *The Office* (2005–2013) all struggled in their first season to attract viewers. But savvy network executives gave the shows a second and even a third season to rise to the top. In the mid-2000s, that thinking changed and some programs were cancelled after the third or fourth week. But now with more shows than ever on the schedule, the benchmark of what is considered a successful show is lower than before. Hence, programs like *The Muppets* (2015–present) and *Scream Queens* (2015–present) were renewed for 2016 despite low numbers of viewers. It takes very poor audience numbers to cancel a show. For example, *Wicked City* (2015) was the first cancellation of the 2015 season, and only after it failed to grab even 4% of the audience. But more remarkable, the cancellation was not announced until the 2015 season was 9 weeks in, which marks the longest cancellation-free streak in recent years.

Network executives understand that in today's crowded lineup, it takes the audience time to find a new show or get into the habit of watching a show that is getting the buzz. In the days of only three broadcast networks (ABC, CBS, NBC), viewers chose which of three programs to watch. Television viewers of today comb through hundreds of programs, which appear on a multitude of broadcast and cable networks at any given time, plus even more shows archived by online services.

Syndicators

Television syndicators are companies that distribute television programs to local television stations, cable television networks, and other media outlets. Local television affiliates and independents stations both depend on syndicated material to fill out their broadcast schedules. There are two basic types of syndicated programs: first run and off network.

First-run syndicated programs are produced for television stations but are not intended for prime-time network airing. First-run syndicated shows mostly fill the morning, afternoon, and early-evening time slots. Examples of **first-run syndicated programs** include *Jeopardy!*, *Wheel of Fortune*, and *Judge Judy* (syndicated by CBS Television Distribution), as well as *Maury* (NBC Universal Television Distribution) and *The Ellen Degeneres Show* (Telepictures Productions Inc.).

Off-network syndicated programs are those that once aired or are still shown as regularly scheduled programs on one of the broadcast networks. For example, even though new episodes of *Seinfeld* (1989–1998) are no longer being produced, many local broadcast stations and cable networks buy the rights to old episodes from a syndicator. The stations air the show at whatever time of day it will draw the largest number of viewers in the market.

Prime-time network programs usually air once a week, but once they have been syndicated to a local station, they often run 5 days a week (a practice known as **stripping**). To have enough episodes for syndication, prime-time programs need to stay on the air for at least 65 episodes. (That way, they can be shown Monday through Friday for 13 weeks.) However, most stations will not pick up a syndicated program unless there are at least 100 to 150 episodes, with 130 being the ideal number (26 weeks of Monday-through-Friday airings). Programs sometimes go into syndication even though new episodes are still being produced for prime-time viewing. For example, in 2001, *Everybody Loves Raymond* had enough episodes for it to go into the off-net syndication market, even though it remained one of the most highly rated network programs on the air. The show was being distributed to local stations by King World Productions at the same time new episodes were being made for prime-time airing.

The off-network syndication market was the brainchild of Lucille Ball and Desi Arnaz, stars of *I Love Lucy* (1951–1957). They were among the first in the industry to envision the financial advantages of rerunning episodes. The major drawback to starting syndication was that many live shows were not recorded at all, and others were preserved on kinescope, a prevideotape process that filmed the show from the video monitor itself and resulted in very poor-quality playback. To their credit, Ball and Arnaz insisted that *I Love Lucy* episodes be shot and preserved on high-quality film rather than kinescope. More than 60 years later, *I Love Lucy* is still a syndication favorite that keeps the audience in stitches. Today, **off-network syndicated programs** are shown on many broadcast and cable networks and are enjoyed by the audience just as much as the first time they were shown.

Local Television Stations

Local television stations produce daily newscasts and programs of interest to the broadcast community, such as local public-affairs shows, weekend travel programs, weather and traffic reports, magazine-style fare, and live sporting events. Local viewers depend on these productions to keep them up on local news and events and informed about the area in which they live.

Hoping to promote competition among competing television stations, the FCC enacted the **prime-time access rule** (*PTAR*) in 1970. First implemented in the 1971 television season, the rule prevented network affiliates in the top 50 markets from programming more than three hours of network shows in prime time (i.e., 7:00 p.m. to 11:00 p.m. Eastern and Pacific; 6:00 p.m. to 10:00 p.m. Central and Mountain). The FCC's goal was to encourage more local programming at television stations and to give independent producers 1 hour of prime time for syndication distribution of their programs outside of the network hold on this lucrative audience. However, most of the stations affected by the rule resorted to finding inexpensive programming rather than producing their own. The result was more game shows and other syndicated fare.

Cable Companies

Cable systems are often required to produce and provide local origination and local-access channels as part of their franchise agreements with cities and local communities. Local-access channels are public/education/government (PEG) channels that are provided free of charge to the local community. Cable services also provide *public-access* channels, which are programmed by local citizens and community groups and government entities. The individuals, organizations, schools, and government agencies involved are responsible for creating their own programs.

Noncommercial Stations

Most noncommercial stations produce their own shows or get programming from other entities. Noncommercial television stations that are affiliated with PBS obtain their programs from independent producers or from the PBS network. PBS–affiliated stations are locally controlled, and its parent organization, Corporation for Public Broadcast, is barred by federal law from producing, scheduling, or disseminating programs to the public. CPB does, however, provide financial support to independent producers to help them create programming for PBS stations. Independent producers create such programs as *Downton Abbey* (2010–2015) for airing on PBS–affiliated stations. Some PBS stations, namely WGBH (Boston), KQED (San Francisco), and WETA (Washington, DC), produce programs for airing on their own local station or throughout the PBS network. For example, *PBS News Hour* and *Washington Week* are produced by station WETA and *Nova* by WGBH, but all three programs are shown on PBS stations nationally.

WHO PAYS FOR PROGRAMS

Producers

When a network or production company produces a television program, it pays the costs. Producing just one episode can cost millions of dollars, but rarely is an episode sold for more than the cost of production. Producers therefore create programs with an eye toward either first-run syndication or network airing and subsequent off-net syndication. They hope their programs are on the air long enough to make it to the off-network syndication market.

The producers will see a profit after a program has been shown on a network and is then sold to a syndicator for future airings. A successful program can command such high prices in syndication that will more than cover the original production cost and reap a huge profit for the producer.

Networks

Television networks pay independent producers for their programs. Specifically, the network actually buys the rights to the program, with the producer maintaining ownership. In some cases, a network will cover the costs of producing its own programs. The network also pays its affiliates (through a contractual relationship) to air the programs. In turn, the network makes money by selling commercial time to advertisers.

Local Television Stations

A local television station that is affiliated with a network is paid by the network to air the network's programs. The amount of compensation depends on the network, the popularity of the program, and the size of the viewing audience. For example, to air a program during prime time, a network could pay its affiliate from several hundred dollars to several thousand dollars per hour. In some cases, such as for a very popular program or game or event, a station may pay the network for the privilege of airing the program, a move known as *reverse compensation*.

Career Tracks: Jay Renfroe, Television Production Co-Owner, Writer, and Producer

FIG. 4.23 Jay Renfroe

What is your job? what do you do?
I own a Los Angeles–based television production company. I create and produce reality and scripted television shows for the broadcast networks (Fox, NBC, ABC, CBS, CW), several cable channels (Discovery, A&E, Lifetime, TLC, Animal Planet, Oxygen, Travel Channel, Turner), and various syndication shows. In the reality world, I am an executive producer, and in the scripted world, I am both an executive producer and a writer.

How long have you been doing this job?
Been in the business for 35 years. I've been a partner in Renegade Entertainment for over 20 years now.

What was your first job in electronic media?
I wrote/directed commercials and industrials in Atlanta.

What led you to the job?
Studied TV/film production in college, did some sports production, had a comedy troupe in Atlanta, produced standup comedy for HBO/Showtime, wrote a play in Los Angeles, which led to writing the screenplay for TriStar based on the play, which led to creating and writing a sitcom for CBS, which led to forming my own company with an old college friend from Florida State University. We're still partners.

What advice would you have for students who might want a job like yours?
Do everything. I was a camera operator, editor, lighting director, theatrical director, writer. Learn every job, and you'll be a better producer. The more material you can create and produce, the more confidence you'll have in yourself as a leader. Learn your own voice and how to communicate your vision to everybody involved in the production. Execution is everything.

In the 1990s, the networks tried to reduce the amount they compensated their affiliates. The networks claimed that affiliate stations could make up for the lost compensation through the sales of additional commercial time that the networks would provide in each program. The affiliates were up in arms over the proposal and rallied against it. The networks backed down. They realized they were faced with growing competition from the new Fox startup and other potential broadcast networks and the availability of syndicated programs as alternatives to network shows.

Local television stations do, however, pay for syndicated first-run and off-network shows, either as cash or as part of a barter agreement with the syndicator. With a **cash purchase**, the station simply pays the syndicator for the right to air the program, and the station in turn sells commercial time. In a **straight barter agreement**, the station gets a program for free, but the syndicator gets to sell a portion of the available commercial time, leaving only a small amount of time for the station to sell. A *cash-plus-barter arrangement* works almost the same way as a straight barter, but the station pays a small amount of cash in exchange for more commercial time.

Cable Companies

Cable operators/providers, such as Comcast, pay cable networks monthly distribution fees based on their numbers of subscribers. Cable operators, in turn, generate revenue by collecting monthly subscription fees from subscribers and by selling commercial time. Some cable programs come from the networks with presold commercial time. In other cases, cable systems barter with cable networks for more commercial time, which they sell locally. Premium cable networks, such as HBO, require a fee-splitting financial relationship. In exchange for carrying a premium cable network, a local cable operator agrees to give the network about half of the fees it collects from its customers, who subscribe to the premium network.

Because cable companies are required by contract to supply local access channels—maybe weather and traffic reports—they are obligated to cover the cost of production. Cable companies also provide public-access channels, but programming is developed by and paid for by the local entities such as school boards, local sports organizations, business clubs, and even individuals who wish to appear on television.

Syndicators

Some syndication companies are part of larger production entities. In these cases, the production arm produces first-run shows for syndication. The syndicator/producer makes back the cost of production by selling the program's broadcast rights to television stations. Syndicators also pay the costs of obtaining the rights to off-network programs, which they then sell to television stations through various cash, straight barter, and **cash-plus-barter** arrangements.

PBS Productions

PBS obtains programs from independent producers and their member stations. Thus, the PBS network/member station programming relationship is opposite the relationship between the commercial networks and affiliates. That is, PBS member stations pay the network for programs and generate revenue through membership drives and from federal, state, and local government funding.

WHAT TYPES OF CHANNELS CARRY WHAT TYPES OF PROGRAMS

BROADCAST NETWORK AFFILIATES

The broadcast television networks vie for the largest possible audience. Up until about 1980, the big three networks (ABC, CBS, NBC) shared approximately 90% of the television-viewing audience. By the late 1990s, new broadcast networks (Fox, PAX, WB, UPN) had hit the screen (ION Media Networks first operated its network as PAX television). The PAX name was changed to Independent Television in 2005 and then to ION Television in 2007. WB and UPN teamed with CBS in 2006 to form the CW Television Network. Additionally, MyNetworkTV is a Fox-owned programming service that airs repeats of broadcast and cable programs such as *Bones* and *Law & Order: SVU*. The Spanish-language networks, Univision and Telemundo, are formidable competitors for the Hispanic audience. With these new broadcast networks and many new cable networks, ABC, CBS, and NBC audience share has shrunk from 90% to about 38%. Cable-supported networks have overtaken them with about 60% audience share, with other cable channels and independent and public stations capturing the remaining viewers.

Prime Time

Airing programs that appeal to a large audience is the key to success for the broadcast networks. With popular programming, the networks can supply large audiences to their advertisers. Even though the broadcast networks' share of the prime-time audience has decreased, they are still the mainstay of the television industry.

Most prime-time programs are designed to appeal across the demographic spectrum—that is, to all ages and education levels. That means a viewer does not have to be an attorney to understand *Law & Order* or be a doctor to like *Grey's Anatomy*. Situation comedies and dramas are the most common forms of prime-time entertainment. Prime time also includes theatrical and made-for-television movies, one-time specials (such as the Emmy, Oscar, and Grammy Award shows), and made-for-TV miniseries.

Non–Prime Time

Networks concentrate their efforts on prime-time programs even though they are actually less profitable than those shown at other times of the day. Non–prime-time programs are generally less expensive to produce and contain 5 to 7 more commercial minutes than prime-time programs. Granted, non–prime-time commercials do not sell for nearly as much as those aired during prime time, but the additional minutes coupled with lower production costs maximize revenue.

Daytime and late-night programs are different from their prime-time counterparts and are driven by audience size and composition. For instance, fewer viewers tune in during non–prime-time hours. Also, the daytime audience is less diverse and made up mainly of children, stay-at-home parents, senior citizens, students, and shift workers. Children's shows, after-school specials, soap operas, talk shows, and game shows are most likely to be aired during the daytime hours. Late-night television attracts more male viewers than female. Males gravitate toward late-night entertainment/talk, sports, comedies, and movies.

CABLE NETWORKS

Many cable programs specialize in particular subjects, such as golfing, home and garden, or history. In contrast to the broadcast networks, which target a large, mass audience, the cable networks target smaller niche or specialty audiences. Thus, a cable network such as the Food Network appeals to viewers interested in cooking and so attracts a much smaller audience than an ABC sitcom, which might appeal to millions of viewers. Other cable networks, such as USA, air mostly old programs from the broadcast networks, which appeal to a broad audience.

There are hundreds of cable networks that focus on specific topics, such as sports (ESPN), music (VH1, MTV), golfing (Golf Channel), weather (The Weather Channel), movies (AMC, Bravo, Turner Classic Movies), news (CNN, MSNBC), science and nature (Discovery), history (The History Channel), animals (Animal Planet, PetsTV), travel (The Travel Channel), and ethnic culture (BET, Univision), to name just a few.

FYI: Top 10 Cable Entertainment Networks by Total Viewers—2014

USA	2.14 million
TNT	1.99 million
History	1.83 million
TBS	1.82 million
FX	1.40 million
Discovery	1.38 million
HGTV	1.34 million
AMC	1.33 million
A&E	1.26 million
ABC Family	1.15 million

Source: 50 top cable entertainment channels of 2014, 2014

Premium and Pay-per-View Channels

Premium channels like HBO, Cinemax, and Encore differ from standard cable channels by the fee structure: an extra charge is added to an existing cable subscription. Premium channels have historically specialized in showing movies and special pay-per-view sporting events like boxing. But premium channels have expanded their line-ups and now offer television dramas and serials and other types of shows. Some premium channels also produce their own shows, like HBO's popular *Game of Thrones* (2011–present) and Showtime's *Dexter* (2006–2013).

Local-Origination and Public-Access Channels

Local-origination channels provide programs such as local news, weather and traffic, program guide, local sporting events, and real estate listings. The local cable provider generally produces these shows as a way to inform subscribers about local events.

Public-access programs run the gamut from city council meetings, PTA meetings, public roundtables, and religious sermons to some guy playing a guitar in a local coffeehouse. As such, public-access programs often serve as an electronic soapbox for ordinary citizens to speak their minds. These shows are generally produced by an independent agent or company not affiliated with the local cable provider. The 1992 film comedy *Wayne's World*, starring Mike Myers and Dana Carvey, is about two loser guys with a public-access show that make it to the big time.

ZOOM IN 4.17

- Read through a local television program guide (online or printed), and identify each channel as a broadcast network affiliate, an independent station, a local-access channel, a cable network, or a premium channel.
- Examine the types of programs shown on each type of channel.

HOW PROGRAMS ARE SCHEDULED

SEASONS

Broadcast networks historically introduced new programs and new episodes of programs in 'seasons.' The big introduction began in September after the Labor Day weekend, which coincided with when the car companies would bring out the year's new models. Not surprisingly, the automakers were some of biggest spenders among television advertisers. Television seasons usually adhered to a 30- to 39-episode year with summer reruns.

Television seasons shortened a bit starting in the late 1960s. Most shows aired 24 to 26 episodes from September through March with reruns from April through August. Since the 1980s, competition from the growing number of cable channels and a stronger emphasis on Nielsen ratings extended the fall season until May, which is one of the 'sweep' months in which Nielsen collects audience viewing data. Programs still typically consist of 20 to 26 episodes but take a hiatus during the winter holidays. Sometimes shows with low ratings are replaced mid-season. *Batman* (1966–1968) was the first mid-season replacement. ABC took a chance when it debuted *Batman* in January, but it turned out to be such a big hit that the networks took to the idea and began replacing their ratings duds with new shows before the season ended.

Although September is still the official new season kickoff month, the full season is sometimes split into two separate units of 10 to 13 weeks. A new show might initially only consist of 10 episodes with which to attract a sizable audience. Successful shows are 'picked up' for the remainder of the season and new episodes shot, while less popular ones are replaced by other programs.

After years of losing audience share to cable, broadcast network executives are searching for new strategies to retain and draw new viewers. The traditional television season is sometimes ignored, and instead new shows premiere throughout the year, and instead of the typical 22-episode season, shows air in bursts of 8-, 10-, and 13-week episodes. AMC's acclaimed *Mad Men* is a perfect example of a 'season-free' program. Making its debut in July 2007 instead of the usual September, its first two seasons and the fourth season ran 13 episodes from July through October, its third season ran August to November, and then the fifth season shifted to March to June, with season six airing April through June. To heighten the drama surrounding long anticipated final episode, *Mad Men's* seventh season was split into two seven-episode 'seasons' airing from April to May of 2014 and 2015.

PROGRAMMING STRATEGIES

The big three television networks use various programming strategies to attract viewers to their channels. In the pre–cable-network, pre–remote-control days of television, these strategies worked more effectively because there were only three networks, and viewers were more comfortable sticking with one channel than getting up and walking to their sets to change to another channel. A network's goal is to control audience flow, or the progression of viewership from one program to another, and to keep viewers from changing to another channel.

- **Tentpoling:** A popular program is scheduled between two new or poorly rated programs. The theory is that viewers will tune to the channel early in anticipation of watching the highly rated program and thus see at least part of the less popular preceding show. Presumably, viewers will stay tuned to the same channel after the conclusion of the popular program to catch the beginning of the next not-so-popular show.
- **Hammocking:** A new or poorly rated show is scheduled between two successful shows. After watching the first program, the network hopes that viewers will stay tuned to the same channel and watch the new or poorly rated show while waiting for the next program to begin.
- **Leading in:** The idea is to grab viewers' attention with a very strong program, anticipating that they will watch that popular show and then stay tuned to the next program on the same channel.
- **Leading out:** A poorly rated program is scheduled after a popular show with the hope that the audience will stay tuned to the same channel.
- **Bridging:** A program is slotted to go over the starting time of a show on a competing network. For example, the season finale of a reality show could be scheduled from 8:00 p.m. to 9:30 p.m. to compete with another network's reality program that is scheduled to start at 9:00 p.m.
- **Blocking:** A network schedules a succession of similar programs over a block of time—for example, four half-hour situation comedies scheduled over 2 hours. The network hopes to attract sitcom lovers and then keep them watching the whole block of shows.
- **Seamless programming:** One program directly follows another, without a commercial break or beginning or ending credits. Some programs use a split-screen technique, in which program credits and closing materials blend in with the start of the following program.
- **Counterprogramming:** One type of program, such as a drama, is scheduled against another type of a program, say, a sitcom, on another network. Counterprogramming works especially well if a strong program of one type is scheduled against a struggling show of another type.
- **Head-to-head programming:** This strategy is the opposite of counterprogramming; that is, two popular shows of the same genre are pitted against each other. For example, one network may schedule its highly rated reality show against another network's highly rated reality show to make viewers choose one show over the other.
- **Stunting:** A special program, such as an important sporting event or a holiday show, is scheduled against a highly rated, regularly scheduled show on another network. Although the network may capture new viewers only for its special program, it is still drawing viewers away from the competition.
- **Repetition:** Used mostly by cable networks. Repetition involves scheduling a program such as a movie to air several times during the week or even during the day.
- **Stripping:** Normally used for syndicated programs, stripping occurs when a program is shown at the same time 5 days a week. For instance, *Seinfeld* (1989–1998), which originally aired 1 day per week in prime time, is stripped in off-network syndication—episodes are aired 5 days per week in non–prime-time hours.

ZOOM IN 4.18

Look through a television program schedule for the current season, and find examples of each type of programming strategy. Then check out program ratings at http://tvbythenumbers.zap2it.com/

SEE IT LATER: RADIO AND TELEVISION

MARKETING SYNDICATED PROGRAMS

National Association of Television Program Executives (NATPE) was first organized in 1963 as a way to bring together television program executives faced with a rapidly changing television industry. NATPE is now more than 4,000 members strong and holds an annual convention at which writers, program creators, and producers market their shows to syndicators, stations, networks and other program distributors. The NATPE convention is considered the primary venue for buying and selling television programming. Industry executives schmooze and wheel and deal for licensing and distribution rights.

But as television has changed, so has NATPE. New to the convention floor are media technology companies that demonstrate new ways to deliver programming online in direct competition with traditional television stations. With more networks producing their own shows and distributing them to their affiliates, affiliates have less need to purchase programs from syndicators. Moreover, as television station ownership rules are relaxed, program directors are often buying the rights to syndicated shows for station groups rather than individual stations, which makes it easier for the syndicators, who pitch their products to fewer decision makers and cover more stations with a single transaction.

As a result of the changing marketplace, many syndicators have recently pulled out of the NATPE convention. They claim that with station consolidation, there are fewer program buyers and ultimately not enough to justify the cost of attending the convention. Although attendance at the NAPTE conventions started dropping in the mid-2000s, the convention still gets about 6,000 registrants. Just where NATPE and the business of buying and selling programs will go remains to be seen, but as the industry changes, so must the traditional ways of doing business.

ZOOM IN 4.19

Learn more about the **National Association of Television Program Executives (NATPE)** at www.natpe.org

---

CABLE TELEVISION

Perhaps the biggest change in television programming is the rise of award-winning cable-originated programs. Historically, cable channels retransmitted broadcast programs to an audience that could not receive broadcast network over-the-air shows. Later, cable networks produced their own programs, which were mostly coverage of sporting events or low-budget shows targeted for niche audiences. Shows about cooking, home decorating, fashion, dating, and so on abounded on cable networks, while the broadcast networks (ABC, CBS, NBC, Fox) excelled in top-rated narrative programs. Until about the late 1990s, broadcast network shows received almost all Emmy Award nominations and wins. HBO's *The Larry Sanders Show* (1992–1998) was one of the first cable-originated shows to win multiple Emmy awards. But the popularity of HBO's *The Sopranos* (1999–2007) cemented the reality that cable-originated shows could compete head to head with broadcast programs. In its 8-year run *The Sopranos* won 21 Emmy awards and 5 Golden Globes. Until 2008, the only winners of an Emmy for top dramatic series were

broadcast network and HBO programs. Then came AMC's *Mad Men* (2007–2015), which took the highly coveted honor of best dramatic series at the 2008, 2009, 2010, and 2011 Emmy Award ceremonies. *Breaking Bad* (AMC, 2008–2013) racked up 58 Emmy nominations and 16 wins, including the top prize for Outstanding Dramatic Series in 2013 and 2014. Cable networks are basking in the glory of many hit shows. HBO's *Game of Thrones* (2011–present), *Boardwalk Empire* (2010–2014), *Silicon Valley* (2014–present), *True Blood* (2008–2014), *Girls* (2012–present), and *VEEP* (2012–present), FX's *Louie* (2010–2015), IFC's *Portlandia* (2011–present), Showtime's *Homeland* (2011–present), and *Dexter* (2006–2013), and even PBS's *Downton Abbey* (2010–2016), give the big four broadcast networks fierce competition for accolades and audience.

The wildly successful *Orange Is the New Black* (2013–present), *House of Cards* (2013–present), and *Arrested Development* (2013) have catapulted Netflix into the world of television program production, yet Netflix is not a cable network but an on-demand, streaming service. Future television production and programming will not be limited to typical production houses and the broadcast networks. Cable networks and Internet companies are discovering that they too can successfully draw a large following of viewers.

SUMMARY

Early radio was quite the magical medium. For the first time, music floated into homes. People could enjoy a concert from the comfort of their living room instead of having to go to a theater. Dramas, comedies, soap operas, quiz shows, music, and many other types of programs filled the radio airwaves.

Television emerged in the late 1940s and took its place as a mass medium in 1948, when the numbers of sets, stations, and audience members all grew by 4,000%. Radio listeners gravitated to television especially as radio programs transitioned to the screen. Anthologies and dramas were popular in the 1950s, but as new production techniques and more portable cameras made outside location scenes possible, the audience's taste moved to more realistic and action-packed shows.

As television gained in popularity, radio's audience shrank. To survive as a medium, radio needed to find a way to live alongside television, and it did so by focusing on music instead of programs. Stations played one or two types of music, or formats, be it rock 'n' roll, classical, middle-of-the-road, easy listening, jazz, or country. Station formatting depends on the market and the competition.

With the growth of cable television, the broadcast networks' share of the audience declined. Looking for ways to maintain their audience, television programmers devised programming strategies to control audience flow from one program to another.

As the Internet, DVRs, and other new media technologies emerge, media consumers gain increasing control of their own viewing schedules and become their own programmers. The networks' best-laid plans may be thwarted as consumers find ways to avoid being tied to programming schedules and serving as a captive audience for advertisers. Media consumers have much to look forward to in the coming years as new programs, new delivery systems, and new ways of watching 'television' emerge.

BIBLIOGRAPHY

50 top cable entertainment channels of 2014. (2014, December 23). *The Wrap*. Retrieved from: www.thewrap.com/from-usa-to-ifc-the-top-50-cable-entertainment-channels-of-2014/

66th Primetime Emmy nominees and winners. (2014, September). *Television Academy of Arts & Sciences*. Retrieved from: www.emmys.com/awards/nominees-winners

About Meet the Press. (2003). *MSNBC*. Retrieved from: www.msnbc.com/news/102219.asp

About PBS. Corporate Facts. (2010). *PBS.org*. Retrieved from: www.pbs.org/about pbs/aboutpbs_corp.html

About Premiere. (2015, n.d.). *Premiere Networks*. Retrieved from: www.premiere radio.com/pages/corporate/about.html

Ahmed, S. (2001). History of radio programming. Retrieved from: web.bryant. edu/,history/h364proj/fall_01/ahmed/Other_radio_programs.htm

Ahrens, F. (2006, August 20). Pausing the panic. *The Washington Post*, p. F1.

American family. (2002). *PBS*. Retrieved from: www.pbs.org/lanceloud/american

Amos 'n' Andy Show. (2003). *Museum of Broadcast Communications*. Retrieved from: www.museum.tv/archives/etv/A/htmlA/amosnandy/amosnandy.htm

Arango, T. (2009). Broadcast TV faces struggle to stay viable. *The New York Times*, pp. A1, A15.

Atkins, L. (2003, October 3). Finally, liberal radio. *The Philadelphia Inquirer*. Retrieved from: www.philly.com/mld/inquirer/news/editorial/6919581.htm

Barstow, D., & Stein. R. (2005). Is it news or public relations? Under Bush, the lines are blurry. *The New York Times*, pp. 1, 18, 19.

Bellis, M. (2010). History of the television remote control. *About.com*. Retrieved from: www.inventors.about.com/od/rstartinventions/a/remote_control.htm

Bibel, S. (2015, February 2). Super Bowl XLIX is most-watched show in U.S. television history with 114.4 million viewers. *TVByTheNumbers*. Retrieved from: http://tvbythenumbers.zap2it.com/2015/02/02/super-bowl-xlix-is-most-watched-show-in-u-s-television-history/358523/

Bird, J. B. (2003). Roots. *Museum of Broadcast Communications*. Retrieved from: www.museum.tv/archives/etv/R/htmlR/roots/roots.htm

Broadcasting and cable yearbook. (2010). Newton, MA: Reed Elsevier.

Brooks, T., & Marsh, E. (1979). *The complete directory to prime time network TV shows*. New York: Ballantine Books.

Butler, J. G. (2001). *Television: Critical methods and applications*. Cincinnati: Thomson Learning.

Cable seizes primetime share in 6/07. (2007). *Cable Advertising Bureau*. Retrieved from: www.thecab.tv/main/whyCable/ratingstory/08ratingsnapshots/cable-seizes-primetime-sh.shtml [August 28, 2009]

Campbell, R. (2000). *Media and culture*. Boston: St. Martin's Press.

Captain Kangaroo. (2003). *The Fifties Web*. Retrieved from: www.fiftiesweb.com/tv/captain-kangaroo.htm

Carter, B. (2006, January 25). With focus on youth, 2 small TV networks unite. *The New York Times*. Retrieved from: www.nytimes.com/2006/01/25/business/25network.html [August 30, 2009]

Carter, B. (2008, September, 23). A television season that lasts all year. *The New York Times*, p. B3.

Costello, C. (2009, October 21). Talk radio: Do we need a new fairness doctrine? *CNN.com*. Retrieved from: amfix.blogs.cnn.com/2009/10/21/talk-radio-do-we-need-a-new-fairness-doctrine/#more-7659 [February 7, 2010]

Digital: News sources for Americans by platform. (2012, September 27). *The Pew Research Center*. Retrieved from: www.journalism.org/media-indicators/where-americans-get-news/

Dizard, W., Jr. (2000). *Old media, new media*. New York: Longman.

Dominick, J. R. (1999). *The dynamics of mass communication*. New York: McGraw-Hill.

Eastman, S. T., & Ferguson, D. A. (2002). *Broadcast/cable/web programming* (6th ed.). Belmont, CA: Wadsworth.

Eggerton, J. (2005). Survey says: Noncom news most trusted. *Broadcasting & Cable*. Retrieved from: www.broadcastingcable.com/article/158913-Survey_Says_Noncom_News_Most_Trusted.php

Enger, J. (2009, August 25). Share your *Guiding Light* memories. *The New York Times*. Retrieved from: artsbeat.blogs.nytimes.com/author/jeremy-egner/ [August 26, 2009]

Famous weekly shows. (1994–2002). *Old Time Radio*. Retrieved from: www.old time.com/weekly

Farhi, P. (2009, May 17). Click, change. *The Washington Post*, pp. E1–E8.

For "SNL," doubts follow a banner year. (2009, September 8). *The Washington Post*, pp. C1, C6.

Garvin, G. (2009, July 9). When Casey Kasem started counting down, America rocked together. *The Washington Post*, p. C10.

Gertner, J. (2008, January). A clicker is born. *The New York Times Magazine*, pp. 34–35.

Goel, V. (2014, September 9). Twitter and Facebook wield little influence on TV watching. *International The New York Times*, pp. 18, 20.

Golden years. (1994–2002). *Old Time Radio*. Retrieved from: www.old-time.com/golden_age/index.html

Gomery, D. (2003). Movies on television. *Museum of Broadcast Communications*. Retrieved from: www.museum.tv/archives/etv/M/html/moviewsontel/movie sontel.htm

Grant, A. E., & Meadows, J. H. (2002). *Communication technology update*. Boston: Focal Press.

Gross, L. S. (2003). *Telecommunications*. Boston: McGraw-Hill.

Guiding Light. (2009). *Wikipedia*. Retrieved from: en.wikipedia.org/wiki/Guiding_Light [August 26, 2009]

Gundersen, E. (2002, November 22). Uncovering the real Osbournes. *USA Today*, p. E1.

Hare, B. (2014, May 20). 'The Maya Rudolph Show:' What's the verdict. *CNN.com*. Retrieved from: www.cnn.com/2014/05/20/showbiz/tv/maya-rudolph-show-whats-the-verdict/index.html

Head, S. W., Sterling, C. H., & Schofield, L. B. (1994). *Broadcasting in America*. Boston: Houghton Mifflin.

Hilliard, R., & Keith, M. (2001). *The broadcast century and beyond* (3rd ed.). Boston: Focal Press.

History of the Batman. (2010). *Batman-on-Film.com*. Retrieved from: www.batman-on-film.com/historyofthebatman_batman66.html [February 6, 2010]

How do public broadcasters obtain programming? (n.d.). *Corporation for Public Broadcasting*. Retrieved from: www.cpb.org/aboutpb/faq/programming.html

Infrared remote controls. (2010). *Howstuffworks.com*. Retrieved from: www.how stuffworks.com/search.php?terms5remote1control [February 2, 2010]

Ingram, B. (2003). Captain Kangaroo. *TV Party*. Retrieved from: www.tvparty.com/lostterrytoons.html

Java, J. (1985). *Cult TV*. New York: St. Martin's Press.

Jenkins, D. (2002, December 19). From Ozzie to Ozzy. *About.com*. Retrieved from: classictv.about.com/library/weekly/aa121902a.htm

Johnson, P. (2002). Fox News enjoys new view—From the top. *USA Today*, pp. 1A–2A.

Johnson, T. J., & Kaye, B. K. (2002). Webelievabilty: A path model examining how convenience and reliance on the web predict online credibility. *Journalism & Mass Communication Quarterly*, 79(3), 619–642.

Kaye, B. K., & Johnson, T. J. (2003). From here to obscurity: The Internet and media substitution theory. *Journal of the American Society for Information Science and Technology*, 54(3), 260–273.

Kinescope. (n.d.). *The Museum of Broadcast Television*. Retrieved from: http://www.museum.tv/eotv/kinescope.htm

Klein, S. (2000). It's a myth that Internet journalism is less accurate. Retrieved from: www.content-exchange.com

Koblin, J. (2015, November 19). The ax falls, but slowly. *The New York Times*, pp. A1, A4.

Lim, D. (2011, April 17). Reality TV originals in drama's lens. *The New York Times*, Television, pp. 24–25.

Massey, K. B., & Baran, S. J. (1996). *Television criticism*. Dubuque, IA: Kendall/Hunt.

Matsa, K. E. (2014, January 28). Local TV audiences bounce back. *Pew Research Center*. Retrieved from: www.pewresearch.org/fact-tank/2014/01/28/local-tv-audiences-bounce-back/

Masterpieces and milestones. (1999, November 1). *Variety*, p. 84.

McGinn, D. (2002, November 11). Guilt free TV. *Newsweek*, pp. 53–59.

Media and Communications Trends. (2009, October 15). *Nielsen Media*. Retrieved from: en-us.nielsen.com/etc/content/nielsen_dotcom/en_us/home/ insights/webinars_registration

Mershan, P. W. (2003, February 28). *Amos 'n' Andy* spurred controversy, and many collectibles. *News Journal*. Retrieved from: www.mansfieldnewsjournal.com/news/stories/20030228/localnews/1078705.html

News audiences increasingly politicized. (2004, June 8). *Pew Research Center*. Retrieved from: people-press.org/report/?pageid5838 [September 1, 2009]

Nielsen. (2014, February 3). Super Bowl XLVIII draws 111.5 million viewers, 25.3 million tweets. Retrieved from: http://www.nielsen.com/us/en/insights/news/2014/super-bowl-xlviii-draws-111-5-million-viewers-25-3-million-tweets.html

Nielsen TV ratings. (2010). *Nielsen*. Retrieved from: en-us.nielsen.com/rankings/insights/rankings/television [February 10, 2010]

NPR fact sheet. (2014, November). *National Public Radio*. www.npr.org/about/press/NPR_Fact_Sheet.pdf

NPR's growth during the last 30 years. (2003). *National Public Radio*. Retrieved from: www.npr.org/about/growth.html

Online papers modestly boost newspaper readership. (2006, July 30). *Pew Research Center*. Retrieved from: people-press.org/report/?pageid51066 [August 26, 2009]

Over and history. (2002). *National Association of Television Program Executives*. Retrieved from: www.natpe.org/about

Pareles, J. (2012, February 1). A smooth operator in the name of soul. *The New York Times*, pp. C1, C8.

Parsons, P. R., & Frieden, R. M. (1998). *The cable and satellite television industries*. Boston: Allyn & Bacon.

Pew Research Center. (2000). *Internet Sapping Broadcast News Audience: Investors Now go Online for Quotes, Advice*. Retrieved from: www.people-press.org/media00rpt.htm

Peyser, M., & Smith, S. M. (2003, May 26). Idol worship. *Newsweek*, pp. 53–58.

Poindexter, P. (2012). Too busy for news: Unlimited time for social media. In *Millennials, news, and social media: Is news engagement a thing of the past?* (pp. 53–69). New York, NY: Peter Lang Publishing.

Powers, L. (2011, February 7). 10 most watched TV shows ever. *The Hollywood Reporter*. Retrieved from: www.hollywoodreporter.com/blogs/live-feed/10-watched-tv-shows-97180

Premiere Radio Networks. (2009, August 26). *Premiere Radio Networks.com*. Retrieved from: www.premiereradio.com/category/view/talk.html [August 26, 2009]

Remote control. (2010). *The Great Idea Finder*. Retrieved from: www.ideafinder.com/history/inventions/remotectl.htm [February 2, 2010]

Seipp, C. (2002). Online uprising. *American Journalism Review, 24*(5), 42.

Sandomir. R. (2009, July 19). Amid blizzard, Cronkite helped make sports history. *The New York Times*, Sports, pp. 1, 8.

Schlosser, J. (2002, January 28). Is it check-out time at NATPE? *Broadcasting & Cable*. Retrieved from: www.tvinsite.com/broadcastingcable

Sesame Street Awards (2014). *IMDb*. Retrieved from: www.imdb.com/title/tt0063951/awards

Stanley, A. (2008, September 21). Sitcoms' burden: Too few taboos. *The New York Times*, pp. MT 1, 6.

State of the media: Audio Today. (2014). *Nielsen*, Retrieved from: http://www.nielsen.com/content/dam/corporate/us/en/reports-downloads/2014%20Reports/state-of-the-media-audio-today-feb-2014.pdf

Stelter, B. (2009, August 1). Voices from above silence a cable TV feud. *The New York Times*, pp. A1, A3.

Stelter, B. (2009, August 24). Reality TV star, a killing suspect, is found dead. *The New York Times*. Retrieved from: www.nytimes.com/2009/08/24/arts/television/24real.html [August 30, 2009]

Stelter, B. (2013, April 28). Two classics of the soaps are heading to the Web. *The New York Times*, pp. B1, B4.

Stelter, B. (2013, September 21). Emmys highlight a changing TV industry. *The New York Times*, pp. C1, C4.

Stelter, B., & Carter, B. (2009, November 20). A daytime network franchise bets on her future with cable. *The New York Times*, pp. A1, A3.

Sterling, C., & Kittross, J. (2002). *Stay tuned: A history of American broadcasting*. Mahwah, NJ: Erlbaum.

Stuever, H. (2014, July 11). Same ol' Emmy drama. *The Washington Post*, p. C1.

Syndicated TV ratings. (2016, April 27). *TV by the Numbers*. Retrieved from: http://tvbythenumbers.zap2it.com/2016/04/27/syndicated-tv-ratings-april-11-17-top-8-shows-repeat/

Top 20 cable programming networks. (2009). *National Cable & Telecommunications Association*. Retrieved from: www.ncta.com/StatisticChart.aspx?ID528 [February 1, 2010]

Top 25 syndicated programs. (2009). *Television Bureau of Advertising*. Retrieved from: tvb.org/nav/build_frameset.asp?url5/rcentral/index.asp [February 1, 2010]

TV ratings. (2009). *Broadcasting & Cable*. Retrieved from: www.broadcastingcable.com/channel/TV_Ratings.php [February 2, 2010]

Vanderbilt, T. (2013, April). The new rules of the hyper-social, data-driven, actor-friendly, super-seductive platinum age of television. *Wired*, pp. 90–103.

Walker, J. R., & Ferguson, D. A. (1998). *The broadcast television industry*. Boston: Allyn & Bacon.

What is NPR? (2010). *National Public Radio*. Retrieved from: www.npr.org/about/

Weber, B. (2012, April 19). TV Emperor of rock 'n' roll and New Year's Eve dies at 82. *The New York Times*, pp. A1, A24.

Wolbe, T. (2014, February 13). Can we save AM radio? *The Verge*. Retrieved from: www.theverge.com/2014/2/13/5401834/can-we-save-am-radio

Wolk, D. (2005, November 17). College radio. *Slate.com*. Retrieved from: www.slate.com/id/2130587/ [August 26, 2009]

Wyatt, E. (2009, August 2). TV contestants: Tired, tipsy and pushed to the brink. *The New York Times*, pp. A1, A14.

Yahr, E. (2009, August 14). "Family Guy" channels controversy onstage. *The Washington Post*, p. C6.

Interconnected by the Internet 5

Contents

Although many people believe that the Internet is a 1990s invention, it was actually envisioned in the early 1960s. Over the last 50 or so years, the Internet has gone from having the technological capability of sending one letter of the alphabet at a time from one computer to another to a vast system in which trillions of messages are sent around the world every day.

This chapter covers how the Internet works and discusses how the various Internet resources are used: the Web, email, YouTube, blogs, electronic mailing lists, newsgroups, chat rooms, instant messaging, social network sites, Twitter, and podcasting. Social media and blogs are also discussed in more detail and from a social perspective in Chapter 9. This chapter focuses primarily on the benefits and challenges of online information and how it affects traditional media, especially radio and television. An in-depth look at new content delivery systems, such as Spotify and Hulu, is included in Chapter 6. The Internet has improved the lives of many people but has caused problems for others. Understanding how a technology developed and made its way into everyday life is the first step to guiding where it is going.

SEE IT THEN

HISTORY OF THE INTERNET AND THE WORLD WIDE WEB

In the early 1960s, scientists approached the U.S. government with a formal proposal for creating a decentralized communications network that could be used in the event of a nuclear attack. By 1970, **ARPAnet (Advanced Research Projects Agency Network)** was created to advance computer interconnections.

The interconnections established by ARPAnet soon caught the attention of other U.S. agencies, which saw the promise of using an electronic network for sharing information among research facilities and schools. While disco music was hitting the airwaves, Vinton Cerf, later known as 'the father of the Internet,' and researchers at Stanford University and UCLA were developing packet-switching technologies and transmission protocols that are the foundations of the Internet. In the 1980s, the National Science Foundation (NSF) designed a prototype network that became the basis for the Internet. At the same time, a group of scientists in the **European Laboratory of Particle Physics (CERN)** was developing a system for worldwide interconnectivity that was later dubbed the World Wide Web. Tim Berners-Lee headed the project and was dubbed 'the father of the World Wide Web.'

ZOOM IN 5.1

See Tim Berners-Lee on YouTube talking about how the Internet was created:

www.youtube.com/watch?v=yF5–6AcohQw

CHARACTERISTICS OF THE WORLD WIDE WEB

For many years, the Internet was the domain of scientists and researchers. It came into widespread use only in 1993 with the advent of the easy-to-use World Wide Web. Since then, the Internet has become a tremendously important part of our daily media diet. Most online users have come to think of the Internet and the Web as synonymous, but the Web is the part of the Internet that brings graphics, sound, and video to the screen.

WHAT IS THE INTERNET?

Simply stated, the **Internet** is a worldwide network of computers. Millions of people around the globe upload and download information to and from the Internet every day. The Internet is a digital venue that connects users to each other and to vast amounts of information. Like radio and television, the Internet is a **mass medium** in the respect that it reaches a broad range of users.

By definition, before any medium can be considered a *mass medium*, it needs to be adopted by a critical mass of users. In the modern world, about 50 million users seems to be the benchmark. The Internet became a mass medium at an unprecedented speed. Radio broadcasting (which began in an era with a smaller population base) took 38 years to reach the magic 50 million mark, and television took 13 years. The Internet surpassed 50 million regular U.S. users sometime in late 1997 or early 1998, only about 5 years after the World Wide Web made it user friendly.

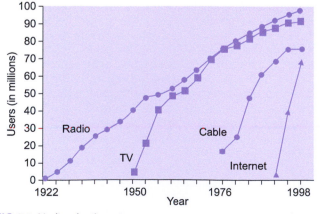

FIG. 5.1 Media adoption rates
Source: Shane, 1999

The Internet is distinguished from other mass media, such as radio and television, by its inclusion of text, graphics, video, and sound into one unique medium. Whereas newspapers are print only, radio audio only, and television audio and video, these properties converge on the Internet. Convergence is generally defined as the "coming together of all forms of mediated communication in an electronic, digital form, driven by computers" (Pavlik, 1996, p. 132). Another researcher defined **convergence** as the "merging of communications and information capabilities over an integrated electronic network" (Dizard, 2000, p. 14).

FYI: The Word 'Internet'

The word *Internet* is made up of the prefix *inter*, meaning 'between or among each other,' and the suffix *net*, short for *network*, which is 'an interconnecting pattern or system.' An *inter-network*, or *internet* (small *i*), refers to any 'network of networks' or 'network of computers,' whereas *Internet* with a capital *I* is the specific name of the computer network that provides the World Wide Web and other interactive components (Bonchek, 1997; Krol, 1995; Yahoo! Dictionary Online, 1997).

TECHNOLOGY

The **World Wide Web** is a technologically separate and unique medium, yet it shares many properties with traditional media. Both its similarities and differences have made it a formidable competitor for the traditional mass media audience.

When comparing traditional media, each can be distinguished by unique characteristics. Radio is convenient and portable and can be listened to even while the audience is engaged in other activities. Television is aural and visual and captivating; print (magazines, newspapers) is portable and can be read anytime, anywhere. The Web has some of these same advantages. For example, people can listen to online audio while attending to other activities, they can read archived information anytime they please, and they can sit back and be entertained by graphics and video displays. In addition, the Internet provides online versions of print media, which can be read electronically or even printed to provide a portable version. The *New York Times* at www.nytimes.com, *Rolling Stone* at www.rollingstone.com, and *Elle Magazine* at www.elle.com are just a few example of online versions of print media. The Internet also offers benefits not offered by traditional media: two-way communication through email, chat, and other interactive tools.

Although the Internet's proponents highly tout this medium, it falls short of traditional media in some ways. Users must have access to a computer, smartphone, or other **Wi-Fi**-enabled device (such as a tablet) to access online material. Many cannot afford a computer or other device, and there are many places where it is inconvenient or just not possible to get an Internet connection. Unlike free over-the-air radio broadcasting, access to the Internet requires a subscription to an **Internet service provider (ISP)**, such as Earthlink, a cable company like Comcast, a phone company like Verizon, or a public wired or wireless connection.

CONTENT

The Web is unique because it displays information in ways similar to television, radio, and print media. Radio delivers audio, television delivers audio and video, and print delivers text and graphics. The Web delivers content in all of these forms, thus blurring the distinction among the media.

The Web's big advantage over traditional media is its lack of space and time constraints. The length of radio and television content is limited by available airtime, and print is constrained by the available number of lines, columns, or pages. These limitations disappear online. News and entertainment on the Internet are not physically confined by seconds of time or column inches of space but are restricted only by editors or Web page designers.

Although the amount of content is largely unlimited, the speed of online delivery is constrained by bandwidth, which is the amount of data that can be sent all at once. Think of bandwidth as a water faucet or a pipe. The circumference of the faucet or pipe determines the amount of water that can flow through it and the speed at which it flows. Similarly, bandwidth determines the speed of information flow and thus affects how quickly content appears on the screen. Web designers might choose to reduce the amount of content to increase speed, for example. Bandwidth is becoming less of a concern, however, now that fast broadband connections are becoming commonplace.

HOW THE INTERNET WORKS

The Internet operates as a **packet-switched network**. It takes bundles of data and breaks them up into small packets or chunks that travel through the network independently. Smaller bundles of data move more quickly and efficiently through the network than larger bundles. It is kind of like moving a home entertainment system from one apartment to another. The DVD player might be packed in the car and the large-screen television in the truck. The home entertainment system is still a complete unit, but when transporting, it is more convenient to move each part separately and then reassemble all of the components at the new place. The Internet works in almost the same way, except it disassembles bits of data rather than a home entertainment system and reassembles the bits into a whole unit at its destination point.

The bits of data that make up an email message, Web page, or image, flow through interconnected computers from their points of origin to the destination. The sender's computer is the origination point, known as the **client**. The message bits leave the client computer and travel in separate packets to a **server**. A server is basically a powerful computer that provides continuous access to the Internet. From there, packets move through routers. A **router** is a computer that links smaller networks and sorts each packet of data until the entire message is reassembled, and then it transmits the electronic packets either to other routers or directly to the addressee's server. The server holds the entire message until an individual directs his or her client computer to pick it up.

Servers and routers deliver online messages through a system called **transmission control protocols/Internet protocols (TCP/IP)**, which define how computers electronically transfer information to each other on the Internet. TCP is the set of rules that governs how smaller packets are reassembled into an IP file until all of the data bits are together. Routers follow IP rules for reassembling data packets and data addressing so information gets to its final destination. Each computer has its own numerical **IP address** (which the user usually does not see) to which routers send the information. An IP address usually consists of between 8 and 12 numbers and may look something like *166.233.2.44*.

Because IP addresses are rather cumbersome and difficult to remember, an alternate addressing system was devised. The Domain Name System (DNS) basically assigns a text-based name to a numerical IP address using the following structure: *username@host.subdomain*. For example, in *MaryC@anyUniv.edu*, the user name, *MaryC*, identifies the person who was issued Internet access. The @ literally means 'at,' and *host.subdomain* is the user's location. In this example, *anyUniv* represents a fictitious university.

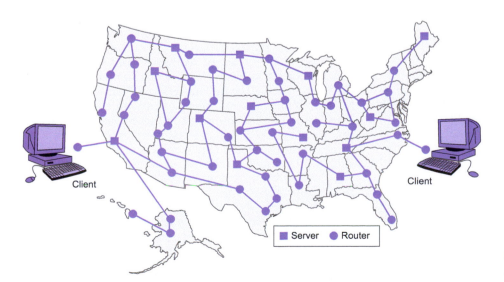

FIG. 5.2 How the Internet works
Source: Kaye and Medoff, 2001

The top-level domain (TLD), which is always the last element of an address, indicates the host's type of organization. In this example, the top-level domain, *edu*, indicates that *anyUniv* is an educational institution.

The Internet Corporation for Assigned Names and Numbers (ICANN) assigns top-level domains. The first top-level domains created for the Internet were .com (commercial), .org (organization), .net (network), .int (international), .edu (education), .gov (government), and .mil (military). Over the years, other domain names were added such as .biz (business), .museum (museum), .coop (cooperatives), and .xxx (pornographic), and domains for countries and regions such as .ec (Ecuador), .it (Italy), .pl (Poland), .gb (Great Britain), and .eu (European Union). But the largest expansion came after 2013 when ICANN began accepting applications for custom domain names. By early 2015, there were just over 700 top-level domain names.

FYI: Top-Level Domains (2014)

Most Popular New Top-Level Domains added in 2014

.today	(current events and issues)
.guru	(expert advice)
.solutions	(technical support)
.photography	(photography)
.tips	(sharing knowledge about any subject)
.email	(email)
.technology	(new technology)
.directory	(directory or lists)
.clothing	(retailers, boutiques, outlets, designers)
.solar	(solar power industry)

Sources: Kerner, 2014; Skiffington, 2014

Prior to the creation of the Web, Internet content could be retrieved only through a series of complicated steps and commands. The process was difficult, time-consuming, and required an in-depth knowledge of Internet protocols. As such, the Internet was of limited use. In fact, it was largely unnoticed by the public until 1993 when undergraduate Marc Andreessen and a team of University of Illinois students developed Mosaic, the first Web browser. Different from the Internet in which data were retrieved by entering cumbersome commands, Mosaic was based on a system of clickable, intuitive hyperlinks. Mosaic caught the attention of Jim Clark, founder of Silicon Graphics, who lured Andreessen to California's Silicon Valley to enhance and improve the browser. With Clark's financial backing and Andreessen's know-how, Netscape Navigator was born. This enhanced version of Mosaic made Andreessen one of the first new-technology, under-30-year-old millionaires. Once Netscape Navigator hit the market, the popularity of Mosaic plummeted. Since Mosaic and Netscape Navigator, other browsers, such as Netscape Communicator, have come and gone as new and improved ones, such as Firefox and Chrome, have taken their place.

ELECTRONIC MASS MEDIA ONLINE

So far, this chapter has focused on the history of the Internet and on what the Internet is and how it works. As online technology improved, radio and television saw the potential for increasing their audience and revenue by delivering their content online. Stations and networks slowly developed online counterparts to their traditional over-the-air and cable-delivered content.

THE RISE OF INTERNET RADIO

Almost all local radio stations have established a Web site, which typically provides news and community information, promotes artists and albums, and streams over-the-air programs and music. These Web sites lend credence to the media industry's concern that Internet users may one day discover that they no longer need radios. Instead, they will just access radio programming over the Internet.

The rise of Internet radio somewhat mirrors the development of over-the-air radio. In the early days of radio, amateur (ham) operators used specialized crystal sets to transmit signals and voice to a limited number of listeners (usually other ham operators) who had receiving sets. Transmitting and receiving sets were difficult to operate, the reception was poor and full of static, and the sets themselves were large and cumbersome, leaving only a small audience of technologically advanced listeners who knew how to operate them. In the early days of radio, the technologically adept were the first to gravitate toward the new medium. In the 1990s, the technologically savvy again took the lead, but this time, they paved the way for *cybercasting*. In its early years, the Internet was accessed largely through a modem, which allowed digital signals to travel from computer to computer via telephone lines. Just as primitive radios and static-filled programs once kept the general public from experiencing the airwaves, bandwidth limitations and slow computer and modem speeds kept many radio fans from listening via the Internet. For example, using a 14.4-Kbps modem, *Geek of the Week*—a 15-minute audio-only program—took almost 2 hours to download, which was considered very fast in the mid-1990s.

Small- to medium-market radio stations and college stations generally led the way to the Internet. As their success stories quickly spread throughout the radio industry, other stations eagerly set up their own music sites. Real-Networks, the provider of **RealAudio** products, was the first application to bring real-time audio on demand over the Internet. Since the introduction of RealAudio in 1994, thousands of radio stations have made the leap from broadcast to online audiocast. By using RealAudio technology, AM and FM commercial radio, public radio, and college stations captured large audiences and moved from simply providing prerecorded audio clips to transmitting real-time audio in continuous streams. Streaming technology pushes data through the Internet in a continuous flow, so information is displayed on a computer before the entire file has finished downloading. Streamed audio and video selections can be played as they are being sent, so there is no waiting for the entire file to download.

Since RealAudio's introduction, several other companies have developed audio-on-demand applications and

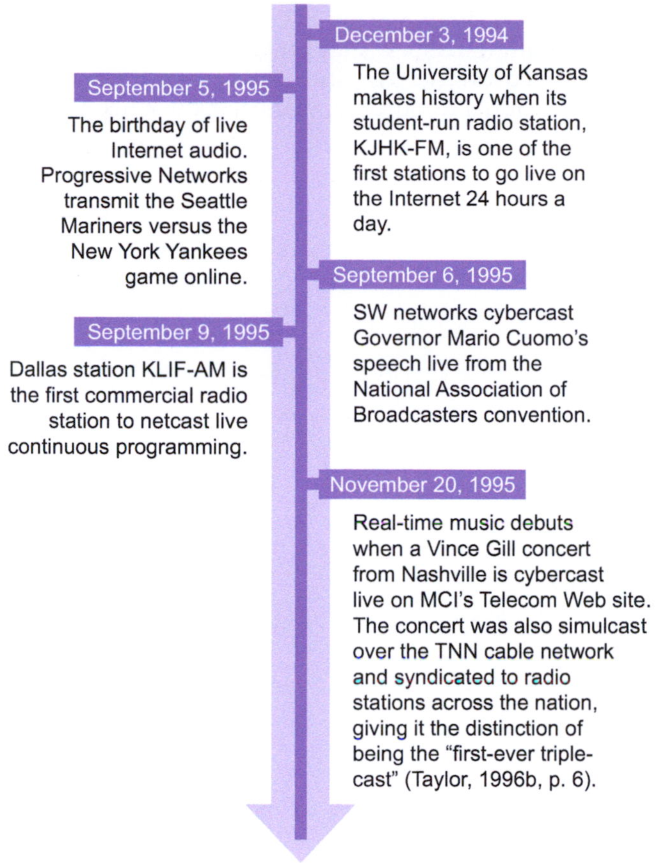

September 5, 1995

The birthday of live Internet audio. Progressive Networks transmit the Seattle Mariners versus the New York Yankees game online.

September 9, 1995

Dallas station KLIF-AM is the first commercial radio station to netcast live continuous programming.

December 3, 1994

The University of Kansas makes history when its student-run radio station, KJHK-FM, is one of the first stations to go live on the Internet 24 hours a day.

September 6, 1995

SW networks cybercast Governor Mario Cuomo's speech live from the National Association of Broadcasters convention.

November 20, 1995

Real-time music debuts when a Vince Gill concert from Nashville is cybercast live on MCI's Telecom Web site. The concert was also simulcast over the TNN cable network and syndicated to radio stations across the nation, giving it the distinction of being the "first-ever triple-cast" (Taylor, 1996b, p. 6).

FIG. 5.3 Highlights of early Internet radio

new protocols for increasing bandwidth for even faster streaming. Online audio has come a long way since the early days of the Internet. Computer and Internet technology has become less expensive, easier to use, and able to deliver high-quality sound.

In 1999, online music was transformed when college student Shawn Fanning created Napster, the first software for finding, downloading, and swapping **MP3 (Moving Picture Experts Group, Audio Layer 3)** music files online. Young adults' love for music made the MP3 format and MP3 player the hottest trend since the transistor radio hit the shelves in the 1950s. At first, music file-sharing sites seemed like a good idea, but they quickly ran into all kinds of copyright problems and found themselves knee deep in lawsuits. Although it has always been legal for music owners to record their personal, store-bought CDs in another format or to a portable player, sharing copies with others who have not paid for the music is considered piracy and copyright infringement.

In the early 2000s, the recording industry launched a vigorous campaign against so called 'music pirates,' who shared or downloaded music files online for free. In 2003, **RIAA (Recording Industry Association of America)** sued 261 music lovers, targeting excessive pilferers. In some cases, the RIAA held individuals liable for millions of dollars in lost revenue, sometimes equaling up to $150,000 per song. After making its point, the RIAA worked out settlements in the $3,000 to $5,000 range and instituted the Clean Slate Amnesty program for those who wanted to avoid litigation by issuing a written promise to purge their computers of all files and never download music without paying again.

Napster was at the center of the illegal music download imbroglio. After building a clientele of about 80 million users but facing several years of legal wrangling, Napster went offline in July 2001. Napster made its comeback in late 2003, this time as a legal site. Napster has since merged with Rhapsody, a music streaming service with a monthly fee.

The question in the early 2000s was whether Internet users would pay for what they used to get for free. One study reported that only about one-quarter of Internet users who downloaded music in the past would be willing to pay to do so in the future, and another study claimed that only about one-third of college students would pay more than $8.50 per month to download music. Yet another report found that people who download free music do so to sample it, and if they like it, they'll buy it. But now paying for downloading music has become commonplace and, for many, preferable to buying a CD.

TELEVISION'S MIGRATION TO THE WEB

Early television programs such as *Amos 'n' Andy*, *Life of Riley*, *Guiding Light*, *You Bet Your Life*, and *The Lone Ranger* all originated as radio programs, as did many other shows televised in the 1950s. Given the ability of television to bring so much more life to a program than radio ever could, producers transformed radio shows into exciting, dynamic television programs.

Many ardent fans have long hailed television as the ultimate form of entertainment. Television can be watched anytime, anywhere, and choosing what to watch is as easy as pushing a button or flicking a switch. Yet despite the popularity of watching television, the viewing public is always looking for new forms of entertainment.

Over-the-air television was once the primary medium for news and entertainment. Although cable was established early in the life of television, it did not take hold with viewers until the 1970s. More recent technologies, especially satellite, gave rise to newer means of program delivery, and the Web itself was hailed as the 'television of the future.'

In the mid-1990s, the Web was touted as the up-and-coming substitute for television. The physical similarities between a television and a computer monitor, coupled with the promise that online content would soon be as plentiful and exciting as that on television, led people to believe that the Web would soon replace television. The novelty of the Web also drew many television viewers out of plain curiosity. Between 18 and 37% of Web users were watching less television than they had before, and the Internet was cutting more deeply into time spent with television than with other traditional media. On the other hand, the Web may not have affected the time viewers spent with television except perhaps among those individuals who watched little television in the first place.

As the 1990s came to a close, the Web could deliver some short, full-motion videos at best, but it did not evolve into a substitute for television as predicted. Users crossed into the 21st century still hoping that their computers would soon be technologically capable of receiving both Web content and television programs.

September 1994

CBS connects to Prodigy, a proprietary online service.

February 1995

The Weather Channel debuts on CompuServe; a forum that provides an outlet for discussing the weather and the channel's coverage.

March 1996

WRNN-TV, a New York–based cable regional news network, launches a Web site with live online video reports.

March 1996

Lifetime is one of the first cable networks to launch a Web site.

September 1996

NFL games are shown live online. Highlights from previous games, player profiles, and other background information is accessible while watching live game action in an onscreen show.

October 1996

American Cybercast launches the first TV-type network online and the first online soap opera, *The Spot*.

November 1998

CNN, ABC News, MSNBC, and Fox offer real-time election night coverage, providing up-to-the-minute results for voters.

FIG. 5.4 Highlights of early Internet television

SEE IT NOW

INTERNET USERS

The Internet has come a long way since 1993, when there were about 14 million users (0.3% of the world population). The number of users increased tenfold in the first 10 years of its existence, and by mid-2016, there were about 3.5 billion users (48.6% global penetration), including 279 million in the United States (86.7% of the population), with most going online every day. The U.S., however, does not have the highest Internet penetration rate; that honor goes to the United Kingdom with 89.9%. Other countries close to the U.S. percentage of Internet users are Germany (86.7%), Japan (86.0%), and France (85.7%).

Estimates of the number of hours users spend online vary widely, but most research indicates that individuals spend about 3.5 to 5 hours per day online, which constitutes about 40% of their daily media use. Although male users originally dominated the Web, women now use it almost as much. Generally, online users tend to be younger than age 65, highly educated, and more affluent than the U.S. population at large, and they tend to live in urban and suburban areas.

Going online is often the first activity of the morning, and many daily routines revolve around the Internet. Completing routine tasks, such as making airline, hotel, or dinner reservations, contacting friends and family, looking up information about any topic, keeping up with the latest news, finding recipes, shopping, and playing games, are conducted online. The Internet has evolved into a multidimensional resource that has become a necessity.

FYI: Demographics of Internet Users (2013)

Age	
18–34	81.2%
35–44	83.3%
45–64	80.6%
65+	64.3%
Gender	
Male	79.4%
Female	77.6%
Education	
less than high school	53.7%
high school graduate	69.7%
some college/associate degree	82.4%
Bachelor's degree or higher	91.5%

Source: U.S. Census Bureau, 2014.

FYI: Undergraduate Students and Technology

98% use the Internet

96% have a smart/cell phone

93% have broadband access

88% have a laptop

86% use social network sites

84% have an iPod or other MP3 player

58% have a game console

9% have an ebook reader

5% have a tablet computer

Source: Pew Internet

A WIRELESS WORLD

To encourage settlement of the American West in the mid-1880s, prominent newspaper editor Horace Greeley urged, "Go west, young man, go west." If he was alive today, he would probably exhort us to venture into an even newer territory with "Go wireless, young man, go wireless."

Wireless technology, now commonly called Wi-Fi, became accessible to consumers when Apple introduced its AirPort Base Station in 1999. Base station technology gave way to wireless routers that sent signals to any computer via a wireless card or transmission receiver that needed to be connected to a computer.

Most computers now come equipped with built-in wireless receiving capabilities. The ease and convenience of Wi-Fi has made it the preferred mode of connecting to the

Internet. Wi-Fi makes it possible to pick up Internet signals from almost anywhere. These signals cannot be seen or felt, but in many cafés, coffeehouses, airports, universities, and other public places, Internet signals are bouncing off the walls and through the walls. The distance wireless signals reach depends on many factors such as outdoor terrain and number of walls or other obstacles between the router and the device. Like a cell phone, Wi-Fi operates as a kind of radio. Newer 4G (fourth-generation) wireless technologies significantly improve data transmission speed and range over 3G wireless, and newer 5G promises yet greater speed and capabilities. Yet other new wireless technologies, such as Bluetooth, are meant for very short-range connections, such as between a computer and a printer.

Wireless is not perfect. The speed of data transmission fluctuates, it can interfere with satellite radio, and the signals are sometimes blocked by walls and other solid objects. The best wireless signal is obtained in a large, unobstructed space, like a large office that has few ceiling-to-floor walls. Despite these issues, once wireless has been set up, it is quick and easy to use and compatible with most computers and mobile devices.

FYI: Internet Connections (percentage of U.S. homes, 2013)

- Cable modem 42.8%
- Mobile broadband: 33.1%
- No Internet 25.6%
- DSL connection: 21.2%
- Fiber optic: 8%
- Satellite: 4.6%
- Dial-up connection: 1%

(Figures exceed 100%—some homes have multiple connections)

Source: United States Census Bureau, 2014

INTERNET RESOURCES

WORLD WIDE WEB

The sheer number of online users attests to the large variety of online content. As astonishing as it seems, there are about 650 million to 1 billion Web sites on the Internet, up from 1 in 1991, 10 in 1992 and 130 in 1993. Moreover, each site contains multiple pages, totaling up to an astounding 14.3 trillion pages of content.

As Tim Berners-Lee pointed out during an interview on PBS's *Nightly Business Report*, there are more Web pages than there are neurons in a person's brain (about 50–100 billion). Further, Berners-Lee claims that more is known about how the human brain works than about how the Web works in terms of its social and cultural impact.

Web information comes in text, graphics, and audio and video formats. The multiformat presentation capability distinguishes it from other parts of the Internet. Point-and-click browsers, first developed in the early 1990s,

make it easy to travel from Web site to Web site. Gathering information is as easy as clicking a mouse. The Web is a gateway to other online resources. Email, electronic mailing lists, newsgroups, chat rooms, blogs, and social network sites (SNS) are mostly accessed through a Web site. For example, Facebook is a Web site, and blogs, such as Instapundit, are Web sites.

The Web has changed and along with it the information and entertainment worlds. The Web is altering existing media use habits and the lifestyles of millions of users who have grown to rely on it as a source of entertainment, interactivity, and information.

ZOOM IN 5.2

To see Web pages from the old days, go to WayBack Machine, http://archive.org/web/web.php, a service that brings up Web sites as they were in the past. The archives go as far back as 1996.

Navigating the World Wide Web

Web browsers such as Firefox and Chrome are based on **Hypertext Markup Language (HTML)**, a Web programming language. Hypertext is "non-linear text, or text that does not flow sequentially from start to finish" (Pavlik, 1996, p. 134). The beauty of hypertext is that it allows non-linear or nonsequential movement among and within documents. Hypertext is what lets users skip all around a Web site, in any order they please, and jump from the beginning of a page to the end and then to the middle, simply by pointing and clicking on hot buttons, links, and icons.

FYI: Hypertext Markup Language (HTML)

HTML is the World Wide Web programming language that basically guides an entire document or site. For Instance, it tells browsers how to display online text and graphics, how to link pages, and how to link within a page. HTML also designates font style, size, and color.

Specialized commands or tags determine a document's layout and style. For example, to center a document's title—say, How to Plant a Containerized Rose in Houston—and display it as a large headline font, the tags <HTML><HEAD> <TITLE> are inserted before and after the title, respectively, like this:

<center><H1><I>How To Plant a Containerized Rose In Houston </TITLE> </HEAD></HEAD>

The first set of commands within the brackets tells the browser to display the text centered and in headline bold font. The set of bracketed commands containing a slash tells the browser to stop displaying the text in the designated style.

The HTML source code for most Web pages can be viewed by users. In Firefox, click on the Tools pull-down box in the browser's tool bar and then click on the Web Developer option. Then click on Page Source.

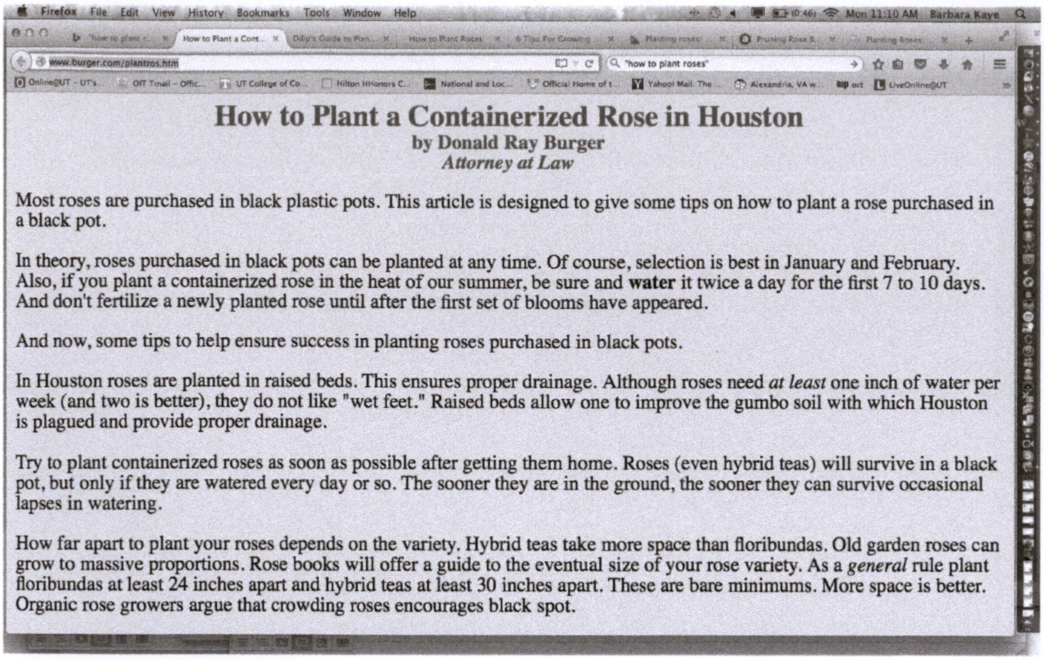

How to Plant a Containerized Rose in Houston
by Donald Ray Burger
Attorney at Law

Most roses are purchased in black plastic pots. This article is designed to give some tips on how to plant a rose purchased in a black pot.

In theory, roses purchased in black pots can be planted at any time. Of course, selection is best in January and February. Also, if you plant a containerized rose in the heat of our summer, be sure and **water** it twice a day for the first 7 to 10 days. And don't fertilize a newly planted rose until after the first set of blooms have appeared.

And now, some tips to help ensure success in planting roses purchased in black pots.

In Houston roses are planted in raised beds. This ensures proper drainage. Although roses need *at least* one inch of water per week (and two is better), they do not like "wet feet." Raised beds allow one to improve the gumbo soil with which Houston is plagued and provide proper drainage.

Try to plant containerized roses as soon as possible after getting them home. Roses (even hybrid teas) will survive in a black pot, but only if they are watered every day or so. The sooner they are in the ground, the sooner they can survive occasional lapses in watering.

How far apart to plant your roses depends on the variety. Hybrid teas take more space than floribundas. Old garden roses can grow to massive proportions. Rose books will offer a guide to the eventual size of your rose variety. As a *general* rule plant floribundas at least 24 inches apart and hybrid teas at least 30 inches apart. These are bare minimums. More space is better. Organic rose growers argue that crowding roses encourages black spot.

FIG. 5.5 Simple text-based Web site http://www.burger.com/plantros.htm
Photo courtesy of Donald Burger

```
<HTML><HEAD> <TITLE>How to Plant a Containerized Rose in Houston by Donald Burger, Houston, TX</TITLE>
</HEAD></HEAD><BODY bgcolor="#FFE4E1"><font size=5 color=brown><center>How to Plant a Containerized
Rose in Houston</font><BR><font size=3 color=brown>by Donald Ray Burger<BR>Attorney at Law</font>
</center>

<p>Most roses are purchased in black plastic pots. This article is designed to give some tips on how to plant a rose
purchased in a black pot.

<P> In theory, roses purchased in black pots can be planted at any time. Of course, selection is best in January
and February. Also, if you plant a containerized rose in the heat of our summer, be sure and water it twice a day
for the first 7 to 10 days. And don't fertilize a newly planted rose until after the first set of blooms have appeared.

<P> And now, some tips to help ensure success in planting roses purchased in black pots.

<P> In Houston roses are planted in raised beds. This ensures proper drainage. Although roses need at least one
inch of water per week (and two is better), they do not like "wet feet." Raised beds allow one to improve the
gumbo soil with which Houston is plagued and provide proper drainage.

<P> Try to plant containerized roses as soon as possible after getting them home. Roses (even hybrid teas) will
survive in a black pot, but only if they are watered every day or so. The sooner they are in the ground, the sooner
they can survive occasional lapses in watering.

<P>How far apart to plant your roses depends on the variety. Hybrid teas take more space than floribundas. Old
garden roses can grow to massive proportions. Rose books will offer a guide to the eventual size of your rose
variety. As a general rule plant floribundas at least 24 inches apart and hybrid teas at least 30 inches apart. These
are bare minimums. More space is better. Organic rose growers argue that crowding roses encourages black spot.

<P>To plant the rose one needs a Stanley utility knife (of the type used for cutting cardboard boxes), a shovel and
a ruler. Place the pots in the desired locations, using the ruler to check spacing. Dig the holes one at a time, not all
at once. This allows minor adjustments in spacing as one proceeds.

<P> Use the rose in the pot to determine how deep to make the hole. The bud union of the rose should be about
two inches above ground level. The bud union is that knot on the stem where the rose was grafted onto the root
stock.
```

<P> There is no bud union on miniatures and old garden roses because they are grown on their own roots. Plant these roses at the same level they are growing at in their pots. Go by the soil level in the pot, not by the rim of the pot.

<P> Make the hole a little bigger than the black pot for ease of planting. Once you are satisfied with the hole, determine which side of the plant is the "front." Do this by rotating the plant while it is still in the pot. Usually one is looking for a fan effect, with a majority of the cane angling back to the rear of the bed.

<P>Now to the tricky part.

<P> If one simply turns the pot upside down and tries to shake the rose out there is a real risk of dislodging the dirt and damaging the fine feeder roots. The utility knife is the tool to use to avoid this.

<P> Take the knife and slowly and carefully cut the bottom out of the pot, using the drainage holes on the side of the pot as a guide. Cut as close to the bottom as possible. Toss this plate-shaped piece of black plastic aside. Next, starting at the bottom cut half way up the "back" side. Place one hand on the dirt at the bottom of the pot and carefully place the pot in the hole with the "front" facing the front of the bed and complete the "back" cut. Then, again starting at the bottom of the pot, cut up the "front" side of the pot. This allows one to carefully ease the black plastic out of the hole without disturbing the plant. Fill in the gaps with the soil you removed to make the hole. Water lightly. Continue with the rest of the roses. That's it.

<P>If you join the Houston Rose Society you can get their list of recommended roses for Houston. <a href="hrs. htm">For information on how to join click here.</a>

<p> The American Rose Society publishes a yearly Handbook for Selecting Roses that gives a numerical evaluation of each rose. Many of the nurseries and bookstores offer rose booklets that discuss selecting roses. I like the ones by Sunset and HP Books.

<P> Also, consulting rosarians from the Houston Rose Society can be found at the better nurseries in town each spring. They will be happy to help you with your selections.

<P>Happy rose growing.

<P>Written by Donald Ray Burger
Attorney at Law

<P><a href="mailto:burger@burger.com"><FONT COLOR="#003300">mail comments to burger@burger. com</a></font size>

<p><a href="http://www.burger.com/roseindx.htm">[Go Back to My Rose Page]</a>

<p><a href="http://www.burger.com/garden.htm">[Go Back to MyGarden Page]</a>

<P><A HREF = "http://www.burger.com">[Go Back to MyHome Page]</A>

FIG. 5.6 HTML source code for the "How to Plant Containerized Roses in Houston" Web page
Photo courtesy of Donald Burger

All Web browsers operate similarly, yet each has its own unique features and thus markets itself accordingly. The competition among browsers has always been stiff, and market share fluctuates. For example, in the mid-1990s, Netscape Navigator led the U.S. market, but by 2003 Internet Explorer (IE) commanded a 90% share of users, which left Netscape with only about a 7% share and Mozilla and other browsers fighting for the remaining market.

A decade and a half into the new millennium, the two pioneering browsers, Mosaic and Netscape Navigator, are now defunct, having been replaced with more sophisticated ones such as Firefox, and newcomer Google Chrome, which is still the most widely used browser. IE's market share has slipped to 31.6% and Firefox is a distant third capturing about 8% of Internet users, with the rest of the online market using several other less popular

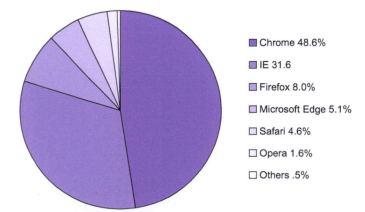

- Chrome 48.6%
- IE 31.6
- Firefox 8.0%
- Microsoft Edge 5.1%
- Safari 4.6%
- Opera 1.6%
- Others .5%

FIG. 5.7 Browser market share
Source: Web Browser Market Share for June 2016. CP Entertainment, 2016

browsers. While still trying to become the number-one browser in the U.S., Google Chrome dominates globally with 62% of the South American market and 49% of the Asian market.

America Online was also a popular way to access online content, but it was more than just a browser. America Online was a proprietary online content provider. Sometimes the term 'walled garden' was used to describe America Online's business model because only subscribers could access certain services. In 1997, almost half of all U.S. homes with Internet access had it through America Online. In 2000, America Online merged with Time Warner, and several years later its name changed to AOL. Through the years, AOL shed its model as a proprietary service and became an Internet portal, similar to Yahoo!. AOL is now a news and information site and offers free access to the Internet, email accounts, and other online resources.

ELECTRONIC MAIL

Electronic mail, or email, is one of the earliest Internet resources and one of the most widely used applications. The first known email was sent from UCLA to Stanford University on October 29, 1969, when researchers attempted to send the word 'login.' They managed to send the letter L and then waited for telephone confirmation that it had made it to Stanford. They then sent the letter O and waited and waited until it arrived. Then they sent the letter G, but due to a computer malfunction, it never arrived. Just as the letter S was the first successful transatlantic radio signal, the letters L and O made up the first successful email message.

Email has come a long way since that first attempt. About 200 billion email messages fly through the global cyberspace every day from more than 4 billion accounts. Approximately 56% of all emails are not wanted (43% spam, 13% other nonessential). Business users send and receive an average of 141 emails per day, accounting for about 70% of all email traffic. The number of personal emails is decreasing as consumers turn to social networking, instant messaging, texting, chatting, and other forms of digital communication.

ZOOM IN 5.3

Contact this spam-fighting organizations and Web sites for more information:

- Coalition Against Unsolicited Commercial Email (CAUCE): www.cauce.org

---

The rise of email and the creation of online marketing and billing, business-to-business email, and new systems for collecting and verifying online signatures have reduced reliance on the United States Postal Service (USPS). The volume of mail handled by the USPS has been declining in recent years. From 2000 to 2009, volume declined from 208 billion pieces of mail to 177 billion and to 155.4 billion in 2014. Because there are now more emails sent each day than letters and packages through the USPS each year, the U.S. Postmaster General had asked Congress to allow the USPS to cut delivery from 6 days a week to 5 days to save money in fuel, vehicle wear and tear, and employee wages and benefits. Congress had planed on addressing the issue in 2017. However, after USPS struck an agreement with Amazon to deliver its packages on Sundays, enacted cost savings measures in operations, and enjoyed a 10% increase in volume of holiday packages and mail, it realized a net profit in 2015 and is set to expand rather than cut back on services.

YOUTUBE

YouTube differs from other interactive Web sites in that it is specifically used for sharing video. Created in 2005 and acquired by Google in October 2006, YouTube is now one of the most widely used sites. Each month, slightly more than 1.3 billion unique users watch more than 6 billion hours of videos and short clips that range from professionally produced movie and television shows and musical performances to amateur content such as homemade videos of cats playing and people doing stupid things and talking about whatever. Every minute, 300 hours of video are uploaded to the site for open-access viewing.

BLOGS

Blogs (originally dubbed Weblogs) are a type of Web site that allows users to interact directly with the blog host (blogger). It is hard to know exactly when the first blog came online, but the term was coined in 1997. Blogs have been part of the Internet landscape since the late 1990s, but they became popular shortly after September 11, 2001. These diary-type sites were an ideal venue for the outpouring of grief and anger that followed the terrorist attacks on the United States. Blogs are now exceedingly popular, and the number of blogs has proliferated to an estimated 250 million.

Blogs are places where online intellectuals, the digital and politically elite, and everyday people meet to exchange ideas and discuss war and peace, the economy, politics, celebrities, and myriad other topics without the interference of traditional media. As such, bloggers are part of a tech-savvy crowd who often scoop the media giants and provide more insight into current events than the traditional media.

Blogs are free-flowing journals of self-expression in which bloggers post news items, spout their opinions, criticize and laud public policy, opine about what is happening in the online and offline worlds, and connect visitors to essential readings. A blogger may be a journalist, someone with expertise in a specific area such as law or politics, or an everyday person who enjoys the exchange of opinions. Blog readers (those who access a blog) post their comments about a current event, issue, political candidate, or whatever they want to the blogger. These comments are often accompanied by links to more information and analysis and to related items. The blogger posts the blog readers' comments and links and adds his or her own opinions and links. Blog readers are attracted to the freewheeling conversation of blogs and to the diverse points of view posted by the blogger and other blog readers.

Moving beyond the typical blog are videologs, also known as vlogs. Vlogs are kind of mini-video documentaries. Commentaries, rants, and raves are all presented as full-motion video. Anyone equipped with a digital video camera and special software can produce his or her own vlog.

FIG. 5.8 Anyone can write a blog.
Photo courtesy of iStockphoto. © AlexValent, image #3220999

Career Tracks: Glenn Reynolds. Blogger. InstaPundit.com

FIG. 5.9 Glenn Reynolds

What is your job?
I'm Professor of Law at the University of Tennessee. I also write a twice-weekly column for *USA Today*. Before coming to Tennessee I went to Yale Law School, clerked for a federal judge on the U.S. Court of Appeals for the Sixth Circuit, and practiced law at the Washington, DC, office of a Wall Street law firm.

What gave you the idea of starting your blog?
In 2000/2001 I was operating some music Web sites and reading some of the very first blogs—Kausfiles, Andrew Sullivan, Virginia Postrel, etc.—when I ran across the Blogger.com site, which Ev Williams was then running from his basement, I believe. I had been using a program called Dreamweaver, which was very labor intensive. With Blogger you just logged on to a Web site, typed in your

post, and hit 'publish.' It was so easy, I decided to give blogging a try. Also, I was teaching Internet law and wanted to keep up with new things. I didn't expect it to last so long, or to have such a large audience.

Describe your blog?
Today, Instapundit gets around 500,000–600,000 page views on a typical weekday. It consists mostly of short posts on things that interest me. Often it's just a link with a brief description, sometimes I'll make a few pithy observations, and sometimes I'll produce a lengthy post incorporating reader comments. It just depends. I tend to put politics, war, and the 'heavier' stuff in the morning, with science, technology, and pop culture in the afternoon, though that's more of a guideline than a rule. Basically, I post whatever I find interesting. I've been bringing in guest bloggers occasionally for years as well.

How is PJ Media connected to Instapundit?
PJ Media is a company I started with a couple of other bloggers back in 2005. I occasionally write for their site, and we share hosting and advertising. They also have editors who will keep an eye on my blog when I'm busy and fix typos, etc.

What do you do on a daily basis to keep your blog going?
On a typical day, the blog will have about 30–50 posts. Since about ten years ago, I've been able to write posts in advance and set a time when they will self-publish, a technological improvement that has made my blogging life much easier. I usually set up a skeleton for the day the night before, so that if I get busy there will still be new stuff appearing every hour. Then I add breaking news and other new stuff that comes up throughout the day as I have time. I get a lot of my links from my readers, and a lot from Twitter, where I follow a lot of people from different circles. Sometimes I travel to events—conferences, protest rallies, etc.—and

report on them via my blog, posting photos and videos that I shoot and edit myself.

What advice do you have for students who might want to start a blog?

Don't start a blog if you can't handle criticism. On the Internet, everyone's a critic. Fortunately, I'm pretty thick skinned. Focus on stuff that interests you, not on copying blogs that you like. The Internet audience is vast, so if you're good at blogging even on a niche target you can build up a big audience. Cultivate a personal relationship with your readers—that's what distinguishes a blog from a Big Media outlet. And don't be afraid to mix up your interests: The joke about my blog is 'come for the politics, stay for the nanotechnology.' Specialization is good, but too much specialization is for insects.

Also, bear in mind that your blog is a marketing tool for your skills. I was writing the occasional op-ed for newspapers before I started blogging (many law professors do that sort of thing), but I've gotten a lot of higher-level writing opportunities because editors know my general take on things, know that I can write, and know that I can write fast. Plus, when I publish a piece and then post a link on my blog, it brings traffic of its own. In today's click-conscious environment, that's a big advantage that comes with having your own audience.

ELECTRONIC MAILING LISTS

Electronic mailing lists are similar to email, in that messages are sent to electronic mailboxes for later retrieval. The difference is that email messages are addressed to individual recipients, whereas electronic mailing list messages are addressed to the electronic mailing list's address and then forwarded only to the electronic mailboxes of the list's subscribers.

Electronic mailing lists connect people with similar interests. Most lists are topic specific, which means subscribers trade information about specific subjects, like college football, gardening, computers, dog breeding, and television shows. Mailing lists are commonly used in the workplace to connect personnel within a department, or managers, or others with similar titles and duties. Many clubs, organizations, special-interest groups, classes, and media use electronic mailing lists as a means of communicating among their members. Most electronic mailing lists are open to anyone; others are available only on a subscribe-by-permission basis. Electronic mailing lists are often referred to generically as listservs; however, LISTSERV is the brand name of an automatic mailing list server that was first developed in 1986.

NEWSGROUPS

Similar to electronic mailing lists, newsgroups bring together people with similar interests. Web-based newsgroups are discussion and information exchange forums on specific topics, but unlike electronic mailing lists, participants are not required to subscribe, and messages are not delivered to individual electronic mailboxes. Instead, newsgroups work by archiving messages that users access at their convenience. Think of a newsgroup as a bulletin board hanging in a hallway outside of a classroom. Flyers are posted on the bulletin board and left hanging for people to sift through and read.

CHAT ROOMS

A chat room is another type of two-way, online communication. Chat participants exchange live, real-time messages. It is almost like talking on the telephone in that a conversation is going on, but instead of talking, messages are typed back and forth. Chat is commonly used for online tech or customer product support. Chat is very useful for carrying on real-time, immediate-response conversations.

INSTANT MESSAGING

Instant messaging (IM) is another way to carry on real-time typed conversations. Different from chat-room conversation, which can occur among anonymous individuals, instant messaging takes place among people who know each other. Instant messaging is basically a private chat room that alerts users when friends and family are online and available for conversation. Because users are synchronously linked to people they know, IM has a more personal feel than a chat room and is more immediate than email. Some IM software boasts video capabilities to create a more realistic face-to-face setting.

'Instant messaging' differs from 'text messaging' in that IM is Internet-based on computer-to-computer connections, whereas texts are sent to and from cell phones and other handheld devices. Instant messaging services were once the domain of America Online (AOL) and Microsoft, but Skype, Facebook, and Google have since added IM applications.

SOCIAL NETWORK SITES

Millions of online users are drawn to social network sites (SNS) as a means of keeping in touch with friends and family and building a network of new 'friends' based on shared interests and other commonalities, such as politics, religion, hobbies, and activities. Although social network sites hit their stride in the late-2000s, they have been in existence since the late 1990s. SNS such as sixdegrees.org, AsianAvenue, BlackPlanet, and LiveJournal were the precursors to the second wave of SNS, which includes MySpace and Facebook. About three-quarters of online users have a profile on at least one of the estimated thousands of SNS. But of all SNS, about 200 have emerged as the most popular, with Facebook leading the pack. The number of social network users over the age of 50 almost doubled between 2009 and 2010, but they are still more likely to network with friends and family using email, whereas those between the ages of 18 and 29 are just as likely to use both SNS and email. The growing popularity of social network sites is signaling a shift in how consumers are using the Internet.

ZOOM IN 5.4

Go to any search service, find a list of blogs, newsgroups, and social network sites, and click on some you are unfamiliar with. Join some you find interesting.

www.ebizmba.com/articles/blogs

www.bloggingfusion.com

www.google.com/press/blogs/directory.html

http://traffikd.com/social-media-websites/

FYI: The Internet of Things

The Internet of Things (IoT) is one of the newest buzz phrases. The term refers to the interconnections of everyday objects through the Internet. For example, smartphone-programmable thermostats, washers and dryers, and lights, cars with built-in sensors, medical and fitness tracking devices are IoT technologies. Intel chips that drive IoT generated 2.1 billion in revenue in 2014, and IoT gadgets were all over the 2015 Consumer Electronics Show.

TWITTER

Twitter took only a few years to infiltrate the social network world. Twitter has moved beyond connecting small groups of friends to being a way to reach millions of followers at once. Celebrities, politicians, and companies use Twitter to promote themselves, their ideas, and their products to their followers. Although messages, known as 'tweets,' are limited to 140 characters, they are long enough to make a point, and it is thrilling to be part of a conversation that includes the rich and famous. Twitter is defined as a microblog because of the limited length of a tweet. But microblog or not, there is nothing micro about the size of some users' networks, which reach hundreds of thousands or even millions of followers. The downside of Twitter is that once a message is out there, it cannot be taken back. So one wrong word or misstatement can and probably will come back to haunt the author. Some users have lost their jobs or have been socially ostracized because of something they said on Twitter. Although proponents love the collection of voices, critics claim that most tweets are nothing but useless babble that wastes time.

FYI: Most Popular Online Activities (2012)

Percentage of Users

Sending or reading email	92%
Using search engines to find information	89%
Search for a map or driving instruction	86%
Research a product or service	81%
Check the weather	80%
Look for health or medical information	73%
Get travel information	73%
Read news and current affairs	73%
Shopping	71%
Visit a local, state, or federal government site	66%
Make travel reservations	64%
Surf the Web for fun to pass the time	62%

Source: Pew Internet and American Life Project

THE WORLD WIDE WEB AND THE MASS MEDIA

Many Internet users are abandoning traditionally delivered radio and television for Internet-delivered programming. Audio Web sites and services like iTunes offer an array of musical choices well beyond local over-the-air radio, and it is so much more convenient to watch video clips or full-length programs online than to wait for a television program to come on.

One of the main concerns about online content does not regard delivery but the nature of the content itself. Decentralized information dissemination means that online materials are often not subjected to traditional methods of source checking and editing. Thus, they may be inaccurate and not very credible.

When listeners tune to radio news or viewers watch television news, they are generally aware of the information source. They know, for instance, that they are listening to news provided by National Public Radio (NPR) or watching the ABC television network. In addition, broadcast material is generally written and produced by a network or an independent producer or credentialed journalist. Audiences rely on these sources and believe them fairly credible. But mainstream media suffer from low public opinion. About three-quarters of people think the news media are influenced by powerful groups and favor one side and that reporters try to cover up their mistakes.

Internet users, especially novices, cannot be sure that what they read and see online is credible and accurate, especially if it is posted by an unfamiliar source. Many Web sites are hosted by reliable and known sources, such as CNN and NBC. However, anyone can produce a Web page or post a message on a blog, a social network site, bulletin board, or chat room, bringing into question source credibility. Such **user-generated content (UGC)** now accounts for more online information than that produced by journalists, writers, and other professionals. Internet users must grapple with the amount of credence to give to what they see and read online. Thus, it is a good idea to sift through cyber information very carefully. It is the user's responsibility to double-check the veracity of online information. Users should be cautious of accepting conjecture as truth and of using Web sources as substitutes for academic texts, books, and other media that check their sources and facts for accuracy before publication.

ONLINE RADIO TODAY

Almost all of the nation's 15,400 broadcast radio stations have some sort of Internet presence. Online radio offers so much more to its audience than broadcast stations, which are limited by signal range and audio-only output. Internet radio delivers audio, text, graphics, and video to satisfy a range of listener needs.

Early online radio was difficult to listen to and hard to access, and it took an excruciatingly long time to download a song. Moreover, the playback was tinny and would often fade in and out. Few Internet users had computers powerful enough to handle audio, and few stations were streaming live content. In the late 1990s, online audio was still so undeveloped that only 5 to 13% of Web users reported that the amount of time they spent listening to over-the-air radio had decreased in favor of online audio.

Although some radio station Web sites retransmit portions of their over-the-air programs, other audio sites produce programs solely for online use and are not affiliated with any broadcast station. Many audio sites offer very specific types of music. A site may feature, for example, Swedish rock, Latin jazz, and other hard-to-find music. There are hundreds of audio programs available on search services and radio program guides that give the time and URL of a live online program or a long list of music-oriented Web sites.

Radio station Web sites are promotional by nature, but most also offer local news and entertainment and have the capability of transmitting audio clips or delivering live programming. However, doing so has hit a snag. Streamed audio is a great way for consumers who do not have access to a radio or who live outside the signal area to listen to their favorite station. But after many complaints from record labels, artists, and others, the Library of Congress implemented royalty fees, requiring Webcasters to pay for simultaneous Internet retransmission. The fees vary from per-song costs to a percentage of gross revenues, depending on the online pricing model. Whatever the exact amount charged, labels and artists claim this fee is too low, while online music providers say it is too high.

The advantage of both over-the-air radio and online radio is that they can be listened to while working or engaging in other activities. But online radio has a distinct advantage over AM/FM radio—users can listen to stations from all over the world, and they can choose from a huge selection of music genres. If none of the local radio stations play a listener's favorite music, he or she can go online and find many stations that do. So Internet users may not be turning their backs on music radio but rather abandoning the old over-the-air delivery for online delivery, which provides clear audio at convenient times and can be set to play their favorite types of music.

Over-the-air broadcast radio stations are concerned that their listening audience is abandoning them for online music. However, despite strong growth in online radio listenership, over-the-air AM/FM listeners remain loyal. Although some studies claim that radio listening is decreasing among Internet users, others claim that radio is gaining favor among those who go online. Digital delivery adds to audio listenership, but it is not a substitute for over-the-air delivery, except perhaps for adults aged 18 to 24, half of whom said they spend less time listening to broadcast radio in favor of online radio. Local news and information is the strongest draw to broadcast radio, and it is the main reason half of U.S. adults tune in. Research indicates a trend in which broadcast radio is relied on for nonmusic programming, whereas digital radio is used more for music, especially for learning about new tunes. Indeed, about two-thirds of online users say they learn more about new songs and bands from the Internet than from the radio, and almost one-half often look at the media player/computer to see the name of the song or artist.

The audience for streaming Internet radio continues to grow as sound quality improves and as listeners embrace newer, more portable playback devices such as laptops, smartphones, and tablets, and customizable services such as Spotify and Pandora. There were about 93 million Internet radio listeners (30% of the U.S. population) in 2010, and just 5 years later, that number rose to almost 170 million users (52%).

> ### FYI: U.S. Audio Listening: 28 Hours per Week
>
> Broadcast radio (AM and FM): 16 hours
> Downloads, vinyl, CDs, tapes: 5 hours 30 minutes
> Streaming services: 3 hours 20 minutes
> Satellite radio: 2 hours and 10 minutes
> Podcasts: 30 minutes
> Other (e.g., audiobooks): 30 minutes
>
> Source: Stutz, 2014

Total U.S. music revenue in 2013 hovered at about $7 billion, about half of what it was before digital music changed the industry forever. As of 2013 there were about 400 licensed digital music services, such as iTunes, Rhapsody, and Groove Music, which account for about $1.6 billion in digital track sales (1.3 billion units) and $1.2 billion in digital album sales (118 million units). But digital downloads are giving way to custom streaming services. The portion of revenue generated by streaming jumped from 3% in 2007 to 21% in 2013 to 33% in 2015. Unfortunately, the revenue generated by online downloads and streaming does not make up for the downturn in CD sales, which have been decreasing since 2000.

File sharing and downloading, both legal and illegal, spawned a new way to obtain and listen to music and other audio files. Gone are the days when a song or audio clip could take up to an hour to download and then be so garbled that it was not worth the effort. Cybercasting, either by live streaming or by downloading, is a great way for little-known artists who have a hard time getting airplay on traditional stations to gain exposure. Additionally, listeners can easily sample new music, and links are often provided to sites for downloading and purchasing songs and CDs. Downloading music from online sites has boomed in recent years, especially since the advent of the iPod and other portable digital music devices.

And the Recording Industry Association of America is still going after music pirates. Since the first round of lawsuits, RIAA has filed more than 30,000 more. In the summer of 2009, a 32-year-old Minnesota mother of four and a Boston University graduate student were both fined for illegally downloading music. The woman was fined $80,000 for each of 24 songs she illegally downloaded, for a total judgment of $1.9 million—for songs that only cost 99 cents each; the graduate student was fined $675,000 for illegally downloading and sharing 800 songs between 1999 and 2007. Both fines were later settled for lesser amounts.

The RIAA is serious about getting its message out: "Importing a free song is the same as shoplifting a disc from a record store" (Levy, 2003b, p. 39). The RIAA's legal and educational efforts have paid off; just over half of those who regularly downloaded music illegally claimed that the crackdown has made them less likely to continue to pirate music. Prior to the initial lawsuits, only about 35% of people knew file sharing was illegal; now about double that percentage is aware of what constitutes illegal downloading. Further, the percentage of U.S. Internet users who downloaded music illegally dropped from 29% to 9% from 2000 to 2010. Although some online users might not think there is much harm in illegally downloading a song here and there, the scope of the problem shows otherwise—illegal downloads cost the U.S. economy $12.5 billion annually.

> ### FYI: Challenges of Internet Radio
>
> - With a slower connection, the delivery may be choppy, with dropped syllables and words.
> - High treble and low bass sounds are diminished as the data squeeze into available bandwidths.
> - Without external computer speakers, quality of sound is often like listening to AM radio or FM mono radio, at best.
> - A high-speed cable modem or direct Internet access with increased bandwidth is needed to hear audio of FM stereo quality and even CD-quality sound.
> - Downloading audio files can still take longer than a user is willing to wait.
> - The number of simultaneous listeners is often limited.
> - A music player, tablet, or smartphone is needed for portability.

> ### FYI: Benefits of Internet Radio
>
> - Web audio files can be listened to at any time.
> - Netcasts can be listened to from anywhere in the world, regardless of their place of origin.
> - Online radio can be heard and seen. Song lyrics, rock bands in concert, and news items can be viewed as text, graphics, or video.
> - Online radio supports multitasking, or the ability of users to listen to an audio program while performing other computer tasks and even while surfing the Web.
> - Online music services customize individual playlists to a listener's preferences.

> ### ZOOM IN 5.5
>
> Get both sides of the story about file sharing. First, go to www.riaa.com to learn the point of view of a music-licensing agency. Then go to www.eff.com for information from organization that does not consider online file sharing a crime.

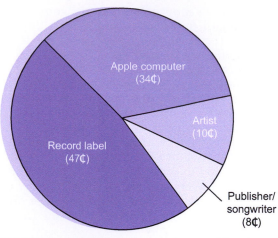

FIG. 5.10 99 cents per download. So where does your 99 cents go when you pay to download a song from a legal music site? Here's an example from Apple's iTunes.
Source: Spors, 2003

FYI: Top Digital Song Download Sales 2015

Artist, Song	Total Sold
Mark Ronson featuring Bruno Mars, "Uptown Funk"	5,529,000
Ed Sheeran, "Thinking Out Loud"	3,976,000
Wiz Khalifa featuring Charlie Puth, "See You Again"	3,801,000
Adele, "Hello"	3,712,000
Maroon 5, "Sugar"	3,343,000
Walk the Moon, "Shut Up and Dance"	2,986,000
Fetty Wap, "Trap Queen"	2,730,000
OMI, "Cheerleader"	2,698,000
The Weeknd, "The Hills"	2,586,000
Taylor Swift featuring Kendrick Lamar, "Bad Blood"	2,580,000

Source: Caulfield, 2016

FYI: Adele Sets a Record

Adele's "25" holds the record for the most downloads in a calendar year with 2.31 million sales in 2015. This record is especially amazing considering that the album was not released until mid-November 2015.

Podcasting

In addition to music, podcasts are another type of audio file that has caught the ears of online listeners. Podcasts are audio-only programs produced by radio and television stations and networks, talk show hosts, celebrities, professors, and anyone who thinks he or she has something interesting to say about topics such as movies, music, popular culture, cooking, and sports. Many corporate media Web sites—such as npr.org, espn.com, and abc.go.com—offer hundreds if not thousands of podcasts.

One of the first times that the term 'podcasting' was mentioned was in an issue of the British newspaper *The Guardian* in 2004. The term is a combination of the acronym 'personal on demand' (POD) and the word 'broadcasting.' According to Jason Van Orden, author of Promoting Your Podcast, a podcast is "a digital recording of a radio broadcast or similar programme, made available on the Internet for downloading to a personal audio player" (Van Orden, 2008).

The purpose of a podcast is not just to transfer readable information to audio format but to provide users with updated and sometimes live online audio. Different from a typical audio file, a podcast is created through **Real Simple Syndication (RSS)** feed, a standardized format used to publish frequently updated materials, such as news headlines and blog entries. Technically, a podcast is an audio file embedded in an RSS feed. Podcasts are subscribed to through a software program, known as a podcatcher, that syncs podcasts to a computer-connected MP3 player. Podcasts free listeners from media, business, and personal schedules, and they can be archived on a computer or burned onto a CD.

With podcasting as the new downloading rage, the question arises as to how media organizations consider people who download music and other audio programs. If someone subscribes to the podcasts of National Public Radio's *Talk of the Nation*, should they be considered a radio listener or a podcast listener? Perhaps the consumption of audio content is what matters rather than the mode of delivery. As the number of podcasts and podcasters surge, the media have much to consider; they need to redefine themselves and rethink their methods of content delivery and how to measure their online and mobile audiences.

It is impossible to know the definitive number of podcasts, but it is estimated there are about 250,000 unique podcasts in over 100 languages, of which about 115,000 are recorded in English. Apple reports that in 2013, the number of podcasts downloaded through iTunes reached 1 billion. Class lectures, interviews, tutorials, stories, training, and speeches are often podcast. Some of the most common podcast topics include, music, technology, comedy, education, and business.

About 33% of the U.S. population has listened to a podcast. With about 320 million people in the United States, that percentage translates into 105 million listeners—an incredibly high number considering that podcasting first appeared online only in 2004. Slightly more males (54%) than females (46%) listen to podcasts, and just over one-half (52%) of the podcast listening audience is between the ages of 12 and 34. Further, podcast listeners have a higher-than-average annual income.

One of the most critically acclaimed podcasts to date first appeared in October 2014. *Serial* delves into police investigation of the 1999 death of an 18-year-old woman in Baltimore, MD. The victim's boyfriend was found guilty and sentenced to life plus 30 years despite his pleas of innocence. Each of the 12 episodes dramatically unravels the story of the woman's strangling and painstakingly retraces each step of the investigation, including scrutinizing witnesses' testimony and analyzing court documents. The gripping audio documentary becomes especially compelling as the boyfriend's guilt comes into question—was he wrongly convicted? *Serial* has caught the ears of about 40 million listeners, including lawyers for Project Innocence, who will be retesting DNA samples found on the victim's body.

FYI: Top 10 Podcasts 2015 (by number of downloads)

Fresh Air (NPR)

Serial

Stuff You Should Know

This American Life

Planet Money

Radiolab

Freakanomics Radio

The Joe Rogan Experience

The Nerdist

The Dave Ramsey Show

Source: Eadiccio, 2016

USING TELEVISION, USING THE WEB

The Internet caught on largely because it is used the same way television is watched—sitting in front of a screen that displays text, graphics, audio, and video. Switching from Web site to Web site is, in some ways, similar to changing television channels. When Internet users switch from one Web site to another, they do so by typing in a **uniform resource locator (URL)**, by simply clicking on a link, or by clicking on their browser's Back and Forward buttons, which function like the up and down arrow keys on a television remote-control device. Even the lingo of Web browsing is borrowed from television. Commonly used terms such as 'surfing' and 'cruising' are used to describe traversing from one Web site to another and are also used to describe television channel-switching behavior.

Now, new combo computer monitors/televisions further integrate the utility of the Internet and television. Many televisions (smart TVs) come with built-in Wi-Fi, a Web browser, icons that link to Hulu and Netflix, and connections to DVD players, DVRs, and computers. On some televisions, the picture is shown in one window and computer functions such as email in another window on the same screen. Television and the Internet are functionally combined in many ways and inseparable in the minds of many viewers.

FYI: Top Multiplatform Properties

Monthly Rankings: January 2016

Property	Unique Visitors
Google Sites	245.2 million
Facebook	207.6 million
Yahoo! Sites	205.4 million
Microsoft Sites	202.9 million
Amazon Sites	183.0 million
AOL Inc.	169.9 million
Comcast NBC Universal	161.0 million
CBS Interactive	156.4 million
Apple Inc.	139.8 million
Mode Media	139.2 million

Source: comScore, 2016

Audience Fragmentation

Traditional broadcast television programs were created to appeal to millions of viewers. Then cable television came along and altered programming by introducing narrowcasting, in which topic-specific shows were specifically created to appeal to smaller but more interested and loyal viewers. The Internet has taken narrowcasting a step further by cybercasting information targeted to smaller audiences. In this way the Web can be thought of as a "personal broadcast system" (Cortese, 1997, p. 96).

Television industry executives are worried that the Web is further fragmenting an already fragmented audience. In the early days of television, viewership was mostly shared among three major broadcast television networks, which were and still are fiercely competitive. (Even a small gain in the number of viewers means millions of dollars of additional advertising revenue). Cable television, which offers hundreds of channels, has further fragmented the viewing audience. Now, as viewers increasingly subscribe to cable and satellite delivery systems and turn to the Web as a new source of information and entertainment, the size of television's audience is eroding further—and, with it, potential advertising revenue.

To help offset audience loss and to retain current viewers, most television networks have established Web sites on which they promote their programs and stars and offer insights into the world of television. Many new and returning television shows are heavily promoted online with banner ads, a Web site, and sometimes blogs, Twitter, bulletin boards, and chat rooms. As one television executive said, "The more they talk about it, the more they watch it" (Krol, 1997, p. 40). Television program executives view the Internet as a magnet to their televised fare.

Selecting Television Programs and Web Sites

Wielding a television remote-control device gives viewers the power to create their own patterns of television channel selection. Some viewers quickly scan through all the available channels; others slowly sample a variety of favorite channels before selecting one program to watch. Television viewers tend to surf through the lowest-numbered channels on the dial (2 through 13) more often than the higher-numbered channels. Unlike television, however, the Web does not have prime locations on its 'dial,' so one Web site does not have an inherent location advantage over another. However, sites with short and easy-to-remember domain names may be accessed more frequently than their counterparts with longer and more complicated URLs.

As users become adept at making their way around the Web, customized styles of browsing are emerging. Some users access only a set of favorite sites; others are not loyal to any particular site. When viewers sit down to watch television, they usually grab the remote control and start pushing buttons. But instead of randomly moving from one channel to another, most viewers have developed a favorite set of channels they go through first. Most viewers' channel repertoire consists of an average of 10 to 12 channels, regardless of the number of channels offered by their cable or satellite systems.

Selecting a television program or Web site also depends on the mood and goals. Sometimes viewers select television programs and Web sites instrumentally; they are looking for specific information. They pay attention to the content and purposely switch channels or move from site to site because they have a goal in mind. At other times, the television or Internet is turned on out of habit or to pass the time. Viewers ritualistically surf through the channels or click on links, paying attention to whatever seems interesting. The Web generally requires more attention than television, but many sites are designed for users to kick back and become a 'Web potato' instead of a 'couch potato.' Longer video and audio segments and text are designed to minimize scrolling and clicking and to keep users glued to the screen for longer periods of time.

Television-viewing behaviors have transferred to using the Web. As users become more familiar with online content, they develop their own repertoire of favorite sites that they link to regularly. These preferred sites are easily bookmarked for instant and more frequent access.

Similar to how television adapted its programs from radio, many Web sites reproduce material that has already appeared in traditional media. For instance, many news sites and other media-oriented sites are made up largely of text taken directly from the pages of newspapers, magazines, brochures, radio and television scripts, and other sources. In some cases, however, materials are adapted more specifically to the Web. The text is edited and rewritten for visual presentation and screen size, and short, summary versions may be linked to longer, detailed ones. Bold graphic illustrations, audio and video components, and interactive elements also enliven Web pages and give them a television-like appearance.

Television-oriented sites are still typically used to promote televised fare and are among the most popular Web sites, excluding search services. Each of the big three television networks (ABC, NBC, CBS) and dozens of cable networks tried out the Web for the first time in 1994, and all now have their own Web sites.

FYI: Top U.S. Online Video Content Properties (January 2016)

Property	Total Unique Viewers
Google Sites (including YouTube)	233.9 million
Facebook	174.0 million
Yahoo! Sites	84.2 million
Vimeo	47.1 million
CBC Interactive	45.3 million
Comcast NBC Universal	45.0 million
VEVO	43.6 million
Warner Music	39.4 million
Maker Studios	38.9 million
Microsoft sites	36.5 million

Source: comScore, 2016

Is the Web Stealing Television's Viewers?

Just as radio's audience was encroached on by broadcast television and, in turn, broadcast television's viewers were drawn to cable, many fear that the Web is slowly attracting users away from television. Time spent on the Web is time that could be spent watching television. The advertising industry is especially concerned that when consumers watch programs online, they do not see the commercials that ran in the original broadcast.

Until about the mid-2000s, television was a bit protected from Web pillaging for several reasons. Online technology could not deliver the same clear video and audio/video syncing as television, not all television programs were available online in their entirety, and television-quality original Web programs were scarce. Some Web sites, however, did offer short episodes (Webisodes) of online-only programming.

But that was yesterday. Now, given all of the advances in full-motion video technology, almost every television show is posted on Web sites such as Hulu.com and YouTube.com. Given the Internet's vast storage capabilities, it is getting easier to go online and watch any program without being tied to a television schedule or a to a cable or satellite company. Viewers who have dropped their cable service and watch television programs solely online are known as '**cord-cutters**,' and those who have never

bringing down the barriers that once separated a television set from the Internet and are instead setting up a symbiotic relationship that makes it unnecessary to have both a traditional television set and a separate computer with Internet access.

FIG. 5.11 With the Internet, you can watch online video anywhere.
Photo courtesy of iStockphoto. © adamdodd, image #4481014

subscribed are called 'cord nevers.' Reports vary on how many television viewers fall into either category. About 15% of U.S. viewers have cut the cable cord, joining the 9% who have never subscribed to cable or satellite. But yet 15% of cable customers say they will never cut the cord. Generally, about 48% of adults watch more programs online than through a cable or satellite service, but that percentage jumps to about 70% for 18- to 35-year-olds.

The cable industry is very concerned about reaching Millennials. Between 13% and 19% of viewers 18 to 35 get all of their television from Netflix, Hulu, and other such services, compared to about 5% of viewers over 35. Further, Millennials strongly influence household cable decisions. Only about 6.5% of U.S. households have cut the cable cord compared with 12.4% of households that include at least one person 18 to 34 years old. Interestingly, 40% of cord cutters and cord nevers do not subscribe to home broadband/Internet services, meaning they watch streamed programs on their smartphone or tablet.

Millions of viewers are watching television online every day. The top video sites by time spent are Netflix, YouTube, Hulu, and MegaVideo. Viewers are most likely to watch full-length movies on Netflix than on the other services, and thus they spend about 10 hours per month using the service, whereas they spend a little over 2 hours per month watching video on the other sites. Compared to about 135 hours a month watching traditional television, online viewing still has a way to catch up, and no one knows for sure if the gap will narrow and/or how long it will take. Even though Internet technology is the catalyst for changing the culture of television viewing, people love television, and so far it has proven resilient against the lure of online delivery. New smart television sets with built-in Internet and links to Netflix and YouTube are

FYI: A Sampling of Radio and Television Online

Radio Station Locator	www.radio-locator.com
Radio Tower	www.radiotower.com
National Public Radio	www.npr.org
Radio Free Europe	www.rferl.org
Voice of America	www.voa.gov
ABC Television	www.abcnews.com
BBC iplayer Radio	www.bbc.co.uk/radio
BBC Radio iplayer Radio	www.bbc.co.uk/radio1
CBS Television	www.cbs.com
CBS Sports	www.cbssports.com/
FOX	www.fox.com
NBC	www.nbc.com
ION Television	www.iontelevision.com
The CW	www.cwtv.com
CNN Interactive	www.cnn.com
Comedy Central	www.cc.com
Discovery Channel	www.discovery.com
E! Online	www.eonline.com
ESPN SportsZone	espn.go.com
HBO	www.hbo.com
MSNBC	www.msnbc.com
The Learning Channel	www.tlc.com
Lifetime	www.mylifetime.com
MTV	www.mtv.com
Nick-at-Nite	www.nickatnite.com
PBS	www.pbs.org
SyFy	www.syfy.com
TV Guide Online	www.tvguide.com
Newslink	www.newslink.org/broad.html
10News.com (San Diego)	www.thesandiegochannel.com
LasVegasNow	www.lasvegasnow.com
KPIX-TV	www.cbs5.com
WBIR-TV	www.wbir.com
WBOC-TV	www.wboc.com
WFLA-TV	www.wfla.com

The Web and Television News

The Web is an ideal venue for reporting the news in that it is free from the constraints of time and space. Most news organizations gather more information than they have the time to air or the space to print, but they can post an unlimited amount of information online. Online news is sometimes written as a summary with a link to an in-depth version as well as related stories. Late-breaking news can be added almost instantaneously, and stories can be updated and amended as needed. For example, TMZ.com was the first news outlet to announce Michael Jackson's death. It even posted the Los Angeles coroner's office report of Jackson's death 6 minutes before anyone else. Critics claim that the posting was premature and TMZ just got lucky that Jackson did indeed die, but the organization countered that it has an inside track because of its large network of reliable sources.

Online news reports are richly presented in audio, text, video, and graphic formats. Whereas radio is bound to audio, television to audio and video, and newspapers to text and graphics, on the Web traditional media cross over into other forms of presentation. Radio news on the Web is presented visually, television news with text, and newspaper stories with audio. The characteristics that distinguish radio and television news presentation from one another blur on the Web.

The Web has become increasingly important to both television networks and affiliate stations as an alternative means of distributing around-the-clock and up-to-the minute information. Television stations work with Internet companies to help develop and maintain their online news presence. For example, CNN partnered with Internet Broadcasting Systems for access to local news content and locally placed advertisements.

Online media are having a difficult time distinguishing themselves from their competitors; therefore, media sites are turning to brand awareness as a strategy for getting their site selected over another, such as cbssports.com over espn.com. The media aim to transfer their strong brand names, such as NBC and CNN, to the online environment. Marketers speculate that early users of the Internet were not brand sensitive but rather tried out many different sites before returning to the ones they liked the best. Users who are more Web tentative tend to be brand loyal

FIG. 5.12 The Internet keeps us connected to the latest news and events. *Photo courtesy of iStockphoto. © sjlocke, image #9652079*

and stick to known sites. In other words, a Web surfer who regularly watches CNN on television may be more likely to access cnn.com than another Web news site.

Mainstream media aside, online-only sites such as Huffington Post and Reddit are highly respected sources of news and information. Reddit is positioned as a news and social network site on which registered users post content. Reddit is a giant message board that is based on free speech and gives readers the authority to rate submissions with an 'up' or 'down' vote. Items with the most 'up' votes appear most prominently. Reddit resonates with the public—about 231 million unique visitors make 8.2 billion page views a month. Many such online news sites are the major source of news for Millennials. Although adults of all ages rely on various sources of news—mainly television and newspapers, both digital and not—along with the Internet, Millennials are the least interested in mainstream media.

Regardless of whether news is generated by radio or television, online content delivery is evolving as the Web gains recognition as a distinct medium. Traditional ways of reporting and presentation are giving way to dynamic and interactive methods that hold promise for engaging and drawing new audiences to the Web.

PRIVACY

The right to privacy is an ongoing concern that refers not only to the privacy of behavior behind closed doors but also to the right to privacy of personal information—namely, information about how people spend their money and their time. Generally, people do not want personal information like their phone numbers, addresses, credit card numbers, and medical information released without their consent.

Privacy is particularly relevant in the context of the Internet. Companies and other organizations track online surfing by placing cookies on a Web site visitor's computer. **Cookies** reveal what sites have been visited, monitor online purchases and product preferences, and collect geographic and demographic information. Moreover, because the Internet is commonly used for commerce between individuals, such as eBay auction transactions, as well as between individuals and commercial entities, private information is sometimes given (or sold) to third parties. For example, credit card numbers intentionally given to one person or company could be stolen and used surreptitiously by someone else.

Some groups have endorsed encryption technology, which scrambles information and makes it unintelligible to third parties. While this seems like a simple enough way to guarantee privacy, the U.S. government has restricted the availability of powerful encryption programs. In fact, the government believes that it should have access to personal data to protect citizens against terrorist plots and that encryption technology hinders its information-gathering efforts.

Although many online users object to government scrutiny, they do not seem to mind that Google, Yahoo!,

FIG. 5.13 In 1993, cartoonist Peter Steiner created for *The New Yorker* magazine a one-panel cartoon now known as "On the Internet Nobody Knows You're a Dog." The cartoon illustrated that Internet users at the time had almost total anonymity. The cartoon became the most reproduced in the magazine's history. The lower cartoon above, created by cartoonist Tom Toles, parodies the original to show how much the Internet had changed in seven years from "nobody knowing you're a dog" to "everybody knowing you're a dog," even your breed, hobbies, friends, online activities, and other personal information (Cavna, 2013).
Toles ©2000 The Buffalo News. Reprinted with permission of Universal Press Syndicate. All rights reserved.

Facebook, and other Internet giants collect and store millions of bits of private information for commercial and marketing purposes. The importance of data collection is underscored by its financial impact on Internet commerce. Online companies utilize consumer data to sell targeted ads, which in turn lead to the sale of products and services. Without the ability to collect data, the cycle would break and revenue would drop precipitously. One study found that at least one-half of the Internet's economic value stems from consumer data.

Another privacy issue is the use of an individual's image without his or her permission. Celebrities and public officials relinquish much of their right to privacy when they gain popularity or become elected to public office. Private citizens are entitled to more protection, however. Recently, videos of individuals in partying mode have been marketed widely with great success (e.g., *Girls Gone Wild*). The individuals shown in the videos do not receive profits, raising the issue of whether they granted permission to use their images. Thus far, the courts have held that individuals give up some of their right to privacy when they are in public places. The wide availability of video cameras has made many people aware of the fact that their public behavior may become everyone else's entertainment.

FYI: Video Gone Wild

Watch out when posting video on YouTube. If you will be embarrassed if others see it, do not do it. Look what happened to the two Domino's Pizza workers who foolishly sneezed on and otherwise tainted food they were preparing for delivery and captured it all on video. The workers claim they were just fooling around, but once the video hit YouTube, disgusted viewers spread the word all over the Internet. In the end, the workers were fired and the pair was arrested. Additionally, Domino's reputation was harmed, and it filed civil charges against the former employees.

See the original video at www.youtube.com/watch?v=oMO_uysMOXU

Baby Boomers and GenXers are probably the last two generations to know what it means to have private information kept private, whereas Millennials have grown up with their lives an open book for anyone to read. But after years of privacy breaches, hijacked identities, and embarrassing social media photos, a security uprising is brewing. No matter what generation they are from,

FIG. 5.14 College students are the most avid of computer and Internet users.
Photo courtesy of iStockphoto. © track5, image #4758492

online users are embracing 'disappearing' message apps, like SnapChat, and they are pressuring social media like Facebook to allow them greater control over what information is public and what is for friends only. Users are diligently setting up encryption tools, creating stronger passwords, enabling additional authentication tools, and just being more careful about what they post about themselves and what they upload to the cloud. Even though 86% of online users have taken steps to secure their privacy, 60% of them believe that online anonymity is an impossible dream. Moreover, the more people disclose about themselves online the more they desire privacy. The paradox of what they do versus what they say they want is likely attributable to getting little back for what they have made known about themselves.

Some relief is on the way, at least for Europeans. The European Court of Justice ruled in 2014 that people have a right 'to be forgotten' by the Internet. The Court asserted that search engines like Google must erase reputation-harming search results generated by inputting someone's name. The ruling limits the Internet's ability to archive personal information indefinitely, especially if it is incorrect or harmful, and it places the right to privacy above the public's right to know. It should be clear, however, that the ruling does not directly require that information be taken off of a site; it applies only to the search results and the links that direct users to the information.

Legal experts claim that such a ruling in the U.S. would be regarded as censorship and clash with the First Amendment. In today's world, it is impossible to remove or correct information from sites that are not in a user's direct control. Anyone can say anything, post false data, or paint a misleading picture, but there is almost nothing a victim can do about it. Unfortunately, it is doubtful that the U.S. will adopt a European type of a 'right to be forgotten.'

SEE IT LATER

The Internet is changing so rapidly that what is written here in the 'See It Later' section of this chapter may very well be more appropriate for the 'See It Now' section by the time the book is published. It is not that the publishing process is slow—it is more that digital technology and digital culture move at lightning speed.

The traditional media are struggling to keep up with the Internet. As radio listeners and television viewers come to prefer online delivery, the media must offer online content. But just putting up content is not enough; the media have to figure out how to survive in the digital age.

Radio has survived broadcast television, cable television, records, cassettes, CDs, and music videos, but can it survive the Internet? Even though the audience for over-the-air radio is still huge, broadcasters are not doing very well financially. As advertisers are moving online, so must broadcasters. The future is "broadcasting to an audience of one" (Internet is future of the radio, 2009). Radio must take full advantage of digital technology that allows users to be their own programmers. MP3 players and streaming are merely new ways to deliver audio and should not be viewed as the radio receiver's enemies but rather as ways to reach an audience. Indeed, those who listen online are more apt to tune in to over-the-air radio, which indicates that radio listeners are not abandoning radio but are merely turning to its online counterparts.

Because nearly three-quarters of all radio listening occurs in vehicles, broadcast radio has pretty much been the only way to tune in. CDs, MP3 players, and satellite radio compete for drivers' ears as well. Almost one-quarter of people use their cell phone to listen to music while in the car, and about 55% of 18- to 24-year-olds depend on their iPod/MP3 player. And now, Web radio is calling shotgun. It is still not easy to connect to Web-streamed audio, but it can be done with an Internet-connected smartphone that is hooked through car speakers and by installing audio software, such as Pandora or Slacker. But too much Web listening runs the risk of using up the monthly cell phone gigabyte quota. Common use of in-car Web radio is still years away, though about 17% of drivers listen to it rather than AM/FM radio.

Since Mosaic hit the market in 1993, there has been much talk about the potential for viewing television online. Now the talk is over, as most programs can be viewed online. It used to be that television was a family or group activity during which everyone would squeeze together on the sofa and sit back and watch and make comments about the show. As the number of multiple-television households grew and with the proliferation of cable channels, television viewing became more of a solo activity and more individualized. But still, watching programs when they aired gave viewers something to talk about the next day. The Internet and other video delivery systems are contributing to a culture in which a shared television viewing experience may become an activity of the past.

Network television must find a way to retain an audience and turn a profit. Offering online full episodes containing commercials and making them available for downloading on mobile devices are positive steps in attracting an audience and revenue. Perhaps offering more behind-the-scenes clips, unedited news and interviews, and making it easier for viewers to help create news stories would make networks, stations, and online sites more engaging and interesting. Perhaps instead of thinking of television as an imperiled medium, it should be thought of as one with new and exciting ways of delivering content.

The younger Web-savvy generation enjoys interacting online and feeling like they are part of the action. Blogs and vlogs are instrumental in creating the interactive culture that marks the online world. Users are no longer content as receivers of news—they want to create news and report on events. During the 2009 protests in Iran challenging the election results, a young protester, Neda Agha-Soltan, was killed. She became a symbol of unrest as millions of viewers watched in horror as her death was captured on cell phones and video cameras and posted on YouTube as it happened. The day Michael Jackson died, several witnesses videoed the EMTs putting him into the ambulance and his arrival at the hospital before the major news outlets were on the scene. And because news travels in seconds across the Internet, thousands of mourners gathered at the hospital within minutes of Jackson's death. These are examples of how individuals became reporters rather than just viewers. The difficulty for the television networks is verifying the user-generated content—anyone can shoot a fake video. Television news organizations need to strategize how to deal with these types of situations and set information policies regarding privacy, surveillance, ethical standards, verifying sources, and protecting anonymity.

As laptops once freed us from our desktop computers, mobile connectivity by cell phone and other handheld devices holds promise to unfetter us from our laptops. Mobile devices are quickly becoming the primary way to connect to the Internet. Their portability and low cost make them the optimal ways to stay socially in touch and linked to online media and the real world.

FIG. 5.15 The Internet is a combination of television, film, radio, newspaper, magazines, and other media.
Photo courtesy of iStockphoto. © Petrovich9, image #6030817

NET NEUTRALITY

Net neutrality and what it means to the Internet's future is one of the hottest issues of the day. Internet neutrality is basically the idea that "Internet should be an impartial conduit for the information that flows through it" (Scola, 2014, p. B2), and everyone should have equal and unfettered access to online content ensured by regulations that high-speed Internet service providers cannot charge extra or slow or block the flow of certain types of content.

Online content providers such as Netflix, Google, and Yahoo! use huge amounts of broadband data to efficiently and effectively deliver content, especially video and graphics. But the Internet service providers, cable companies, and cell phone carriers that provide broadband access to the content companies are demanding that they be allowed to increase charges for certain amounts and types of access. As an executive for SBC Communications so elegantly proclaimed, "For a Google, or a Yahoo, or a Vonage to expect to use these pipes for free is nuts!" (Scola, 2014, p. B2). On the other hand, content providers and consumer groups protest that such a fee increase, and the right to slow bandwidth for some content would hinder access and go against the principle of a free Internet.

In 2005, the Federal Communications Commission (FCC) issued the Internet Policy Statement, which listed four principles to keep the Internet 'open.' According to the 2005 policy:

1. Consumers can access all lawful Internet content.
2. Consumers can use applications and services that they choose as long as they are lawful.
3. Consumers can connect devices to the Internet as long as the devices do not harm the network.
4. There should be free and open competition among network providers, application and service providers, and content providers.

The net neutrality controversy centers on the last point. The fear is that some Internet providers may block some Internet applications or content, especially from their competitors, thus creating a two-tiered system of Internet content haves and have-nots, depending on how much they can afford to pay. For example, if a couple of young entrepreneurs want to start a Web-based video service that requires high-speed bandwidth, under net neutrality they could simply set up their site and be in business. But without neutrality rules, an Internet Service Provider could charge them the same amount of money for high-speed-quality service that they charge large established sites, thus competitively disadvantaging the newcomers. But opponents claim that ISPs should have the ability to provide tiered services, such as giving companies the option of paying more for high-quality data transmission at higher speeds than their competitors.

While ISPs are leaning on the FCC to reconsider its policy, the FCC is trying to uphold its policy and have it etched as law, but in January 2014, the courts said otherwise. But the FCC and neutrality advocates have not given up. Traditionally, the FCC has stayed away from mandating how ISPs should operate, fearing that too much interference would hinder the growth of broadband. But the changing nature of the Internet and fierce competition among ISPs has altered the FCC's position. The FCC and supporters of neutrality would like to see cable companies established as 'common carriers,' as are telephone companies and other utilities, so they cannot filter or interrupt online content. Common carrier status would allow the FCC to uphold and enforce its net neutrality rules.

In late February 2015, the FCC voted to regulate the Internet as a utility. Tom Wheeler, FCC commission chairman, justified the agency's stance as protecting innovators and consumers and preserving "the Internet's role as a core of free expression and democratic principles" (Ruiz & Lohr, 2015, p. B1). Cable and telecommunication companies and some members of Congress immediately objected to what they deem government interference with enterprise. A few months earlier, a Congressional committee floated net neutrality legislation that would have prohibited ISPs from blocking or slowing content but would have allowed them 'reasonable network management.' Further, the rules would have blocked the FCC from classifying broadband as a common carrier or utility and would have stopped it from creating any new neutrality policies.

Opponents of the FCC utility ruling have not given up and are looking into legal ways to curb it, even hoping that the Supreme Court will intervene. There is much at stake, as shown by the almost 4 million public comments sent so far to the FCC. Clearly, the controversy continues and could take years to settle.

SUMMARY

The history of the Internet is longer than most people think. It dates back to the 1960s, when scientists were experimenting with a new way to share information and keep connected in times of crisis. In its early days, the Internet was largely limited to communicating military, academic, and scientific research and was accessed by using complicated commands. In 1993, the first Web browser, Mosaic, came onto the scene, and the Internet quickly caught the public's attention. Since then, it has become the most quickly adopted new medium in history.

Without a doubt, the Internet has changed our lives tremendously. We no longer have to passively sit and absorb whatever news and information the traditional media want to send our way, but we can select what we want to know. Online technologies have sprouted new ways of gaining news and information and having an individual voice. Despite these unique qualities, the Internet also has many of the same properties as the traditional print and broadcast media. It delivers audio, video, text, and graphics in one package.

Not only has the Internet changed the way we receive and provide information, but it is also altering our traditional media use behaviors. Millions of people around the world log onto the Internet on a regular basis. For some, the Internet will always be a supplement to radio and television, but for others, the Internet may become the medium they turn to first for news, information, and entertainment.

Online radio delivery has attracted many users who prefer clicking a button to hear their favorite audio over tuning in to an over-the-air station. For audio providers, online radio is relatively easy to set up and inexpensive to maintain. Cyber radio is the ideal medium for those who want to reach a global audience. People have been enamored with television and the act of watching television since the 1940s, which means it will be hard, if not impossible, to tear them away from it. Instead of abandoning television, new program delivery systems provide new ways to watch—on a PC, laptop, tablet, or smartphone. Television content still rules; the only difference is what type of screen it is watched on.

Clearly, the Internet is quickly catching up with radio and television when it comes to news delivery. In many ways, the Internet has surpassed radio and television when it comes to providing in-depth news. The Web is not constrained by time and space as are the traditional media. News is posted immediately and updated continuously, as the situation warrants. Moreover, Weblogs and vlogs redistribute news and information delivery from established media into the hands of everyday people.

Broadcast radio and television networks and stations are competing among and between themselves and with the Internet, often with their own online counterparts. Television and radio must compete with the Web for a fragmented audience and precious advertising dollars. The traditional media are concentrating their efforts on designing Web sites that draw viewers away from their online competitors. But at the same time, they have to be sure that they do not lure viewers to the Web at the expense of their over-the-air fare. Even though research differs on whether the Internet is taking time away from radio and television, even a short amount of time spent online is time taken away from the "old-line media" (Dizard, 2000).

The Internet is still a relatively new medium, and so no one knows for sure what form it will end up taking as it keeps changing and adapting to technological innovations and diverse social and cultural needs. But what we do know is that it has had an enormous impact on the radio and television industries. If the prognosticators are right, the Web will eventually merge with radio and television, offering both conventional radio and television fare and Web-based content through a single device. But new net neutrality laws could change everything from how much we have to pay for content, the speed and quality of content, and the type of content we receive to the Internet business model in general.

BIBLIOGRAPHY

Anderson, J. Q., & Rainie, L. (2008, December 14). The future of the Internet III. *Pew Internet & American Life Project*. Retrieved from: www.pewinternet.org/Reports/2008/The-Future-of-the-Internet-III.aspx [October 9, 2009]

AOL. (2009). *Wikipedia*. Retrieved from: en.wikipedia.org/wiki/American_Online

As Web sites go, which are the click magnets? (2014, September 28). *ComScore, The Washington Post Magazine*. p. 7.

Audio by the numbers: The state of the news media. (2012). *Pew Research Center*. Retrieved from: www.stateofthemedia.org/2012/audio-how-far-will-digital-go/audio-by-the-numbers/

Berkman., R. I. (1994). *Find it online*. New York: McGraw-Hill.

Berniker, M. (1995, October 30). RealAudio software boosts live sound, music onto the Web. *Broadcasting & Cable*, p. 67.

Berr, J. (2014, November 20). Has the cord-cutter threat to pay TV been exaggerated? *CBS Money Watch*. Retrieved from: www.cbsnews.com/news/has-the-cord-cutters-threat-to-pay-tv-been-exaggerated/

Blood, R. (2001, September 7). Weblogs: A history and perspective. *Inkblotsmag.com*. Retrieved from: rebeccablood.net/essays/weblog_history.html

Bollier, D. (1989). How Americans watch TV: A nation of grazers. *Channels Magazine* [Special Report], pp. 41–52.

Bonchek, M. S. (1997). *From broadcast to netcast: The Internet and the flow of political information*. Doctoral dissertation, Harvard University, Cambridge, MA. Retrieved from: www.esri.salford.ac.uk/ESRCResearchproject/papers/bonch97.pdf

Broadcasting and cable yearbook. (2010). Newton, MA: Reed Elsevier.

Bromley, R. V., & Bowles, D. (1995). The impact of Internet use on traditional news media. *Newspaper Research Journal*, 16(2), 14–27.

Browser market share. (2010, January). *Net Market Share*. Retrieved from: marketshare.hitslink.com/browser-market-share.aspx?qprid50 [February 8, 2010]

Cai, X. (2002, August). *Are computers and traditional media functionally equivalent?* Paper presented at the annual meeting of the Association for Education in Journalism and Mass Communication, Miami, FL.

Calunod, J. (2016, July 13). Web browser market share wars for June 2016. CP Entertainment, Retrieved from: www.christianpost.com/news/web-browser-market-share-wars-for-june-2016-google-chrome-vs-microsoft-internet-explorer-vs-mozilla-firefox-vs-edge-vs-safari-166387/

Campbell, F. B. (2014, December 26). The slow death of 'do not track.' *The New York Times*, Op Ed, 4.

Caulfield, K. (2016, January 5). Adele's "25" rules as Nielsen music's top album of 2015 in U.S. *Billboard*. Retrieved from: www.billboard.com/articles/columns/chart-beat/6835248/adele-25-top-album-of-2015-in-us

Cavna, M. (2013, July 31). Nobody knows you're a dog. *The Washington Post*. Retrieved from: https://www.washingtonpost.com/blogs/comic-riffs/post/nobody-knows-youre-a-dog-as-iconic-internet-cartoon-turns-20-creator-peter-steiner-knows-the-joke-rings-as-relevant-as-ever/2013/07/31/73372600-f98d-11e2-8e84-c56731a202fb_blog.html

Chmielewski, D. (2009, January 2). Music downloads hit 1 billion mark. *Chicago Tribune*. Retrieved from: archives.chicagotribune.com/2009/jan/02/business/chi-fri-digital-music-milestone-jan02 [October 25, 2009]

Cohen, N. (2009, July 13). How the media wrestle with the Web. *The New York Times*. Retrieved from: www.nytimes.com/2009/07/13/technology/internet/13link.html [July 14, 2009]

Cole, J. I. (2001). The UCLA Internet report 2001: Surveying the digital future. Retrieved from: www.ccp.ucla.edu

College students fastest growing smartphone segment. (2009, October 22). *News Center*. Ball State University. Retrieved from: www.bsu.edu/news/article/0,1370,-1019-63096,00.html [January 27, 2010]

comScore. (2012, August 17). ComScore releases July 2012 U.S. online video rankings. Retrieved from: www.comscore.com/Insights/Press-Releases/2012/8/comScore-Releases-July-2012-US-Online-Video-Rankings

comScore. (2016). ComScore ranks the Top 50 U.S. digital media properties for January 2016. Retrieved from: www.comscore.com/Insights/Market-Rankings/comScore-Ranks-the-Top-50-US-Digital-Media-Properties-for-January-2016?ns_campaign=comscore_general&ns_source=social&ns_mchannel=social_post&ns_linkname=link_name&ns_fee=0

comScore. (2016, February 24). comScore releases January 2016 U.S. desktop online video rankings. Retrieved from: www.comscore.com/Insights/Rankings/comScore-Releases-January-2016-US-Desktop-Online-Video-Rankings

Cord cutters alert: 5% of TV broadband users watch all their TV online. (2013, August 1). *Gigaom Research*. Retrieved from: https://gigaom.com/2013/08/01/five-percent-cord-cutters/

Cord-cutters grew by 44% in the past four years. (2014, April 21). *Experian*. Retrieved from: http://press.experian.com/United-States/Press-Release/cord-cutters-grew-by-44-percent-in-the-past-four-years-with-7-6-million-households.aspx

Cortese, A. (1997, February 24). A way out of the Web maze. *Business Week*, pp. 95–101.

Davidson, J. (2016, February 15). At USPS, a net profit—but few cheer. *The New York Times*, p. A15.

Demographics of Internet users. (2010). *Pew Research Center*,. Retrieved from: www.pewinternet.org/Static-Pages/Trend-Data/Whos-Online.aspx [January 27, 2010]

Digital audio usage trends: A highly engaged listenership. (2011). *Parks Associates*. Retrieved from: www.iab.net/media/file/TargetSpotInc_DigitalAudioUsageTrends_WhitePaper2011.pdf

Digital video awareness grows; YouTube still dominates. (2009, September 11). *SeekingAlpha.com*. Retrieved from: seekingalpha.com/article/160949-digital-video-awareness-grows-youtube-still-dominates [October 28, 2009]

Dizard, W., Jr. (2000). *Old media, new media*. New York: Longman.

Eadicicco, L. (2016, December 9). The 10 most popular podcasts of 2015. *Time*. Retrieved from: http://time.com/4141439/podcasts-most-popular-year-2015/

Eager, B. (1994). *Using the World Wide Web*. Indianapolis, IN: Que Corporation.

Email Statistics Report, 2014–2018. (2014, April). *The Radicati Group, Inc.* Retrieved from: www.radicati.com/wp/wp-content/uploads/2014/01/Email-Statistics-Report-2014–2018-Executive-Summary.pdf

Email vs. social networks. (2010, September 13). *Pew Research Center*. Retrieved from: www.pewresearch.org/daily-number/email-vs-social-networks/

Fahri, P. (2009). TMZ earns second look with scoop on Jackson. *The Washington Post*, pp. C1, C12.

Ferguson, D. A., & Perse, E. M. (1993). Media and audience influences on channel repertoire. *Journal of Broadcasting & Electronic Media, 37*(1), 31–47.

First U.S. Web page made its debut 10 years ago. (2001, December 13). *Knoxville News-Sentinel*, p. C1.

Fitzpatrick, A. (2013, February 26). Custom top-level domains will be live this year. *Mashable*. Retrieved from: http://mashable.com/2013/02/26/top-level-domains-2013/#

Flatow, I. (2009, October 30). Happy birthday, Internet. Transcript. *NPR.com*. Retrieved from: www.npr.org/templates/story/story.php?storyId5114319703. [November 4, 2009]

Fowler, G. A. (2002, November 18). The best way . . . Find a blog. *Wall Street Journal*, p. R8.

Frauenfelder, M., & Kelly, K. (2000). Blogging. *Whole Earth*, 52–54.

Freese, D. D. (2002). New media, new power. *Tech Central Station*. Retrieved from: www.eddriscoll.com/archives/2002_12.php

Friend, E. (2009, June 18). Woman fined to tune of $1.9 million for illegal downloads. *CNN.com*. Retrieved from: www.cnn.com/2009/CRIME/06/18/minnesota.music.download.fine/index.html [October 25, 2009]

Fung, B. (2015, March 18). GPO: FCC's IG is looking into net neutrality. *The Washington Post*, p. A13.

Fung, B. (2015, December 22). Survey: Many cord-cutters don't have broadband. *The Washington Post*, p. A12.

Getting serious online. (2002). *Pew Research Center*. Retrieved from: www.pewinternet.org/reports

Gamerman, E. (2014, November 13). 'Serial' podcast catches fire. *The Wall Street Journal*. Retrieved from: www.wsj.com/articles/serial-podcast-catches-fire-1415921853

Golbeck, J. (2007). The dynamics of Web-based social networks: Membership, relationships, and change. *First Monday, 12*, 11.

Greenspan. R. (2003, July 24). Blogging by the numbers. *Click Z Network*. Retrieved from: www.clickz.com/big_picture/applications/article.php/2238831

Gross, G. (2015, January 20). Republican Net neutrality bill allows 'reasonable' network management. *InfoWorld*. Retrieved from: www.infoworld.com/article/2871945/government/republican-net-neutrality-bill-allows-reasonable-network-management.html

Haring, B. (1997, August 11). Web sites strive to brand their names into users' minds. *USA Today*, p. 4D.

Heeter, C. (1985). Program selection with abundance of choice. *Human Communication Research, 12*(1), 126–152.

Heeter, C., D'Allessio, D., Greenberg, B., & McVoy, S. D. (1988). Cableviewing behaviors: An electronic assessment. In C. Heeter & B. Greenberg (Eds.), *Cableviewing* (pp. 51–63). Norwood, NJ: Ablex.

Helft, M. (2009). Google to add captions, improving YouTube Videos. *The New York Times*, p. B4.

Hesse, M. (2009). Web series are coming into a prime time of their own. *The New York Times*, pp. E1, E12.

Hotz, R. L. (2009). A neuron's obsession hints at biology of thought. *The Wall Street Journal*, p. A14.

How old media can survive in a new world. (2005, May 23). *The Wall Street Journal*. Retrieved from: online.wsj.com [September 29, 2009]

How to make music radio appealing to the next generation. (2005, December 8). *USC MediaLab*. Retrieved from: www.frankwbaker.com/mediause.htm [October 26, 2009]

Human Brain. (2009). *Wikipedia*. Retrieved from: en.wikipedia.org/wiki/Human_brain [October 25, 2009]

Husted, B. (2002). And the beat goes on. *Atlanta Journal-Constitution*, pp. P1, P4.

ICANN. Registry Listing. (2010). *Internet Corporation for Assigned Names and Numbers*. Retrieved from: www.icann.org/en/registries/listing.html [January 27, 2010]

Indvik, L. (2011, March 25). Internet piracy is on the decline. *USA Today*. Retrieved from: http://content.usatoday.com/communities/technologylive/post/2011/03/us-internet-piracy-is-on-the-decline-1/#.VtnbHinN_xY

Internet 2008 in numbers. (2009, January 22). *Royal Pingdom.com*. Retrieved from: www.bizreport.com/2009/06/americans_greatly_increasing_time_spent_online_1.html [October, 25, 2009]

Internet 2012 in numbers. (2012, January 16). *Royal Pingdom.com*. Retrieved from: www.royal.pingdom.com/2013/01/16/internet-2012-in-numbers

Internet 2013 in numbers. (2014. January 14). *Dubai Chronicle*. Retrieved from: www.dubaichronicle.com/2014/01/31/internet-2013-numbers/

Internet and American life. (2000). *Pew Research Center*. Retrieved from: www.pew-internet.org/reports [January 14, 2002]

Internet eats into TV time. (1996). *St. Louis Post Dispatch*, pp. 6C, 8C.

Internet history. (1996). *Silverlink LLC*. Retrieved from: www.olympic.net/poke/IIP/history.html

Internet is future of the radio. (2009, September 11). *InfoZine*. Retrieved from: www.infozine.com/news/stories/op/storiesShowPrinter/PrintId/37442/?goto-5print [October 27, 2008]

Internet radio's audience turns marketers heads. (2013, February 6). *eMarketer*. Retrieved from: www.emarketer.com/Article/Internet-Radios-Audience-Turns-Marketer-Heads/1009652

Internet users. (2015, January 19). *Internet Live Stats*. Retrieved from: www.internetlivestats.com/internet-users/

Irvine, M. (2010, February 3). Is blogging a slog? Some young people think so. *Excite News*. Retrieved from: apnews.excite.com/article/20100203 [February 4, 2010]

Itzkoff, D. (2009, August). Graduate student fined in music download case. *The New York Times*, p. A3.

Jesdanun, A. (2001, October 14). In online logs, Web authors personalize attacks, retaliation. *Florida Times-Union*. Retrieved from: www.Jacksonville.com/tu-online/stories/101401/bus_7528493.html

Johnson, D. T. J., & Kaye, B. K. (2002). Webelievabilty: A path model examining how convenience and reliance on the Web predict online credibility. *Journalism & Mass Communication Quarterly, 79*(3), 619–642.

Johnson, T. J., & Kaye, B. K. (2004). Wag the blog: How reliance on traditional media and the internet influence perceptions of credibility of weblogs among blog users. *Journalism & Mass Communication Quarterly, 81*(3), 622–642.

Johnson, T. J., & Kaye, B. K. (2011). Can you teach a new blog old tricks? How blog users judge credibility of different types of blogs for information about the Iraq War. In B. K. Curtis (Ed.), *Psychology of trust* (pp. 1–25). New York: NOVA Publishers.

Judge cues Napster's death music. (2002, September 3). *Wired News*. Retrieved from: www.wirednews.com

Kang. C. (2014, September 26). Listening got easier, and podcasters found profits. *The Washington Post*, A1, A2.

Kaye, B. K. (1998). Uses and gratifications of the World Wide Web: From couch potato to web potato. *New Jersey Journal of Communication, 6*(1), 21–40.

Kaye, B. K., & Johnson, T. J. (2002). Online and in the know: Uses and gratifications of the Web for political information. *Journal of Broadcasting & Electronic Media, 46*(1), 54–71.

Kaye, B. K., & Johnson, T. J. (2003). From here to obscurity: The Internet and media substitution theory. *Journal of the American Society for Information Science and Technology, 54*(3), 260–273.

Kaye, B. K., & Johnson, T. J. (2004). Blog day afternoon: Weblogs as a source of information about the war on Iraq. In R. D. Berenger (Ed.), *Global media go to war* (pp. 293–303). Spokane, WA: Marquette.

Kaye, B. K., & Medoff, N. J. (2001). *The World Wide Web: A mass communication perspective*. Mountain View, CA: Mayfield.

Kerner, S. M. (2014, September 2). 10 most popular new top-level domains. *eweek*. Retrieved from: www.eweek.com/cloud/slideshows/10-most-popular-new-top-level-domains.html

Knight, K. (2009, June 26). Americans greatly increasing time spent online. *Biz Report.com*. Retrieved from: bizreport.com/2009/06/americans_greatly_increasing_time_spent_online_1.html [October 25, 2009]

Krol, E. (1995). *The whole Internet: User's guide & catalog*. Sebastopol, CA: O'Reilly & Associates Publishing.

Krol, C. (1997). Internet promos bolster networks' fall programs. *Advertising Age*, p. 40.

Kurtz, H. (2002, July 31). How weblogs keep the media honest. *The Washington Post*. Retrieved from: www.washingtonpost.com/ac2/wp-dyn/A25354–2002Jul31?language5printer

Leichtman Research Group. (2015, November). Over 80% of U.S. Households get broadband at home. Retrieved from: www.leichtmanresearch.com/press/120315release.html

Let's make a deal. (2001, March 5). *Newsweek*, p. 63.

Levy, S. (1995, February 27). TechnoMania. *Newsweek*, pp. 25–29.

Levy, S. (2002, July 15). Labels to Net radio: Die now. *Newsweek*, p. 51.

Levy, S. (2003a, March 29). Blogger's delight. *MSNBC*. Retrieved from: www.msnbc.com

Levy, S. (2003b, September 22). Courthouse rock? *Newsweek*, p. 52.

Lewis, R. (2014, March 18). Music industry revenue in 2013 stayed flat at $7 billion, RIAA says. *Los Angeles Times*. Retrieved from: www.latimes.com/entertainment/music/posts/la-et-ms-music-industry-revenue-riaa-report-streaming-digital-20140318-story.html

Lin, C. A. (2001). Audience attributes, media supplementation, and likely online service adoption. *Mass Communication and Society*, 4, 19–38.

Listserv. (2003). *Webopedia*. Retrieved from: www.webopedia.com/TERM/L/Listserv.html

Luckerson, V. (2013, February 28). Revenue up, piracy down: Has the music industry turned a corner? *Time*. Retrieved from: http://business.time.com/2013/02/28/revenue-up-piracy-down-has-the-music-industry-finally-turned-a-corner/

Madden, M. (2008a). Podcast downloading 2008. *Pew Internet & American Life Project*. Retrieved from: www.pewinternet.org/Reports/2008/Podcast-Downloading-2008.aspx [October 9, 2009]

Madden, M. (2008b). The audience for online video shoots up. *Pew Internet & American Life Project*. Retrieved from: www.pewinternet.org/Reports/2009/13-The-Audience-for-Online-VideoSharing-Sites-Shoots-Up.aspx [October 9, 2009]

Magna on demand quarterly. (2008, September). *TVbytheNumbers*. Retrieved from: tvbythenumbers.com/2008/09/11/latest-dvr-and-on-demand-trends-report/5058s [January 30, 2010]

Major music industry groups announce breakthrough agreement. (2008, September 23). *Recording Industry Association of America*. Retrieved from (RIAA): www.riaa.com/newsitem.php?id5C9C68054-D272-0D33-6EDBDF08022C7E3A&searchterms5&termsinclude5royalties,%20music,%20downloads&termexact5 [October 26, 2009]

Mandese, J. (1995). Snags aside, nets enamored of 'net. *Advertising Age*, p. S20.

Marc Andreessen, co-founder of Netscape. (1997). *Jones Telecommunications and Multimedia Encyclopedia Homepage*. Retrieved from: www.digitalcentury.com/encyclo/update/andreess.htm [January 21, 1998]

Markus, M. L. (1990). Toward a "critical mass" theory of interactive media. In J. Fulk & C. Steinfield (Eds.), *Organizations and communication technology* (pp. 194–218). Newbury Park, CA: Sage.

Media and communications trends—ways to win in today's challenging economy. (2009, October 15). *Nielsen*. Retrieved from: www.en-us.nielsen.com/etc/medialib/nielsen_dotcom/en_us/documents/pdf/webinars.Par.33055.File.pdf [January 30, 2010]

Murphy, K. (2014, October 4). We want privacy, but can't stop sharing. *The New York Times*, Op Ed 4.

Music downloads. (2008, August). *Wikipedia*. Retrieved from: en.wikipedia.org/wiki/Music_download [October 25, 2009]

Nixon, R. (2013, November 15). Postal services reports improved $5 billion loss. *The New York Times*. Retrieved from: www.nytimes.com/2013/11/16/us/postal-service-trims-its-losses.html?_r=0

Noack, D. R. (1996, June). Radio stations are blooming on the Internet as the Net becomes a radio medium. *Internet World*. Retrieved from: www.webweek.com

Nocera, J. (2009, August 7). It's time to stay the courier. *The New York Times*. Retrieved from: www.nytimes.com/2009/08/08/business/08nocera.html [October 26, 2009]

Number of Internet users. (2015). *Internet Live Stats*. Retrieved from: www.internetlivestats.com/internet-users/

Online listening shows big gains. (2009, February 10). *Convergence*. Retrieved from: www.radioink.com/cw/Article.asp?id51161496&spid525728 [October 26, 2009]

Our mission. (2015). *Facebook newsroom*. Retrieved from: http://newsroom.fb.com/company-info/

Outing, S. (2002, July 18). Weblogs: Put them to work in your newsroom. *PoynterOnline*. Retrieved from: www.pointer.org

Palser, B. (2002). Journalistic blogging. *American Journalism Review*, 24(6), 58.

Parks, M. (2015, January 2). He came, Reddit conquered, now what? *The New York Times*, pp. C1, C2.

Patterson, T. (2009, May 8). Is the future of TV on the Web? *CNN.com*. Retrieved from: www.cnn.com/2008/SHOWBIZ/TV/05/01/tv.future/index.html [October 28, 2009]

Pavlik, J. V. (1996). *New media technology: Cultural and commercial perspectives*. Boston: Allyn & Bacon.

Pegoraro, R. (2009, May 31). Web radio hits the road. *The New York Times*, p. G2.

Perse, E. M., & Dunn, D. G. (1998). The utility of home computers and media use: Implications of multimedia and connectivity. *Journal of Broadcasting & Electronic Media*, 42, 435–456.

Petrozzello, D. (1995, September 11). Radio 1 listeners: A match made on the Internet. *Broadcasting & Cable*, p. 34.

Podcast. (2009). *Wikipedia*. Retrieved from: en.wikipedia.org/wiki/Podcasting [October 25, 2009]

Podcast Alley. (2009). Retrieved from: www.podcastalley.com [November 1, 2009]

The podcast consumer. (2012. May 29). *Edison Research*. Retrieved from: www.slideshare.net/webby2001/the-podcast-consumer-2012

Postal service stretched thin. (2010, February 18). *The Washington Post*, p. B3.

Progressive networks launches real audio player plus. (1996, August 18). *Real Audio*. Retrieved from: www.realaudio.com/prognet/pr/playerplus.html

Quittner, J. (1995, May 1). Radio free cyberspace. *Time*, p. 91.

Radio facts and figures. (2014). *News Generation*. Retrieved from: www.newsgeneration.com/broadcast-resources/radio-facts-and-figures/

Radio fans tuning online. (2001, April 9). *NUA Surveys*. Retrieved from: www.nua.ie/surveys [January 14, 2002]

Radio sets new record for number of stations on-air. (2015, January 16). *Inside Radio*. Retrieved from: www.insideradio.com/article.asp?id=2852119#.VLmLmyfeNFU

Rafter, M. (1996b, January 24). RealAudio fulfills Web's online sound promise. *St. Louis Post Dispatch*, p. 13B.

Rainie, L. (2008, December 14). The next future of the Internet. *Pew Internet & American Life Project*. Retrieved from: www.pewinternet.org/Commentary/2008/December/The-Next-Future-of-the-Internet.aspx#

Rainie, L. (2009, October 2). The unfinished symphony: What we don't know about the future of the Internet. *Speech given to the Internet Governance Forum—USA*. Retrieved from: www.pewinternet.org/Search.aspx?q5the%20unfinished%20symphony [October 12, 2009]

Ranie, L., & Cohn, D. (2014). Census: Computer ownership, internet connection varies widely across U.S., *Pew Research Center*. Retrieved from: www.pewresearch.org/fact-tank/2014/09/19/census-computer-ownership-internet-connection-varies-widely-across-u-s/

Richtel, M. (2001, November, 29). Free music is expected to surpass Napster. *The New York Times*. Retrieved from: www.nytimes.com/2001/11/29/technology/29MUSI.html [November 29, 2001]

Richtel, M. (2002, September 3). Napster says it is likely to be liquidated. *The New York Times*. Retrieved from: www.nytimes.com

Roberts, A. (2014, December 23). The 'Serial' podcast by the numbers. *CNN.com*. Retrieved from: www.cnn.com/2014/12/18/showbiz/feat-serial-podcast-btn/index.html

Royalties. (2009). *Wikipedia*. Retrieved from: en.wikipedia.org/wiki/Royalty_fee#Music_royalties [October 25, 2009]

Rubin, A. M. (1984). Ritualized and instrumental television viewing. *Journal of Communication*, 34(3), 67–77.

Ruiz, R. R., & Lohr, S. (2015, February 27). F.C.C. votes to regulate the Internet as a utility. *The New York Times*, pp. B1, B2.

Schonfeld, E. (2012, January 8). How people watch TV online and off. *Tech Crunch*. Retrieved from: http://techcrunch.com/2012/01/08/how-people-watch-tv-online/

Schwartz, J. (2001, October 29). Page by page history of the Web. *The New York Times*. Retrieved from: www.nytimes.com/2001/10/29/technology/ebusiness

Scola, N. (2014, June 15). 5 myths about net neutrality. *The New York Times*, p. B2.

Scope of the problem. (2015). *Recording Industry Association of America*. Retrieved from: http://riaa.com/physicalpiracy.php?content_selector=piracy-online-scope-of-the-problem

Scott, M. (2014, October 8). Tim Berners-Lee, Web creator, defends net neutrality. *The New York Times*. Retrieved from: http://bits.blogs.nytimes.com/2014/10/08/tim-berners-lee-web-creator-defends-net-neutrality/

Seals, T. (2013, July 1). TV cord-cutters, cord-nevers total 19 percent of US adults. *Cable Spotlight*. Retrieved from: http://cable.tmcnet.com/topics/cable/articles/2013/07/01/344134-tv-cord-cutters-cord-nevers-total-19-percent.htm

Segailer, S. [Executive producer] (1999). *Nerds 2.0.1: A brief history of the Internet*. Television broadcast. Public Broadcasting Service.

Seipp, C. (2002). Online uprising. *American Journalism Review*, 24(5), 42.

Self, C. (2003, September 22). Admitted copyright offenders avoid suits. *Daily Beacon*, p. 1.

Semuels, A. (2009, February 24). Television viewing at all-time high. *Los Angeles Times*. Retrieved from: articles.latimes.com/2009/feb/24/business/fi-tvwatching24 [September 1, 2009]

'Serial' season one. (2014). *Serial*. Retrieved from: http://serialpodcast.org/

Shane, E. (1999). *Selling electronic media*. Boston: Focal Press.

Shifting Internet population recasts the digital divide debate. (2003, April 16). *Pew Internet and American Life*. Retrieved from: www.pewinternet.org/ releases

The size of the world wide web. (2015, January 19). *World Wide Web Size*. Retrieved from: www.worldwidewebsize.com/

Skiffington, C. (2014, July 30). Top 10 most popular new top-level domains. *1 & 1*. Retrieved from: http://blog.1and1.com/2014/07/30/top-10-most-popular-new-top-level-domains/

Smith, A., Rainie, L., & Zickuhr, K. (2011, July 19). College students and technology. *Pew Research Internet Project*. Retrieved from: www.pewinternet.org/2011/07/19/college-students-and-technology/

Smith, C. (2016, February 19). By the numbers: 60+ amazing Reddit statistics. *DMR*. Retrieved from: http://expandedramblings.com/index.php/reddit-stats/

Social networking fact sheet. (2014, January). *Pew Research Internet Project*. Retrieved from: www.pewinternet.org/fact-sheets/social-networking-fact-sheet/

Spors, K. K. (2003, October 26). Musicians get little from 99-cent songs. *Knoxville News-Sentinel* [*Wall Street Journal* Supplement], p. WSJ3.

Stelter, B. (2009, March 10). Serving up television without the TV set. *The New York Times*, pp. C1, C3.

Stelter, B. (2009, June 27). A web site scooped other media on the news. *The New York Times*, p. A10.

Stelter, B. (2009, November 11). TV news without the television. *The New York Times*, pp. B1, B10.

Stelter, B. (2011, September 26). Pew Media study shows reliance on many outlets. *The New York Times*, pp. B1, B10.

Stelter, B., & Helft, M. (2009, April 17). Deal brings TV shows and movies to You-Tube. *The New York Times*, pp. B1, B2.

Stempel, G., III, Hargrove, T., & Bernt, J. P. (2000). Relation of growth of use of the Internet to changes in media use from 1995 to 1999. *Journalism & Mass Communication Quarterly, 77*, 71–79.

Stone, B. (2004, January 12). OK with pay for play. *Newsweek*, p. 10.

Streitfeld, D. (2014, May 14). European curt lets users erase records on the Web. *The New York Times*, p. A1, A3.

Stross, R. (2009, February 8). Why television still shines in a world of screens. *The New York Times*, pp. D1, E4.

Students would pay for Napster. (2001, April 11). *NUA Surveys*. Retrieved from: www.nua.ie/surveys [January 14, 2002]

Stutz, C. (2014, June 19). The average American listens to four hours of music each day. *Spin*. Retrieved from: www.spin.com/2014/06/average-american-listening-habits-four-hours-audio-day/

Taylor, C. (1996a). Tube watchers, unite. *MediaWeek*, pp. 9–10.

Taylor, C. (1996b). Zapping onto the Internet. *MediaWeek*, p. 32.

Taylor, C. (1997, July 5). Net use adds to decline in TV use; radio stable. *Billboard*, p. 85.

TCP/IP green thumb: Effective implementation of TCP/IP networks. (1998). *LAN Magazine, 8*(6), 139.

Tierney, J. (2009, March 1). Industry supports proposed USPS delivery cut-backs. *MultiChannel Merchant*. Retrieved from: multichannelmerchant.com/mag/0301-industry-supports-usps-cutbacks/ [October 25, 2009]

Total number of web sites. (2015). *Internet Live Stats*. Retrieved from: www.internetlivestats.com/total-number-of-websites/

Totally wired campus. (2009, November 12). *Alloy Media 1 Marketing*. Retrieved from: www.harrisinteractive.com [January 27, 2010]

TV v. Net: Not a zero-sum game. (1998). *Wired*. Retrieved from: www.wired-news.com

Twitter. (2009). *Wikipedia*. Retrieved from: en.wikipedia.org/wiki/Twitter [October 25, 2009]

United States Postal Service. (2015). *Size and Scope*. Retrieved from: https://about.usps.com/who-we-are/postal-facts/size-scope.htm

United States Postal Service. (2015, November 15). *U.S. Postal Service Expects to Deliver more than 15 Billion Pieces of Holiday Mail and Packages this Year*. Retrieved from: https://about.usps.com/news/national-releases/2015/pr15_059.htm

U.S. Census Bureau. (2014, November). Computer and Internet use in the United States. Retrieved from: www.census.gov/content/dam/Census/library/publications/2014/acs/acs-28.pdf

U.S. Census Bureau. (2015, January 18). U.S. and world populations clocks. Retrieved from: www.census.gov/popclock/

Van Orden, J. (2008). *How to Podcast*. Retrieved from: www.how-to-podcast-tutorial.com/what-is-a-podcast.htm [October 24, 2009]

Vogt, N. (2015). Podcasting: Fact sheet. *Pew Research Center*. Retrieved from: www.journalism.org/2015/04/29/podcasting-fact-sheet/

Walker. R. (2009, October 18). The song decoders. *The New York Times Magazine*, pp. 49–53.

Weekly Internet radio listenership jumps from 33 to 42 million. (2009, April 16). *About.com: Radio*. Retrieved from: http://radio.about.com/b/2009/04/16/weekly-internet-radio-listenership-jumps-from-33-to-42-million.htm [October 26, 2009]

What are the most popular online activities? (2012). *Pew Internet American Life Project*. Retrieved from: http://abhisays.com/internet/what-are-the-most-popular-online-activities.html

What is AudioNet? (1997, February 4). *AudioNet*. Retrieved from: www.audionet.com/about

Who music theft hurts. (2015). *Recording Industry Association of America*. Retrieved from: http://riaa.com/physicalpiracy.php?content_selector=piracy_details_online

Why we do what we do. (2015). *Recording Industry Association of America*. Retrieved from: http://riaa.com/physicalpiracy.php?content_selector=piracy-online-why-we-do-what-we-do

Wood, M. (2014, September 8). Younger generation still values privacy. *International The New York Times*, p. 15.

World Internet users. (2016, July 10). *Internet Usage Stats*. Retrieved from: http://www.internetworldstats.com/stats.htm

Yahoo! Dictionary Online. (1997). Retrieved from: www.zdnet.com/yil/content.

YouTube press room statistics. (2015). *YouTube*. Retrieved from: www.youtube.com/yt/press/

Zittrain, J. (2014, May 15). Don't force Google to forget. *The New York Times*, p. A25.

Digital Devices: Up Close, Personal, and Customizable 6

Contents

The primary focus of this book is electronic media and the way they are used to reach a mass audience. So far this book has covered the origins of radio and the subsequent development of broadcast television, cable television, and satellite communications.

Chapter 4 focuses on radio and television programming and how it is created and distributed to a mass audience, and Chapter 5 is about the Internet. But this chapter is about how new communication technologies have changed the way mass media operate. Media traditionally sent professionally crafted messages to a mass audience. But using new communication tools, the everyday person can become a media outlet and construct 'user-generated content' (UGC) and send it to a mass audience, even to individual recipients. To avoid being left behind, traditional media have had to embrace new models of content distribution that include UGC and online enterprise.

This chapter begins with a historical look at personal communication devices, such as smartphones, tablets, and smart watches, what they are and how they work. It then examines the new world of customizable radio and television, in which listeners and viewers choose what they want to listen to or watch and design their own media use styles.

FIG. 6.1 ZITS by Jerry Scott and Jim Borgman
© Zits Partnership. Reprinted with permission of King Features Syndicate.

But in addition to providing simple descriptions of these new technologies, this chapter discusses their cultural implications and how they affect media content and distribution and personal lifestyles.

SEE IT THEN

POINT-TO-POINT/ONE-TO-ONE COMMUNICATION

LANDLINE TELEPHONE

The invention of the **telegraph** in 1835 ushered in the era of electronic communication and, more importantly, an almost instantaneous way to communicate one-to-one (point-to-point). A person could write a short message to another person and send it long distance instead of by Pony Express, train, or other ground transportation. The telegraph was the precursor to telephone voice transmission, which was invented by Alexander Graham Bell in 1876. By 1915, about 30% of U.S. households had at least one telephone installed, by 1945 one-half of all homes had a phone, by 1960 that percentage rose to three-quarters, and by the 1980s landline penetration was at its highest in about 95% of all homes.

These **'landline' phones** are connected by wire to outside telephone transmission lines strung from pole-to-pole to a telephone company. Telephone voice transmission still dominates electronic person-to-person communication.

Advances were soon made to the telephone, and it became a way to send pictures over telephone lines, a process known as **facsimile** or telefax. The term 'fax,' as it is called today, literally means a copy or reproduction. Although the technology for 'printed telegraph' transmissions has been in existence since the late 1800s, it was not until the 1920s when scientists at AT&T improved the technology and in the mid-1940s when Western Union built a compact machine did 'faxing' become a tool for business communication. By the 1970s and 1980s, desktop versions of fax machines made their way to the consumer market and began appearing in offices and homes.

At about the same time, the telephone answering machine started being marketed to consumers for home use. The answering machine made it possible for callers and receivers to leave and retrieve voice messages. The answering machine freed people from having to sit at home waiting for a call, and it also changed interpersonal relationships. Some callers preferred to leave a message rather than talk to someone, and receivers would screen calls to know who was calling without having to answer. Outgoing messages became informative and entertaining: "If you're calling about the red Toyota, sorry, it's already been sold"; "If your call is about good news or money, leave a message. If not, send me a letter."

CELL PHONES

The first prototype of a cell phone was developed in 1947 as a way to communicate from cars. This innovation was limited, however, because it was very expensive, and the Federal Communications Commission allowed only 23 simultaneous conversations in the same geographical area. And because of the FCC's apathy in allocating airwave space for wireless one-to-one communication, the introduction of wireless phones to the public was further delayed. Then in 1978, AT&T and Bell Labs introduced a **cellular** system and held general consumer test trials in Chicago. Within a few years, other companies were also testing cellular systems and pressuring the FCC to authorize commercial cellular services.

The first portable or **mobile** telephone units were the size and weight of a brick. They were called transportables or luggables. Cell phones were commercially available starting in 1984, and by the mid-1990s they were gaining in popularity. Initially, the cost of a cell phone was too high for most consumers, but by 1990 there were 5.3 million cell phone subscribers.

FYI: First Mobile Telephone Call

The first mobile telephone call was made on April 3, 1973. Martin Cooper, an employee of Motorola, placed the first call using a prototype of what would become the company's DynaTac cell phone. Cooper developed the phone, which weighed 2.5 pounds, was 9 inches long, 5 inches deep, and 1.75 inches wide. The cost was $3,995. Although no one remembers what words were said or the length of the first call, it was a historic step in the birth of the cell phone industry.

Source: Seltzer, 2013

FIG. 6.2 Dr. Martin Cooper with the first portable handset
© Ted Soqui/Corbis

ZOOM IN 6.1

Watch an ad for the DynaTAC cell phone from the 1980s: www. youtube.com/watch?v=0WUF3yjgGf4

PERSONAL DIGITAL ASSISTANTS

A personal digital assistant (PDA) was a handy device that freed people from the burden of lugging around a three-ring binder type of daily organizer. Apple's Newton was the first organizing and messaging handheld PDA. Introduced in 1993, Newton was an immediate hit; sales soared to 50,000 units in the first 10 weeks the product was on the market. But users quickly became disillusioned with the Newton's poor handwriting recognition, complexity of use, size, and expense. The Palm Pilot debuted in 1996 as a lightweight, small, easy-to-use organizer with enough memory to store thousands of addresses and notes. Its simple interface caught on with the public, and the name Palm was almost synonymous with PDA, even though several other manufacturers made PDAs. Hewlett Packard acquired Palm but eventually discontinued its production.

A PDA was basically an electronic calendar that stored contact information (addresses, phone numbers, email addresses) and sent reminders for appointments and special events, homework assignments, and projects. It also contained simple spreadsheets for tracking expenses and other numerical information. All those functions fit onto one small, palm-sized electronic device. Although some later PDAs retrieved, stored, and sent information to other PDAs and computers, they were generally self-contained, meaning that stored information could not be readily shared. PDAs eventually gave way to multifunctional and Internet-connected smartphones and tablets.

THE INTERNET

In the early days of the Internet, users had to connect their desktop computers to a modem or router to receive email and Web pages. The Internet was not mobile, nor was it nearly as extensive in terms of content and applications as it is today. The 1990s was the era of the Internet, a time when businesses were considering whether establishing a Web site would be a good return on investment, media were wondering if anyone would go online to listen to or watch their content, and individual users were thinking of ways that the Internet could make their lives easier and their social circles wider yet more intimate.

The Internet was still thought of as a separate medium, albeit one that had the capability of presenting text, graphics, audio, and video. Although consumers quickly adopted the Internet, it was still not used as a replacement for print, radio, or television news and entertainment. It was an extra and alternative source of information. Although online radio was becoming more popular, until the 2000s, few full-length television programs were available. Consumers still watched news, information, and entertainment from the traditional television set.

FYI: Fifth Generation—5G

The term 5G stands for 'fifth-generation' wireless service. The first generation (1981) was comprised of early analog cell phones and services. The second generation (1992) provided digitized phones and services. The third generation (2001) provided more efficient delivery and multimedia capabilities. Smartphones depend on 3G capabilities. The fourth generation (4G) (2012) provides mobile broadband Internet access to smartphones and other mobile devices through long-term evolution (LTE) technology that has even faster connection speeds and puts listeners in more control of their media consumption. The fifth generation (5G), due in 2020, will provide still faster service and facilitate device-to-device communication.

NEWS AND ENTERTAINMENT

News had not become a prominent part of broadcast television until about the 1960s. News events were shot on film and then developed and edited, a time-consuming process that often prevented stories from being ready in time for the next newscast. Anchors would report a breaking story and then lure viewers to the next newscast with a "film at eleven" promise. Beginning in the late 1970s and becoming widespread in the 1980s, film gave way to battery-powered portable video cameras and quick editing procedures. Stories shot in the field were processed more speedily and on the air within a short period of time. As portable video cameras became common at networks and television stations in the 1980s, news gathering and reporting became more timely and sophisticated. And 6 p.m. news producers would have the video ready to air instead of making the anchors say, "We will have that video for you tonight at 11:00 p.m."

Cable News Network (CNN) started operating in 1980 and, thanks to satellite news gathering (SNG), could easily fill the airwaves 24/7. Using newer satellite technology, on-the-scene reporters could shoot live video with portable cameras and uplink it to a satellite that would send it to the studio. During the 1991 Gulf War, CNN reporters relied heavily on satellite delivery to show the world what was happening with up-to-the-minute accounts. The broadcast networks had a hard time competing with CNN because they did not have nearly the same number of news bureaus and reporters in the field, and their on-air news time was limited to several half-hour segments throughout the day. The networks had to preempt their entertainment programs to show Gulf War events in more detail. CNN's reputation as a premier source for news was solidified.

In the pre-Internet days, news was the domain of professional journalists and news organizations. The common person on the street did not have a way to disseminate eyewitness accounts or commentary. Consumers turned to television, radio, and newspapers to learn about the world. Television and radio entertainment programs were professionally produced for a mass audience. High-level productions were transmitted over the air or via cable or satellite. Media consumers were just that, consumers, and producers were producers. It was not until new digital technologies emerged that the lines between content producers and content consumers were blurred. Two effects of this change are the rise of citizen journalists (ordinary people photographing and reporting on news events) and sites like YouTube where consumers provide the content.

SEE IT NOW

POINT-TO-POINT OR ONE-TO-ONE COMMUNICATION

CELL PHONES

The term 'mobile telephone' has given way to the more familiar 'cellular' or 'cell phone.' 'Mobile' describes how it is used, and 'cellular' describes how it works. A cell phone is really a type of two-way radio. Basically, a city or county is divided into smaller areas called cells, usually a few miles in radius. Each cell contains a low-powered radio transmitting/receiving tower that covers the cell area. Collectively, the cell towers provide coverage to the entire larger area, be it a city or county or some other region. The size of each cell varies according to geographic terrain, number of cell phone users, demand, and other criteria.

ZOOM IN 6.2

How cell phones work: www.explainthatstuff.com/cellphones.html

FYI: Dos and Don'ts of Cell Phone Etiquette

Do respect those who are with you. When you are engaged face-to-face with others, give them your complete and undivided attention. If a call is important, apologize and ask permission before accepting it.

Do be a good dining companion. No one wants to be a captive audience to a third-party cell phone conversation or to sit in silence while their dining companion talks with someone else. Always silence and store your phone before being seated. Never put your cell phone on the table.

Do let voicemail do its job. When you are in the company of others, let voicemail handle nonurgent calls.

Do keep arguments under wraps. Nobody can hear the person on the other end. All they are aware of is a one-sided screaming match a few feet away.

Do respect the personal space of others. When you must use your phone in public, try to keep at least 10 feet between you and others.

Do exercise good international calling behavior. The rules of cell phone etiquette vary from country to country.

Don't yell. The average person talks three times louder on a cell phone than they do in a face-to-face conversation.

Don't ignore universal quiet zones such as the theater, church, the library, classrooms, and funerals.

Don't make wait staff wait. Whether it is your turn in line or time to order at the table, always make yourself available to the server. Making servers and other patrons wait for you to finish a personal phone call is never acceptable.

Don't talk and drive. Permissible but only as a 'hands-free' call.

Don't forget to filter your language. A rule of thumb: If you would not walk through a busy public place with a particular word or comment printed on your T-shirt, do not use it in cell phone conversations.

Source: The do's And don'ts of cell phone etiquette, 2013

When a call is made, it is picked up by a cell tower and transmitted over an assigned radio frequency. When the caller travels out of the cell area while still talking, a switching mechanism transfers the call from that cell area to another cell area and to the corresponding radio frequency. Most of the time, callers do not notice the switch, but sometimes a call might be dropped because of spotty coverage.

By-mid 2002, more than 135 million Americans were cell phone customers—18 times the number of users than in 1992. American teens constituted 32 million of the 135 million consumers and spent $172 billion on cell phones and accessories. To keep up with the demand, in July 2002 the Bush administration allotted space in the radio spectrum for the use of wireless communication services. The number of users reached 262 million in 2008. The U.S. Commerce Department's National Telecommunications and Information Administration worked closely with the FCC and the cellular industry to make even more spectrum space available at the end of 2010 to continue to meet consumers' wireless voice and data communication needs.

Today more than 90% of U.S. residents have a mobile telephone, and of those, about two-thirds are smartphones. Additionally, about 41% of households are wireless only, with no landline. In May of 2014, The International Telecommunication Union estimated there are nearly 7 billion mobile subscriptions worldwide (6% more than the year before). This amount is equivalent to 95.5% of the world population. The number of mobile phone subscriptions is not the same as the number of mobile users. There are around 4.5 billion mobile users because many people have subscriptions for several devices. Mobile use continues to rise. Portio Research predicts that the number of mobile subscribers worldwide will reach 8.5 billion by the end of 2016.

% of U.S. household with and without a working landline telephone*

● Landline phone   ● Cell phone only

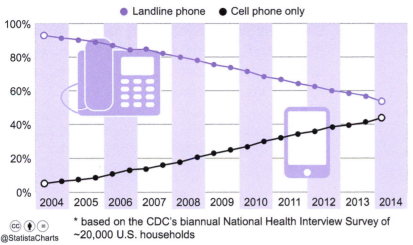

* based on the CDC's biannual National Health Interview Survey of ~20,000 U.S. households

@StatistaCharts

FIG. 6.3 Landlines are a dying breed.
Photo courtesy of The National Center for Health Statistics

FROM CELL PHONES AND PERSONAL DIGITAL ASSISTANTS TO SMARTPHONES AND TABLETS

What was once known as a personal digital assistant has merged with a cell phone to become a smartphone. It used to be that cell phones were just handy conversation devices. But cell phones have quickly morphed into smartphones/mobile devices with multiple uses—for example, 63% of mobile phone owners use their phone to go online. Even newer digital devices, known as tablets, combine the functionality of a PDA, cell phone, and laptop.

A tablet is a general-purpose computer contained in a touch-screen panel. Early tablet computers required a stylus, whereas modern tablets use a finger-sensitive touchscreen, a pop-up keyboard, and an optional stylus. In 2010, Apple launched the first tablet, the iPad. The iPad is a lightweight mobile device that works by touching icons on the screen. At a quarter-inch deep and about 1.5 pounds, it is easy to carry around and conduct online activities while on the go. Even television news anchors use iPads to read Twitter posts live on air as viewers send in comments.

In early 2015, it was announced that the worldwide tablet market had experienced its first decline since 2010, but still about 42% of U.S. adults own a tablet.

Tablets are generally more expensive than smartphones and can only be used to make Skype or Facetime calls. Some manufacturers are sizing newer smartphones between smaller older models and tablets. For example, the iPad screen measures 9.7 inches diagonally, but Samsung's Galaxy Mega 6.3 smartphone boasts a 6.3-inch display, and the Apple iPhone 6 Plus measures 5.5 inches in comparison to older iPhones, with about 3.5-inch displays. The tablet market is crowded with many competing brands. Because tablets are small and lightweight, there are times when they are more convenient to use than a laptop, but the lack of internal storage and a built-in keyboard keep them from becoming a true substitute for a computer.

Whether it is an iPhone, a 'Droid, or some other brand, it is probably a combination phone, camera, computer, video

player, calculator, address book, organizer, and alarm clock. Users send and retrieve email, access the Web, play games, text, download music, store photos and movies, and get driving directions. Smartphones and tablets both rely on function-specific applications (apps). Mobile applications are to smartphones/mobile devices what software programs are to computers. There is an app for almost anything a user needs to know or needs to do. There are apps to read the weather forecast, to find a recipe or the closest Starbucks, to keep up with the stock market, to reserve a rental car, to check flight times, to watch videos, and to engage in many other activities while on the go.

The app craze began in 2008, and users were quick to download new apps as they came along. In 2009, the average number of different apps used per month was 22, 2 years later 23.2, and by 2014 26.7. After the initial app rush in 2008 and 2009, the number of different apps used each month has remained fairly steady, indicating that there might be an upper limit to the total number of apps users need to access. Although the number of apps used has not increased dramatically, the amount of time users spend with apps went up by about 65% from 2012 to 2014. At the end of 2012, users were on their apps about 23 hours per month, but 2 years later time spent shot up to 37 hours and 28 minutes.

FIG. 6.4 Text message meeting
Photo courtesy of iStockphoto. © Michael DeLeon, image #930020

FIG. 6.5 The mobile phone generation
Photo courtesy of iStockphoto. © elcor, image #3648770

FYI: What Smartphone Owners Do With Their Phones

*68% occasionally follow breaking news events.

*67% share pictures, videos, or commentary about events happening in their community.

*67% occasionally depend on it for turn-by-turn navigation while driving.

*62% have looked up information about a health condition.

*56% occasionally learn about community events or activities.

*57% do online banking.

*44% look up real estate listings or other information about a place to live.

*43% look up information about a job.

*40% look up government services or information.

*30% take a class or get educational content.

*25% occasionally get public transit information.

*18% have submitted a job application.

*11% occasionally reserve a taxi or car service.

Source: Smith, 2015

Some Features are Popular With a Broad Spectrum of Smartphone Owners; Social Networking, Watching Video, and Music/Podcasts are Especially Popular Among Young Users

% of smartphone owners in each age group who used the following features on their phone at least once over the course of 14 surveys spanning a one-week period

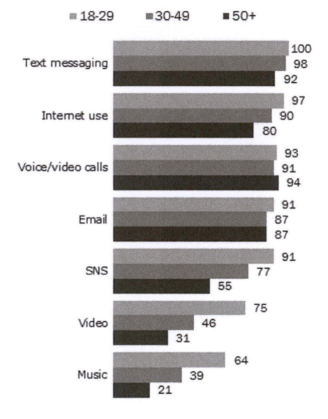

Pew Research Center American Trends Panel experience sampling survey, November 10-16 2014.

Respondents were contacted twice a day over the course of one week (14 total surveys) and asked how they had used their phone in the preceding hour (besides completing the survey). Only those respondents who completed 10 or more surveys over the course of the study period are included in this analysis.

PEW RESEARCH CENTER

FIG. 6.6 U.S. Smartphone Use in 2015 *Source: Smith, 2015*

FYI: Smartphone Phone Activity by Age Group

Age Group	18–29	30–49	50–64	65+
Send/receive text messages	97%	94%	75%	35%
Access Internet	84	72	45	19
Send/receive email	73	60	43	17
Download apps	77	59	33	14
Talk	92	92	92	91
Social media	74	69	67	58
Camera	54	50	50	56
Get directions	68	57	37	19
Listen to music	80	59	26	08
Video chat	40	24	10	03
GPS	43	40	34	35
News	18	17	15	13
Play games	52	49	50	53

Source: Smartphone usage and demographics, 2014

MOBILE PHONES: BENEFICIAL OR IRKSOME?

In an episode of the old sitcom *Friends*, Ross was getting married (again), this time in England to Emily. As his friend Joey was about to escort Emily's mother down the aisle, her cell phone started to ring. She answered it and handed the phone to Joey. Much to Ross's annoyance, Joey not only took the call but held up the phone throughout the ceremony so a friend back in the States could listen to the vows.

Although this incident was just a funny scene in a television sitcom, such use of mobile phones is common today. Stories about mobile phones ringing during speeches given by dignitaries and heads of state pervade the media. In fact, in response to these increasing interruptions, former president George W. Bush banned cell phones from his staff meetings. No event or location is sacred from mobile phones. They ring during funerals, in theaters, during concerts, and in classrooms, often despite regulations and urgings to turn off the ringer. And of course, the unfortunate practice of talking or texting on a mobile phone while driving is particularly dangerous as well, even more risky than driving while intoxicated. Even using voice-activation systems, like Siri, can be just as distracting as using a handheld cell phone.

FYI: Cell Phones, Texting, Driving

Young adults confident they can drive safely and text	77%
Young drivers who see parents use cell phone while driving	48%
12- to 17-year-olds in car while driver texting	48%
Adults who send/receive text messages while driving	27%
Collision risk when texting	23%
Drivers who surf the Web while driving	20%
Teens who text and drive outside their lane	10%

Source: DWI: Driving While Intoxicated, n.d.

In the 20 years since cell phones made their way to the consumer market they have had a strong social and cultural impact. In some respects, cell phone users have become tethered to their friends and family, as mobile phones are often used to keep tabs on each other. Some people are so connected that they have to let someone know where they are every minute of the day. Moreover, families and friends come to expect constant and immediate contact and may become alarmed if there is no answer.

There was a time when answering machines gave people a new way to project their self-image and impress others with a clever outgoing message. Many people thought of their answering machine as a barometer of popularity and could not wait to get home and see how many messages they had received. Whether people used answering machines to filter calls, charm others with witty messages, untether themselves from their landlines, or build self-esteem, they quickly came to depend on them. Like answering machines, cell phones serve many purposes. They can make users feel popular, important, and independent yet needed.

Opinions vary regarding proper mobile phone etiquette. At one time when cell phones were rare and reserved mostly for professionals who could afford one, when a cell phone would ring, others would give the recipient admiring glances. But now public mobile phone use is regarded as annoying, intrusive, and irritating yet acceptable and even normal. Think of how many times inane conversations take place: "Honey, do you want me to buy Twinkies or Ding Dongs?" It almost seems that cell phones have taken away the ability to make even the most mundane decisions without consultation. Worse, when on a cell phone, people speak at a volume that is three times louder than when in face-to-face conversation, so what they say reverberates for all to hear. Many wonder whether the mobile phone frenzy is a technological step forward or a step backward, but it seems to have become a style of communicating that is now a permanent part of our culture.

Despite the trivial uses of cell phones, the advantages can outweigh the disadvantages. Cell phones help users keep in touch about important matters and give comfort when physically isolated. Moreover, they are convenient and easy to use and, in many cases, less costly than making long-distance calls over traditional landlines. They can also be credited with saving lives in emergencies.

Voice-to-voice calls, however, are not the only way cell phones are used to communicate. **Text messaging** has become a more popular means to get in touch. Users send or receive about 40 texts per day on average. Six billion SMS (short message service) text messages are sent every day, and more than 2.2 trillion are sent each year in the United States. Globally, 8.6 trillion text messages are sent each year. Although there are many advantages to texting, critics claim that it is out of control. Some teenagers send and receive more than 3,000 text messages per month, which equates to about 100 texts per day. A 13-year-old from California racked up 14,528 texts in one month. Parents, teachers, and psychologists are concerned that adolescents often text late into the night, during school, and as an escape from homework and other activities.

Digital technology has created a new world of communication. Because of smartphones and tablets, communication among friends and strangers takes place 24/7, and users are unchained from desktop computers, landline phones, paper and pens, and even geographic locations. But along with instant communication comes cultural changes. Interpersonal relations, the way information is gathered and disseminated, the way media are used, how situations are analyzed, problems solved, and decisions made are all influenced by technology.

ZOOM IN 6.3

How wireless technology works: www.wireless-technology-advisor.com/how-does-wireless-technology-work.html

How home networks work: www.cnet.com/how-to/home-networking-explained-part-1-heres-the-url-for-you/#!

PHONE CALLS OVER THE INTERNET (VOIP)

Plain old telephone service (POTS) is suffering from competition of cell phones. Dealing yet another blow to standard landline phone services is Internet telephony, which is a way to make phone calls over the Internet through systems known as **VoIP (voice over Internet protocol)** and **IP telephony**.

Internet telephony is a packet-switching system—similar to the Internet. Packet switching is much more efficient than the cellular phone system and the POTS circuit-switching system, in which circuits are open and dedicated to the call. With packet switching, the connection is kept open just long enough to send bits of data (a packet) back and forth between the caller and the receiver, so connection time is minimized and there is minimal load on the computer. In the transmission space taken up by one POTS call, three or four Internet calls can be made. Basically, Internet telephony digitizes voice and compresses it so it will flow as a series of packets through the Internet's limited bandwidth.

- *Computer or mobile device to telephone.* Free phone calls from a computer or a mobile to a telephone number—whether to a landline, cell phone, or smartphone. To make a free call using an Internet phone, software needs to be downloaded from an Internet phone provider Web site.

- *Computer or mobile device to computer or mobile device.* This Internet method is probably the easiest way to connect, and long-distance calls are free. In most cases, all that is needed is an Internet connection, speakers, a microphone, and software, such as Skype or Facetime. Both the caller and the recipient must be using the same application or be on the same Web site. The calls can be either voice only, voice-to-video, or video-to-video.

Making a call over the Internet has become a very popular way to connect with friends and family, not only because the calls are free but also because of the video capabilities. Seeing someone while talking is better (in most cases) than just hearing a voice. Parents delight in seeing their children and grandkids, couples separated by distance feel closer, and employers save travel costs by interviewing candidates over an Internet call.

Placing an international call via the Internet is much easier than connecting telephone-to-telephone, plus it is free. The use of Skype for international calls is growing at a faster rate than the use of a traditional telephone company (telco). In other words, Skype calls are replacing international telephone-to-telephone calls. In 2013, Skype captured an additional 36% of the international calling market. Users spent a total of 214 billion minutes on international Skype-to-Skype calls.

VoIP calling has so many advantages that the combined business and residential VoIP services will grow to be worth $88 billion in 2018. These companies will be worth more because they will become a more essential service in the future. The number of residential and small office/home office (SOHO) subscribers to hosted VoIP services is expected to grow from 955 million in 2013 to 1.2 billion worldwide by the end of 2018.

As mobile VoIP usage increases and more people make the switch to this technology and lower monthly phone bills, innovative new services continue to be developed. One of the most recent developments is VoWi-Fi, or VoIP Wi-Fi, which is a wireless version of VoIP. VoIP innovations provide new ways of one-to-one communication.

As technological changes lead to social changes, social changes lead to behavioral changes, such as dropping a landline phone for a cell phone or eschewing Verizon, AT&T, Sprint, T-Mobile, or other carrier that charges for calls in favor of free Internet calls.

A WIRELESS WORLD OF NEWS AND ENTERTAINMENT

Restaurants, coffee shops, hotels, and universities are just some of the many places that provide wireless Internet service for their customers. Data transfer speeds are increasing, and connecting is easy. Wi-Fi–enabled laptops, smartphones, and tablets connect to the Internet without the annoyance of cords.

Wireless capability has already changed the way people access information. Mobile device users download the latest news while taking a walk, check their email while ordering lunch, and collect research for their homework assignment while working out in the campus gym. In late 2014, Apple launched its Apple Pay system that lets users pay for many goods and services just by waving their cell phone in front of an electronic reader. Apple Pay digitizes and replaces the credit or debit magnetic stripe card transaction at credit card terminals. The system has seen steady increase in users and support from banking partners and retail stores. Apple Pay is also available in Europe.

AUDIO/SATELLITE RADIO

Audio delivery via the Internet is supported by faster connections and improved bandwidth compression—less bandwidth needed per service. Online audio has clear advantages over traditional radio. Online listeners do not have to depend on local stations to hear their favorite music, and they can connect to hometown news. Online listeners select their favorite types of music and listen to it in the order they prefer. On a portable player, such as an iPod, users can download hundreds of hours of music and create their own playlists and organize the songs according to preference.

Two satellite radio companies, XM and Sirius, took the chance in the early 2000s that over-the-radio listeners would pay to listen to commercial-free, or at least commercial-low, radio delivered by satellite. It took a few years for the concept to catch on, and both companies suffered financially, but deals made with automakers to provide in-car receivers and contracts with big-name stars like Howard Stern boosted the companies' fortunes. In 2008, the companies merged becoming SiriusXM. The service offers close to 200 commercial and commercial-free stations, ranging from news, talk, sports, and old-time radio shows to an array of music such as genre-oriented jazz, soul, alternative, and garage band channels, decade-categorized channels, even ones dedicated to particular bands like the Grateful Dead channel. No matter where the almost 30 million subscribers are located, they can listen to whatever type of music they like, and they do not have to worry about driving out of a station's broadcast range; the signal stays with them. A jazz fan driving

FIG. 6.7 Downloading music onto a portable player
Photo courtesy of iStockphoto. © abalcazar, image #7238674

through rural Nebraska might have trouble picking up an over-the-air local station that plays that type of music, but SiriusXM's jazz channel will tag along for the ride across the country.

SiriusXM faces a serious challenger—Internet radio. Company executives fear that drivers will not want to pay for the satellite service when they can just as conveniently tune to Spotify, Pandora, Beats, or other customizable online audio service. Even though 75% of all vehicles manufactured in the U.S. come with SiriusXM installed, the connected car, with the Internet integrated into the dashboard, could prove a formidable competitor.

TELEVISION

Program providers such as the broadcast networks, cable channels, and even some independent production houses realize that using the Internet for delivery of their products is technically viable and beneficial. Also, the broadcast networks are pushing audiences to their Web sites to encourage them to view episodes of their favorite shows. Many online television sites merely serve as promotional vehicles for their broadcast and cable counterparts. However, about 90% of television programs are now available in their entirety on the networks' sites. Given the Internet's vast storage capabilities, it is easy to go online and watch any program at any time without being tied to a television schedule. The current television era is a complex media environment that is forcing changes to the traditional broadcast models of program delivery.

Over-the-Top Content

Over-the-top content (OTT) is a relatively new term that refers to the delivery of audio and video programming over the Internet, bypassing or going 'over the top' of cable and satellite providers. OTT describes third-party content providers such as Hulu and Netflix. Many viewers watch their favorite programs, such as *Family Guy*, *Grey's Anatomy*, and *South Park*, newscasts, and movies for free or for a monthly fee on Netflix, YouTube, and other OTT services. About four in ten households now have a subscription to a video-on-demand OTT service, such as Netflix or Amazon Prime Instant Video, up from 35% in 2013.

Netflix provides on-demand Internet streaming to 70 million subscribers in North America, South America, and parts of Europe, and it delivers DVDs in the U.S. for a flat-rate fee of about $8 to $10 per month. Netflix started out as a service that streamed reruns first shown on broadcast or cable channels. But Netflix has since scored big with original programming like Emmy-winning *House of Cards*, *Daredevil*, and *Jessica Jones*, and with new episodes of the cancelled Fox sitcom *Arrested Development*. Netflix is going back to the future with its new family-oriented *Fuller House*, a revival of the old ABC sitcom *Full House* (1987–1995). It is hoped that the show, which stars several of the actors from the original series (minus the Olsen twins), will attract family viewership and add to the Netflix subscription list.

Amazon is also producing original programs, such as *The Man in the High Castle*, for viewing on its Amazon Video network. The successes of Netflix and Amazon have compelled traditional television networks to follow their lead and stream their own programs. CBS did just that when it notably live streamed for free the Super Bowl in 2015 and 2016 on cbssports.com. And in the second year, it streamed the same commercials as they were shown on television. About 5% of those who planned to watch the game in 2016 said they would do so online, compared to 57% via cable, 28% on satellite, and 9% with an antenna. Although most fans watched the game on television, streaming the game opened viewing to cord-cutters and **cord-nevers** and created promotional buzz, which might have increased viewership.

HBO fired up HBO Now, a stand-alone, untethered-from-cable, video-streaming/OTT service in April 2015. HBO Now streams programs such as *Game of Thrones* and *Veep*, which are also shown on HBO cable, but because HBO Now bypasses cable it is an OTT service. HBO Now is aimed at cord-cutter and cord-nevers who have an Internet connection but eschew cable. HBO Now has been slow to gain an audience and one year since its debut, its 800,000 subscriber number is not even close to Netflix's tens of millions.

Apple TV is poised to become the "biggest gateway to online video—the new Comcast for the Internet" (Kang, 2015a, A14). Apple TV is a digital media player and pay service that hooks to a television via an HDMI cord to deliver digital programing from OTT providers and broadcast and cable networks. Although on the surface the service sounds like it is a good alternative to cable television, it could be that viewers end up trading cable-bundled channels for Apple TV–bundled channels but in the long run do not save any money. Consumer demand for Apple TV seems dependent on consumer frustration and disappointment with cable and satellite television packages and prices. But Apple TV is also moving into the programming business with its first original program starring Dr. Dre. The debut has not yet been set, but hopes are high that the show and others that may follow will propel Apple to a major force in the television industry.

Google also has its eye on television of the future and is poised to lead the move from cable or Apple TV–type boxes and players to digital delivery integrated within a smart television set. For now the company is watching and waiting for a pending FCC decision that could force cable companies to release some control of their set-top boxes.

YouTube is a bit different from other program distribution sites in that videos are created and uploaded by ordinary online users, politicians, corporations, and schools but not necessarily by professional program and movie producers. The videos range from short, homemade ones of silly things cats do to full-length, professionally created ones. Most videos are free, but starting in 2013 some cost 99 cents per viewing. More than 1 billion users watch about 6 billion YouTube videos each month—that is almost 1 hour for every person on Earth. YouTube reaches more U.S. adults ages 18 to 34 than any cable network.

Premiers of new television series are important to networks and to the studios that produce them. Broadcast networks generally debut new programs during late September and early October. A new programming strategy is to release the first episode online days, weeks, and even months before its actual television premiere. An early online preview of Starz's *Outlander* drew almost 1 million viewers. The Disney Channel now regularly debuts almost all of its shows online, on its official WATCH Disney streaming app. In the autumn of 2014, three different broadcast networks placed the full versions of some new shows online in August, well before their TV premieres. NBC's *A to Z* (premiered Oct. 2), ABC's *Selfie* (premiered Sept. 30), and Fox's *Red Band Society* (premiered Sept. 17) were posted on Hulu and their networks' respective Web sites before the series started on television. But the strategy of online previewing is only as good as the program, and in the case of the three shows above, all were cancelled within weeks of their television premiere, but they could still be seen on the Internet.

Other types of online programs include Webisodes, which are either streamed or downloaded onto a computer. Webisodes are mini-programs—some series consist of only a few episodes, each of which may only be a few minutes long. Webisodes such as *Convos with my 2-year-old*, *2040*, and *Bee and Puppycat* have a following of their own and are considered a new entertainment genre. Webisodes are streamed from sites like YouTube and Vimeo, and many are produced by media companies such as Warner Bros., NBC, and Sony. Popular Webisodes often develop into longer series with help from crowd-funding, like Kickstarter, to raise money for production. Webisodes have their own award ceremonies, plus of course fans, critics, and blogs.

Celebrities and companies often create Webisodes as part of their digital marketing campaigns. For example, Feed Us is a company that develops Web-based content-management applications. Instead of a hard, direct pitch about the features and benefits of its products, Feed Us produced humorous Webisodes about two of its technicians who answer customer problems over the phone. *Funny or Die* is a site that includes original and user-generated Webisodes. Viewers vote for their favorite

Webisodes to stay on the site, while less appealing ones are booted off.

Since the changeover to digital broadcasting, viewers experience the advantages of television sets with built-in wireless Internet capabilities. Viewers are beginning to wonder if cable and satellite subscriptions are necessary if they can get their favorite programs and movies directly from the Internet but watch them on the television screen. Many young adult viewers have given up their expensive cable or satellite subscriptions and now only use an Internet connection, their computers, and a flat-screen monitor to provide them with video entertainment. And flat-screen monitors come in various sizes and clarity. The newest '4K' display is super high definition, with 4,000-pixel horizontal resolution. In other words, the picture is almost as clear as being there in person.

Nielsen reported in 2012 that the number of American households with a television declined 2 years in a row, as viewers transitioned to the Internet. Further, between 2013 and 2015, the number of households with Internet service but no television increased 12%, but those households represent only 3% of the total.

As entertainment television moves online, so does news and information programming. Media sites stream live newscasts so people away from their televisions can see the latest happenings. For example, on the day Michael Jackson was buried, the Small Business Administration's Internet connections had slowed dramatically because so many employees were watching the live newscast of the funeral online. Workers without access to a television set got to see the day's events online exactly as shown on the television networks. Online news delivery, however, is cutting into broadcast and cable news television audiences. For example, CNN's viewership is declining, and in January of 2015, it lost 39% of prime-time audience from the same period in 2013.

The Internet, Wi-Fi, and smartphones, combined with access to social media sites where content is controlled by users, are compelling media to make as big changes in news delivery as when CNN ushered in satellite news gathering in 1991.

Place Shifting/OTT

Dish Network—Sling TV: Sling TV is an OTT provider that is also ushering in a new way of watching television called *placeshifting*. Sling TV is a streaming device that lets users remotely watch cable and satellite channels on an Internet-connected computer or mobile device. With Sling TV, viewers never miss their favorite shows while away from home and without a television. Sling TV works through an app that can be installed on multiple devices and works with streaming media players, smart televisions, and Android and Apple devices. A major downside to Sling TV is the absence of the broadcast networks.

But the upside is that a viewer who is vacationing in Tahiti but wants to watch a college bowl game can do so his or her laptop through Sling TV. Sling TV's deal with ESPN to stream its games live might just be what it takes to get diehard cable subscribers to drop bundled cable channels and 'appointment television' in favor of personalized, anywhere, anytime viewing.

> ### FYI: Binge Viewing
>
> Binge watching is viewing two or three episodes of a program at one sitting. This cultural change in the way television is watched has been aided by services such as Netflix and Hulu. Slightly more than 6 in 10 Netflix viewers binge watch at least once every few weeks.
>
> *Source: Thanks to DVR and streaming services, 2013*

Time Shifting/DVR

Personalized television viewing is more than just having a wide selection of programs that fulfill individual needs but also includes choosing when to watch. Online 24/7 streaming provides individual scheduling, as do digital video recorders (DVRs). The DVR has long replaced the old-fashioned videotape using VCR, but the purpose is the same—to record programs for later viewing. A DVR digitally records and stores hundreds of hours of programming that viewers can watch whenever they want, and they can fast-forward through commercials. Some DVRs record up to four shows at once, even while channel-surfing back and forth. If a viewer is hungry but engrossed in a live program, he or she can pause the show, make a snack, and then come back and pick up the program where it left off. With a DVR, viewers have ultimate control over their viewing experience. They can fast forward, reverse, pause, and use instant replay or slow motion and never miss a scene.

DVR viewing is popular. Viewers spent 14 hours and 20 minutes per month watching time-shifted television in 2014, up from 13 hours and 12 minutes in 2013. About

FIG. 6.8 Television has come a long way
Photo courtesy of iStockphoto. © fredrocko, image #6954585

48% of U.S. households have at least one DVR, and the television is on 15% longer than in homes without DVRs. DVR growth is slowing, however, as viewers turn to streaming services.

Streaming services have benefitted from the decline in cable and satellite television subscriptions, which have dropped about 1.2% between 2010 and 2016. Although this small percentage might not seem like much, it translates into about 1 million fewer subscribers. Moreover, viewers are watching 12 fewer minutes of television. Almost 13% of households now have at least one smart television that can stream Internet programs, and some households are getting rid of their television sets altogether. Yet despite all of the online viewing and movement way from cable and satellite services, television is still tops when it comes to entertainment. Viewers watch about 135 hours of television per month compared to about 13 hours and 44 minutes on the Internet.

FYI: DVR Influence on Ratings

DVR Viewing increases TV Shows' Ratings (May 2015 for 18–49 age demographic)

Wayward Pines (Fox)	110%
Marvel's Agents of S.H.I.E.L.D. (ABC)	92%
Blacklist (NBC)	81%
Modern Family (ABC)	70%
Scandal (ABC)	70%
Grey's Anatomy (ABC)	64%

Source: Kondolojy, 2015

FYI: DVR Use

76% of homes use DVR, Netflix, video on demand

47% of homes use DVR

66% of homes using DVR have more than $75,000 annual income

33% of homes using DVR have less than $30,000 annual income

Sources: 76 percent of homes have DVR, 2015; DVRs leveling off, 2013

FYI: Most Time-Shifted Programs of 2015 (by percentage change in viewers)

1.	Fargo (FX)	+240.3%
2.	True Detectives (HBO)	+220.7%
3.	Better Call Saul (AMC)	+215.2%
4.	Bates Motel (A&E)	+201.8%
5.	Game of Thrones (HBO)	+195.9%

Source: Pennington, 2015

FYI: Binge Viewing—Percentage of Internet Users Who Watch Time-Shifted TV (2013)

Age group	18–29	30–39	40–54	55+
Video-on-demand	47%	45%	41%	35%
Hulu, Netflix	71	60	33	19
TiVo/other	27	46	40	36
Amazon	17	18	8	6
Time shifters who have binged viewed	78	73	58	48

Sources: Hinckley, 2014; Thanks to DVR and streaming services, 2013

FYI: Methods Used to Binge View (%)

Primary

Online subscription service	41%
Network/cable Web site	15%
DVR	15%
Rent DVD	14%

Have Used

Online subscriptions service	63%
Network/cable Web site	53%
DVR	44%
Rent DVD	51%

Source: Thanks to DVR and streaming services, 2013

FYI: Aereo

Aereo was a short-lived way to time-shift television programs. Aereo started in March of 2012 and worked by leasing a remote antenna to its subscribers. The antenna picked up broadcast television for live or recorded viewing. The antenna was what made Aereo different from Internet streaming services. In other words, it was an over-the-air service, not an Internet based one.

Aereo filed for bankruptcy in November of 2014 after several broadcast networks sued for copyright infringement. The broadcasters asserted that Aereo needed to pay royalty fees just like cable companies. Aereo's defense was that it was obtaining the signals legally and simply making them available to the public. However, the court ruled in favor of the networks. Aereo was available in New York, Atlanta, and Boston and had about 80,000 subscribers before it was forced to shut down.

OTHER PERSONAL DIGITAL DEVICES

GOOGLE TRANSLATE

Google is adding a captioning system to many of its videos that translates English audio into text in 90 different languages, thus expanding its market of non–English speakers and the hearing impaired. The captioning system allows users to search for particular words within videos, and it automatically tags photos with a description, such as "Two pizzas sitting on top of a stove." This technology shows the future of artificial intelligence when machines will do the work users now spend time doing.

E-BOOKS

The point of having an e-reader is that hundreds of books can be downloaded onto one device, which is similar in size to a typical paperback novel. On newer e-readers, users can underline, highlight text, and write notes in the margins, and some apps work with e-readers to display photo galleries and 3D models. Even though most people still prefer a printed book to a digital one, e-reader use is on a steady climb—about 32% of U.S. adults own an e-reader.

Amazon and the e-book maker Kindle have formed a quid-pro-quo partnership. Amazon provides the digitized books that are downloaded onto a Kindle. This union is a good example of the changing media environment, in which new models of monetization are being developed as new devices become available.

CELL PHONE/WRISTWATCH

Just when the world thought multiuse mobile devices could not get any smaller or any more portable, here comes the cell phone/wristwatch. In early 2003, Microsoft introduced smart personal object technology (SPOT) software, which is driven by tiny but powerful microchips that were the basis for the first 'smart watches.'

The Apple Watch is the latest big consumer hit. Worn on the wrist like an ordinary watch, the Apple Watch face displays the time amid app icons that pop open at a touch. The watch is paired to the wearer's iPhone and uses both Wi-Fi and Bluetooth to transfer data back and forth. Unless a wearer has super-sharp eyesight, the Smart Watch might be bit awkward to use, but otherwise it is a convenient, multiuse gadget that is sure to turn heads.

Even before its release to the consumer market in spring 2015, there were already more than 4,000 Apple Watch apps to do everything from receiving news, weather, traffic reports, stock updates and recording sleep activity to sending and receiving texts, getting phone calls, listening to music, setting calendar dates and reminders, and using Siri. Apple Watch also monitors fitness. Like a Fitbit Flex and other such devices, it tracks physical activity such as number of steps taken, calories burned, and distance traveled.

GOOGLE GLASS

Not every new device is successful. Some have a limited life span. Google Glass, the high-tech eyewear, debuted with much fanfare in 2013. Worn like a pair of glasses, Google Glass was promoted as the wearable Internet. Put them on and have online connectivity anywhere. Google Glass functions with voice commands and by touching the glasses in different spots. The consumer cost was about $1,500, much more than a smartphone. Sales did not live up to expectations, and so in January of 2015, the consumer version was taken off the market.

WEBCAMS

A Webcam is a simple, low-cost video camera that transmits live images through the Internet. Video from older Webcams was slow and jerky because cameras could only capture an image once every 5 or 10 seconds because of bandwidth limitations. But newer Webcams operate at 30 frames per second, the same speed as a television camera. Online video looks just as sharp and flows just as easily as what is on television.

Webcams are also great devices for communicating with family members in the service or with a boyfriend or girlfriend studying abroad. Users mug for the camera and send all kinds of goofy images back and forth. A Webcam helps ease loneliness by visually connecting people to each other.

Critics contend that Webcams appeal to people who are voyeuristic and that except for parents and a few close friends, most people really do not care about peeking into the boring lives of others. However, Webcams also have more serious uses. For example, they are used for Internet conferencing, as public relations tools (such as a city or university campus streaming live Webcam images for potential visitors), and for online courses. Some radio station Web sites entertain users by focusing Webcams on their disc jockeys at work, and some local news stations have installed Webcams on reporters' desks so they can tape bulletins without going into the studio.

The Internet abounds with live Webcams turned on funny pets, a live colony of ants, street musicians, and just about anything or anybody imaginable. One guy even has a Webcam inside his beer refrigerator. A person does not have to be a famous rock star with a reality program to be live on the screen. With a Webcam setup, anyone can let the world into his or her life.

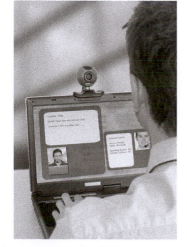

FIG. 6.9 Internet users connect using Webcams.
Photo courtesy of iStockphoto. © andresr, image #2248379

ZOOM IN 6.5

Check out these Webcam links:

- WebCam World: www.webcamworld.com
- EarthCam: www.earthcam.com/usa/louisiana/neworleans/bourbonstreet
- Hermosa Beach, California: www.hermosawave.net/webcam/
- Animal Planet Cam: www.apl.tv/

A WORLD OF DIGITAL DEVICES

More than just fancy gadgets, smartphones, tablets, and watches serve the purpose of keeping us connected to information sources and to other people. As their functional utility increases, the personal gratifications keep people hooked. These devices become more than just tools for telling time or keeping a schedule. Rather, they keep us involved in the world. Our own personal sphere widens as we let in more and more information, which has led some people to wonder just how much we can absorb before reaching an emotional and psychological limit.

As convenient as mobile devices may be, there is the issue of feeling overwhelmed by an excess of information. Critics question the need for having such immediate and often trivial information at our fingertips and are also concerned about our overreliance on these devices. It is especially annoying to be with someone who keeps looking at the device or, worse, using it instead of conversing. Nothing like having a friend whip out his or her device just for the pleasure of proving you wrong about some subject. And nothing is private. Every time a smartphone is used to download an application, view a Web site, or engage in other activities, marketers are tracking movements and tailoring ads to match the user's interests and demographic profile.

Online users are bombarded with commercial messages, pelted with sound bites, and bored to death with the tedious details of friends' and family members' lives. Every beep of the mobile device means more information. At some point, people begin to question how much information they really need and to wonder when keeping in touch crosses the line to being under surveillance.

FIG. 6.10 Studying on the campus green with a laptop
Photo courtesy of iStockphoto. © quavando, image #4178977

SEE IT LATER

CORPORATE CHANGES

A growing trend is the convergence of various data and voice networks. Worldwide telephone, cable television, wireless communication, and computer data networks are becoming less of a standalone system; that is, they are converging into a powerful unified network based on the Internet protocol packet-switching system, which is versatile and can transmit any kind of information quickly and at low cost.

Rather than having people make calls to verify transactions, place orders, or move money from businesses to banks and so on, computers are doing it over high-speed broadband wires. And as the amount of information transferred continues to grow, the need for more bandwidth will become more acute as well.

As is the case for other delivery systems, telecommunications companies will continue to consolidate. Mergers, such as the one between Cingular and AT&T in 2004, formed huge communication companies, leading this part of the industry to an oligopoly similar to those in the cable and satellite industries. However, regulators are showing some concerns. AT&T's 2013 attempt at acquiring T-Mobile for $1.2 billion was rejected by U.S. regulators.

In early 2015, the FCC held an auction, in which it sold the rights to use parts of the electromagnetic spectrum to accommodate the growing use of digital media. Companies like AT&T, Verizon, and T-Mobile bid about $44 billion for these frequencies that are needed to further develop wireless networks, cell phones, satellite television, and other electronic media.

LIFESTYLE CHANGES

Someday almost everyone in the world will own a smartphone. Few people leave their homes without their smartphone in tow and some even sleep with it under their pillow. The smartphone has become our communication hub and security blanket. In Japan, a young female store clerk used a smartphone to snap a photo of a man who had fondled her on a commuter train. She then sent the photo to the police, who promptly arrested the perpetrator. Smartphones were on the scene of the Boston Marathon bombing in 2013, and Internet-posted videos of the aftermath helped investigators piece together the sequence of events leading up to and after the bombs exploded. Smartphones recorded the peaceful marches and riots in Ferguson, Missouri, in 2014 after a police shooting of an African American man. In April of 2015, the *New York Times* was provided video taken with a smartphone of a police officer in South Carolina shooting an apparently unarmed man after the man was stopped for traffic violation. New apps are being developed to specifically monitor police activity. For example, Cop Watch is an iPhone app that automatically begins recording when the icon is tapped and automatically uploads the video to YouTube when the recording is stopped. Smartphones and such apps are already changing law enforcement procedures.

The new capabilities of smartphones may bring some legal and ethical problems as well. The managers of athletic clubs and other facilities with locker rooms worry about lawsuits stemming from secret smartphone photo snapping. There is concern that laws may be passed that require smartphones to include 'kill' switches that make it possible to disable the phone remotely. The kill switch could help stop cell phone theft, since the phone could be shut down if stolen, but phones that could also be shut down by government officials would limit citizen journalists from recording important events.

Smartphones are sure to become smarter and will eventually become miniature handheld or wearable computers. Google was the first to come up with wireless eyeglasses that delivered Internet content directly into the wearer's retinas. But consumers thought the glasses looked silly, and so Google took it off the market. Google Glass or some other type of wearable computer might one day be back in favor, and how these miniature take-anywhere eyeglasses and wearable computers will change us culturally is yet to be seen.

Television and audio remain an important part of life. Although DVR growth seems to have stalled out, large-screen televisions, HDTV, 4K screens, and Internet audio are still exciting. But it always comes down to the same question: "What's next?"

Perhaps it will be the *hypersonic sound system (HSS)*, which takes an audio signal from any source (television, stereo, CD, or computer), converts it to ultrasonic frequency, and directs it to any target up to 100 yards away. For example, one roommate could watch television while the other blasted the stereo, and neither would hear what the other is hearing. Likewise, a car full of passengers could each listen to his or her own music without hearing what others have on, or a nightclub could have several dance floors, each playing a different type of music and none interfering with the others.

FIG. 6.12 Modern home entertainment system
Photo courtesy of iStockphoto. ©3alexd, image #5267037

ZOOM IN 6.6

How HSS Works
Check out this Web site to watch the inventor of HSS talk about how HSS works: www.ted.com/talks/woody_norris_invents_amazing_things?language=en

FIG. 6.13 Teen technophile
Photo courtesy of iStockphoto. © stray_cat, image #11186952

The next big device could be Melomind, a headband that measures brainwaves in the same way as an electroencephalogram (EEG) and uses the output to create a playlist of relaxing music. Melomind pairs with a mobile app that actually plays the tunes.

THE GRID AND THE INTERNET OF THINGS

Perhaps the future is 'the grid.' Imagine riding in a car that is automatically steering you to your destination while also scanning your stock portfolio in 3D, transmitting your vital statistics to your physician as part of your annual exam, sending engine performance data to your mechanic, and keeping an eye on your favorite television program. Also, you are wearing an outfit that was coordinated by microchips embedded in your body.

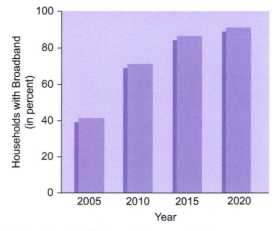
FIG. 6.11 Broadband—U.S. household penetration forecast
Source: Vanston, 2002

The Internet of Things (IoT) describes **machine-to-machine (M2M) communication** that is possible because of cloud computing and networks of data-gathering sensors, which can be placed in any 'thing'—device, person, or animal. A 'thing,' in the Internet of Things, can be a person with a heart monitor implant, a farm animal with a biochip transponder, an automobile that has built-in sensors to alert the driver when tire pressure is low, or any other natural or manmade object that can be assigned an IP address and is provided with the ability to transfer data over a network. Products built with M2M communication capabilities are often referred to as being smart.

Wired cement provides this example of how IoT will be used in the future. Engineers will soon use smart cement to build bridges. The smart cement will be equipped with sensors to monitor stresses, cracks, ice, and warps, and will alert engineers to fix problems before a catastrophe arises. Smart cement will also alert drivers to road problems; the sensors in the concrete detect trouble spots and communicate the information to cars via the wireless Internet. Once a car knows there is a hazard ahead, it will instruct the driver to slow down, and if the driver does not, then the car will slow down automatically. This example is just one of the ways that sensor-to-machine and machine-to-machine communication can take place.

IoT devices are all the rage as consumers discover not only the 'cool' factor but the convenient functionality. The number of IoT devices grew by more than 20% from 2013 to 2014, and it is predicted that by 2020, there will be about 40.9 million of them in the U.S. alone.

No device is completely secure, and security breaches will become more frequent and disrupting as more devices are added to IoT. In 2014, a Hewlett Packard study found that common smart items such as televisions, Webcams, thermostats, sprinkler systems, home alarms, garage door openers, baby monitors, security cameras, and surgical robots have already been hacked.

FYI: The Internet of Things: Security Problems

80% of devices raised privacy concerns.

80% of devices failed to require passwords of sufficient complexity or length.

70% of devices did not encrypt communication to the Internet and local network.

60% of devices displayed concerns with Web interface.

60% of devices did not use encryption when downloading software updates.

Source: Hewlett Packard, 2014

More advanced than any current technology, the grid is a "linkage of many servers into a single system in which complex computing tasks are parceled out among various machines." The grid is sort of a super-supercomputer that

acts as a "universal translator between previously incompatible computer systems. It can also turn information into visual representation of, or solution to, a problem" (Foroohar, 2002, p. 34J). Big businesses are clearing the way for the grid, which they consider to be one of the most advantageous and time- and money-saving inventions of the past 200 years. By embedding microchips in inanimate objects (and even in humans), companies can link to each other, to consumers, to governments, and to other institutions.

The largest grid computing network, accessing 170 computer centers in 42 countries, is used by the physicists as CERN (Conseil Européen pour la Recherche Nucléaire). It allows scientists to analyze more than 26 million gigabytes of data produced by the Large Hadron Collider (LHC) each year. Physicists access CERN's grid using mobile devices like iPhones and iPads to submit work, monitor results, and control processes.

ZOOM IN 6.7

To visit the LHC grid, go to http://wlcg.web.cern.ch/

Another computing grid developed by IBM and University of California, Berkley, is already using the computing power of mobile devices. The World Community Grid (WCG) now allows Google's Android devices to work on the grid. In 2013, researchers began using the combined power of volunteers' handsets to search for new drugs to fight HIV. According to its Web site, the World Community Grid enables anyone with a computer, smartphone, or tablet to donate their unused computing power to advance cutting-edge scientific research on topics related to health, poverty, and sustainability. Through the contributions of more than 650,000 individuals and 460 organizations, the World Community Grid has supported 25 research projects to date, including searches for more effective treatments for cancer, HIV/AIDS, and neglected tropical diseases. Other projects are looking for low-cost water filtration systems and new materials for capturing solar energy efficiently.

ZOOM IN 6.8

Visit the World Community Grid at https://secure.worldcommunitygrid.org/about_us/viewAboutUs.do

---

There are currently several other grid computing systems in operation around the world that are being used by groups of scientists to help them with research, including the search for extraterrestrial life and identifying potential new drugs. Developers hope that as more computers join the systems and they link together, a worldwide grid will be created, giving access to resources around the world at the touch of a button. For home users, it would mean no longer having to upgrade their computer every couple of years to ensure it is powerful enough to run the current software.

The grid has the potential of creating a real-life version of *The Minority Report*, in which the hero, played by Tom Cruise, is flooded with personalized ads and followed at every move by retina tracking devices. He reaches his breaking point and discovers that the only way to escape the grid is to have a black-market eyeball transplant. Privacy advocates fear such intrusion. Others claim, however, that consumers will always have the ability to control the information collected about and transmitted to them.

Even more formidable than the grid is artificial intelligence. Science fiction writers and scientists have long imagined a world in which computers transcend biology. Artificial intelligence is already being used to automate and replace some human functions. There are computers that can learn, answer questions, and solve problems. But the hopes are that computers will one day be self-aware and be of superhuman intelligence, and the thoughts and information in our brains can be downloaded or transferred to a computing environment.

While artificial intelligence is being developed, one elderly Microsoft researcher is moving data from his brain onto a computer. Video equipment, cameras, and audio recorders are his constant companions. They capture his every move, his conversations, and his experiences. Plus, he takes pictures of all of his receipts, event tickets, and other records. He is digitizing his life as an e-memory. It could be that someday everyone's life history will be online and searchable. Some people find this possibility very frightening, but others do not seem to mind at all.

In the end, it may all come down to what the famous journalist Edward R. Murrow said in his last public speech in 1964:

> *The speed of communications is wondrous to behold. It is also true that speed can multiply the distribution of information that we know to be untrue. The most sophisticated satellite has no conscience. The newest computer can merely compound, at speed, the oldest problem in the relations between human beings, and in the end the communicator will be confronted with the old problem, of what to say and how to say it. (Kendrick, 1969, p. 5)*

SUMMARY

This chapter covered new communication technologies and how traditional media have evolved to survive increased competition and changing audience needs. Although it is doubtful that traditional media as we know them today will ever become obsolete, they may have to carve out their own niches to remain competitive in the modern world of personal communication devices. Industry analysts insist that radio and television

FIG. 6.14 Actor Tom Cruise tries to escape the grid in the movie *The Minority Report*.
Photo courtesy of Twentieth Century Fox/ Photofest. © Twentieth Century Fox

FIG. 6.15 Pearls Before Swine comic
Source: From Washington Post 2/15/15 (PEARLS BEFORE SWINE (c) 2015 Stephan Pastis. Reprinted by permission of UNIVERSAL UCLICK FOR UFS. All rights reserved.

will continue to fulfill entertainment and information needs, but it could be that the needs served in the future will be very different from those served today. In all likelihood, traditional media will remain important sources of information and entertainment that appeals to the masses, but they will yield to newer communication technologies for the delivery of personalized news and services.

New communication technologies are creating a world in which the mass media are becoming more personal. Users determine which types of news stories they want to receive, what types of information are relevant to their lives, and when and where they want to be exposed to information. Commercial messages are tailored to personal characteristics and lifestyles. The electronic media still reach the masses, but the messages are personalized.

Streaming, video on demand, and services like Netflix, Hulu and Sling TV have changed the way the audience watches television. Young viewers tend to stream television programs at times that fit their busy schedules, and they tend to binge watch multiple episodes in one sitting. These new ways of viewing will change the way advertisers and programmers reach their audiences.

Some of the biggest changes will happen because of the convergence of devices, accessed through wireless networks and grids. Our lives will continue to be influenced by technological change. Concerns about privacy and hacking are only two of the issues that will need to be resolved. Even more important to consider are the changes that personal technological devices are making in each of our lives. What we can say for sure is that the media environment will continue to change as technology changes and people adapt that technology to create a better lifestyle.

FYI: Paradoxes of Technology

A famous French philosopher, Jacques Ellul, discussed the paradoxes of technology. Four are listed below. Consider how your use of technology has brought both good and bad effects to your life.

- All technical progress exacts a price; it adds something and subtracts something.
- All technical progress raises more problems than it solves and tempts us to see the problems as technical in nature and to seek technical solutions to them.
- The negative effects of technological innovations are inseparable from the positive; they are not neutral.
- All technological innovations have unforeseeable effects.

BIBLIOGRAPHY

76 percent of homes have DVR, Netflix or use video on demand. (2015, January 4). Retrieved from: www.benton.org/headlines/76-percent-homes-have-dvr-netflix-or-use-video-demand

About You Tube. (2015, January). *YouTube*. Retrieved from: www.youtube.com/yt/about/

Adoption of new technology. (2008, February 18). *Visualizing Economics*. Retrieved from: http://visualizingeconomics.com/blog/2008/02/18/adoption-of-new-technology-since-1900

Albrecht, C. (2008, September 15). Study: DVRs in 27% of homes. *newteevee.com*. Retrieved from: newteevee.com/2008/09/15/study-dvrs-in-27-percent-of-tv-homes/ [January 30, 2010]

Americans and their cell phones. (2006). *Pew Internet & the American Life Project*. Retrieved from: www.pewinternet.org/Reports/2006/Americans-and-their-cell-phones/2-Methodology.aspx?r51 [January 12, 2010]

Analysis of primary TV watching habits: Insight Report. (May 2014). *Biz Civic Science*. Retrieved from: biz.civicscience.com

Anderson, M., & Caumont, A. (2014, September 14). How social media is reshaping news. *Pew Research Center*. Retrieved from: www.pewresearch.org/fact-tank/2014/09/24/how-social-media-is-reshaping-news/

Anderson, N. (2007, August 13). PDA sales drop by 40 percent in a single year, vendors bolt for exit. *ARS Technica*. Retrieved from: arstechnica.com/business/news/2007/08/pda-sales-drop-by-40-percent-in-a-single-year-vendors-bolt-for-exit.ars [November 1, 2009]

Anjarwalla, T. (2010, July 9). Inventor of cell phone: We knew someday everybody would have one. *CNN*. Retrieved from: www.cnn.com/2010/TECH/mobile/07/09/cooper.cell.phone.inventor/index.html

App watch: Facebook's 11 million farmers. (2009, August 31). *The Wall Street Journal*. Retrieved from: blogs.wsj.com/digits/2009/08/31/app-watch-facebooks-11-million-farmers/tab/print/ [January 16, 2009]

Apple iPad. (2010). *Apple.com*. Retrieved from: www.apple.com/ipad/design/ [January 28, 2010]

Auletta, K. (2014, February 3). Outside the box: Netflix and the future of television. *The New Yorker*, Retrieved from: www.newyorker.com/magazine/2014/02/03/outside-the-box-2

Benner, K., & Sisaro, B. (2016, February 13). Apple TV and Dr. Dre are said to be planning an original TV show. *The New York Times*, p. B6.

Berquist, L. (2002). Broadband networks. In A. E. Grant & J. H. Meadows (Eds.), *Communication technology update* (pp. 257–267). Oxford: Focal Press.

Best for last? (2009, August 3). *The New York Times*, pp. C1, C8.

Blumberg, S. J., & Luke, J. V. (2014). Wireless substitution: Early release of estimates from the National Health Interview Survey, January–June 2014. Retrieved from: www.cdc.gov/nchs/data/nhis/earlyrelease/wireless201412.pdf

Bond, P. (2015, June 3). The Internet of things (You can Sue about). *Forbes*. Retrieved from: www.forbes.com/sites/danielfisher/2015/06/03/the-internet-of-things-you-can-sue-about/

Brown, D. (2002). Communication technology timeline. In A. E. Grant & J. H. Meadows (Eds.), *Communication technology update* (pp. 7–46). Oxford: Focal Press.

Burger, A. (2014, April 16). Infonetics VoIP forecast $88 billion in service revenues in 2018. *Telecompetitor*. Retrieved from: www.telecompetitor.com/infonetics-voip-forecast-88-billion-in-service-revenues-in-2018/

Burrus, D. (2014). The Internet of Things is far bigger than anyone realizes. *Wired*. Retrieved from: www.wired.com/2014/11/the-internet-of-things-bigger/

By the numbers: 70 amazing Netflix statistics and facts. (2016, January 21). *DMR*. Retrieved from: http://expandedramblings.com/index.php/netflix_statistics-facts/2/

Cauley, L. (2008, May 14). Consumers ditching land-line phones. *USA Today*. Retrieved from: www.usatoday.com/money/industries/telecom/2008-05-13-landlines_N.htm [November 1, 2009]

Chang, R. S. (2009, June 25). Texting is more dangerous than driving drunk. *The New York Times*. Retrieved from: http://wheels.blogs.nytimes.com/2009/06/25/texting-is-more-dangerous-than-driving-drunk

Chen, B. X. (2016, February 4). How to watch the Super Bowl when you don't have cable. *The New York Times*, p. B7.

Chen, B. X., & Wingfield, N. (2012, January 19). Apple unveils app and tools for digital textbooks. *The New York Times*, p. B2.

Cell phone-only households eclipse landline-only homes. (2009, May 6). *The Tech Chronicles*. Retrieved from: www.sfgate.com/cgi-bin/blogs/techchron/detail?blogid519&entry_id539683 [January 21, 2009]

Cisco VNI service adoption forecast 2013–2018 white paper. (2013). *Cisco*. Retrieved from: www.cisco.com/c/en/us/solutions/collateral/service-provider/vni-service-adoption-forecast/Cisco_VNI_SA_Forecast_WP.htm

Clifford, S. (2009, March 11). Advertisers get a trove of clues in smartphones. *The New York Times*, pp. A1, A14.

The dangers of texting while driving. (n.d.). *Federal Communications Commission*. Retrieved from: www.fcc.gov/guides/texting-while-driving

DeSilver, D. (2014, July 8). CDC: Two of every of every 5 US households have only wireless phones. *Pew Research Center*. Retrieved from: www.pewresearch.org/fact-tank/2014/07/08/two-of-every-five-u-s-households-have-only-wireless-phones/

Dolan, B. (2009, April 1). @CTIA: 1 Trillion text messages in 2008. *MobiHealth News.com*. Retrieved from: mobihealthnews.com/1109/ctia-1-trillion-text-messages-in-2008/ [November 3, 2009]

The do's and don'ts of cell phone Etiquette. (2013, July 23). Retrieved from: www.nydailynews.com/life-style/good-mobile-manners-article-1.1406873

Dugan, M., & Smith, A. (2013, September 16). Cell Internet use 2013. *Pew Research Center*. Retrieved from: www.pewinternet.org/2013/09/16/cell-internet-use-2013/

DVRs leveling off at about half of all TV households. (2013, December 6). *Leichtman Research*. Retrieved from: www.leichtmanresearch.com/press/120613release.html

DWI: Driving While Intoxicated. (n.d.). Retrieved from: www.textinganddriving safety.com/texting-and-driving-stats/

Dwyer, J. (2009, August 23). Hello oven? It's phone. Now let's get cooking! *The New York Times*, p. A26.

Ericsson mobility report. (2014). *Ericsson*. Retrieved from: www.ericsson.com/mobility-report

Ericsson Mobility Report: 90 Percent will have a mobile phone by 2020. (2014, November 18). *Ericsson*. Retrieved from: www.ericsson.com/news/1872291

Federal Motor Carrier Safety Administration. (2009). *Driver distraction in commercial vehicle operations (FMCSA-RRR-09–045)*. Washington, DC.

Fischer, S. (2015, May). 20 Ways to make Free Internet Phone Calls. *About.com*. Retrieved from: http://freebies.about.com/od/computerfreebies/tp/free-internet-phone-calls.htm

Foroohar, R. (2002, September 15). A new way to compute. *Newsweek*, p. 34J.

Freierman, S. (2009, January 19). Popular demand. *The New York Times*. Retrieved from: query.nytimes.com/gst/fullpage.html?res59505E4D6123FF93AA25752 C0A96F9C8B63&scp51&sq5popular&st5nyt [January 30, 2010]

Freudenrich, C. C. (2002). How personal digital assistants (PDAs) work. *How stuff works.com*. Retrieved from: www.howstuffworks.com/pda.htm

Fung, B. (2016, February 15). Google is considering a role in the future of television. *The Washington Post*, p. A13.

Gibbs, N. (2012, August 16). Your life is fully mobile. *Time*. Retrieved from: http://techland.time.com/2012/08/16/your-life-is-fully-mobile/

Gil, L. (2015, May 31). What you can do with Apple watch when you paired iPhone is out of Range. *Mac Rumors*. Retrieved from: www.macrumors.com/how-to/apple-watch-with-iphone-out-of-range/

Global mobile statistics 2014. (2014). *Part A: Mobile Subscribers; Handset Market Share; Mobile Operators*. Retrieved from: mobiforge.com/research-analysis/global-mobile-statistics-2014-part-a-mobile-subscribers-handset-market-share-mobile-operators#subscribers

Gray, R. (2013, August 14). How CERN's Gird may place the power of world's computers in your hands. *The Telegraph*. Retrieved from: www.telegraph.co.uk/technology/news/10242837/How-CERNs-Grid-may-place-the-power-of-the-worlds-computers-in-your-hands.html

Gross, D. (2010, January 18). Red Cross text donations pass $21 million. *CNN.com*. Retrieved from: www.cnn.com/2010/TECH/01/18/redcross.texts/index.html?hpt5T2 [January 19, 2010]

Hafner, K. (2009, May 26). Texting may be taking a toll. *The New York Times*. Retrieved from: www.nytimes.com/2009/05/26/health/26teen.html [May 26, 2009]

Harwell, D. (2015, February 12). For HBO Now, 800,000 is a disappointing number. *The Washington Post*, p. A17.

Harwell, D. (2016, November 17). Holiday ads moving from TV to digital. *The Washington Post*, p. A16.

Haiti text donation to Red Cross pass $5 million. (2010, January 14). *Associated Press/Yahoo! News*. Retrieved from: news.yahoo.com/s/ap/20100114/ap_on_hi_te/us_tec_haiti_text_donations [January 20, 2010]

Hesse, M. (2009, October 19). Worldwide ebb. *The Washington Post*, pp. C1, C8.

Hesse, M. (2010). Your dullest details, now part of history. *The Washington Post*, pp. A10, A1.

Hewlett Packard. (2014). The Internet of things research study. Retrieved from: www8.hp.com/h20195/V2/GetPDF.aspx/4AA5-4759ENW.pdf

Hinckley, D. (2014, March 5). Average American watches 5 hours per day, Report shows. *New York Daily News*. Retrieved from: Time shifting www.nydailynews.com/life-style/average-american-watches-5-hours-tv-day-article-1.1711954

Holson, L. (2014, July 3). Social media's vampires: They text by night. *The New York Times*. Retrieved from: www.nytimes.com/2014/07/06/fashion/vamping-teenagers-are-up-all-night-texting.html?_r=0

Holson, L. M. (2008, March 9). Text generation gap: UR2 Old (JK). *The New York Times*, pp. B1, B9.

Home landlines losing ground to cell phones. (2009, May 21). *Cell phone Shop*. Retrieved from: blog.cell phoneshop.com/2009/05/home-landlines-losing-ground-to-cell.html [January 21, 2009]

Homes with only wireless telephones continues to grow. (2013, December 19). *National Center for Health*. Retrieved from: NCHStats.com/category/wireless-phone-usage

Hosted PBX reseller markets seeing rapid growth. (2013, February 25). *PR Newswire*. Retrieved from www.bizjournals.com/prnewswire/press_releases/2013/02/25/CG65326

Houston, S. M. (2014, December 31). The 10 best web Series of 2014. *Paste*. Retrieved from: www.pastemagazine.com/articles/2014/12/the-10-best-web-series-of-2014.html

Hulu. (n.d.). Retrieved from: www.hulu.com

Inside these lenses, a digital dimension. (2009, April 26). *The New York Times*, p. 4.

The Internet of Things really is things, not people. (2015). *Deloitte*. Retrieved from: www2.deloitte.com/global/en/pages/technology-media-and-telecommunications/articles/tmt-pred-the-iot-is-things-not-people.html

It's history. (2003, March 17). *Newsweek*, p. 14.

ITU releases 2014 ICT Figures. (May 5, 2014). Retrieved from: www.itu.int/net/pressoffice/press_releases/2014/23.aspx

Janisch, T. (2007, March 13). See how they run: Radio ads and Google. *Wisconsin Technology News*. Retrieved from: wistechnology.com/articles/3768/ [November 3, 2009]

Jarvinen, A. (2009). Game design for social networks: Interaction design for playful dispositions. *Proceedings of the ACM Siggraph Video Game Symposium*, New Orleans, pp. 95–102.

Johnston, S. J. (2002). Entering the 'W' zone. *InfoWorld*, pp. 1, 44.

Kang, C. (2014, October 17). CBS jumps on online streaming bandwidth. *The Washington Post*, p. A13.

Kang, C. (2015a, January 6). Decision to stream ESPN may be TV game-changer. *The Washington Post*, pp. A1, A14.

Kang, C. (2015b, March 19). Cord cutters look to Apple for the joy of no TV bundles. *The Washington Post*, pp. A1, A14.

Kelly, H. (2012, December 3). OMG, the text message turns 20; but has SMS peaked? *CNN*. Retrieved from: www.cnn.com/2012/12/03/tech/mobile/sms-text-message-20/

Kendall, B., & Hagey, K. (2014, June 25). Supreme Court rules Aereo violates broadcasters' copyright. *The Wall Street Journal*. Retrieved from: www.wsj.com/articles/supreme-court-rules-against-aereo-sides-with-broadcasters-in-copyright-case-1403705891

Kendrick, A. (1969). *Prime Time: The Life of Edward R. Murrow*. New York: Little, Brown and Company. p.5.

Kim, E. (2015, April 15). The number of Americans paying for traditional TV peaked in 2012. *Business Insider*. Retrieved from: www.businessinsider.com/decline-of-us-tv-subscribers-2015-4

Koblin, J. (2016, February 26). All for the family. *The New York Times*, pp. B1, B6.

Kondolojy, A. (2015, June, 1). Modern Family leads adults 18–49 viewership growth. *TV By The Numbers*. Retrieved from: http://tvbythenumbers.zap2it.com/2015/06/01/modern-family-leads-adults-18-49-viewership-growth-wayward-pines-tops-percentage-gains-in-live-7-ratings-for-week-33-ending-may-17/410944/

Kowlke, M. (2014, February 28). What global VoIP usage can teach us. Retrieved from: http://ip-communications.tmcnet.com/articles/371863-what-global-voip-usage-teach-us.htm

Krane, J. (2003, January 9). Microsoft's Gates touts consumer gadgets. *Excite.com*. Retrieved from: http://apnews.excite.com/article/20030109/D7OER2Q00.html

Landline phones are a dying breed. (2015, February 24). *Statista*. Retrieved from: www.statista.com/chart/2072/landline-phones-in-the-united-states/

Layton, J. (2009). How to go completely mobile. *HowStuffWorks.com*. Retrieved from: electronics.howstuffworks.com/how-to-tech/how-to-go-completely-mobile.htm [November 3, 2009]

Levy, S. (2001, December 10). Living in a wireless world. *Newsweek*, pp. 57–58.

Levy, S., & Stone, B. (2002). The Wi-Fi wave. *Newsweek*, pp. 38–40.

Manjoo, F. (2016, January 14). Why media titans would be wise not to overlook Netflix. *The New York Times*. Retrieved from: www.nytimes.com/2016/01/14/technology/why-media-titans-need-to-worry-about-Netflix

Markoff, J. (2009, May 24). The coming superbrain. *The New York Times*, pp. 1, 4.

McCue, T. J. (2012, December 27). Google voice stays free in 2013 but VOIP is $15 billion industry. *Forbes*. Retrieved from: www.forbes.com/sites/tjmccue/2012/12/27/google-voice-stays-free-in-2013-but-voip-is-15-billion-industry/

Metz, R. (2014, November 26). Google glass is dead; long live smart glasses. *MIT Technology Review*. Retrieved from: www.technologyreview.com/featured story/532691/google-glass-is-dead-long-live-smart-glasses/

Meyers, D. (2011, October 11). Infonetics ups its VoIP service forecast to $76 billion by 2015. *Infonetics*. Retrieved from: www.infonetics.com/pr/2011/1H11-VoIP-and-UC-Services-Market-Highlights.asp

Miller, C. C. (2009, October 23). The cell refuseniks, an ever-shrinking club. *The New York Times*, pp. B1, B5.

Miller, C. C. (2010, May 6). In expanding reach, Twitter loses its scrappy start-up status. *The New York Times*, p. B7.

Mirkinson, J. (2014, January 28). CNN ratings plunge to near record lows. *The Huffington Post*. Retrieved from: www.huffingtonpost.com/2014/01/28/cnn-ratings-january-2014_n_4682699.html

Mitchell, A., & Rosenstiel, T., & Christian, L. (2012). The state of the news media 2012. *Pew Research Center*. Retrieved from: www.stateofthemedia.org/2012/mobile-devices-and-neconsumption-some-good-signs-for-journalism/what-facebook-and-twitter-mean-for-news/

Mobile technology fact sheet. (2014, January). *Pew Research Center*. Retrieved from: www.pewinternet.org/fact-sheets/mobile-technology-fact-sheet/

More than 80 percent of Americans with a DVR can't live without it according to NDS survey. (2008, September 3). *NDS*. Retrieved from: www.nds.com/press_releases/NDS_DVR_Survey-US_030908.html [January 20, 2010]

Moscaritolo, A. (2015, February 2). Tablets see first-ever yearly decline. *PC Magazine*. Retrieved from: www.pcmag.com/article2/0,2817,2476199,00.asp

Murchu, I. O., Breslin, J. G., & Decker, S. (2004). Online social and business networking communities. *Digital Enterprise Research Institute*. Technical Report 2004-08-11.

Nielsen reports DVR playback is adding to TV viewing levels. (2008, February 14). *Nielsen*. Retrieved from: en-us.nielsen.com/main/news/news_releases/2008/0/nielsen_reports_dvr [January 30, 2010]

O'Grady, M. (2012, June 19). SMS usage remains strong in the U.S.: 6 billion SMS messages are sent each day. *Michael O'Grady's Blog*. Retrieved from: blogs.forrester.com/michael_ogrady/12-06-19-sms_usage_remains_strong_in_the_us_6_billion_sms_messages_are_sent_each_day

Over the top television set to change TV consumption in the home. (2016). *BCi*. Retrieved from: www.bci.eu.com/over-the-top-tv/over-the-top-television-ott-tv/

Pachal, P. (2015, February 5). The next Google glass will probably look nothing like the first one. *Mashable*. Retrieved from: mashable.com/2015/02/05/google-glass-redesign/

Pegoraro, R. (2009, January 28). *The New York Times*, pp. A1, A18.

Pennington, G. (2015, December 8). Here are 2015's top rated TV shows, according to Nielsen. *St. Louis Post Dispatch*. Retrieved from: www.stltoday.com/entertainment/television/gail-pennington/here-are-s-top-rated-tv-shows-according-to-nielsen/article_ab1f8852-0808-51f0-9be7-0615ef7ca1ce.html

Poggi, J. (2016, April 15). The CMO's guide to over-the-top TV. *Advertising Age*. Retrieved from: http://adage.com/article/media/cmo-s-guide-top-tv/298013/

Poniewozik, J. (2016, February 27). To a 'fuller house' we go, marching down memory lane. *The New York Times*, pp. C1, C2.

Portio research mobile Factbook 2013. (2013). Retrieved from: www.portio.com

Press, G. (2014, August 2014). Internet of Things by the numbers: Market estimates and forecasts. *Forbes*. Retrieved from: www.forbes.com/sites/gilpress/2014/08/22/internet-of-things-by-the-numbers-market-estimates-and-forecasts/#555e72ba2dc9

Quenqua, D. (2011, September 19). Text messaging levels off among U.S. adults. *ClickZ*. Retrieved from: www.clickz.com/clickz/news/2110274/text-messaging-levels-adults

Richtel, M. (2014, October 7). Voice activation systems distract drivers, study says. *The New York Times*, B4.

Rohit. (2011, October). VoIP Industry to grow to $76.1 Billion in 2015. *DID for Sale*. Retrieved from: www.didforsale.com/voip-industry-to-grow-to-76-1-billion-in-2015

Seltzer, L. (2013, April 3). Cell phone inventor talks of first cell call. *Information Week*. Retrieved from: www.informationweek.com/wireless/cell-phone-inventor-talks-of-first-cell-call/d/d-id/1109376

Selyukh, A., & Nayak, M. (2015, January 29). U.S. wireless spectrum auction raises record $44.9 billion. *Reuters*. Retrieved from: www.reuters.com/article/2015/01/29/us-usa-spectrum-auction-idUSKBN0L227B20150129

Seward, Z. M. (2013, April 3). The first mobile phone call was made 40 years ago today. *The Atlantic*. Retrieved from: www.theatlantic.com/technology/archive/2013/04/the-first-mobile-phone-call-was-made-40-years-ago-today/274611

Sisario, B. (2016, February 21). SiriusXM fights to dominate the dashboard of the connected car. *The New York Times*, pp. B1, B6.

Sling TV review. (2015, January 26). *Consumer Reports*. Retrieved from: www.consumerreports.org/cro/news/2015/01/sling-tv-first-look/index.htm

Small crowd, young and old watch Super Bowl on Web. (2016, February 4). *The New York Times*. Retrieved from: www.nytimes.com/reuters/2016/02/04/arts/04reuters-nfl-superbowl-streaming.html

Small, Talk. (2001). *News week*, p. 72.

Smartphone: So many apps so much time. (2014, July 1). *Nielsen*. Retrieved from: www.nielsen.com/us/en/insights/news/2014/smartphones-so-many-apps—so-much-time.html

Smartphone usage and demographics. (2014, September 11). *Media Horizons*. Retrieved from: www.mediahorizons.com/blog/item/416-smartphone-usage-and-habits-by-demographics

Smith, A. (2009, April 15). The Internet's role in campaign 2008. *Pew Internet & American Life Project*. Retrieved from: www.pewinternet.org/Reports/2009/6-The-Internets-Role-in-Campaign-2008/3—The-Internet-as-a-Source-of-Political-News/5—Long-tail.aspx?r51 [April 19, 2009]

Smith, A. (2012, November 30). The best (and worst) of mobile connectivity. *Pew Research Center*. Retrieved from: www.pewinternet.org/2012/11/30/the-best-and-worst-of-mobile-connectivity/

Smith, A. (2015, April 1). U.S. smartphone use 2015. *Pew Research Center*. Retrieved from: www.pewinternet.org/2015/04/01/us-smartphone-use-in-2015/

So many apps, so much more time for entertainment. (2015, June 6). *Nielsen*. Retrieved from: www.nielsen.com/us/en/insights/news/2015/so-many-apps-so-much-more-time-for-entertainment.html

Solsman, J. E. (2015, May 7). Sling TV may add broadcast networks but won't force you to use them. *CNET*. Retrieved from: www.cnet.com/news/sling-tv-may-add-broadcast-networks-but-wont-force-you-to-buy-them/

Sophy, J. (2014, January 19). Will most international calls be on Skype someday? *Small Business Trends*. Retrieved from: http://smallbiztrends.com/2014/01/someday-international-calls-might-skype.html

Steel, E. (2014a, June 29). Stung by Supreme Court, Aereo suspends service, *The New York Times*, p. A21.

Steel, E. (2014b, October 17). Cord-cutters rejoice: CBS joins web stream, *The New York Times*, pp. A1, A17.

Steel, E. (2015, January 5). Dish network unveils Sling TV, a streaming service to rival cable (and it has ESPN). *The New York Times*. Retrieved from: www.nytimes.com/2015/01/06/business/media/dish-network-announces-web-based-pay-tv-offering.html?_r=5

Steel, E. (2016, February 11). HBO Now has 800,000 paid streaming subscribers, Time Warner says. *The New York Times*, p. B3.

Sterling, C. (2013, October). Numbers don't lie: Impressive stats on the VOIP Industry. Retrieved from: https://virtualphonesystemreviews.com/numbers-dont-lie-impressive-stats-voip-industry/

Sterling, G. (2014, January 16). Pew: 50 percent in U.S. now own a tablet or an e-reader. *Marketing Land*. Retrieved from: marketingland.com/nielsen-time-accessing-internet-smartphones-pcs-73683

Sterling, G. (2014, February 11). Nielsen: More time on Internet through smartphones than PCs. *Marketing Land*. Retrieved from: marketingland.com/nielsen-time-accessing-internet-smartphones-pcs-73683

Stelter, B. (2013, July 18). Netflix does well in 2013 primetime Emmy nominations. *The New York Times*, p. C1,

Stevens, T. (2007, November 14). 82% of Americans own cell phones. *Switched.com*. Retrieved from: www.switched.com/2007/11/14/82-of-americans-own-cell-phones/ [November 3, 2009]

St. George, D. (2009, February 22). 6,473 texts a month, but at what cost? *The Washington Post*, p. A17.

Stone, B. (2009, August 10). Breakfast can wait. The day's first stop is online. *The New York Times*. Retrieved from: www.nytimes.com/2009/08/10/10morning.html [August 13, 2009]

Sutter, J. D. (2009, September 29). Microsoft researcher converts his brain into "e-memory." *CNN*. Retrieved from: www.cnn.com [September 29, 2009]

Svahn, M. (2009). Processing play: Perceptions of persuasion. Paper presented to the Digital Games Research Association (DiGRA) symposium. September 1–4.

Tanaka, J. (2001, September 17). PC, Phone home! *Newsweek*, pp. 71–72.

Thanks to DVR and streaming services, binge TV viewers abound. (2013, April 18). *eMarketer*. Retrieved from: www.emarketer.com/Article/Thanks-DVR-Streaming-Services-Binge-TV-Viewers-Abound/1009823

Thing (in the Internet of Things). (2015). *WhatIs.com*. Retrieved from: wgatus.techtarget.com/definition/thing-in-the-Internet-of-things

Thompson, C. (2008, September 7). I'm so totally close to you. *The New York Times Magazine*, pp. 42–47.

Tierney, J. (2009, May 5). Ear plugs to lasers: The science of concentration. *The New York Times*, p. D2.

Time shifting/DVD usage. *Cable Nation*. Retrieved from: www.thecab.tv/cross-plat form-timeshifting.php

Total audience report. (2014, December 3). *Nielsen*. Retrieved from: www.nielsen.com/us/en/insights/reports/2014/the-total-audience-report.html

Tsukayama, H. (2014, June 29). Aereo hits 'pause' button on service, will issue refunds to customers. *The Washington Post*, p. A8.

Tsukayama, H. (2015, January 3). At tech show, expert, bigger and smarter. *The Washington Post*, p. A10.

Tyson, J. (2002). How telephony works. *How Stuff Works*. Retrieved from: www.howstuffworks.com/iptelephony.htm

Vanston, L. K. (2002). *Residential broadband forecasts*. Austin, TX: Technology Futures, Inc.

Vinyals, O., Toshev, A., Bengio, S., & Erhan, D. (2014, November 17). A picture is worth a thousand (coherent) words: Building a natural description of images. *Google Research Blog*. Retrieved from: googleresearch.blogspot.com/2014_11_01_archive.html

Watch 2014 Fall TV preview online. (2014). *Hulu*. Retrieved from: www.hulu.com/browse/picks/2014-fall-tv-preview

Weiss, T. (2015, May 20). Tablet market continues to shrink around the world: Report. *Eweek*. Retrieved from: www.eweek.com/mobile/tablet-market-contin ues-to-shrink-around-the-world-report.html

What is telephony? (2002). *About.com*. Retrieved from: http://netconference.about.com/library/weekly/aa032100a.htm

Wortham, J. (2010, January 15). Burst of mobile giving adds millions in relief funds. *The New York Times*. Retrieved from: www.nytimes.com/2010/01/15/tech nology/15mobile.html?scp56&sq5%22haiti%22%20%22twitter%22&st5cse [January 17, 2010]

Advertising: From Clay Tablets to Digital Tablets 7

Contents

Advertising plays an important role in the U.S. economy and fulfills many consumer needs. Advertising is so prolific and we have become so accustomed to seeing and hearing promotional messages that we do not always notice them. We will watch a string of television commercials, yet 5 minutes later, most of us are not able to remember what products or brands the commercials were advertising. We will flip through magazines and look at the full-color ads, but when asked what products we saw, there is a good chance we will not remember. The point is that advertising can have a subtle effect on us. We do not always realize the influence advertising has on our purchasing decisions.

Most people tend to think of advertising as a modern-day phenomenon when actually it has been around for thousands of years. The origins of advertising can be traced back to ancient Babylon, where tradesmen inscribed their names on clay tablets. In medieval England, tavern owners distinguished their establishments with creative names and signs. Even back then, merchants recognized the need to get the word out about their products or services. Although the advertising of long ago was nothing more than written names or figures drawn on signs, these methods of promotion led the way to modern advertising. Today, advertising is a very complex business that employs principles of psychology, sociology, marketing, economics, and other sciences and fields of study for the end purpose of selling goods and services.

Getting products and services to customers is a multistep process called marketing. Marketing is the overall function that includes product and service pricing, distributing, packaging, and advertising. Marketing is "the process of planning and executing the conception, pricing, promotion, and distribution of ideas, goods, and services to create exchanges that satisfy individual and organizational objectives" (Vanden Bergh & Katz, 1999, p. 155). Before a product or service can be sold, it must first be marketed. In other words, it must be packaged, priced, and distributed to sellers or directly to buyers. Products and services must also be advertised to potential customers. Thus advertising is a function of marketing.

Traditional industry models define *advertising* as "nonpersonal communication for products, services, or ideas that is paid for by an identified sponsor for the purpose of influencing an audience" (Vanden Bergh & Katz, 1999, p. 158). More simply, advertising is any "form of nonpersonal presentation and promotion of ideas, goods, and services usually paid for by an identified sponsor"

155

(Dominick, 1999, p. 397) or "paid, mass mediated attempt to persuade" (O'Guinn, Allen, & Semenik, 2000, p. 6).

In today's world, these definitions are fluid, as newer personal media individualize and deliver advertising messages directly to a consumer. Advertising is not limited to the typical radio or television commercials that flood the airwaves but also includes awareness-creating endeavors such as T-shirts, ball caps, pens, and coffee mugs imprinted with brand, store, or Web site logos, free products, free service trials, product demonstrations, sponsorships of 5k runs, and tweeting customers a product related message. Advertising is no longer 'nonpersonal' and is not always 'mass mediated.' Web sites are a combination of advertising and direct selling. Web sites track users' online movements to customize ads. Social network sites, blogs, and Twitter also individualize ads and even send them to 'friends,' thinking that people who hang out together are probably interested in the same products and brands. New online and personal technologies have transformed advertising into an "art of engagement" (Othmer, 2009).

This chapter begins with an overview of the origins of advertising in about 3000 BCE and moves to the contemporary world of advertising on radio, television, cable, and the Internet. This chapter also describes the purpose and function of advertising agencies and addresses advertising regulation. This chapter ends with a look at the criticisms aimed at contemporary advertising.

FYI: Four Ps of Marketing

1. Product
2. Price
3. Place
4. Promotion
   4a. Advertising

ZOOM IN 7.1

Close your eyes and think about all of the promotional messages in the room where you are sitting. Now open your eyes and look around the room. How many advertising messages do you count? Be sure to include clothing, posters, coffee mugs, pens, pencils, calendars, and other everyday items imprinted with logos. How many of the advertisements did you remember before you looked? Do you tend to notice most of the advertising messages around you, or do you tend not to pay any attention?

SEE IT THEN

ADVERTISING: 3000 BCE TO 1990

The earliest form of advertising has been traced back to about 3000 BCE. Clay tablets inscribed with merchants'

names have been unearthed near what was the city of Babylon in ancient Mesopotamia. Ancient Egyptians advertised on papyrus (a form of paper) that was much more portable than clay tablets, and the ancient Greeks used town criers to advertise the arrival of ships carrying various goods. The ancient Romans hung stone and terra-cotta signs outside their shops to advertise the goods sold inside.

It was Johannes Gutenberg's invention of the printing press around 1450 that gave people the idea of distributing *printed* advertisements. Toward the end of the 1400s, church officials were printing handbills and tacking them up around town. The first printed advertisement for a product is thought to have appeared in Germany around 1525; it promoted a book about some sort of miracle medicine.

The growing popularity of newspapers in the late 1600s and early 1700s led to the further development of print advertising. In 1704, the *Boston Newsletter* printed what is thought to have been the first newspaper ad, a promise to pay a reward for capturing runaway slaves. Benjamin Franklin, one of the first publishers of colonial newspapers, endorsed advertising and increased the visibility of ads placed in his papers by using larger type and more white space around the ads. Until the mid-1800s, most newspaper ads were in the general form of what is known today as classified ads, or simply lines of text.

Advertising took on a new role in the 1800s with the Industrial Revolution. Countless new machines were invented, leading to huge increases in manufacturing. People moved away from their rural farms and communities into the cities to work in factories. The cities' swelling populations and the mass production of goods gave rise to mass consumption and a mass audience. Advertising provided the link between manufacturers and consumers. City dwellers found out about new products from newspaper and magazine ads rather than from their friends and family.

Manufacturers soon realized the positive effect advertising had on product sales. They also realized they needed help designing ads, writing copy, and buying media space in which to place the ads. Volney B. Palmer filled the void by opening the first advertising agency in Boston in 1841. Palmer primarily contracted with newspapers to sell advertising space to manufacturers. About 30 years later, another promotions pioneer, Francis Ayer, opened the first full-service agency in Philadelphia. Ayer's agency produced and placed print ads in newspapers and magazines. As new agencies opened and advertising as a business took off, there came new insights into consumerism and eventually new advertising strategies and techniques. For example, by 1860, full-color magazine advertisements flourished thanks to new cameras and linotype machines.

EARLY RADIO ADVERTISING

A new form of advertising was born with the advent of radio. At first, radio was slow to catch on with the public, because there were not very many programs on the air.

One of the first radio broadcasts was of the 1921 heavyweight boxing championship between Jack Dempsey and Georges Carpentier. As each punch was thrown in Hoboken, New Jersey, on-site telegraph operators tapped out the action to station KDKA in Pittsburgh, Pennsylvania, where telegraph operators translated the signals and vocally reported what was happening over the airwaves. Such live broadcasts piqued interest in radio, and as more people purchased radio sets, the demand for programming increased, and broadcasters were left grappling with how they were going to finance the endeavor.

In the early 1920s, stations and ham (amateur) radio operators were airing programs but not generating any revenue for their efforts. Setting up a radio station meant buying expensive transmitters, receivers, and other equipment. Plus, broadcasters and other personnel had to be paid salaries. Given these costs, there was a collective call among broadcasters to figure out a way to generate income, because they knew this gratuitous service could not go on indefinitely. In 1924, *Radio Broadcast* magazine held a contest with a $500 prize for the person who could come up with the best answer to the question, "Who is going to pay for broadcasting and how?"

Although the idea of commercial radio was undergoing serious discussion, radio advertising was largely considered in poor taste and an invasion of privacy. Additionally, radio had to comply with the Communications Act of 1934, which stated that there must be a recognizable difference between commercials and program content. That is, a station is required to disclose the source of any content it broadcasts for which it receives any type of payment. Many consumers and broadcasters were resistant to over-the-air commercialism and claimed that radio should not be used to sell products. Then-Secretary of Commerce Herbert Hoover, who later became president of the United States, claimed that radio programming

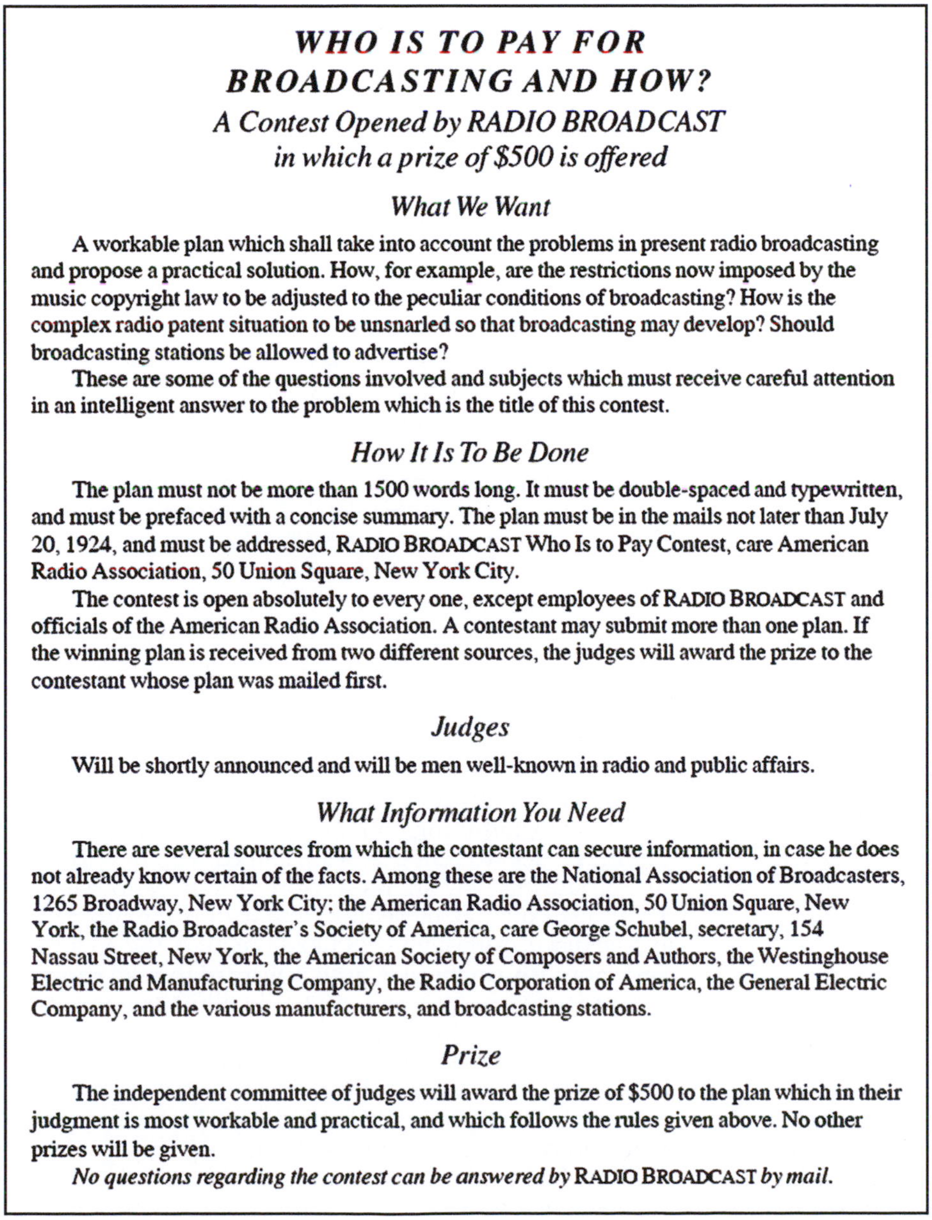

WHO IS TO PAY FOR BROADCASTING AND HOW?
A Contest Opened by RADIO BROADCAST in which a prize of $500 is offered

What We Want

A workable plan which shall take into account the problems in present radio broadcasting and propose a practical solution. How, for example, are the restrictions now imposed by the music copyright law to be adjusted to the peculiar conditions of broadcasting? How is the complex radio patent situation to be unsnarled so that broadcasting may develop? Should broadcasting stations be allowed to advertise?

These are some of the questions involved and subjects which must receive careful attention in an intelligent answer to the problem which is the title of this contest.

How It Is To Be Done

The plan must not be more than 1500 words long. It must be double-spaced and typewritten, and must be prefaced with a concise summary. The plan must be in the mails not later than July 20, 1924, and must be addressed, RADIO BROADCAST Who Is to Pay Contest, care American Radio Association, 50 Union Square, New York City.

The contest is open absolutely to every one, except employees of RADIO BROADCAST and officials of the American Radio Association. A contestant may submit more than one plan. If the winning plan is received from two different sources, the judges will award the prize to the contestant whose plan was mailed first.

Judges

Will be shortly announced and will be men well-known in radio and public affairs.

What Information You Need

There are several sources from which the contestant can secure information, in case he does not already know certain of the facts. Among these are the National Association of Broadcasters, 1265 Broadway, New York City; the American Radio Association, 50 Union Square, New York, the Radio Broadcaster's Society of America, care George Schubel, secretary, 154 Nassau Street, New York, the American Society of Composers and Authors, the Westinghouse Electric and Manufacturing Company, the Radio Corporation of America, the General Electric Company, and the various manufacturers, and broadcasting stations.

Prize

The independent committee of judges will award the prize of $500 to the plan which in their judgment is most workable and practical, and which follows the rules given above. No other prizes will be given.

No questions regarding the contest can be answered by RADIO BROADCAST *by mail.*

FIG. 7.1 Ad from *Radio Broadcast* magazine, May 1924

should not be interrupted with senseless advertising. Such anti-advertising sentiment even extended to some station owners.

But then in 1922, AT&T–owned radio station WEAF came up with the idea of **toll broadcasting**—payment for using airtime. On August 28, a Long Island, New York, real estate firm paid $50 for 10 minutes of time to persuade people to buy property in the New York area. Although toll advertising may seem like a modern-day infomercial, AT&T did not consider these toll messages 'advertising' but simply courtesy announcements, because the prices of the products and services were never mentioned.

ZOOM IN 7.2

- Learn more about WEAF's first commercial and listen to a short clip of the spot at www.npr.org/2012/08/29/160265990/first-radio-commercial-hit-air-waves-90-years-ago
- Listen to old-time radio commercials within the Old Time Radio site: www.old-time.com/commercials/index.html

---

Radio advertising helped boost product sales. The Washburn Crosby Company (now known as General Mills) saw the sales of Wheaties cereal soar after introducing the first singing commercial on network radio in December 1926. In geographic areas in which the Wheaties jingle was aired, the cereal became one of the most popular brands, but in the areas in which the commercial did not air, sales were stagnant.

ZOOM IN 7.3

Listen to the first singing commercial at www.oldtimeradio-fans.com/old_radio_commercials

Click on Wheaties. The spot starts with an announcer talking and then ends with the first singing jingle.

By the late 1920s, the initial resistance to advertising gave way to the increasing cost of operating a radio station, and eventually broadcasters and the public endorsed the idea of advertising-supported radio, though largely in the form of sponsorships.

SPONSORED RADIO

In 1923, the Browning King clothing company bought weekly time on station WEAF to sponsor the Browning King Orchestra. Whenever the name of the orchestra was announced, so was the company's name. But in keeping with WEAF's anticommercial sentiment, the announcers were careful not to mention that Browning King sold clothing. Other companies and then later advertising agencies took the lead from Browning King and assumed the production of radio programs in turn for being recognized program sponsors. The **sponsor identification rule (Section 317)** of the Communications Act of 1934 protected listeners from commercial messages coming from unidentified sponsors by requiring broadcasters to reveal sponsors' identities.

At first, sponsored radio seemed like a good idea for the audience and the broadcasters. Programs, usually 15 minutes in length, were produced by an ad agency in partnership with the sponsoring company. The ad agency benefited by being paid for its creative work, the company benefited by gaining brand recognition and hopefully sales, and the audience benefited by being made aware of a product but without being subjected to blatant promotional messages.

Rather than running a commercial per se, a sponsor's name would either be part of a short narrative read by an announcer at the beginning, middle, and end of a program, or it would be mentioned in the script. For example, in *Oxodol's Own Ma Perkins*, the sponsor's laundry products were woven into the storyline when characters did the clothes washing. The American Tobacco Company sponsored *The Lucky Strike Hit Parade*, a program that played the best-selling records. Every time the announcer said the name of the program, the audience would hear the name of the brand Lucky Strike. By naming the program after Lucky Strike, brand recognition increased, and smokers who listened to the program were presumably likely to purchase Lucky Strike cigarettes.

Despite the benefits of sponsorship, radio stations began to resent that they had no control over program content. Ad agencies and advertisers were fully responsible for program content and even had complete control over performers. The ultraconservative Rexall Drugs would not let their spokesman, entertainer Jimmy Durante, appear on a campaign program that was soliciting votes for Democratic president Franklin D. Roosevelt. Sponsors also bought particular time slots for their programs; thus, radio stations had little control over when programs aired. Consequently, at hours of peak listenership, the stations were sometimes forced to air unpopular programs when they would have preferred to air shows that would draw large audiences.

By the mid-1940s, stations and executives were tired of ceding program control to ad agencies and advertisers, so CBS radio network owner William S. Paley devised a plan for programming and advertising. He set up a programming department that was charged with developing and producing new shows. In turn, the network would recoup its expenses by selling time within the programs to advertisers. Although CBS liked the plan, the ad agencies vigorously opposed it, as they wanted to keep control over programs and advertisers. Eventually, programming became the network's responsibility, but the agencies that bought time within programs still controlled casting and scheduling. Then, as radio shows moved to television, radio program sponsorships and agency control over programs diminished. Network radio programs gave way to individual station programming, and advertising was increasingly sold not as sponsorships but as commercials within and between programs. The number of the advertising spots increased substantially between 1965 and 1995, as did advertising revenue.

ZOOM IN 7.4

To listen to old radio sponsorships for programs such as *Little Orphan Annie* (sponsored by Ovaltine) and *Your Hit Parade* (sponsored by Lucky Strike), go to www.old-time.com/weekly

EARLY TELEVISION ADVERTISING

The public got its first glimpse of television at the 1939 New York World's Fair, but television did not immediately catch on. World War II interrupted television set manufacturing and program creation and transmission, stalling consumers' adoption of television until about 3 years after the war. In 1948, the economy was booming, American consumers were very enthusiastic about television, and the industry exploded with new stations, new programs, new sets, and new viewers, realizing growth of more than 4,000% in that year. Television quickly became a mass medium and promised to become as popular as radio, which left industry executives struggling with how to make television a financially lucrative medium.

SPONSORSHIP

At first, television advertising was based largely on radio's sponsorship model. Advertisers and their agencies produced sponsored programs such as *Kraft Television Theater* (1947–1958) and *Texaco Star Theater* (1948–1953). However, television programs were much more expensive to produce than radio programs, and agencies and advertisers were spending more than they were generating in increased product sales. For example, the Kudner Agency spent about $8,000 per week producing *Texaco Star Theater*, an amount that quadrupled 3 years later, and Frigidaire spent about $100,000 for each Bob Hope special it sponsored. The return on investment (ROI) did not always satisfy the sponsors. The high cost of television sponsorship kept all but a few of the largest and most financially stable companies off the air. It soon became apparent that the model for television advertising had to change.

SPOT ADVERTISING

NBC television executive Pat Weaver (father of *Alien* and *Imaginary Heroes* star Sigourney Weaver) extended William Paley's idea of selling ad time within radio programs to television and came up with what is known as the **magazine style** of television advertising. This approach later became known as **spot advertising**. Weaver had figured out that television could make more money by selling time within and between programs to several sponsors than by relying on one sponsor to carry the entire cost. His idea was similar to the placement of magazine advertisements between articles. Instead of advertisers purchasing television program sponsorships, they would purchase advertising time in 1-minute units. Weaver's plan also promoted the production of programs by network and independent producers, keeping advertisers out of the business of programming.

With Weaver's plan, advertisers found it much less expensive to purchase 1 minute of time rather than the entire 15 or 30 minutes of a program, and they did not have to concern themselves with program content. Affordable airtime brought many more advertisers (especially smaller, lesser-known companies) to television for the first time, much to the chagrin of larger, wealthier sponsors and ad agencies, which were concerned about losing their broadcast dominance.

The Bulova watch company was the first advertiser to venture to television and also the first company to purchase spot radio time (as opposed to a program sponsorship). Starting in 1926, the United States ran on Bulova time, with its well-known radio commercial announcements: "At the tone, it's 8:00 p.m. B-U-L-O-V-A. Bulova watch time." Bulova later adapted its radio spot to television. On July 1, 1941, Bulova paid $9 to a New York City television station for a 20-second ad that aired during a Dodgers versus Phillies baseball game. The Bulova commercial showed a watch face with the current time but without the audio announcement. Bulova later kept the close-up of the watch face but added a voice announcing the time.

Advertising and television were a highly successful pairing. Advertisers had a new and popular mass medium that brought product and audience together. For the first time, consumers could view products and product features from their own living rooms. For instance, they could watch a brand of laundry soap remove stubborn stains from a blouse or toothpaste whiten dull, yellowed dentures.

Yet television advertising did not have a very smooth beginning. In the early years, many television programs and commercials were produced live, but unrefined production techniques left viewers watching coarse black-and-white images. Soap sponsors touting the whitening power of their products discovered that on black-and-white television, viewers could not tell the difference between 'white' and 'whiter.' Commercial actors had to hold up a white shirt next to a blue one for contrast and pretend that it had been washed with a competing soap.

Live product demonstrations did not always go off as planned, either, embarrassing the advertiser and the product spokesperson. Aunt Jenny, a character on the *Question Bee*, dripped beads of perspiration from the hot studio lights onto the chocolate cake she had just freshly baked using Spry vegetable shortening. To make matters worse, she licked some cake off the knife blade and then cut more slices with the same knife. Spry was not happy, to say the least. Neither was *Variety* magazine, which called the whole business of television 'unsanitary.'

In another bungled demo, a hand model demonstrated live Gillette's new automatic safety razor, except it was not so automatic. The new disposable blade, which was supposed to twist open for easy and safe replacement, became stuck, and the television audience watched as the

FIG. 7.2 Andy Griffith (Sheriff Taylor) and Clint Howard take a break from filming the *Andy Griffith Show* to sell Jell-O brand pudding.
Photo courtesy of CBS Photo Archive/ Getty Images

FIG. 7.3 Westinghouse commercial with Betty Furness, 1956
Photo courtesy of Photofest

Corn Flakes commercial, an announcer was asked to tout the cereal while eating it, but he could not speak clearly with his mouth full, so he figured that with the help of the camera, he would just pretend he was eating. Unfortunately, the cameraman zoomed out too far, and viewers watched in amazement as the announcer discarded spoonful after spoonful of cereal over his shoulder instead of putting it in his mouth. Refrigerator doors stuck shut, can openers would not open cans, and spokespeople holding up one brand but trumpeting another are just some of the other debacles of live television commercials. Despite these bloopers, advertisers continued to promote their products on the airwaves, and television executives cheered as they watched advertising revenues rise rapidly throughout the 1950s.

ZOOM IN 7.5

Watch classic television commercials at these sites:

- Classic TV ads: www.youtube.com/results?search_query= old+commercials&oq=old+commercials/classic.htm
- Saturday morning commercials (1960–1970): www.tvparty. com/vaultcomsat.html

hands desperately struggled to unstick it, to no avail. That was the last live product demonstration for Gillette. From then on, it prerecorded its commercials. In a Kellogg's

Although sponsors and agencies were gradually giving up control over production, scripts, and stars, the decade was still rife with sponsored programs, such as *The Dinah Shore-Chevy Show* (1951–1957) and *The Colgate-Palmolive Comedy Hour* (1950–1955). Program sponsorship, however, took a severe hit with the quiz show scandal that rocked the television industry in the late 1950s. At the time, television quiz shows were the most popular programs on the air. Sponsors and their advertising agencies produced most of the quiz shows, with a few produced by the networks. The competition for viewers was fierce, as there were many quiz shows on the air, and they were often on at the same time on different channels. To keep popular contestants on the air (and thus to keep viewers watching the shows), sponsors started to secretly give well-liked contestants the answers to the questions before the game. One contestant, who had lost to a competitor who had been given the right answers, came forward and exposed the quiz shows as fraudulent.

When the scandal made the headlines, the public was outraged at the deception and blamed the networks, even though it was the sponsors who had coerced the networks to cheat under threat of losing sponsorship. The networks figured that if they were going to be held responsible for televised content, then they should take over programming from the advertising agencies. The networks pushed for control of program content and for when shows would be on the air. The networks had the public on their side, and later, the influential *Advertising Age*—the advertising industry's major trade publication—strongly urged advertising agencies to regain their reputations by leaving the production of television programs to the television industry. Further, the publication pushed for separating sponsors and programs by switching entirely to the magazine-style method of advertising sales. Sponsors and agencies gradually gave up control over production, scripts, and stars, and instead of one advertiser sponsoring an entire program, various advertisers bought commercial time throughout the show.

Throughout the 1960s, most television commercials were sold in 60-second units known as *spots*. Because programs were no longer produced and sponsored by single advertisers, spots within and between programs were available to a number of advertisers. Thus, many different commercials featuring different brands and products were shown during the course of one program, initiating a more competitive marketplace. Consider that when an advertiser sponsors a program, that brand is the only one promoted during the entire program. But with spot advertising, a variety of products and brands are promoted during the course of a show. For many advertisers, this was the first time they had faced strong competition for the audience's attention. They had to come up with creative ways to make their product or brand stand out from the others. Slogans, jingles, and catchy phrases started to make their way into commercials.

> ### ZOOM IN 7.6
>
> Think of a product brand that you like and one that you do not like—maybe running shoes, colas, or sports cars. What's your image of each brand, and why do you prefer one instead of the other?

A NEW LOOK

Television advertising took on a slightly new look in 1971 after the federal government banned commercials featuring tobacco products. Cigarette companies had been among television's biggest advertisers until they were forced to transfer their advertising dollars from the airwaves to print, leaving broadcasters scrambling to fill unsold time. Television networks quickly discovered that many other companies simply could not afford to buy commercial time in 60-second blocks, but they could buy 30-second units. This arrangement proved profitable for the networks, as they could sell two 30-second spots for more money than they could one 60-second spot. The effect on an hour of television programming was that programs became infiltrated with shorter spots from more advertisers.

In 1965, about 70% of all commercials were 60 seconds in length. That percentage decreased throughout the late 1960s. Four years after the 1971 ban on tobacco advertising, only 11% of commercials were 60 seconds in length and almost 80% were 30-second spots. By 1985, almost 90% of all commercials were 30 seconds in length and only 2% ran for a full minute. Throughout the 1980s, the

FIG. 7.4 Cindy Crawford drinks a Pepsi in a commercial, 1991.
Photo courtesy of Getty Images Entertainment

FIG. 7.5 Brooke Shields selling StarKist tuna, 1983
Photo courtesy of Adam Scull, Globe Photos, Inc.

length of commercials began to vary and included 10-, 15-, 20-, and 45-second spots. The 15-second spots especially caught favor, and by 1990, they accounted for about one-third of all commercials, and by 2008 about 4 out of 10 spots were 15 seconds in length. Although they are rare, sponsorships do still occur. For example, Ford Motor Company was the sole sponsor of the movie *Schindler's List* when it made its television debut in 1997, and in the early 2000s Kleenex aptly sponsored several 'tearjerker' movies, such as *An Officer and a Gentleman* and *Steel Magnolias*. *The Biggest Loser* has been sponsored by Wrigley's Extra sugarless gum, and BMW's early sponsorship of AMC's *Mad Men* limited commercials to brand mentions at the beginning and end of an episode plus a few breaks, which contained only BMW spots and network promos.

CREATIVE STRATEGIES

With more commercial spots on the air, advertisers needed to be creative to remain competitive. Faced with new creative challenges, many commercials took on a more narrative approach. Rather than just showing a product and reciting its features, commercials became more like 30-second mini-movies that tell a story with characters and plots that build a product or **brand image**. This approach, called *image advertising*, goes beyond simply promoting a product; instead, it attempts to set a product or brand perception in the consumer's mind. For example, Whole Foods and Chipotle have both cultivated a brand reputation committed to organic food and sustainability.

TARGET ADVERTISING

Throughout the 1970s and into the early 1980s, the three broadcast television networks (ABC, CBS, and NBC) shared about 90% of the viewing audience and television advertising dollars. As more homes hooked up to cable television in the 1980s, broadcast television found a new rival. Cable offered advertisers many new and specialized channels that attracted niche audiences. Rather than pay a large amount of money to reach a mass audience, advertisers could easily target their consumers on specialty cable networks. For example, it is more efficient and effective for Ping golf clubs to target golfers by running commercials on the Golf Channel or ESPN instead of on a broadcast network where only a small percentage of viewers may be interested in golf equipment. The growth of cable television's audience came at the expense of the broadcast networks, whose share of the viewing audience and advertising revenues decreased.

CABLE ADVERTISING

Cable television was originally developed to retransmit broadcast programs to geographic areas that were unable to receive clear antenna reception. A cable company received broadcast signals and redistributed them via cable. Later, original cable programming and channels were established, most of which charged for delivery. In other words, in addition to the cost of hooking up to a cable service to receive broadcast programs, subscribers also either paid a fee to receive a cable network, such as HBO or Showtime, or they paid to watch a program special, such as a boxing match.

Advertiser-supported cable network programming appeared by the mid-1980s. Cable service providers then began offering programming packages. Instead of paying for each channel received or for an individual program, subscribers pay a base monthly fee for a set of cable and broadcast channels. The cable and broadcast networks and the local cable service sell commercial time within these advertiser-supported programs.

As more cable channels dedicated to particular interests became available, the television viewing audience became more fragmented, and each cable and broadcast network found itself competing for a diminishing share of the overall audience. When there were only three broadcast networks, each could be certain that about one-third of the audience was watching its programs and commercials. But with each new cable channel, all the existing channels saw their audience shrink.

Fragmented audiences and more channel choices have contributed to a decline in advertising effectiveness. With the touch of a button, a viewer can easily switch to another channel during a commercial break, which hampers a network's ability to deliver a steady audience to its advertisers. The proliferation of new cable channels has eroded the broadcast networks' share of the audience to about 42%, while cable's proportion has grown to about 54%. The share fluctuates—cable share increases in the summer, and broadcast programs are tops in fall and spring.

When audience share increases or decreases, so do advertising dollars, which is why the cable and broadcast industries are in such heavy competition. Even though cable television still has a long way to go before its advertising revenues catch up with those of the networks, every dollar it makes is a dollar less for the broadcast networks.

SEE IT NOW

ADVERTISING: 1990 TO PRESENT

FROM A BUSINESS PERSPECTIVE

Clearly, no business today could survive without using some form of promotion. Some companies and professional services might rely entirely on word-of-mouth referrals, while others might use a combination of referrals, the Yellow Pages, and ballpoint pens inscribed with the company's name. Other businesses might choose to create full-blown media advertising campaigns. Regardless, most businesses rely on some type of advertising.

Advertising plays several crucial roles in an entity's marketing efforts. It identifies a target audience, differentiates products, and generates revenue by increasing sales. Advertising tells the audience about the benefits and value of a product or service. Advertising also differentiates products by creating a *brand image* for each one. A brand image is the way a consumer thinks about a product or service. A brand image is different for each product and makes the product stand out from its competitors. Sports drinks, in general, are very similar to one another, but there may be a big difference in consumers' minds between Gatorade and PowerAde. For instance, a consumer might believe that one sports drink is more effective at rehydrating than the other, even though any real differences may be indistinguishable.

Advertising generates revenue by persuading people to purchase products and services. But persuading people to buy a product is psychologically complex; the process includes building a positive brand image and relies on generating brand loyalty, the repeated purchase of a product.

FYI: U.S. Advertising Revenue by Medium (2014) (in billions)

Internet Total	$49.5
Nonmobile	$37.0
Mobile	$12.5
Broadcast TV	$40.5
Cable TV	$25.2
Radio	$17.2
Newspapers	$16.7
Magazine (Consumer)	$12.8
Out-of-Home	$8.4
Videogame	$1.0
Cinema	$.8
Total	$172.1 billion

Source: Interactive Advertising Bureau, 2015

FYI: Top U.S. Advertisers by Ad Spending: 2014 (in billions)

1.	Proctor & Gamble	$3.13
2.	Comcast Corp.	$2.18
3.	General Motors	$1.94
4.	AT&T	$1.79
5.	L'Oreal	$1.58
6.	Verizon Communications, Inc.	$1.55
7.	News Corp	$1.39
8.	Toyota	$1.35
9.	Berkshire Hathaway, Inc.	$1.28
10.	Chrysler Group LLC	$1.20

Source: Total advertising spending by company, 2015

FROM A CONSUMER PERSPECTIVE

Although advertisements may be annoying, they are also beneficial. Advertisements serve social and economic purposes. Advertising is also educational; after all, it is the way consumers learn about new products and services, sales, and specials. Advertising benefits the economy by promoting free enterprise and competition. The results of these forces are product improvements, increased product choices, and lower prices.

Advertisements also serve a social function in that they reflect popular culture and social values and give people a sense of belonging. Spots often include the hottest celebrities, the latest trends, and the most popular music. Products are advertised within the cultural environment. For instance, after the 9/11 terrorist attacks on Washington, DC, and New York City, many commercials contained shots of the American flag and other symbols of national unity.

Buzz marketing, which has been defined as "the transfer of information from someone who is in the know to someone who isn't" (c.f. Gladwell, 2003), is a contrived version of word-of-mouth endorsement. True word-of-mouth advertising is traditionally a highly trusted and very effective type of communication. 'Word of mouth' implies that someone who has used a product or service is giving their honest opinion about it of their own volition. With buzz marketing (also known as *viral marketing*), a company pays people to pass themselves off as ordinary consumers using a product and to promote its features and benefits regardless of whether they believe in it. For example, Vespa hired good-looking, hip young men and women to cruise around Southern California hot spots on its scooters. As admirers asked about the scooters, the paid endorsers touted the product and even handed out the address and phone number of the nearest Vespa dealer.

ADVERTISING AGENCIES

Advertising agencies have long been the hub of the advertising industry. Agencies are hotbeds of creativity—the places where new ideas are born and new products come to life. Agencies are where many minds come together—account representatives, copywriters, graphic artists, video producers, media planners, researchers, and others who collaborate to devise the best campaigns possible.

Advertising agencies are not all alike. There are large agencies that serve national advertisers, and there are small agencies that serve locally owned businesses. Some agencies have offices around the world, some have offices around the nation, and others have an office or two in some city or town. Some agencies have staffs of thousands, and others are one- or two-person shops. Although agencies exist to serve advertisers, how they do that is not the same from agency to agency. Different types of agencies serve different functions.

FULL-SERVICE AGENCIES

Full-service agencies basically provide all of the advertising functions needed to create an advertising campaign. Agency employees plan, research, create, produce, and place commercials and advertisements in various media. They often provide other marketing services as well, such as promotions, newsletters, and corporate videos. Some full-service agencies are very structured; each group or department focuses on its strengths, and projects are basically moved down the line in each step of the process. Other agencies are more collaborative, often grouping employees with different fields of expertise together on a project. For instance, to create a campaign, an account representative may team with copywriters, graphic artists, researchers, planners, and media buyers.

CREATIVE BOUTIQUES

Some advertisers have an in-house staff that plans, researches and buys media time and space but needs help with the actual creation of a campaign. These advertisers contract with a creative boutique rather than with a full-service agency. Creative boutiques focus specifically on the actual creation of ads and campaigns and are therefore staffed with copywriters, graphic artists, and producers. Advertisers benefit by hiring a group of people with expertise in creative work.

MEDIA-BUYING SERVICES

Some advertisers have in-house creative departments that write and produce their commercials and advertisements, but they might not know the best media in which to place their spots and print ads. These advertisers depend on a *media-buying service*. Once the advertisements and commercials have been produced, they must be placed in the most effective media to maximize exposure and sales. Media buyers are experts in media placement. They know which media are best for which products as well as which media will help their clients achieve advertising and sales goals.

INTERACTIVE/CYBER AGENCIES

Many advertisers today see the need to expand their advertising to the Internet, DVDs, smartphones, tablets, and other interactive and mobile platforms, so they are turning to media/interactive agencies with expertise in Web design and interactive technology. These *interactive agencies* (or *cyber agencies*) create and maintain client Web sites, create and place banner ads, and produce and distribute other interactive advertising materials. Interactive shops tout expertise in many areas that full-service and other types of agencies cannot provide. Cyber agencies are criticized, however, by their full-service counterparts and others who claim that though these agencies are experts in interactive technology, they are not experts in advertising and marketing.

FYI: One of the First Interactive Agencies

Way back in 1995, Chan Suh and Kyle Shannon came together with the vision of helping companies promote themselves online, and they set up shop with their Mac computer. They called their new interactive enterprise Agency.com. British Airways, one of their first clients, asked them to redesign its Web site. Rather than just give the site a facelift, Agency.com suggested that the company rethink its online business model. As a result, British Airways ended up with an online ticketing system and a site that focused on "new ways to use interactive technology to expand

market share, reduce costs, improve efficiency, and deliver great customer satisfaction" (History, 2004).

Agency.com's success with British Airways led to new online projects with companies such as Compaq, Nike, Sprint, and Texaco. Agency.com quickly became known for its outstanding work, and *Adweek* named it one of the top 10 interactive agencies for 1996. Its mission changed from "helping businesses bring their business online" to "empowering people and organizations to gain competitive advantage through interactive relationships," which they did through a strategy called *interactive relationship management* (History, 2004).

Agency.com caught the attention of Omnicom Group, a marketing communication company, which acquired a large share in Agency.com. By 2011, Omnicom had folded most of Agency.com into its larger operations, and the other parts of the agency merged with Designory.

ZOOM IN 7.7

For a clearer understanding of the differences among creative boutiques, media-buying services, and full-service agencies, visit each of these agency Web sites and read about the types of services it offers:

- Creative boutique—Jugular: http://jugularnyc.com/
- Media-buying service—Revolution Media: http://revolutionmediainc.com/
- Full-service agency—Tombras Group: http://tombras.com/

For more information about advertising agencies and easy access to many agency Web sites, connect to the American Association of Advertising Agencies: www.aaaa.org

ADVERTISING CAMPAIGNS

Airing one or two commercials here and there is not, in most cases, an effective way to sell a product or service. That is why advertising agencies specialize in planning and implementing advertising campaigns. A *campaign* may be comprised of a number of commercials and advertisements for radio, television, the print media, and—in some cases—the Internet that are all tied together using the same general theme or appeal. Successful recent campaigns include the NBA's "where amazing happens" and Apple's "get a Mac."

Strategizing a campaign involves setting advertising and marketing objectives, analyzing a product's uses as well as its strengths and weaknesses, determining the target audience, evaluating the competitive marketplace, and understanding the media market. By using strategies, the creative staff has the fun task of creating a campaign—a series of commercials and print and online ads that follow the same basic theme. Copywriters, graphic artists, and video producers confer, toss around ideas, argue, and change their minds until finally they come up with a campaign that meets the marketing and advertising objectives.

Producing a campaign is often a long, arduous, and stressful process. Difficult-to-please clients sometimes think they know more than the advertising experts. Deadlines come up much too quickly, and the marketplace changes in a flash. Moreover, the advertising business is very competitive, which requires agencies to frequently pitch new advertisers for their business. Just one failed campaign can lose an ad agency a multimillion-dollar account. On the positive side, being part of a successful ad campaign is very satisfying. Agencies and their creative staffs often build client relationships that last for years. Creative staffs are recognized for their excellent work by national and international associations and by receiving Clio Awards, which are the advertising equivalents of Emmy Awards.

FYI: IBM's e-Commerce Campaign

Ogilvy's award-winning campaign for IBM's e-commerce division employed a media mix that included the Web. Ogilvy opened the campaign with multipage advertising in the *Wall Street Journal* that included IBM's Web site URL. It followed up with a flight of television commercials, targeted ads in trade publications, direct-mail packages, and banner ads—all of which directed people to IBM's Web site, where they received the actual selling message.

Ogilvy won a gold Clio and several other prestigious advertising awards for its IBM e-commerce campaign. Additionally, its e-commerce banners won silver and gold Interactive Pencil awards for best interactive campaign and best single interactive banner (Kindel, 1999; Winner of the most, 1999).

ZOOM IN 7.8

For a historical look at the development of Coca-Cola campaigns, go to www.memory.loc.gov/ammem/ccmphtml/colahist.html

ZOOM IN 7.9

Check out the Clio Award winners and learn more about this prestigious award at www.clioawards.com

USING THE MEDIA

Each electronic medium—radio, television, and the Internet—has strengths and weaknesses as a marketing tool. Smart media buyers know which products do best on which medium and in which market. They also know which creative strategies and appeals work best for the different media audiences. It is not possible to say that one medium is *always* better than the others.

RADIO ADVERTISING

Radio commercials are generally 15, 30, or 60 seconds in length. Radio commercials are classified as local, national,

and network spots. One of radio's strengths is the ability to reach a local audience. About 79 cents of every dollar of time is sold to local advertisers who wish to have their message reach the local community. Local restaurants, car dealerships, and stores know that by advertising on the radio, they are reaching the local audience that is most likely to visit their establishments.

National spots are those that air on many stations across the country. For example, McDonald's might buy time on selected radio stations in many different geographic regions. If it has a special promotion going on in the South, it will run its commercials on stations located in that market area.

ZOOM IN 7.10

Next time you listen to the radio, see if you can identify local versus national spots.

A *network buy* is when a national advertiser buys time on a network of radio stations that are affiliated with the same company. Some of the larger radio networks are Clear Channel Communications, which operates or distributes Premiere Networks, Fox Sports Radio and Fox News Radio, Cumulus Media Networks, which includes Westwood One, and Disney, which operates ABC News Radio, ESPN Radio, and Radio Disney.

Radio stations affiliate with a radio network that provides them with programming. When advertisers buy time on a network, they are buying spots within the network's shows that are aired on its affiliated stations. The advertiser benefits from this one-stop media purchase of many stations that reach its target audience.

Advantages of Radio Advertising

- *Local*—Radio spots reach a local audience, the most likely purchasers of local products and services.
- *Flexible*—A radio spot can be sold, produced, and aired within a few days. Copy can quickly be changed and updated. Advertisers may run several different commercials throughout the day.
- *Targets an audience*—Advertisers target selected audiences through station buys. The various program formats make it easy for advertisers to reach their markets. For example, alternative rock stations reach that all-important teen and college-age market.
- *Low advertising cost*—Radio is inexpensive in terms of the number of listeners reached. Radio reaches its audience for generally less money than most television stations or the Internet.
- *High exposure*—Advertisers can afford to buy many spots, so their commercials are heard many times. Through repeated exposure, listeners learn the words to jingles, memorize phone numbers, and remember special deals and other commercial content.
- *Low production costs*—Radio commercials are generally inexpensive to produce. Some spots are simple

announcements and others more elaborate productions, but they are still less expensive to create than television spots.

- *High reach*—Almost everyone in the nation listens to the radio for some time each week. Nearly 75% of all consumers tune in every day, and about 95% listen at least once a week.
- *Portable and ubiquitous*—Radios are everywhere: at work, at sporting events, at the beach, at the gym, on boats, on buses, on trains, in homes, in hospitals, in cars, in bars, and just about everywhere else. Small, lightweight radios are ideal for anywhere, anytime listening.
- *Commercials blend with content*—Commercials with background music and jingles often sound similar to songs, and commercials with dialogue sound similar to talk show conversations.

Disadvantages of Radio Advertising

- *Audio only*—Radio involves only the sense of hearing; thus, listeners are easily distracted by what else they may be doing or seeing. The audio-only format also makes it difficult for listeners to visualize a product.
- *Background medium*—Radio is often listened to while people are engaged in other activities (like working, driving, reading, and eating), so they do not always hear or pay attention to commercials.
- *Short message life*—Radio ads are typically 30 seconds in length, which is not much time to grab attention, especially if a listener is involved in another activity. Also, unlike print, where people can go back to an ad and write down the information, once a radio spot has aired, the information is gone. Missed messages may not be heard again.
- *Fragmented audience*—Most markets are flooded with radio stations, all competing for a piece of the audience. Listeners are fickle, often changing stations many times throughout the day. Fragmentation forces many advertisers to expand their reach by purchasing time on several stations in one market.

TELEVISION ADVERTISING

Television is considered the most persuasive advertising medium. The combination of audio and visual components captures viewers' attention more so than other media do. Plus, almost everyone watches television. Advertisers reap the benefits of an audience of millions. Despite its strong points, broadcast television is not the best advertising outlet for all advertisers. Not everyone needs to reach a large mass audience, and not everyone has the budget to produce television spots. For many advertisers, cable television offers more audience for less money.

FYI: Cost of a 30-Second Spot on Prime-Time Broadcast Network Programs

2014–2015 Season		
Sunday Night Football	NBC	$637,330
Empire	FOX	$521,794
Thursday Night Football	CBS	$462,622
The OT	FOX	$303,200
Big Bang Theory	CBS	$289,621
2013–2014 Season		
Sunday Night Football	NBC	$593,700
American Idol	FOX	$355,946
Big Bang Theory	CBS	$316,912

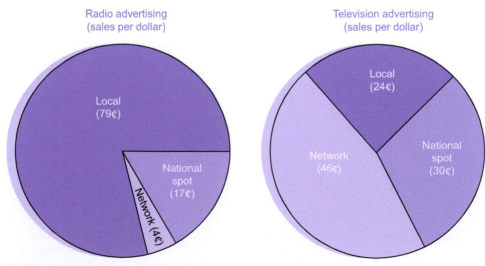

FIG. 7.6 Radio versus television dollars

FYI: Cost of a 30-Second Spot on Prime-Time Broadcast Network Programs (Continued)

The Voice	NBC	$294.038
American Idol Results	FOX	$289,942
2008–2009 Season		
Sunday Night Football	NBC	$339,700
Grey's Anatomy	ABC	$240,462
Desperate Housewives	ABC	$228,851
Two and a Half Men	CBS	$226,535
Family Guy	FOX	$214,750
2005–2006 Season		
American Idol (Wed.)	FOX	$705,000
American Idol (Tues.)	FOX	$660,000
Desperate Housewives	ABC	$560,000
CSI	CBS	$465,000
Grey's Anatomy	ABC	$440,000
2000–2001 Season		
ER	NBC	$620,000
Friends	NBC	$540,000
Will & Grace	NBC	$480,000
Just Shoot Me	NBC	$465,000
Everybody Loves Raymond	CBS	$460,000

Sources: Advertising Age, Prime Time Programs, 2010; Steinberg, 2009, 2015

Television Commercials

It often seems that commercials take up more television time than programs themselves. But in fact, in 2015, commercials on broadcast networks accounted for 17.3% of all programming, and commercials on cable networks took up 20.6% of time. Although some viewers like commercials, others think they are intrusive and click to other channels when they break into a program or appear between shows.

Television commercials are typically 30 seconds in length, but they can run 15, 45, or 60 seconds. Most commercials air in **clusters**, or **pods**, of several commercials between and within programs. Advertisers may be guaranteed that their commercials will not air in the same pod as a competitor's commercial. In other words, a Ford commercial may be guaranteed not to air in the same pod as a Chevrolet commercial. Commercials aired within a particular program are known as **spots**, or **participations**, and commercials that air before and after programs are known as **adjacencies**.

Like radio commercial time, television time is bought as **network, national, and local spots**. Advertisers who want to reach the largest audience possible will place their commercials on broadcast network programs. Automobile, shampoo, toothpaste, and fast-food commercials, for example, most often appear on network programs, which often draw millions of viewers across the country.

These spots are very expensive to purchase, so network time is usually reserved for the limited number of advertisers who can afford it.

Broadcast networks typically sell time in what are called the **upfront market**, the **scatter market**, and the **opportunistic market**. The 'markets' are actually time frames throughout the year. With the upfront market, the networks begin selling time each spring for programs that will air during the fall season. Advertising costs are based on ratings projections—the higher the estimates, the higher the price. Because advertisers are taking a risk by buying ahead, the networks guarantee a minimum audience size. If a program falls short, the network reimburses the advertiser through additional spots. Networks typically strive to sell about 70% to 80% of their inventory up front.

With the scatter market, unsold time is offered to advertisers four times a year for the upcoming quarter. But because advertisers wait until they have a better idea of how a show is doing, their risk is lower. Thus those who buy in the scatter market typically pay a higher rate and typically are not guaranteed a minimum number of viewers. Advertisers also buy opportunistic time that comes up at the last minute. For example, maybe time slots open up in a popular program or a news event spurs a special program that is sure to draw a large audience.

Advertisers also buy national spots in which they place their ads on individual stations in certain markets. For example, in February, a national manufacturer of patio furniture might buy commercial spots on stations throughout Florida. By April, the company might reduce its commercial spending in the Sunshine State and instead buy spots on stations in Georgia and South Carolina, or it may just place its commercials on stations located in beach communities along the East Coast.

Local advertisers are generally hometown businesses that want to reach customers within a single market or geographic region. For a restaurant with one or two locations in Knoxville, Tennessee, it makes sense to buy time on one or two local television stations that reach the greater Knoxville market. Buying nationwide network time would be much too costly and serve no useful purpose, as it is unlikely that anyone from far outside the local viewing area would drive to Knoxville just to eat a meal.

Product Placement

Sometimes, rather than buying commercial time, companies resort to **product placement**, paying to have their product used in or visible within a program scene. Product placement, also known as 'product integration' and 'stealth advertising,' is a subtle but effective way of exposing viewers to products, often without their conscious knowledge or realization. Early programs in which characters mentioned or used a product were a form of product placement. After falling out of favor for a number of years, product placement is making a comeback (especially on cable networks and in movies) as a creative way to set advertisements apart from the typical 30- and 60-second spots and to thwart DVR users from escaping product exposure. Mini Cooper had the right idea when it placed its cars in the 2003 caper *The Italian Job*, about the theft of millions of dollars of gold bricks. Anyone who saw the movie could not help but be awed as a helicopter chased the nimble yet powerful Minis as they wove through traffic, drove down the steps leading to Los Angeles's underground metro train tunnel system, careened through above-ground train yards, and eventually slid inside of a train box car.

Tens of thousands of products are embedded in television shows—35,000 placements in 2008 alone. In 2012, advertisers spent $8.25 billion on product placement. Characters on *House of Cards* interrupt their conversation to swig from bottles of Stella Artois, contestants on the *Biggest Loser* gab about Ziploc bags, *Top Chef* countertops are stacked with Glad products, Liz Lemon on *30 Rock* hypes Verizon cell phone services, and Luke Dunphy on *Modern Family* eats an Oreo cookie.

Apple Computer is one of biggest investors in product placement; its products appeared in 891 program episodes in 2011. In the same year, *American Idol* displayed the most product placements of any program, with 577 occurrences in 39 episodes. *Idol* judges sipped from cups emblazoned with the Coca-Cola label, and viewers were encouraged to vote using their AT&T wireless phones. The show reaps almost $60 million per year in product placement. MSNBC's *Morning Joe* hosts deliver the news while sipping Starbucks drinks, making sure to hold the cups so the viewing audience can see the logo. Critics blast the arrangement, contending that it is a conflict of interest and a breach of journalistic integrity for a news program to tout a company it might report on.

In a test of brand recognition, 46.6% of the participants recognized a brand when exposed to only the commercial, but 57.5% recalled the brand when exposed to a commercial and product placement. Some of the highest recalled recent product placements include Dean Pelton on *Community* giving a speech at a ribbon cutting ceremony where a Subway restaurant is opening, on *Two-and-a-Half Men* when Alan is shown driving a bright-red Porsche, and on the *Good Wife* when David Lee offers Veronica a bowl of M&Ms.

The potential drawback of product placement is that advertisers give up control over how their product will be presented, risking it being shown in an unfavorable light. However, it is in the producers' best financial interest to convey product benefits in a positive manner. Further, scriptwriters might not like being compelled to weave products into storylines, thus subliminally persuading viewers. And many viewers object to this practice of covert marketing.

The Federal Communications Commission (FCC) requires that programmers disclose placement sponsors in a show's credits. But the credits are typically in very small print, and they roll by so quickly that they are almost impossible to read. The FCC is considering new regulations for informing viewers about product placement. The FCC wants to extend the disclosure of product placement to make it more prominent and ban all placements in programs targeting children under the age of 12. The FCC believes that viewers have the right to know when they are being sold a product.

Sponsored Programs

Viewers are becoming weary of the 50-some-year-old advertising format of 30-second spots airing during commercial pods. Moreover, the proliferation of on-demand services is tugging viewers to advertising-free options. Viewers wonder why they should pay for commercial-laden cable or satellite programs when they can pay for ad-free shows or at least those with minimal commercial content on services such as Hulu.

Television executives are scrambling to keep viewers tuned to the broadcast and cable networks, but they know to do so, they must reduce the number of commercials. Fewer commercials, however, means lower revenue—a trade-off that television cannot make if it plans on remaining competitive in the digital era. Advertising executives are turning their attention to the old days of sponsored programs for a solution to their dilemma. Both broadcast and cable advertising strategists are calling for less commercial time in the hopes that a limited supply will increase the cost enough to offset fewer spots. They are also creating 'bonus programming,' such as behind-the-scenes interviews and extended segments, in lieu of the typical spots. Instead of going to commercials at the end of a program, viewers will be treated to what they think is 'extra' programming but is in fact sponsored time. Strategists see this new formulas as a win situation for all; viewers enjoy the extra programming and prefer the subtle promotional messaging over blatant 30-second spots, advertisers have a new and effective way to reach viewers, and television retains viewers who might otherwise have abandoned commercial-dependent programming for ad-free services.

Public Service Announcements

Public service announcements (PSAs) and station promos are also types of on-air radio and television promotions. Public service announcements promote nonprofit organizations, such as the American Lung Association and the United Way, as well as social causes, such as "Friends Don't Let Friends Drive Drunk" and "Click It or Ticket" seatbelt advocacy. Most stations air PSAs at no charge and air them whenever they can fit them into the regular schedule. Radio and television stations also promote their own programs on their own stations. Frequently, a promo airing in the morning will promote an afternoon drive-time show, or an afternoon television promo will alert viewers about a special program coming up later that evening.

Advantages of Advertising on Television

- *Visual and audio*—Television's greatest advantage is its ability to bring life to products and services. The combination of sight and sound grabs viewers, commands their attention, and increases their commercial recall. Viewers often remember the words to commercial jingles, company slogans, and mottos, and—more important—they remember information about the products themselves.

- *Mass appeal*—Television commercials reach a broad, diverse audience. So rather than just target a narrow customer base, advertisers appeal to a large, mass audience.
- *High exposure*—Although the high cost of a television spot generally limits the number of times it can be aired, one exposure will reach many people simultaneously.
- *High reach*—About 99% of U.S. households have at least one television set, and 82% have two or more sets. There are now more televisions in the United States than there are people—116 million television homes, each averaging 2.93 sets, comes to 340 million sets for 322 million people. In an average U.S. household, the television is on for 8.25 hours per day. Adults spend about 5 hours per day, an all-time high, watching television, and children and teens about 3.5 hours a day. Thus, advertisers with products that appeal to a broad audience achieve maximum reach and exposure with television.
- *Ubiquitous*—Television is everywhere. It is rare to go someplace where there is not a television set. Television is more than a medium; it is a lifestyle. People schedule their time around their favorite programs, and they often build social relationships based on common liking of certain programs.
- *Commercials blend with content*—Commercials are cleverly inserted within or between programs so they are not so obvious and blatant. Contemporary commercials are visually exciting and have interesting narratives that capture viewers' attention before they have the chance to change channels or head to the kitchen to make a snack.
- *Variety*—With so many program types, advertisers have many options for commercial placement. For instance, a diaper marketer may choose to place commercials during soap operas, whose viewers are typically work-at-home mothers.
- *Entertaining*—Television is highly entertaining, and this entertainment value spills over into commercials. During certain times, like the Super Bowl and the Emmys, viewers actually find the commercials more exciting than the programs themselves.
- *Persuasive*—Television is the most persuasive commercial medium. Catchy audio, spectacular visual effects, and interesting product demonstrations often get the most skeptical of viewers to try new products.
- *Emotional*—Television makes us laugh, makes us cry, makes us angry, makes us happy, and otherwise engages our emotions. Effective commercials tug at our heartstrings, as we witness the emotional rewards that come from purchasing the advertised product and so are inclined to purchase it ourselves.
- *Prestige*—Many viewers believe that if a company is wealthy enough to buy commercial time and a product is good enough to be advertised on television, then it is good enough to buy. The glamour of television tends to rub off on products.

Disadvantages of Advertising on Television

- *Channel Surfing, Zipping, and Zapping.* These are three terms that advertisers hate to hear. Viewers channel surf when they move all around the television dial, sampling everything that is on at the time. All too often, when a commercial comes on, the channel gets changed. Most viewers are so adept at avoiding commercials that they instinctively know how much time they have to scan other channels and get back to their original program just as the commercial break ends. Advertisers end up paying for an audience that does not even see their commercials.
- Digital video recorders (DVRs) and remote-control devices are commercial avoidance culprits. Viewers merely push a button to fast forward or zip through commercials, and some DVRs blank or zap out the messages all together. Research estimates that about 90% of DVR owners fast forward through commercials. But some makers of DVRs are onto commercial fast forwarders and are thwarting their attempts at avoidance by inserting interactive ads on the screen whenever a viewer pauses a program or fast forwards.
- *Fragmented audience*—With cable and satellite television now offering hundreds of channels, the broadcast television audience has shrunk. Moreover, the average household receives between 172 and 189 channels, which fragments the audience. In other words, channels share a smaller portion of the overall viewing audience. Moreover, of all those channels received, viewers only watch about 17 on a regular basis.
- *Difficult to target*—The mass appeal of broadcast television makes it difficult and expensive to reach a specific target audience. Advertisers often end up paying for wasted coverage—paying for a large audience when they really only wanted to reach a smaller subset of viewers.
- *Portability*—Most television sets cannot be picked up and moved around. With digital transmission, even small battery-operated televisions cannot be operated without a cable or satellite connection, and thus television is not very portable.
- *High cost*—Running commercials on television is very expensive. Dollar for dollar, it is the most expensive medium, especially when considering both production costs and airtime. The average cost of a 30-second commercial on broadcast prime time was $112,000 during the 2014 season.
- *Clutter*—Television commercials are grouped together, either between or within programs. It is common to see three or four commercials in a row, followed by a station promotion, a station identification spot, and then three or four more commercials. Broadcasting and cable nonprogramming time (commercials, station ID, station promotion) takes up an average of 15 minutes in a typical prime-time hour. Early-morning television airs 18 minutes per hour of nonprogramming materials, and daytime television has nearly 21 minutes devoted to nonprogram fare. When so many commercials are cluttered together, viewers tend to pay little attention to any of them, thus hampering message recall.

FYI: The Average Cost of a 30-Second Super Bowl Commercial

It often seems that the best part about watching the Super Bowl is seeing the commercials. Advertisers save their best spots for this contest, and even when the game itself is a real snooze, viewers perk up during the commercials. The Super Bowl delivers a huge audience, so advertisers pay big money to promote their products and services. Below are the average costs of a 30-second spot through the years:

Year	Average Cost
2016	$5,000,000
2015	$4,500,000
2014	$4,000,000
2013	$3,700,000
2012	$3,500,000
2011	$3,000,000
2010	$2,650,000
2009	$3,000,000
2008	$2,700,000
2007	$2,400,000
2006	$2,500,000
2005	$2,400,000
2004	$2,300,000
2003	$2,200,000
2002	$2,200,000
2001	$2,200,000
2000	$2,100,000
1997	$1,200,000
1992	$850,000
1987	$600,000
1977	$125,000
1967	$37,500

Sources: Baumer, 2011; Elliott, 2013, 2014; Super Bowl ads cost average of $3.5M, 2012; Pagels, 2015

ZOOM IN 7.12

Ten Best Super Bowl Ads
View the ads at http://ftw.usatoday.com/2015/01/10-best-super-bowl-commercials-of-all-time

1. McDonald's: "Larry Bird vs. Michael Jordan," 1993
2. Pepsi: "Your Cheatin' Heart," 1996
3. Apple: "1984," 1984
4. Coca Cola: "Mean Joe Greene," 1979
5. Career Builder: "Monkey Office," 2006
6. Budweiser: "Bud Frogs," 2002
7. Nike: "Michael Jordan and Bugs Bunny," 1992
8. Reebok: "Terry Tate: Office Linebacker," 2003
9. Pepsi: "Cindy Crawford", 1992
10. Volkswagen: "The Force," 2011

Source: Scott, 2015

CABLE ADVERTISING

There are several major differences between cable and broadcast television. For example, broadcast television used to be free over the airwaves, but now that over-the-air analog transmission has stopped, viewers have to pay for a cable or satellite connection. Before digital delivery, broadcasting was broadcasting and cable was cable, but now these differences are blurred, at least in the minds of many viewers. But from an industry perspective, the differences still exist. Although both broadcast and cable networks generate revenue through advertising, local cable providers profit from subscriber fees.

Cable television has been around for many years, but only recently has it challenged broadcast television for audience share. For one week in the summer of 1997, basic cable channels for the first time edged out ABC, NBC, and CBS with 40% of the prime-time audience, compared to the networks' 39% share. During the 2001–2002 season, the advertising-supported cable networks drew for the first time a larger prime-time audience than the seven broadcast networks (ABC, CBS, NBC, Fox, UPN, WB, and PAX) combined. Now it is common for the cable networks to outpace the broadcast networks.

Cable offers viewers select program options on which advertisers target niche audiences. Advertisers have the option of buying commercial time on specialty cable networks such as Golf Channel, Home and Garden Television (HGTV), Nick at Nite, MTV, and hundreds of others. These channels are perfect advertising venues for marketers who want to target specific audiences. For example, there is probably not a better place to advertise kitchen appliances or cooking products than on Food Network or gardening supplies on HGTV. Even though advertising-supported cable networks tend to have small audiences, advertisers are attracted to these specialized and often loyal markets.

Most cable buys take place at the network level, wherein advertisers buy time on a cable channel, such as ESPN, and the spots are shown in selected locations or throughout the country. The remaining program time is then sold locally by the cable service. Local cable reps from one service (e.g., Comcast) also team with other cable service reps (e.g., Cox) to sell **interconnects**. Large cities such as New York may have several cable providers, each sending out cable programming to a specific part of the city. Interconnects allow advertisers to purchase local cable time with several providers with one cable buy. In doing so, an advertiser could simultaneously run a commercial on all of New York's cable systems for a larger audience reach.

Advantages of Advertising on Cable Television

- *Visual and audio*—Like broadcast television, cable television's primary strength is its ability to attract viewers through sight and sound.
- *Select audience*—Cable's wide variety of programming and networks attracts small, select target audiences. Rather than spending money on a large broadcast audience, of which only a small percentage may be interested in the product, cable delivers specific consumers to its advertisers.

- *Upscale*—The cable television audience tends to be made up of young, upscale, educated viewers with money to spend, making cable an ideal venue for specialized and luxury items.
- *Variety*—With hundreds of cable networks to choose from, it is easy for an advertiser to match its product with its target audience.
- *Low cost*—With so many cable networks competing for advertisers, they rarely sell all of their available commercial time. The fierce competitive environment also keeps costs down, making cable an attractive buy to many advertisers.
- *Seasonal advantage*—Cable networks have learned to take advantage of the broadcast networks' summer 'vacation.' During those hot months when broadcast television is airing stale reruns, cable is counterprogramming with shows that attract larger-than-normal audiences and pull viewers away from old broadcast shows.
- *Local advantage*—National spot buyers and local businesses can take advantage of cable's low cost and targeting abilities to reach specialty audiences within certain geographic areas.
- *Media mix*—The low cost of cable, coupled with the selective audience it provides, makes it an ideal supplement in the media mix.

Disadvantages of Advertising on Cable Television

- *Zipping, zapping, and channel surfing*—Zipping, zapping, and channel surfing are the enemies of television—both broadcast and cable. These culprits make it easy for viewers to avoid exposure to commercials.
- *Fragmented audience/low ratings*—Cable audiences are fragmented and spread across many cable networks. These small audiences translate into low ratings for cable shows when compared to the ratings of broadcast network shows.
- *Lack of penetration*—Only about 5.5 of 10 U.S. households subscribe to cable television, down from a high of about 75% in the late 1990s. Cable's household penetration is ebbing as more viewers subscribe to satellite services and watch programs on the Internet.
- *Churn*—Cable audience size is affected by *churn*, which is the ratio of new subscribers to the number who disconnect their cable service.

INTERNET ADVERTISING

No other medium is as interactive as the Internet. Typical 30-second radio and television spots do not engage consumers in the same way as clicking on ads, watching and listening to in-banner videos, reading product recommendations, playing games that are disguised ads, or tweeting about recent purchases. In many ways, the consuming public has come to expect to interact with advertisers and other purchasers. Advertisers are up on the advantages of reaching consumers online—so much so that they spend about $49.5 billion a year doing so. More importantly, online advertising expenditures surpassed broadcast television for the first time in 2013. Television advertising is forecast to drop by 12% per year, while online advertising, led by Google and Facebook, is expected to grow at

a rate of 3% per year and hit the $100 billion mark in 2020. But this trend does not mean that television will disappear; rather, the Internet and television will settle as complementary media, with online recognized as the better way to target consumers and television the better way to reach a mass audience.

FYI: First Online Banner Ad

Online advertising was born on October 27, 1994, when HotWired (www.hotwired.com), the online version of *Wired* magazine, posted the first-ever banner ad, which was sponsored by AT&T. The banner was something new and intriguing, and thus it was clicked on by 44% of those who saw it. With no idea of how much traffic HotWired would attract or of its audience's demographic profile, HotWired set a price of $30,000 for a 12-week commitment and was thrilled when other advertisers—MCI, Volvo, and Sprint—signed up.

Since then, the Internet advertising industry has boomed. Ad design and interactive technologies, online media buying and selling, online campaign construction, consumer tracking, and online audience measurement are just a few of the online advertising functions that have come about since the first banner ad.

FIG. 7.7 First banner ad *Source: Edwards, 2013*

FYI: Push/Pull Strategies

Internet advertising includes *push* and *pull* strategies. A *push strategy* means that an ad is pushed, or forced, onto consumers, whereas a *pull* refers to consumers pulling, or seeking, the message. Internet advertising that follows a push strategy is similar to traditional media advertising, in which commercial messages are pushed onto consumers. For example, television commercials are pushed because viewers do not have control over when and which ads to view. Advertisers are taking advantage of Internet technology by pushing new products and product-improvement announcements and other promotional messages through banner ads, email, electronic mailing lists, mobile ads, and so on. Web marketers also use pull strategies, such as establishing Web sites, delivering information through subscription services, and creating links that simply lead consumers to product information at their own convenience.

There is some debate as to what constitutes *Internet advertising*, but it is generally considered to be such when a company pays or makes some sort of financial or trade arrangement to post its logo, product information, or service with the intent of generating sales or brand recognition on someone else's Internet space. For example, when Neiman Marcus pays to place its banner on the *Washington Post* Web site, this is considered Internet advertising. However, when Neiman Marcus sells clothing and other products on its own Web site, it is considered *marketing*. The distinction between online advertising and online marketing is similar to Neiman Marcus buying commercial time on a local radio station as opposed to printing a catalog with ordering information. The former is an ad, and the latter is a marketing endeavor. But these days, the lines are not so clear-cut. Marketers tweet, create Facebook pages, host blogs, use social networks, and send images and text through mobile phones. There is a question as to whether these are marketing or advertising strategies.

Online advertising is known by various names and comes in many forms. Some of the most common types of online ads are described in the paragraphs below. Most online ads can be adapted across platforms for Internet, smartphone, and tablet delivery.

Search Engine Marketing (SEM)

SEM is the most common form of online advertising, accounting for about 41% of all online revenue. SEM optimizes ad targeting with paid placement on search engine results pages. For example, companies like Nike or New Balance would pay to have their ads appear as part of the search results whenever a user enters the term 'athletic shoes.' Directories, such as Zillow.com, which lists homes for sale and rent, are another type of SEM.

Banner Display Ads

Banner ads are the second most common form of online advertising and make up about 19% of all online revenue. Traditional banner ads are static or nonanimated, with little more than an advertiser's logo with some embellishment. To increase consumer interest and to make purchasing easier, a banner usually contains some sort of animation or moving gifs and a link to the advertiser's landing page that relates to the ad or to its homepage. Banners come in many sizes and shapes such as squares and vertical and horizontal rectangles. For example, *tiles* are simply inexpensive small banners usually in the shape of a square.

Rich-Media Ads

The static and animated banner ad has given way to a more exciting visual presence, the rich-media ad. It is not a type of ad per se but describes how an ad is designed. Different from static or animated ads, rich-media banners contain audio and video. Advertisers are concerned that with so many banner ads dancing on Web sites, users might get annoyed at the distraction and choose to ignore them all.

Interstitial/Superstitial

To make sure that their messages are seen, many online advertisers opt for rich-media **interstitials** and **superstitials**. The word 'interstitial' means 'in between.' Thus both interstitials and superstitials appear in a separate browser window in between Web pages or on top of Web page content. An interstitial and superstitial are both rich

media ads, but interstitial is a general term for any type of ad that runs between pages, but the term superstitial is trademarked by Unicast as a descriptor of its rich-media technology. Once referred to as 'polite' ads, because they only played when fully downloaded and initiated by the user, some video superstitials now rudely self-start while a page is loading up, and it is often impossible to turn off the ad until after the page is fully loaded. An interstitial/superstitial that appears on top of Web page content is called a **pop-up**. If it lurks behind the content browser window, it is a **pop-under**, also called the 'evil cousin of a pop-up,' because it surprises users when they close the browser window. Both dazzle the eyes with animation, graphics, interactive transactional engines, and near-television-quality video.

Expanding Rich Media

Pencil pushdown/sliding billboard ads are skinny, full-width ads that expand lengthwise when clicked on. Pencil pushdowns are highly visible, and when clicked and opened, they push down Web site content rather than covering it up. An *expanding banner* looks like a square banner but doubles in size horizontally and covers up adjacent content when the cursor swipes over it.

In-Banner Audio/Video

An **in-banner audio** ad looks like a banner but contains audio that either plays automatically when a user lands on the page or must be initiated by the user. An in-banner video ad is a display ad but has the look and feel of a television commercial, usually 15 seconds in length. Preroll ads are video spots that play before noncommercial video content starts up. For example, YouTube videos often contain a 15-second preroll that comes on before the beginning of a video.

Corner Peel

A **corner peel** ad is a double-image ad in which the top layer peels down from either corner, revealing additional product information. The action is user initiated by running the cursor over a corner tab. The corner peel movement is highly noticeable and, when pulled down, only obstructs a small corner of the Web page content. Clicking on the bottom layer takes a user to the product's Web site.

Hover Ads

Hovers are full-width ads that slide up about one inch from the bottom of a Web page when it is opened. They cover up content and must be clicked on to 'collapse' them.

Floating Ads

This type of ad features an image, perhaps a butterfly, flower, plane, or bird flittering around the screen for a few seconds, and then it 'lands' on the Web page and morphs into a small, very short animated ad that covers up the content and then, 'poof,' disappears on its own. At their most annoying, floaters challenge users to try to nab them with the mouse as they dance inside a browser window and even turn cursors into ads, making it almost

FIG. 7.9 Example of floating ad
The Oklahoman Media Company

impossible to use the page. Frustrated users find themselves playing cat and mouse while desperately trying to sink the floater.

Big Box, Leaderboard, Skyscraper/Tower, Extramercial

These display banners may take several forms. They can be nonanimated (static), contain an animated gif or Flash animation, or contain rich-media elements.

The **big box** is the most popular type of banner. Measuring 300 × 250 pixels, big-box ads are highly visible and flexible in terms of placement on the Web page. Pixels are

tiny dots of color that collectively form an image. On a computer screen, **72 pixels equals 1 inch**.

At 728 × 90 pixels, the **leaderboard** ad spans the width of a Web page and is typically placed between the masthead (the Web page title) and content. Because it is at the top of the page, it is highly visible but does not interrupt the content. Top-of-the-page placement is also called 'above the fold,' because users do not have to scroll to see an ad. When users have to scroll to see an ad, it means it is placed 'below the fold.' The terms come from newspaper advertising in which printed display ads are literally placed above or below the paper's horizontal fold.

The aptly named **skyscraper/tower ad** extends vertically (160 × 600 pixels) along one side of the browser window. The right hand placement makes it visible and high impact but subtle and nonintrusive.

Another type of online advertising is the **extramercial**, which is placed in the 3-inch space to the right of the screen that is usually not visible unless the user scrolls sideways or has his or her monitor sized at 1,024 × 768 pixels or higher. Extramercials have given way to newer formats and are not as commonly used as they were in the 1990s and 2000s.

Wallpaper and Homepage Takeover

These types of ads occupy Web page architecture by actually becoming part of the page. **Wallpaper ads** frame Web content. Wallpaper ads are high visibility and do not block

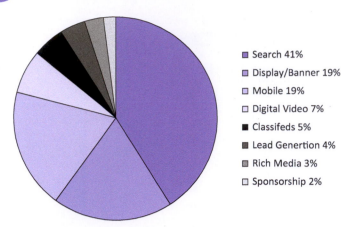

Search 41%
Display/Banner 19%
Mobile 19%
Digital Video 7%
Classifeds 5%
Lead Genertion 4%
Rich Media 3%
Sponsorship 2%

FIG. 7.10 Popular types of Internet ads (2013)

Web page content. When users click on the frame or skin, as it is otherwise known, they are sometimes taken to the product Web site, but not all wallpaper ads are clickable links. *Homepage takeover* ads do as they are named—they block out homepage content. Users are taken aback when they click on a Web site, and instead of seeing the familiar landing page, they get a full-page advertisement that completely covers the content. For example, a few years ago Yahoo! users were awestruck when a few seconds after opening the homepage, a Ford Mustang came careening across the screen, obliterating the site's content. As the Mustang swerved around the screen, it changed colors and morphed into different models to urge visitors to customize their own vehicle on the Ford Web site. A few seconds after the ad finished, the landing page reappeared. A homepage takeover ad can be thought of as a 'road blocking' strategy, in which one advertiser buys all of the ad space on a page.

ZOOM IN 7.13

Check out this Web page for examples of online ads. http://customers.ibsys.com/cs-rich_media/index.html

ZOOM IN 7.14

A 21-year-old British student created what is probably the cleverest use of click-through banners. He came up with the idea to sell click-through logos on one page for $100 each. For their money, advertisers got about a 10 x 10 pixel space, which is very small considering 72 pixels equals 1 inch. It took the student only about 5 months to sell a million dollars' worth of logos. Check out the colorful page at www.milliondollarhomepage.com

Native Ads, Product Placement, Buzz Marketing, and Advertorials/Infomercials

Native Ad. A native ad is the newest blending of commercial messages and editorial content, but the ad mimics the Web site by using the same headline style, type font, layout, and so on. The ads basically look like articles and

draw in even the most experienced online users, who think they are reading real news stories. Thus **native ads** are highly persuasive. For example, BuzzFeed once embedded the headline, "14 People Making the Best of Bad Situation" among other news stories, except the link went to a Volkswagen-created 'story' that tied into its theme, 'Get in. Get happy,' without indicating that it was really an advertisement. Advertisers like to use native ads because they are not subject to '**banner blindness**,' which is when users become so accustomed to seeing banners that they ignore or simply do not see them.

Going native also hinders **ad blockers**, which are software products that prevent advertisements from appearing in a browser window. Because a native product endorsement is written into the content, there is no ad to block as such, and so viewers are exposed to the advertising message. Savvy Web site designers know how to integrate native ads into their content seamlessly. For example, allrecipes.com revamped its site in 2015 and features recipes made with Market Pantry, Kraft, Swanson, and other branded ingredients.

Product Placement

The strategy of placing a product within content has become more common in movies and television over the last several decades, and now products are being placed within Web site content. For example, Target signed on to sponsor NBC's online series *Saturday Night Line*, about people waiting in line for tickets to *Saturday Night Live*.

SnapChat's new form of product placement is imprinting corporate logos on users' photos. For example, a teen who SnapChats the Big Mac he is eating has the opportunity to overlay McDonald's golden arches onto the photo. Teen favorite retailer Hollister sent a SnapChat request to etch its logo on photos archived on phones located near 19,000 high schools in the U.S. and Canada. Hyatt, Burberry, Hallmark, TGI Fridays, Starbucks, and other big retailers have signed on to SnapChat's 'geofilter' program, which makes logos and slogans available to SnapChat users who gladly act as online 'ambassadors' of their favorite brands. As with most product placements in which corporations cede control of their image to promoters, geofiltering has its risks. For example, a user could overlay a logo onto an unflattering photo. But corporations are willing to take the chance of their slogan or logo appearing on a handful of uncomplimentary photos in exchange for potentially millions of positive views.

Buzz Marketing

People are hired by a marketer to pretend they are everyday people who just happens to go online to talk up a well-liked product. These hired promoters present themselves as ordinary consumers to infiltrate chat rooms, blogs, and mailing lists to push a particular brand. Product information may be a sales pitch disguised as a helpful tip or a 'review' that is really a paid-for promotion rather than an unbiased viewpoint. Unsuspecting Internet users might be duped into buying a product or service based on buzz marketing rather than honest word-of-mouth promotion. To curb the obfuscation of paid online endorsements, the Federal Trade Commission is now requiring that bloggers disclose whether they have received any kind of payment or free samples of products they review or endorse.

Advertorials/Infomercials

This type of product promotion appears as part of a Web site's editorial. For example, an online bookstore might promote a particular book with a 'recommendation' that appears to be a Web site editorial but is actually a paid promotion by the book publisher. Similarly, a Web site could offer several recipes that are accompanied by recommended wines, but a user may not know that the wine recommendations are actually paid commercial messages. Product placement and advertorials/infomercials are especially designed to attract the attention of Web users who are adept at ignoring banner ads.

Word of Mouth

Genuine word-of-mouth promotion is very powerful. People buy products that are endorsed by people they trust. For example, when a Harvard University social studies student looked into why the popularity of fringe-rock band Weezer soared even when the band was on a recording hiatus, he discovered that of the 20,000 fans who answered his survey, one-quarter of those who bought the band's 1996 album in 2002 did so because of online word-of-mouth recommendations. Music promotion has exploded online. Musicians promote their music, and music aficionados tout their favorite artists and songs and blast the ones they do not like on such sites as MP3.com, OurWave.com, PureVolume.com, and last.fm.

Social Media Marketing and Advertising

Using social media is another way to generate Web site attention, traffic, and sales. Social media sites, like Facebook, are advantageous for marketers to connect with and establish a 'social' relationship between themselves and consumers. Conversations about products and services instill brand loyalty and leave customers believing that the company cares about them and values their feedback, which in turn leads to increased and repeat sales. Savvy marketers put themselves out on various social media

and keep the conversation going on their blogs, Facebook pages, and Tumblr, Foursquare, Google+, and Twitter accounts. Marketers strive for engagement, the new buzz concept that asserts that the longer a user stays on a Web site or social media page, the more likely he or she is to make a purchase. Further, new software that monitors the conversations and analyzes their general tone cues marketers about brand image and buying intention and helps them identify and target particular consumers who are most likely to purchase or replace an old item.

Advertisers are not only designing attractive and persuasive ads but ads that are compelling and interesting enough to generate social media cachet and get consumers to share them among their contacts. Marketers promote their products through social media and then depend on consumers to spread the word to their own social groups through 'likes,' reposted comments, and tweets. Marketers know that online users trust ads and recommendations that come from people they know more than any other source. Social media is the modern version of word-of-mouth, and the larger the number of social media groups marketers connect with, the larger the number of followers to spread the word to others.

Mobile Advertising

Mobile is the newest way to reach consumers on the go by pushing ads directly to a smartphone or tablet. The sheer number of mobile devices in the world advantages mobile advertising. In the U.S. alone, there are about 140 million tablet users and about 200 million smartphone users. Worldwide projections show tablet users soaring to 1.4 billion and smartphone users to 2.5 billion by 2018.

Measuring the effectiveness of mobile ads is a bit elusive. The **click-through rate** for mobile banners in the U.S. hovers around 0.86%, and in Europe, the Middle East, and Africa, it is up to 1.4%, but the rate only tells part of the story. More difficult to measure is the percentage of users who take action after seeing an ad and how to value the action they take. A user could pull or search for content and then click on a mobile banner linked to a product Web site or landing page, but that is not as valuable an

Internet Ad Revenues by Major Industry Category, 2013

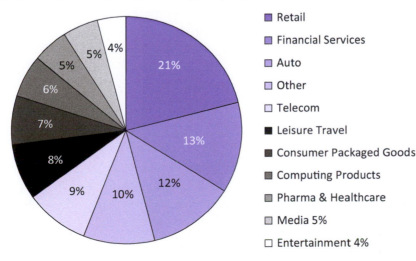

- Retail
- Financial Services
- Auto
- Other
- Telecom
- Leisure Travel
- Consumer Packaged Goods
- Computing Products
- Pharma & Healthcare
- Media 5%
- Entertainment 4%

FIG. 7.11 Internet ads as percentage of spending by major industry categories, 2013

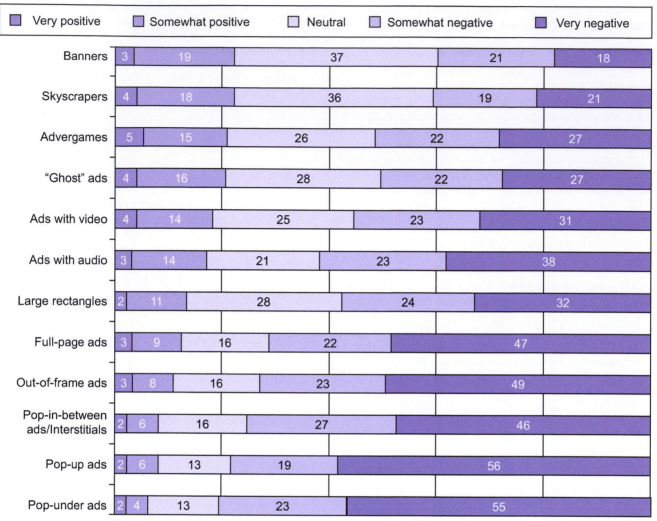

FYI: Digital Media Statistics: Less Intrusive and New Online Formats Perceived Most Positively by Consumers

	Very positive	Somewhat positive	Neutral	Somewhat negative	Very negative
Banners	3	19	37	21	18
Skyscrapers	4	18	36	19	21
Advergames	5	15	26	22	27
"Ghost" ads	4	16	28	22	27
Ads with video	4	14	25	23	31
Ads with audio	3	14	21	23	38
Large rectangles	2	11	28	24	32
Full-page ads	3	9	16	22	47
Out-of-frame ads	3	8	16	23	49
Pop-in-between ads/Interstitials	2	6	16	27	46
Pop-up ads	2	6	13	19	56
Pop-under ads	2	4	13	23	55

FIG. 7.12 Attitudes toward Web ad formats

action as purchasing a product through a mobile device, which people are less likely to do than through their computers. But what mobile device users are likely to do is compare prices (known as 'showrooming') and click on links to phone numbers, maps, and coupons, signaling prepurchasing intention.

The various types of mobile ads include banners placed within apps, top- or bottom-of-page banners, ads placed within mobile games and videos, full-screen interstitials that come up before content appears, and even audio-only jingles. But mobile ads do not have the same visual impact as ads on larger screens, nor are they very interactive, so they are easier to overlook than desktop ads.

As with social media advertising, much mobile advertising depends on consumers to help spread the word. And because much social media and mobile ads are 'wanted' in the sense they are sent to consumers who have expressed interest in a product or brand by going to the Web site, signing up for a newsletter or alert, or 'friending' a company on Facebook or other social medium, they are generally well received. Consumers seem to have accepted the thinking

that if they have to be exposed to ads they may as well be for products they like or are thinking of purchasing. A young woman will probably be happier getting an ad from the shoe company Zappos than one from Just For Men, which sells hair coloring to cover the gray.

FYI: U.S. Ad Spending on Social Media, 2006–2012

2014	$8.5 billion
2013	$6.1 billion
2012	$4.7 billion
2011	$4.2 billion
2010	$3.3 billion
2009	$1.81 billion
2008	$1.43 billion
2007	$920 million
2006	$350 million

Sources: Hoelzel, 2014; Infographic: Rise of social media, 2010; Stambor, 2013

Online Advertising or Online Spying?

A 1993 issue of *The New Yorker* printed what is now one of the most reproduced and infamous cartoons. It is of two dogs looking at a computer screen, and one is saying to the other, "The best thing about the Internet is they don't know you're a dog." In the Internet chapter (Chapter 5) of this book, a 2000 version of the cartoon adds to the earlier one with a second frame in which the computer screen flashes the dog's exact profile. The two cartoons show how quickly online privacy diminished. Throughout most of the 1990s, online users were fairly confident of online privacy; now nothing is private. If a user does not want anyone to know something, he or she should not post it online, and that includes on email, blogs, and Facebook.

Marketing companies, data services, retailers, and manufacturers figured out how to best use online technologies to monitor who uses their sites and who purchases online. Within a few milliseconds of opening a Web site, marketers are scanning the user's activity and sending ads based on past online travels. An online search for a desk lamp and a small flat-screen television for a dorm room will probably yield ads for lamp stores, school supplies, and electronic goods the next time the searcher goes online. Google alone handles about 3 billion search queries per day or about 100 billion per month. The typical U.S. user enters a search term 129 times per month, and each time companies like Yahoo!, Google, Microsoft, Bing, and others are there to process, monitor, track, and collect that information.

But most online users do not know how much and what kinds of data are being collected. The use of cookies is one of the most common methods of collecting online audience data. A *cookie* is a type of software that allows a Web site operator to store information about users and track their movements throughout the Web site. The software essentially installs itself on a Web site visitor's hard drive, often without his or her knowledge or permission. The cookie creates a personal file that the company then uses to customize online information to target that individual. Cookies leave a bad taste because they collect personal data that users do not always intend to give. Data could include where someone lives, past online purchases, sites visited most often, a user's hobbies, music and entertainment preferences, and—with social network sites—even friends' names and their online travels. For the most part, these searches go on without a user's knowledge or permission. But the Internet has become such a big part of everyday life, despite the fact that 85% of users believe that sites should not be allowed to track their online movements.

Marketers tend to think of online monitoring in terms of 'contextual targeting' and 'behavioral targeting.' Contextual marketing refers to placing ads on topic-relevant Web sites—for example, buying an ad for Nike running shoes on a site promoting races and marathons. Thanks to new data collection methods and technology, behavioral targeting has become the new generation of online marketing. Specifically, behavioral targeting is the practice of monitoring and tracking online activity, including searches, for the purpose of delivering ads tailored to a user's interests. Sophisticated models track where a user is traveling within a Web site. For example, if a user is on a general news site and clicks on information about European cities, a travel marketer will then know to target that person with travel ads.

Marketers are finding that online users pay more attention to behavioral targeting ads than to contextual ads by almost a two-to-one margin—65% of online users are more likely to notice an ad directly related to their specific online activity, whereas 39% spot ads of general interest. More important, behavioral targeting ads are clicked on more frequently, and lead to more sales and a higher return on investment than contextually targeted ads.

But it is especially disconcerting when a user's name and image are posted in ways he or she did not expect—in ads. Online services like Google and Facebook make it easy to find and share photos and personal information. For example, Google happily exhibits 'shared endorsements' on its advertising network of about 2 million sites. If a user follows a shoe store or a particular restaurant or rates a new album, that user's name, personal information, and photo could show up in ads for that shoe store, restaurant, or album without his or her explicit permission or knowledge. Although Google and Facebook claim they provide ways to opt out of such endorsements, most people are not aware of this option and do not even know how to go about doing so. Online companies say that they are helping consumers by personalizing online ads, but they are butting heads with privacy advocates who claim they have gone too far in breaching personal security.

All this tracking and personalization, coined 'the long click,' is under fire from advocacy groups, the Federal Trade Commission (FTC), legislators, and Internet users who are concerned that personal information is being collected without direct consent or knowledge. These groups are calling for sites to notify users when information is collected and to offer a way to block the process. Proponents of profiling claim consumers like cookies because they get ads that they are interested in. Although online consumers have the option of limiting or turning cookies off, fewer than 15% do so.

FYI: Ads That Watch You

A poster for domestic abuse at a Berlin bus stop proclaims, "It happens when nobody is looking," and means what it says. When someone is looking at the poster, it shows a happy couple with a man's arm lovingly encircling the woman, but when no one is looking, the image switches to the man raising his fist to strike the woman, who is leaning away and protecting her face with her hands.

A camera with facing-tracking software is recording whether anyone in the bus shelter is looking at the poster, and it measures attention and gender of the onlooker.

The technology is also being used at a German rental car company. When a man sees the poster, he sees an image of a limousine; when a woman sees the same poster, the image changes to a Cabriolet.

Source: Advertisement that watches you, 2009

Bowing to pressure, Google, Yahoo!, and others are offering ways to opt out of contextual targeting and behavioral targeting. Critics object to 'Big Brother' watching and monitoring online activities without users' knowledge or awareness that with most every click, they are giving away information about themselves. Consumers will soon be able to retrieve and remove bits of their own personal data. Consumers, however, will have to dig around to find out exactly how to manage their online profiles. Facebook also announced that it was discontinuing its Beacon program, which alerted users about their friends' online purchases. A widely publicized incident tells of a guy who bought his girlfriend an engagement ring, but Facebook spread the word to all of his friends, including the girlfriend, before he had a chance to propose.

And it is not just the Internet per se that is doing the tracking but also mobile devices and apps. Smartphones and tablets know where people go, what they are doing, whom they talk to, what they buy, and what they like. Online services are tracking those preferences to send consumers hypertargeted, personalized mobile ads. So when people use an app or a Web site to look up running shoes or a resort in Aruba, they could find an ad for Nike or the Hilton the next time they check their smartphone. The downside of mobile advertising, at least for marketers, is that there is not a definitive way to know if mobile ads lead directly to purchasing a product.

Consumers are fending off the onslaught of digital promotions with ad-blocker software, which uses an algorithm to identify ads, block them from appearing on a Web site, and place a blank space where the ad would have appeared without the ad blocker. Users benefit by the absence of all those pesky ads, but Web sites lose out on advertising revenue. Almost 198 million Internet users worldwide have installed ad-blocker software, putting hundreds of millions of advertising dollars at stake. Online businesses and marketers distain ad blockers and have been known to inform such users of ways to support the site and even give instructions on how to selectively unblock the site. Makers of ad-blocker software claim that marketers do not get that users are weary of being pelted with ads, and they are as intent on protecting consumers' privacy as marketers are on monitoring them.

Spam

Commercial messages for Viagra, weight loss, hair loss, body-part enhancement, get-rich-quick schemes, medical cures, and a host of other products and services clog millions of e-mail boxes every day. Email may have started as a promising method of delivering commercial messages, but it has captured the wrath of users, who are up in arms at receiving these unsolicited sales pitches, commonly known as **spam**. According to Internet folklore, there are two origins of the word 'spam.' Some say the term comes from the popular *Monty Python* line, "Spam, spam, spam," which was nothing more than a meaningless uttering (though pronounced with an English accent, of course). Others assert that email spam is akin to the canned sandwich filler: a whole lot of junk but no real meat.

Regardless, spam is commonly thought of as any unsolicited message or any content that requires the user to opt out. Advertisers often justify sending spam by tricking customers into signing up for the information. Sometimes when users are making a purchase, filling out an online poll, or just cruising through a Web site, they inadvertently click on or run their mouse over a link or icon that signals permission to send email messages. Most legitimate businesses offer users a way to opt out of unwanted messages; others make it almost impossible to block out spam.

From a marketing standpoint, sending out promotional material via email is much more efficient than waiting for potential customers to stumble upon a product's Web page or view one of the product's banner ads. Besides, it takes only a few seconds for users to recognize and delete unwanted promotional messages, so marketers figure that recipients are spared any real harm. This kind of thinking can backfire on the advertisers, however. Unsolicited email can be detrimental to advertisers, because customers who are spammed might harbor negative feelings and even boycott these companies' products and services. Despite spam's bad reputation, marketers are spamming full force. About 14.5 billion spam messages are sent globally per day. Estimates claim that about 90% of all email is spam.

FYI: How Spam Works

1. Spammers first obtain email addresses either from low-cost 'spambots' (software that automatically combs the Web), bulletin boards, lists, and other resources or from businesses that sell their customers' personal information.
2. By changing Internet accounts to avoid detection, spammers send out millions of pieces of spam from one or several computers.
3. The spam messages are then sent to stealth servers, which strip away the clues that could identify their origin and add fake return addresses.
4. The spam then gets sent on to unregulated blind-relay servers in Asia, which redirect the spam, making it more difficult to trace.
5. The spam finally travels back to the United States. The circuitous route fools ISPs and spam blockers into thinking spam is legitimate email.

Source: Stone & Lin, 2002

The **Federal Trade Commission** receives about 40,000 complaints and hundreds of thousands of examples of Internet spam each week. In accordance with the 1938 **Wheeler Lea Act**, the FTC is responsible for the regulation of advertising and has the power to find and stop deceptive advertising in any communication medium. The FTC, however, does not have regulatory power over email, but it is politically influential. Although legislation has been introduced to combat this persistent nuisance, federal legislators have been unable to totally stop these cyber intruders, but headway has been made.

Currently, 37 states have passed some type of spam-related law, ranging from Delaware's outright banning of spam with false return addresses to requiring 'remove me' links and demanding subject alerts on sex-related spam. Some states that do not have specific antispamming laws have sued spammers using laws pertaining to deceptive advertising.

ZOOM IN 7.15

Learn more about the FTC at the agency's Web site: www.ftc.gov

Click on the Consumer Protection tab to read about how the agency protects consumers from fraud, deception, and unfair business practices.

On the federal level, Congress passed the **Controlling the Assault of Non-Solicited Pornography and Marketing Act**, or *CAN-SPAM*, in December 2003. CAN-SPAM abolishes the most offensive tactics used by spammers, including forging email headers and sending pornographic materials. Email marketers are now required to have a functioning return address or a link to a Web site that can accept a request to be deleted from the emailing list. The act was updated in 2008 and again with the *M-Spam Act of 2009*, which prohibits spamming on mobile devices. Other help comes from Internet service providers (ISPs) that offer **spam blockers**, special software that filters out spam, and from organizations, such as the Coalition Against Unsolicited Commercial Email and spam.abuse.net, that are fighting for legislation to protect the online community from unwanted email.

Although all this is a good start toward protecting email users, it does not necessarily protect against spam that originates in other countries. Additionally, spammers have all kinds of ways to circumvent the system, including using ordinary subject lines like 'From Me' and 'Returned message.' The battle against spam continues.

FYI: Stopping Spam

ISPs scramble for new ways to outfox spammers' underhanded means of circumventing spam blockers. Here are some examples:

- *ISP's strategy:* Block messages from known spammers.
- *Spammer's strategy:* Set up new email addresses. *Example:* When Mary@offer4U.com is blocked, the spammer simply changes the address to Mary@goodoffer4U.com
- *ISP's strategy:* Use a spam blocker to cross-check the address and verify the sender.
- *Spammer's strategy:* Mask its identity in the email header, making it seem as though the message is coming from someone else.
- *ISP's strategy:* Use antispam software to block messages containing marketing terms.
- *Spammer's strategy:* Alter the spellings of words or add invisible HTML tags to confuse the spam blockers. *Example:* V*I*A*G*R*A, V1AGR@, VI,B.,/B.AGRA

- *ISP's strategy:* Use antispam software to check messages with altered text and to block mail sent to multiple addresses. If the software suspects spam, it requires the sender to access a Web site and enter a displayed number before the message will be delivered. Because there is no receiver to verify computer-generated spam that has been sent to thousands of addresses, it does not get delivered.
- *Spammer's strategy:* Send spam out from many computers through stealth servers, so origin cannot be detected.

Source: Stone & Weil, 2003

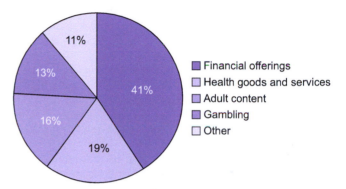

FIG. 7.13 What do spammers sell?

ADVANTAGES OF ADVERTISING ON THE INTERNET

- *Worldwide marketplace*—The Internet serves as a worldwide marketplace, delivering a vast and diverse audience to advertisers. By placing advertisements on the Web, companies can reach out to physically distant customers.
- *Targeting consumers*—The Internet's ability to carry messages to targeted groups is one of its most effective marketing tools. Using special services, marketers deliver targeted advertising to customers based on their IP address or domain name, type of Web browser, and other criteria, including demographic characteristics (age, sex, income) and psychographics, which is the study of consumer lifestyles (interests, activities). The effectiveness of banner ads is increased when they are placed on Web sites with complementary content—such as a banner ad for cookware on a cooking page or a banner ad for a clothing store on a fashion page.
- *Exposure and run time*—Internet ads have longer exposure and run times than ads in traditional media. They are visible for as long as the advertisers post them, they can be accessed any time of day and as often as users wish, and they can be printed and used as paper coupons.
- *Low production costs*—Web advertisements are generally less expensive to produce than ads in traditional media, and the longer exposure and run times make them even more cost efficient. Web ads generally do not require the extensive production techniques

involved in the traditional media; in fact, they can often be designed using digital imaging software. Low production cost is instrumental in attracting a wide range of businesses with small advertising budgets to the Web.

- *Updating and changing ad copy*—Updating and changing the copy and graphics of online ads can be accomplished fairly quickly. These ads can generally be designed and posted within a relatively short period of time.
- *Prestige*—The prestige of online advertising casts a positive image on advertisers and their products.
- *Competition*—The generally lower cost of online advertising allows companies with small advertising budgets to compete with companies with more advertising resources. In this sense, the Web closes the gap between large and small enterprises and places them in the same competitive arena. Online, small businesses are not so small.
- *Quick links to purchases*—Online purchases are made by simply clicking on a banner ad and following the trail of links to an online order form. In many cases, newer interactive banners allow purchases to be transacted directly from the banner, without having to click through a product Web site.
- *Mobility*—Users can check their mobile devices for coupons and specials, for prices, and for products and services from anywhere.

DISADVANTAGES OF ADVERTISING ON THE INTERNET

- *Hidden persuasion*—With most traditional media, consumers are exposed to persuasive messages in an instant, often before they have even had the chance to turn away from the promotion. Although online advertisements intrude on computer screens, the persuasive elements are often at least one click away. A consumer must be interested in the product and must click on the ad before being exposed to the sales message.
- *Banner blindness*—Online users do not pay attention to ads. Consumers see so many ads that they have learned how to screen them out. About 30% to 40% of ads are never even noticed, and only about 1% are clicked on.
- *Creative restrictions*—Online ads are becoming more technologically sophisticated, and many allow on-ad ordering. However, they are still somewhat restricted in creative terms. Many banners are nothing more than the equivalent of a roadside billboard. Advertisers lean toward interstitials and other flashy ads that lure consumers with animation and movement, but these ads may slow Web page downloading to a snail's pace.
- *Fragmentation*—Despite its ability to target an audience, the Web is a highly fragmented medium. Thus, advertisers face the challenge of placing their messages on sites that will draw large enough audiences to make their investments worthwhile. With hundreds of thousands of Web sites and thousands of pages within each site, it is difficult to determine ideal ad placement.

- *Unreliable audience measurement*—Unreliable and unstandardized measurement techniques limit an advertiser's knowledge of how many users are exposed to a message, thus hindering effective advertising buys.
- *Questionable content*—The online audience is already weary of deceptive content and the overcommercialism of the Web, and so they are not very receptive to online advertising. Many users respond unfavorably to these ads and particularly resent advertising popping up all over their screens. More troubling to many online users, especially parents, is the blurring of content and advertising aimed at children. Concerned users are calling on sites to set limits on their advertising and to make it clear when information will be used for marketing purposes.

CRITICISMS OF ADVERTISING

Advertising is highly criticized, not so much for its very nature but because of its content, its negative influences on society, and the types of products it promotes.

ADVERTISING IS MISLEADING AND DECEPTIVE

Critics assert that commercials are often exaggerated and misleading. The FTC regulates deceptive advertising, and even though many ads do not break the law, they do breach ethical standards. Many consumer groups complain about the overuse of exaggerated claims—for instance, basketball players wearing Nike shoes jumping higher than their rivals wearing other brands and Bud Light drinkers having more fun and being surrounded by more gorgeous women than drinkers of other brands. Nike does not directly claim its shoes can make people jump higher, and Anheuser Busch does not state that drinking Bud Light will help men attract beautiful women. Even so, critics claim that people may associate these products with these unlikely outcomes.

Advertising defenders, on the other hand, assert that such depictions are nothing more than harmless puffery: exaggerated claims that a reasonable person knows are not true. Defenders claim that most people know that Nike shoes do not make people jump higher, no matter what an ad might suggest; thus, puffery is harmless to consumers. Critics contend otherwise and claim that puffery is unethical.

The FTC promotes and supports a free marketplace by ensuring that advertising is not untruthful or deceptive with the following three criteria:

- There must be a representation, omission, or practice that is likely to mislead the consumer.
- The act or practice must be examined from the perspective of a consumer acting reasonably in the circumstances.
- The representation, omission, or practice must be a material one ("FTC Policy Statement," 1983).

The FTC issues sanctions against advertisers found guilty of deceptive advertising. Complaints are most commonly settled by requiring the advertiser to sign a consent decree, agreeing to stop misleading advertising practices. An advertiser who breaches the decree may be fined up to $10,000 a day until the deceptive advertising stops. If an advertiser refuses to sign a consent decree, the FTC has the authority to take the matter further by issuing a cease-and-desist order, demanding an end to the deceptive advertising. In this situation, the case would be brought before an administrative law judge, who could impose penalties or overturn it.

If a cease-and-desist order is upheld, the FTC can require the advertiser to run corrective advertising to make up for the previously misleading ad. In 1972, for example, ITT Continental Baking was ordered to run corrective advertising because it had claimed that its Profile bread was beneficial to people watching their caloric intake. In fact, though, Profile bread had about the same calories per ounce as other breads on the market. Ocean Spray was forced to run corrective advertising after claiming that its cranberry juice cocktail beverage had more 'food energy' than other fruit juices. The corrective advertising made clear that the extra food energy came from calories, not protein or vitamins. In 1978, the maker of Listerine mouthwash was required to run corrective ads to state that Listerine did not prevent colds or sore throats, as had been claimed in its advertising. In 2001, the FTC settled with Snorenz, the manufacturer and promoter of an antisnoring mouth spray, for making unsubstantiated product claims. As part of the settlement, the FTC required all future Snorenz promotions to include two disclaimers: to encourage purchasers to see a doctor or sleep specialist and to list the common symptoms of sleep apnea. In 2009, the German pharmaceutical company Bayer, the maker of the oral contraceptive Yaz, aired deceptive television spots. The ads claimed that not only was the drug an oral contraceptive, but it could lessen premenstrual syndrome (PMS) and cure acne. A legal settlement required that Bayer had to submit its ads to the Food and Drug Administration (FDA) for approval. The company also agreed to air a $20 million 'remedial advertising program' and seek the approval of the FTC for any future television ads.

INFOMERCIALS LOOK TOO MUCH LIKE TV PROGRAMS

Prior to the 1980s, commercials longer than 2 minutes in length were considered infomercials and thus were barred from radio and television. The FTC dropped the provision for radio ads in 1981 and for television spots in 1984. Modern-day infomercials are longer than 2 minutes and usually air on early morning, late-night, and weekend television. They are criticized for blurring product endorsements with what could be mistaken for program content. Viewers often think they are watching a program when they are actually watching an infomercial filled with product promotions. Infomercials promote everything from exercise equipment to get-rich-quick real estate schemes. Children are especially vulnerable to sales pitches and may have particular difficulty in distinguishing programs from infomercials, so even though infomercials are allowed on television, the FTC has upheld a ban on those aimed specifically at children.

ADVERTISING ENCOURAGES AVARICIOUSNESS AND MATERIALISM

Many people complain that advertising encourages them to buy items they do not need just for the sake of amassing goods. "Whoever dies with the most toys wins" was a popular bumper-sticker slogan of the 1980s and reflects the type of thinking that encourages greed and competition among friends and neighbors based on the amount of material goods they collect. All too often, people are persuaded to spend money on goods that they cannot afford and do not need, because product promotions have convinced them that their self-worth depends on these purchases. College students often complain they cannot afford to pay for their education or books, yet they purchase television sets and smartphones that are way out of their price range and that they do not really need. Why? They feel these items will make them look cool, or they derive some other sort of personal benefit from having them. No one is immune to the persuasive power of advertising. It is not just college students who are buying more than they need. In 2015, the average U.S. credit card holder was carrying about $3,600 in credit card debt. Many critics insist that runaway debt is the direct result of advertising because it promotes materialism.

ADVERTISING REINFORCES STEREOTYPES

Commercials are under fire for their unrealistic and often demeaning portrayals of women, minorities, and other individuals. Stereotypical images of women mopping floors, men being bosses, smart people being nerds, blondes being dumb, and old people being fools pervade many advertisements. Unfortunately, people's beliefs are shaped by what they see on television, and when they are repeatedly exposed to stereotypes, they come to believe what they see.

Recent pressure on advertisers has brought about some changes in how groups are depicted in commercials and other types of promotional materials. Advertisements that show people in more realistic situations, that include more minorities and women, and that depict them in a more positive light are minimizing the stereotypic portrayals of many groups.

ADVERTISING EXPLOITS CHILDREN

Parents and advocacy groups are concerned about the negative effects of advertising on children. Most children see about 30,000 commercials each year, and marketers spend about $12 billion each year pushing products to children under the age of 12. Critics claim that children cannot interpret the purpose of a commercial, judge the credibility of its claims, or differentiate between program content and a sales message. Critics further contend that advertisers take advantage of children by selling them on products without their knowledge.

ADVERTISING IS INVASIVE AND PERVASIVE

It seems incredible, but individuals are exposed to between 3,000 and 5,000 advertising messages per day, including electronic and print media, billboards, signs, banner ads, T-shirts, labels, and other advertising venues.

Critics claim that advertisements invade our minds and our homes and overwhelm us with trivial information. But research shows that on recall tests, people remember very few of the ads they have seen or heard. Further, people tend to block out much of the advertising they are exposed to and attend to the ads that promote a product or service in which they are already interested. While many critics blast the very existence of commercials, proponents hail them as a necessary component of our free-market system.

FYI: Online Ads That Cannot Be Blocked

Just when we thought our pop-up blockers could save us from annoying ads forever, Apple, Google, and other technology players are working on ways to force users to see ads on any device—computer, phone, television, gaming device, music player—and react to it before it freezes the application. The ads will be programmed to appear randomly and so could potentially interrupt whatever a user is doing. Audio commercials on music players that prompt listeners to push a button to signify attentiveness could be especially intrusive.

It is all coming to being forced to pay for commercial-free information. Even after paying for the device, paying for the apps, paying for the service, an extra fee could be tacked on for commercial-free delivery.

Source: Stross, 2009

ADVERTISING PROMOTES UNHEALTHY BEHAVIORS

In 1971, Congress banned cigarette ads from radio and television and later extended the ban to include all tobacco-related products. Additionally, there had been a decades-old self-imposed ban (the National Association of Broadcasters Code, or NAB) on the broadcasting of hard-liquor commercials. While broadcast television could not accept liquor ads, cable television was able to do so because it is not under the purview of the NAB. Cable had the competitive edge over broadcast for liquor ads until 1996 when the Distilled Spirits Council, an industry trade group, announced support for reversing the ban on broadcast advertising of hard liquor. In effect, this left the decision of whether to accept commercials for hard liquor up to the stations. Seagram's was one of the first advertisers to make its way to the airwaves, and other brands of hard liquor have since followed. The advertising of liquor on broadcast television now accounts for $243 million per year, much to the chagrin of critics, who claim that such advertising encourages drinking, especially among minors.

SEE IT LATER

Although advertising is in many ways beneficial, it also contributes to *information overload*. Consumers are bombarded with ads practically everywhere they look. Television, radio, and the Internet are all packed with ads; the nation's roadways are cluttered with billboards that urge people to pull over and eat, to buy gas, or to listen

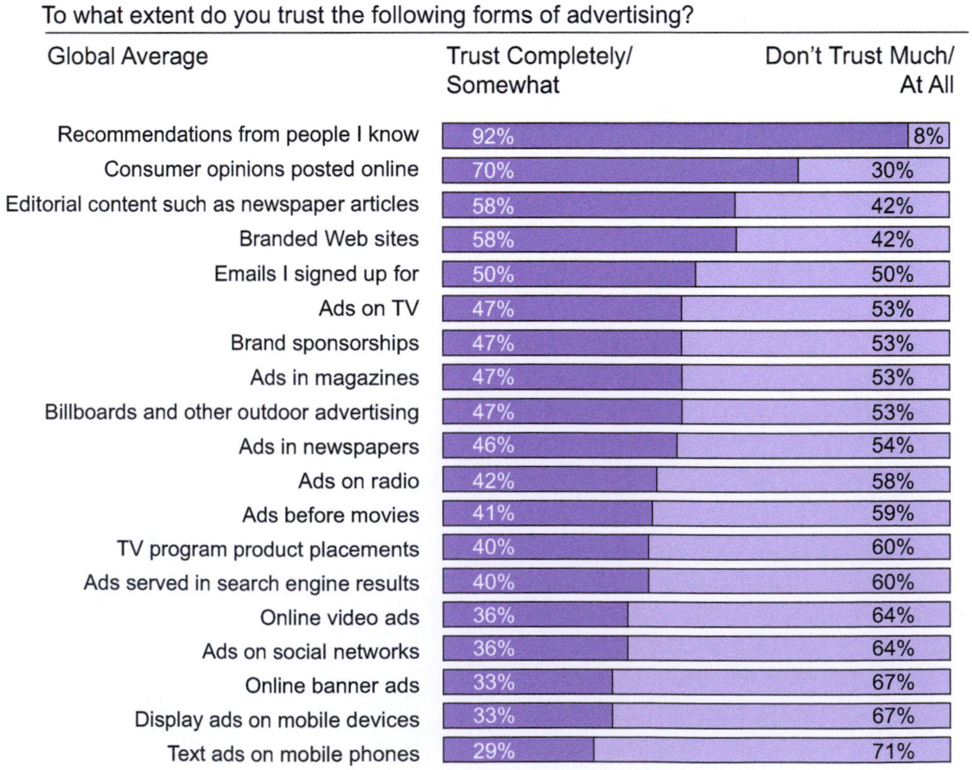

To what extent do you trust the following forms of advertising?

Global Average	Trust Completely/ Somewhat	Don't Trust Much/ At All
Recommendations from people I know	92%	8%
Consumer opinions posted online	70%	30%
Editorial content such as newspaper articles	58%	42%
Branded Web sites	58%	42%
Emails I signed up for	50%	50%
Ads on TV	47%	53%
Brand sponsorships	47%	53%
Ads in magazines	47%	53%
Billboards and other outdoor advertising	47%	53%
Ads in newspapers	46%	54%
Ads on radio	42%	58%
Ads before movies	41%	59%
TV program product placements	40%	60%
Ads served in search engine results	40%	60%
Online video ads	36%	64%
Ads on social networks	36%	64%
Online banner ads	33%	67%
Display ads on mobile devices	33%	67%
Text ads on mobile phones	29%	71%

FIG. 7.14 Trust in advertising
Source: Nielsen: Consumer trust in 'earned' advertising grows in importance, 2012. (c) 2015, of The Nielsen Company, licensed for use herein.

to a particular radio station, and it is almost impossible to buy a product that does not prominently display its brand or logo. Consumers are persuaded to spend their money almost everywhere they look.

Advertisements appear even in the most unusual places. In 2004, horseracing jockeys were permitted to sell logo space on their riding silks. Advertisers paid up to $30,000 for a jockey to wear advertising on his sleeves in the 2004 Kentucky Derby. To avoid the overcommercialization that plagues other sports, the Kentucky Horse Racing Commission has since banned jockeys from wearing sponsored logos within 1 hour of a race.

Although the initial public outcry against the jockeys advertising was minimal, it was loud and strong against Major League Baseball (MLB) promoting the 2004 release of the new *Spider-Man* movie on its bases. As part of a marketing agreement with Columbia Pictures and Marvel Studios, MLB had planned to adorn first, second, and third bases with the red-and-blue Spider-Man logo, but baseball fans shrieked at the idea of commercializing any part of the infield. Although logos are commonly painted on the walls of the outfield, a baseball historian said the infield is regarded as a "magic circle" that is not to be tampered with at the risk of "offending the gods" (Maller, 2004). And so just one day after announcing the *Spider-Man* promotion, MLB backed out of the infield agreement, succumbing to the thousands of calls of protest it received. The MLB still sometimes supports movies, such as when it promoted *G-Force* on its online team schedules in 2009.

From radio to television, clothing, bathroom stalls, and now foreheads, advertising can be found almost anywhere. A London advertising agency paid good-looking, hip college students the equivalent of $7 an hour to paste logos on their foreheads. Advertisers paid about $25,000 for 100 foreheads a week.

ADDRESSABLE ADVERTISING

Digital media is changing the entire advertising landscape, especially as new ways to reach individual consumers are created. Personalized ads are not only being pushed to unwitting consumers through Web sites and apps but also through satellite television services. New 'addressable advertising' is being tested and perfected by DirecTV and Dish Network. The networks troll sophisticated databases to learn about their subscribers and then partner with advertisers to send tailored ads to customers through their satellite-service digital video recorder. Addressable advertising was tested during the 2012 presidential election to deliver targeted ads to voters. For example, if data show that a satellite subscriber is likely to favor a particular political candidate or is an independent but typically votes Democratic, then candidates and other political organizations can partner with the satellite service, which will download an addressable ad that has been created for that specific type of voter and even that particular person. The commercial is automated to appear within any program, but the viewer does not realize that it is being shown just to him or her but not to the viewing audience at large. The downside of addressable advertising is that it is very expensive, and because the ads are invasive, viewers could react negatively. But as targeted ads become more commonplace and acceptable, addressable advertising might turn out to be a cost-effective way to reach consumers.

MOBILE ON THE RISE

Mobile advertising is on the brink of perhaps becoming the ultimate method of consumer targeting. With GPS, the tracking capabilities of apps and browser cookies, users' every move, likes and dislikes, attitudes, lifestyles, and product preferences are collected, sorted, categorized, and analyzed by algorithms that spit out personal consumer profiles, which are sold to marketers who partner with creative teams who design ads that resonate with individual users and compel them to buy products.

There are still several hurdles facing mobile ads. They are still too passive and noninteractive to really grab users' attention, which is mostly due to technological limitations and the small size of a mobile screen. Because the ads are easy to overlook, they are sold for less money and thus are not as lucrative as desktop ads. Plus users are more reluctant to make purchases through mobile devices than through desktop computers.

Despite lingering limitations, some companies are prime for mobile delivery. Facebook, for example, began 2012 with no mobile ad revenue, but by the end of the year it accounted for 5.1% of the global mobile ad market and hit over 25% by the end of 2015. Facebook now makes more than half of its ad revenue from smartphone and tablet ads. Mobile advertising is on the rise, and its power to persuade should not be underestimated. Even newer variations of wearable eyeglasses and other devices could soon be transmitting ads directly into our retinas.

As a research analyst once summed up, "Television is unique, in all forms of advertising, in being able to deliver that emotional power." And as another analyst pointed out, "After all, when was the last time a banner ad made you cry? If you're a greeting card company, it sure better" (Harwell, 2015, p. A16).

SUMMARY

Advertising is not a modern-day phenomenon but has been around since the days of clay tablets. Although it has changed in many ways since its origins, its purposes have remained the same: to get the word out about a product or service and persuade people to buy.

Radio was never meant to be an advertising medium, but high operating costs drove entrepreneurs to come up with a way to raise money for their over-the-air ventures. Stations experimented with toll advertising and later with sponsored programs. By the time television emerged as a new medium, there was not any question that advertising was going to fund its existence. The only question was how best to advertise products on television. Early television experimented with radio's style of sponsored programs. This method was fairly successful until the quiz show scandal exposed the problems of sponsored programs. The television industry then borrowed the magazine concept from print and began selling commercial spots between and during programs.

Most commercials throughout the 1960s were 60 seconds in length, but they began getting shorter toward the end of the decade. The ban on tobacco advertising in 1971 left many networks and stations with unsold commercial time. Stations and networks were happy to find that they could sell two 30-second commercials to two different advertisers for more money than one 60-second spot to one advertiser. Throughout the 1970s, 30-second commercials began to dominate the airwaves, as they still do today.

Advertising benefits both marketers and consumers. Marketers are able to promote their products and services, and consumers get to learn about them. Advertising on each of the electronic media—radio, television, the Internet, and mobile devices—has its own advantages and disadvantages. Some products and services are best served by advertising on network television; some will reach their audiences in a more cost-efficient way on radio or cable television. Advertisers also have the Internet on which to place their ads. Online advertising consists of banner ads, interstitials, superstitials, and extramercials, among other types.

Placing a commercial here and there is usually not as effective an advertising strategy as constructing a campaign: a series of commercials and print ads with the same theme. Most campaigns are created by advertising agencies. There are basically four types of advertising agencies: full-service agencies, creative boutiques, media-buying services, and interactive or cyber agencies.

Effective advertising begins with an understanding of the market. Creative teams understand consumer psychology and what motivates purchasing decisions. Knowing which medium best provides access to the target audience is the responsibility of media market researchers. Measuring a commercial's effectiveness is a crucial part of the advertising process. It provides the feedback needed to reassess and update a campaign.

For all its benefits, there are some downsides to advertising as well. Critics claim that too much commercialism makes us greedy and materialistic. Some ads are said to promote racism and sexism through their stereotypic portrayals of minorities and women. Advertising is blasted for exaggerating product benefits and for misleading consumers into buying items that they are later disappointed in or simply do not need. How children are affected by ads is of particular concern. Most children cannot tell the difference between programming and commercials and do not know if they are being given information or being sold a product. Regardless of how consumers may feel about advertising, without it, the media would not exist, and the economy would be dramatically different.

Given all the new technological developments, critics of advertising scream for relief from the overcommercialized world that it creates. And although some people try to shield themselves from advertising, others have come to accept that it is an everyday part of life. In the future, it is likely that advertising will creep into areas that were once regarded as being above such peddling. Even though public protest may stem the tide, commerce will probably prevail. Advertising will probably become increasingly ubiquitous, increasingly influential, and increasingly controversial.

BIBLIOGRAPHY

2 billion consumers worldwide to get smart(phones) by 2016. (2014, December 11). *eMarketer*. Retrieved from: www.emarketer.com/Article/2-Billion-Consumers-Worldwide-Smartphones-by-2016/1011694

Advertisement that watches you. (2009, December 13). *The New York Times Magazine*, p. 27.

Advertiser expenditure in the largest advertising categories in the United States. (2016). *Statista*. Retrieved from: www.statista.com/statistics/275506/top-advertising-categories-in-the-us/

Ah, the good-old days: The best Super Bowl ads. (2010, February 10). *MSNBC*. Retrieved from: www.msnbc.msn.com/id/35174030 [February 10, 2010]

Ahrens, F. (2006, August 20). Pausing the panic. *The Washington Post*, p. F1.

Amadeo, C. (2015, February 9). Average credit card debt: U.S. statistics. *About News*. Retrieved from: http://useconomy.about.com/od/glossary/g/credit_card_debt.htm

American ad spending on online social networks. (2009, November 10). *TechCrunchies.com*. Retrieved from: techcrunchies.com/american-ad-spending-on-online-social-networks/ [February 3, 2010]

American advertisers increasingly take aim at children. (2009, August 31). *Washington Profile*. Retrieved from: www.washprofile.org/en/node/3396 [September 12, 2009]

Arango, T. (2009, October 6). Soon, bloggers must give full disclosure. *The New York Times*, p. B3.

Average U.S. home now receives a record 118.6 TV channels, according to Nielsen. (2008, June 13). *Nielsen*. Retrieved from: www.marketingcharts.com/television/us-homes-receive-a-record-1186-tv-channels-on-average-4929/ [September 1, 2009]

Barnes, B. (2009, July 27). Lab watches Web surfers to see which ads work. *The New York Times*, p. B1, B6.

Baumer, K. (2011, February 3). Here's a look at the cost of Super Bowl ads through the years. *Business Insider*. Retrieved from: www.businessinsider.com/cost-super-bowl-ads-through-the-years-2011-2

Begun, B. (2002, December 23). Music: How much do you like it? *Newsweek*, p. 9.

Behavioral targeting defined. (2008, June 5). *Digital Media Statistics*. Retrieved from: www.fuor.net/dnn/behavioral1targeting.aspx [January 16, 2010]

Behavioral targeting: I always feel like somebody's watching me. (2009, June 25). *Digital Media Buzz*. Retrieved from: www.digitalmediabuzz.com/2009/06/behavioral-targeting-watching/ [January 12, 2010]

Behavioral targeting: Pushing relevance to maximize advertising spending. (2009, January 26). *Jupiter Research*. Retrieved from: www.forrester.com [January 16, 2010]

Berthon, P., Pitt, L. F., & Watson, R. T. (1996). The World Wide Web as an advertising medium: Toward an understanding of conversion efficiency. *Journal of Advertising Research*, 36(1), 43–54.

Braga, M. (2012, October 17). Web startups look for new ways to make money. *The Globe and Mail*, p. E6.

Bulova story. (2003). *Bulova Watch Company*. Retrieved from: www.asksales.com/Bulova/Bulova.htm

Cable and ADS. (2014, November). *Television Advertising Bureau*. Retrieved from: www.tvb.org/media_comparisons/4729

Campbell., R. (2000). *Media and culture*. Boston: St. Martin's Press.

Can-Spam Act: A compliance guide for business. (n.d.). *Federal Trade Commission*. Retrieved from: www.ftc.gov/tips-advice/business-center/guidance/can-spam-act-compliance-guide-business

Chen, A. (2002, August 19). How to slam spam. *E-week*, pp. 34–35.

Chen, B. X. (2016, February 4). How to watch the Super Bowl when you don't have cable. *The New York Times*, p. B7.

Clifford, S. (2009, April 23). Ads that TiVo hopes you'll talk to, not zap. *The New York Times*, pp. B1, B8.

Cuneo, A. (2002). Simultaneous media use rife, new study finds. *Advertising Age*, pp. 3, 40.

Davis, W. (2009, September 20). Facebook to wind down Beacon to resolve privacy lawsuit. *MediaPost News*. Retrieved from: www.mediapost.com/publications/?fa5Articles.showArticle&art_aid5113848 [September 25, 2009]

Dominick, J. R. (1999). *The dynamics of mass communication* (6th ed.). Boston: McGraw-Hill.

Dominick, J. R., Sherman, B. L., & Copeland, G. A. (1996). *Broadcasting cable and beyond* (3rd ed.). New York: McGraw-Hill.

Dorey, E., & MacLellan, A. (2003). Spread the word—Encourage viral marketing. *DDA Computer Consultants, Ltd.* Retrieved from: www.dda.ns.ca/resource/articles/spreadtheword.html

Edwards, J. (2013, February 13). Behold-The first banner ad ever. From 1994. *Business Insider*, www.businessinsider.com/behold-the-first-banner-ad-ever—from-1994-2013-2

Elliott, S. (2013, January 22). Super Bowl ads will be heavy on colas, beers, and cars. *The New York Times*, p. B3.

Elliott, S. (2014, January 31). High stakes for agencies, and products at Super Bowl. *The New York Times*, p. B6.

Elkin, T. (2003a, September 8). Making the most of broadband. *Advertising Age*, p. 86.

Elkin, T. (2003b, September 22). Spam: Annoying but effective. *Advertising Age*, p. 40.

Ember, S. (2015, September 2). With technology, avoiding both ads and the blockers. *The New York Times*, p. B3.

Ember, S. (2016, February 27). TV networks are recasting the role of commercials. *The New York Times*, p. B3.

Emergence of advertising in America. (2000). *John W. Hartman Center for Sales, Advertising and Marketing History, Duke University.* Retrieved from: www.scriptorium.lib.duke.edu/eaa/timeline.html [June 4, 2001]

Fact book: A handy guide to the advertising business. (2002, September 9). *Advertising Age* [Supplement].

Fahri, P. (2013, February 2). Blurring the line between news and advertising. *The Washington Post*, pp. A1, A9.

Fitzgerald, K. (2002, June 10). Eager sponsors raise the ante. *Advertising Age*, p. 18.

Freeman, L. (1999, September 27). E-mail industry battle cry: Ban the spam. *Advertising Age*, p. 70.

FTC policy statement on deception. (1983). *Federal Trade Commission.* Retrieved from: www.ftc.gov/bcp/policystmt/ad-decept.htm [August 12, 2004].

Garbe, W. (2012, November 19). What is the average number of searches per user per day in the U.S.? *Quora.* Retrieved from: www.quora.com/What-is-the-average-number-of-searches-per-user-per-day-in-the-US

Geuss, M. (2014, May 6). On average, Americans get 189 cable channels and only watch 17. *ArsTechnica.* Retrieved from: http://arstechnica.com/business/2014/05/on-average-americans-get-189-cable-tv-channels-and-only-watch-17/

Giles, C. (1998, October 18). Fighting spammers frustrating for now. *Atlanta Journal-Constitution*, pp. H1, H4.

Gladwell, M. (2003). Alternative marketing vehicles: The future of marketing to one. *Consumer Insight.* Retrieved from: www.acnielsen.com/download/pdp/pubs

Godes, D., & Mayzlin, D. (2002). *Using online communication to study word of mouth communication. Social Science Research Network.* Retrieved from: papers.ssrn.com

Golden Super Bow now grown up form modest beginning. (2016, January, 30). *The New York Times.* Retrieved from: www.nytimes.com/reuters/2016/01/30/sports/football/30reuters-nfl-superbowl-history.html

Gorman, B. (2010, June 4). Basic cable's primetime audience share remains 59%, to broadcast's 39% for 2009–2010. *TV by The Numbers.* Retrieved from: http://tvbythenumbers.zap2it.com/2010/06/04/basic-cables-primetime-audience-share-remains-59-to-broadcasts-39-for-2009–10/53203/

Grant, A. E., & Meadows, J. H. (2012). Communication technology update and fundamentals (13th ed.). New York: Focal Press.

Gross, L. S. (1997). *Telecommunications: An introduction to electronic media* (6th ed.). Madison, WI: Brown & Benchmark.

Hansell, S. (2009a, July 6). A guide to Google's new privacy controls. *The New York Times.* Retrieved from: bits.blogs.nytimes.com/2009/03/12/a-guide-to-googles-new-privacy-controls [January 12, 2010]

Hansell, S. (2009b, July 6). Four privacy protections the online ad industry left out. *The New York Times.* Retrieved from: www.bits.blogs.nytimes.com/tag/behavioral-targeting/?st5cse&sq5nebuad&scp54 [January 12, 2009]

Harwell, D. (2015a, November 17). Holiday ads moving from TV to digital. *The Washington Post*, p. A16.

Harwell, D. (2015b, November 17). SnapChat wants to turn your life into an ad. *The Washington Post*, p. A12.

Head, S. W., Spann, T., & McGregor, M. A. (2001). *Broadcasting in America* (9th ed.). Boston: Houghton Mifflin.

Helft, M. (2009, March 11). Google to offer ads based on Interests, with privacy rights. *The New York Times*, p. B3.

Herlihy, G. (1999, September 26). Deliver me from spam. *Atlanta Journal-Constitution*, p. P1.

Hinckley, D. (2014, March 5). Average American watches 5 hours of TV per day, report shows. *Daily News.* Retrieved from: www.nydailynews.com/life-style/average-american-watches-5-hours-tv-day-article-1.1711954

History. (2004). *Agency.com.* Retrieved from: www.agency.com/our-company/history

Hoelzel, M. (2014, December 2). The social media advertising report. *Business Insider.* Retrieved from: www.businessinsider.com/social-media-advertising-spending-growth-2014–9

How many searches does Google handle per year? (2015, February 11). *New York Law Institute.* Retrieved from: www.nyli.org/how-many-searches-does-google-handle-per-year/

Industry data. (2015). *National Cable & Telecommunications Association.* Retrieved from: www.ncta.com/industry-data

Infographic: Rise of social media. (2010, September 8). *Digital Buzz.* Retrieved from: www.digitalbuzzblog.com/infographic-rise-of-social-media-ad-spending/

Interactive Advertising Bureau. (2015, April). IAB Internet advertising revenue report: 2014 full year results. Retrieved from: www.iab.com/wp-content/uploads/2015/05/IAB_Internet_Advertising_Revenue_FY_2014.pdf

Introduction to electronic retailing. (2001). *Frederiksen Group.* Retrieved from: www.fredgroup.com/ftvu_er101.html [June 16, 2001]

Jockeys can wear ads in Derby. (2004, April 29). *CNN Money.* Retrieved from: www.cnn.money.com

Kang, C. (2008, June 27). Product placement on TV targeted. *The Washington Post.* Retrieved from: www.washingtonpost.com/wp-dyn/content/article/2008/06/26/AR2008062603632_pf.html [January 16, 2010]

Kang, C. (2015, November 10). Online advertising predicted to top TV by 2017 as viewing habits evolve. *The Washington Post*, p. A16.

Kapner, S. (2001, August 31). Kleenex to sponsor movies to cry by on TV. *The New York Times.* Retrieved from: www.nytimes.com/2001/08/31/business/the-media-business-advertising-addenda-kleenex-to-sponsor-movies-to-cry-by-on-tv.html [September 1, 2009]

Kaye, B. K., & Medoff, N. J. (2001a). *Just a click away.* Boston: Allyn & Bacon.

Kaye, B. K., & Medoff, N. J. (2001b). *The World Wide Web: A mass communication perspective.* Mountain View, CA: Mayfield.

Kentucky horse racing commission. (2007, April 6). *Jockey Advertising.* Retrieved from: www.khrc.ky.gov/jockeyadvertising/ [September 12, 2009]

Khermouch, G., & Green, J. (2001, July 30). Buzz marketing. *Business Week.* Retrieved from: www.businessweek.com/print/magazine/content/01_31/b3743001.htm

Kindel, S. (1999). Brand champion. *Critical Mass*, pp. 56–58.

Lawton, C. (2009, May 28). More households cut the cord on cable. *The Wall Street Journal.* Retrieved from: online.wsj.com/article/SB/12347195274260829.html [September 12, 2009]

Lewis, C. (2007, April 27). Study: Ad clutter holds steady. *Multichannel.com.* Retrieved from: www.multichannel.com/article/88621-Study_Ad_Clutter_Holds_Steady.php [September 4, 2009]

Lohman, T. (2009, April 15). 2008 spam was 62 trillion messages. *ComputerWorld.* Retrieved from: www.computerworld.com.au/article/299365/2008_spam_62_trillion_messages_mcafee [September 12, 2009]

Love it or hate it. (2001, September 25). *Pittsburgh Post-Gazette*, p. E-3.

Maller, B. (2004, May 5). Baseball sells out to Spiderman. *Ben Maller.com.* Retrieved from: http://benmaller.com/archives/2004/may/05-baseball_sells_out_to_spiderman.html

Massey, K., & Baran, S. J. (1996). *Television criticism.* Dubuque, IA: Kendall/Hunt.

Massey, K., & Baran, S. J. (2001). *Introduction to telecommunications.* Mountain View, CA: Mayfield.

McCambley, J. (2013, December 12). The first banner ad. Why did it work so well? *The Guardian.* Retrieved from: www.theguardian.com/media-network/media-network-blog/2013/dec/12/first-ever-banner-ad-advertising

McMurry, E. (2013, September 17). MSNBC's *Morning Joe* is no longer 'brewed by Starbucks.' *MediaIte.* Retrieved from: www.mediaite.com/tv/msnbcs-morning-joe-is-no-longer-brewed-by-starbucks/

Media trends track. (2003). *Television Bureau of Advertising.* Retrieved from: www.tvb.org/rcentral/mediatrendstrack/gdpvolume/gdp.asp?c5gdp1

Miller, C. C., & Goel, V. (2013, October 12). Google to sell users' endorsements. *The New York Times*, pp. B1, B2.

Miller, C. C., & Sengupta, S. (2013, October 12). Selling secrets of phone users to advertisers. *The New York Times*, pp. A1, A4.

Mobile banners continue to boast high click rates. (2012, August 27). *eMarketer.* Retrieved from: www.emarketer.com/Article/Mobile-Banners-Continue-Boast-High-Click-Rates/1009299

Murphy, K. (2016, February 21). The ad blocking wars. *The New York Times*, p. SR 7.

Network TV activity by length of commercial. (2008). *Television Bureau of Advertising.* Retrieved from: www.tvb.org/rcentral/mediatrendstrack/tvbasics/26_Net_TV_Activity.asp [September 1, 2009]

Nielsen: Consumer trust in 'earned' advertising grows in importance. (2012, April). *Nielsen.* Retrieved from: www.slideshare.net/ecommercenews/global-trustinadvertising2012

Nielsen reports television tuning remains at record levels. (2001, October 17). *Nielsen.* Retrieved from: en-us.nielsen.com/main/news/news_releases/2007/

october/Nielsen_Reports_Television_Tuning_Remains_at_Record_Levels [September 1, 2009]

Nielsen tops of 2012: Advertising. (2012, December 17). *Nielsen*. Retrieved from: www.nielsen.com/us/en/insights/news/2012/nielsen-tops-of-2012-advertising.html

Number of tablet PC users in the United States from 2010 to 2015. (in millions). *Statista*. Retrieved from: www.statista.com/statistics/199761/forecast-of-tablet-pc-users-in-the-united-states-from-2010-to-2015/

Ogilvy, & Mather. (n.d.). *Ogilvy.com*. Retrieved from: www.ogilvy.com/About/Our-History/David-Ogilvy-Bio.aspx

O'Guinn, T. C., Allen, C. T., & Semenik., R. J. (2000). *Advertising*. Toronto, Canada: South-Western.

Orlik, P. B. (1998). *Broadcast/Cable copywriting*. Boston: Allyn & Bacon.

Othmer, J. P. (2009, August 16). Skip past the ads, But you're still being sold something. *The Washington Post*. Retrieved from: www.washingtonpost.com/wp-dyn/content/article/2009/08/14/AR2009081401629.html

Pagels, J. (2015, January 31). Super Bowl ad rates per viewer are a huge bargain. *Forbes*. Retrieved from: www.forbes.com/sites/jimpagels/2015/01/31/super-bowl-ad-rates-per-viewer-are-a-huge-bargain/

Poggi, J. (2013, October 9). TV ad prices: Football is still king. *Advertising Age*. Retrieved from: http://adage.com/article/media/tv-ad-prices-football-king/244832/

Pogue, D. (2002, June 27). Puncturing Web ads before they pop up. *The New York Times*. Retrieved from: www.nytimes.com/2002/06/27/technology/circuits

Prime time programs and 30 second ad costs. (2010). *FrankWBaker.com*. Retrieved from: www.frankwbaker.com/prime_time_programs_30_sec_ad_costs.htm [February 2, 2010]

Primetime shows with the most product placement. (2012). *CNBC*. Retrieved from: www.cnbc.com/id/45884892/page/1

Product placement move online. (2007, October 3). *eMarketer*. Retrieved from: www.emarketer.com/Article.aspx?R51005444 [September 12, 2009]

Rosen, J. (2012, December 2). Who do they think they are? *The New York Times Magazine*, pp. 40–45.

Rovell, D. (2004, May 7). Baseball scales back movie promotion. *ESPN*. Retrieved from: www.sports.espn.go.com/espn/sportsbusiness/news/story?id5 1796765

Ritchie, M. (1994). *Please stand by*. Woodstock, NY: Overlook Press.

Russell, J. T., & Lane, W. R. (1999). *Kleppner's advertising procedure*. Upper Saddle River, NJ: Prentice Hall.

Scola, N. (2014, August, 23). Tune in to satellite TV? The political ads may be tuned in to you. *The Washington Post*, p. A10.

Scott, N. (2015, January 28). 10 Best Super Bowl commercials of all time. *USA Today*, Retrieved from: http://ftw.usatoday.com/2015/01/10-best-super-bowl-commercials-of-all-time

Semuels, A. (2009, February 24). Television viewing at all-time high. *Los Angeles Times*. Retrieved from: articles.latimes.com/2009/feb/24/business/fi-tvwatching24 [September 1, 2009]

Shane, E. (1999). *Selling electronic media*. Boston: Focal Press.

Shields, M. (2014, November, 14). NBC nabs Target as sponsor of Web video series 'Saturday Night Line." *The Wall Street Journal*. Retrieved from: http://blogs.wsj.com/cmo/2014/11/14/nbc-nabs-target-as-sponsor-of-web-video-series-saturday-night-line/

Sivulka, J. (1998). *Soap, sex and cigarettes*. Belmont, CA: Wadsworth.

Smartphone, tablet uptake still climbing in the U.S. (2013, October 14). *eMarketer*. Retrieved from: www.emarketer.com/Article/Smartphone-Tablet-Uptake-Still-Climbing-US/1010297

Smith, L. (2003). Commercials steal the spotlight. Retrieved from: www.amherst.k12.oh.us/steele/news/record/record.php?articleID5262

Spam statistics and facts. (2015). *SpamLaws.com*. Retrieved from: www.spamlaws.com/spam-stats.html

Stambor, Z. (2013, April 12). Social media ad spending will reach $11 billion by 2017. *Internet Retailer*. Retrieved from: www.internetretailer.com/2013/04/12/social-media-ad-spending-will-reach-11-billion-2017

Stampler, L. (2012, August 7). 12 excellent examples of how Apple product placements rule Hollywood. *Business Insider*. Retrieved from: www.businessinsider.com/apple-product-placements-in-tv-and-movies-2012-8?op=1

Starr, M. (2009, July 22). More TVs than humans. *New York Post*. Retrieved from: www.nypost.com [September 9, 2009]

Steinberg, B. (2009, October 26). "Sunday Night Football" remains costliest TV show. *Advertising Age*. Retrieved from: http://adage.com/article/ad-age-graphics/tv-advertising-sunday-night-football-costliest-show/139923/ [February 2, 2010]

Steinberg, B. (2013, October 10). Increased ad spending on tap for beer, liquor makers. *Variety*. Retrieved from: http://variety.com/2013/biz/news/increased-ad-spending-on-tap-for-beer-liquor-makers-1200709329/

Steinberg, B. (2015, September 29). TV ad prices: Football, 'Empire,' 'Walking Dead,' 'Big Bang Theory,' top the list. *Variety*. Retrieved from: http://variety.com/2015/tv/news/tv-advertising-prices-football-empire-walking-dead-big-bang-theory-1201603800/

Stelter, B. (2009, November 11). TV News without the TV. *The New York Times*, pp. B1, B10.

Stern, L. (2003, May 12). Using your head. *Newsweek*, p. E2.

State laws relating to unsolicited commercial or bulk e-mail (SPAM). (2015, January 9). *National Conference of State Legislatures*. Retrieved from: www.ncsl.org/research/telecommunications-and-information-technology/state-spam-laws.aspx

Stone, B. (2002, October 24). Those annoying ads that won't go away. *Newsweek*, pp. 38J, 38L.

Stone, B., & Lin, J. (2002). Spamming the world. *Newsweek*, pp. 42–44.

Stone, B., & Weil, D. (2003, December 8). Soaking in spam. *Newsweek*. Retrieved from: www.newsweek.com

Story, L. (2008, March 10). To aim ads, Web is keeping closer eye on what you click. *The New York Times*, pp. A1, A14.

Stross, R. (2009, November 15). Apple wouldn't risk its cool over a gimmick, would it? *The New York Times*, p. B1.

Super Bowl ads cost average of $3.5M. (2012, February 6). *ESPN*. Retrieved from: http://espn.go.com/nfl/playoffs/2011/story/_/id/7544243/super-bowl-2012-commercials-cost-average-35m

Super Bowl TV ratings. (2009, January 18). *TV by the Numbers*. Retrieved from: Tvbythenumbers.com/2009/01/18/historical-super-bowl-tv-ratings/11044

Tablet users to surpass 1 billion worldwide in 2015. (2015, January 8). *eMarketer*. Retrieved from: www.emarketer.com/Article/Tablet-Users-Surpass-1-Billion-Worldwide-2015/1011806

Tedeschi, B. (1998, December 8). Marketing by e-mail: Sales tool or spam? *The New York Times*. Retrieved from: www.nytimes.com [February 19, 1999]

Television watching statistics. (2013, December 7). *StatisticsBrain.com*. Retrieved from: www.statisticbrain.com/television-watching-statistics/

Terms for the trade. (2014). *Nielsen*. Retrieved from: www.arbitron.com/downloads/terms_brochure.pdf

Thompson, D. (2013, March). The incredible shrinking ad. *The Atlantic*, pp. 24, 26, 28.

Tops in 2008: Top advertisers, most popular commercials. (2008, December 18). *Nielsenwire*. Retrieved from: www.blog.nielsen.com/nielsenwire/consumer/tops-in-2008-top-advertising [February 2, 2010]

Total advertising spending by company. (2015, May 10). *Statistics Brain*, www.statisticbrain.com/total-advertising-spending-by-company/

Trends in television. (2000). *Television Advertising Bureau*. Retrieved from: tvb.org/rcentral/mediatrendstrack/tv/tv.asp?c5timespent

Tsukayama, H. (2014, July 24). Facebook profits surge, largely from mobile ads, *The Washington Post*, p. A14.

TV costs and CPM trends—Network TV primetime. (2015). *Television Advertising Bureau*. Retrieved from: www.tvb.org/trends/4718/4709

United States Population. (2014, October 19). *World Population Review*. Retrieved from: http://worldpopulationreview.com/countries/united-states-population/

Vanden Bergh, B. G., & Katz, H. (1999). *Advertising principles*. Lincolnwood, IL: NTC Business Books.

Warner, C. (2009). *Media sales*. West Sussex, UK: Wiley-Blackwell.

Waters, D. (2008, March 31). Spam blights e-mail 15 years on. *BBC News*. Retrieved from: news.bbc.co.uk/2/hi/technology/7322615.stm [September 12, 2009]

White, T. H. (2001). United States early radio history. Retrieved from: www.ipass.net/~whitetho/part2.htm [June 4, 2001]

Why Internet advertising? (1997). *MediaWeek*, pp. S8–S13.

Why social media is set to explode. (2013, August 8). *Business Insider*. Retrieved from: www.businessinsider.com/social-media-advertising-set-to-explode-2013-7

Winner of the most industry awards in 1999. (1999). *Ogilvy Interactive*. Retrieved from: www.ogilvy.com/o_interactive/who_frameset.asp

Yarow, J. (2014, May 8). This is the thing that people hate about paying for TV more than anything. *Business Insider*. Retrieved from: www.businessinsider.com/average-number-of-channels-watched-per-household-2014-5

Audience Measurement: Who's Listening, Who's Watching, Who's Surfing?

8

Contents

Let's pretend you own a radio station. You spent all your savings to purchase the station, but to operate it involves many other expenses, including your employees' salaries. How are you going to come up with the money to cover these costs? By selling commercial time. In selling airtime, you are really selling your listeners to your advertisers. And to do so, you must know something about your listeners: how many there are, who they are, when they listen, what music they like, and so on. Finding out this information is what prompted the development of radio audience measurement techniques, which later led to methods for measuring television and Internet audiences.

Knowledge of audience characteristics, including their use of media, is at the core of selling commercial time. This knowledge is translated into various figures, such as **ratings** and **shares**, which are then used to determine the price of commercial time and space. Generally, the program, station, or Web site that draws the largest or most desirable audience charges the most for its airtime or space.

Developing reliable and accurate ways of measuring media audiences is an ongoing process. New computer-based technologies have led to new data collection methods. Techniques for measuring the Internet audience are adapted from those used for researching the traditional mass-media audience. Audience-measuring companies are working on more accurate and reliable ways to discover how listeners, viewers, and online users are using the media and which programs and Web sites they favor.

This chapter begins with an overview of the early methods of obtaining audience feedback and then examines contemporary ways of monitoring radio, television, and Internet audiences. The explanations of these measuring techniques include the mathematical formulas that show how ratings, shares, and other measures are calculated and reported. The chapter then ties ratings to advertising sales by demonstrating how stations and Web sites use audience data to price commercial time and space.

SEE IT THEN

EARLY RATINGS SYSTEMS

In the late 1920s, when radio advertising was just beginning to catch on, stations were stumped as to how much to charge for commercial time. Radio stations were setting charges for time, but advertisers were hesitant to buy unless they had information about the listeners. The void was filled in 1929 when Archibald M. Crossley called on advertisers to sponsor a new way of measuring radio listenership: the **telephone recall system**. With this method, a random **sample** of people was called and asked what radio stations or programs they had listened to in the past 24 hours. Memory failure was the biggest drawback to Crossley's system, as listeners could not accurately remember what stations and programs they had tuned in to.

Crossley's fiercest competitor was C. E. Hooper, who used the **telephone coincidental method** of measuring radio listenership. Using this method, Hooper's staff telephoned a random sample of people and asked, "Are you listening to the radio just now?" The next question was something like, "To what program are you listening?" followed by, "Over what station is the program coming?" (Beville, 1988, p. 11). Many thought Hooper's method was superior to Crossley's because it did not rely on listeners' memories but instead surveyed exactly what they were listening to at the time of the call. Of course, the drawback to Hooper's method was that someone might have listened to a station for hours that day but just happened to have the radio turned off or tuned to another station at the time of the telephone call. Both the Crossley and Hooper methods had their flaws, but at the time, they were the best ways to measure listenership.

In the 1930s, A. C. Nielsen was also working on ways to measure the radio audience. He took a slightly different approach by using an electronic metering device, the *audimeter*, which attached to a radio and monitored the stations that were tuned to and for how long. The audimeter was the precursor to today's audience-metering devices.

In the late 1940s, Hooper supplemented telephone coincidental calling by asking listeners to keep **diaries** of their radio use. American Research Bureau (ARB), which changed its name to Arbitron in 1973, also championed the diary method for measuring radio listenership.

The Crossley, Hooper, and Nielsen companies were the radio-ratings leaders into the late 1940s. Crossley left the ratings business in 1946, and in 1950, Nielsen bought out Hooper and thus eliminated its biggest competitor. Hooper himself, however, met a tragic death 4 years later, when on a duck hunting trip he slipped and fell into a rotating airplane propeller.

As television came to life in the late 1940s, the ratings services—mainly Nielsen and ARB/Arbitron—adapted their radio research methods for the new medium. Although many other ratings services emerged, none was able to overcome the market dominance of these two companies.

SEE IT NOW

GATHERING AUDIENCE NUMBERS

RADIO AND TELEVISION

Media outlets need to know who is listening to their stations and who is watching their programs. Radio station managers must keep tapped into their number of listeners and who listens to what music when. A television station sales manager might want to know what percentage of television households in the market had their televisions tuned to the last game of the World Series. These types of data reflect how consumers use the media and help managers set the price for commercial time.

Nielsen

Arbitron once measured both local radio and television audiences but dropped its less profitable television services in 1994, leaving television audience measurement in the hands of Nielsen. For almost 20 years, Nielsen measured television audiences while Arbitron monitored radio listeners. In September 2013, Nielsen acquired Arbitron and renamed it Nielsen Audio. Nielsen is interested primarily in how the public uses the media—what stations they tune to and what programs and stations they listen to and watch, what Internet sites they use and how they use their mobile devices. If a particular television program has consistently low ratings, it might be moved to another time slot or perhaps dropped. If a radio station is highly rated in the market, management knows that it has a winning format that can command top dollar for commercial time.

ZOOM IN 8.1

For information about Nielsen, visit www.nielsen.com/us

Nielsen geographically segments television viewers and listeners to gain a clearer understanding of audience differences and similarities across the nation. When Arbitron was a separate company from Nielsen, it had developed its own market designation scheme, known as the **area of dominant influence (ADI)**, but it then switched to Nielsen Media Research **designated market areas (DMAs)**. Nielsen divides the United States into 210 DMAs based on geographic location and television-viewing patterns. The largest DMA for 2015 was New York, New York, with 7,368,320 households, and the smallest was Glendive, Montana, with 4,230 households. Market areas are sometimes redrawn if overall viewing behavior shifts because of changes in cable penetration or satellite television subscriptions, and other factors.

Nielsen Audio also breaks out radio markets by DMA. Additionally, it designates 272 distinct geographic market areas as **metro survey areas (MSAs)**, often referred to as *metros*. A MSA is generally composed of a major city or several cities and their surrounding county or counties. In 2015, New York, New York, was the largest MSA (16,278,300 population, ages 12+) and Beckley, West Virginia, the smallest (66,600 population, ages 12+).

ZOOM IN 8.2

To find your city's DMA ranking and to see whether you live in a metered market, visit www.tvb.org/media/file/2015-2016-dma-ranks.pdf

---

Nielsen Audio also measures listenership in 224 designated geographic regions known as **total survey areas (TSAs)**. A TSA is made up of a major city and a larger surrounding area than a metro area. For example, the

MSA population for Knoxville, Tennessee, is 712,300, but Knoxville's TSA population is 1,346,300. Smaller metros help low-powered stations compete in the ratings game against high-powered stations by providing smaller geographic areas of measurement.

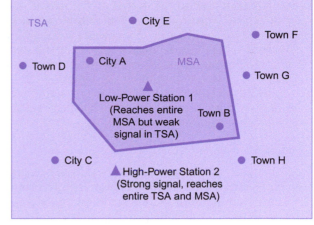

FIG. 8.1 TSA/MSA

FYI: TSA/MSA

Station 1 is a 3,000-watt station whose signal is limited to its metro survey area (MSA). Station 1's ratings are low compared to those of Station 2, a 50,000-watt station whose signal reaches the entire total survey area (TSA).

Station 1 may be very popular throughout the smaller MSA; thus measuring Station 1 within the MSA more accurately reflects its popularity and boosts its ratings. Conversely, measuring Station 1 throughout the TSA will lower its ratings, because its signal does not reach that population.

- TSA550,000 population
- MSA525,000 population
- Station 152,500 listeners
- TSA rating52,500/50,00050.05 or 5%
- MSA rating52,500/25,00050.10 or 10%

RADIO DIARIES

Nielsen Audio collects radio-listening data by sending diaries to a sample of listeners within the top 50 total survey areas. Nielsen surveys most radio markets at least once a year, though larger market areas may be surveyed up to four times per year or even on a continuous basis.

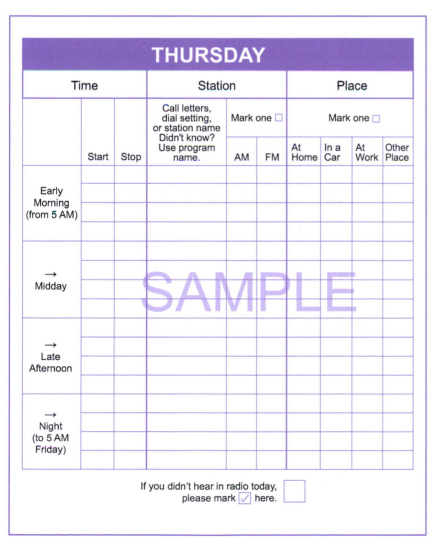

FIG. 8.2 A page from a Nielsen radio diary
Photo courtesy of Nielsen.

Nielsen mails each randomly selected participant a 7-day diary, with each day divided into 15-minute time blocks. In exchange for a small monetary reward (usually less than $5), each participant records when he or she starts and stops listening and the station (identified by call letters and/or dial position) that he or she listens to during each 15-minute time block. Additionally, each participant is asked demographic information, typically age, gender, income, education, and ethnicity.

TELEVISION DIARIES

Nielsen also uses diaries to track television viewership. Nielsen diaries are sent out to sample homes each November, February, May, and July, which are known as **'sweeps'** months or periods. Some larger markets may have additional sweeps during October, January, and March. Television viewers write down the programs and networks they watch in 15-minute blocks over a 1-week period. Viewers also provide demographic and sometimes lifestyle information for the diaries.

In an attempt to hike program viewership—and hence ratings—television and radio station networks use sweeps periods to present special programs, season finales, and much-awaited episodes that resolve a mystery. When these special programs and episodes are aired during sweeps, the ratings may be inflated, because viewers may watch only these heavily hyped episodes and not tune in during the regular season.

Meters

In addition to diaries, Nielsen also uses metering devices to record viewing habits in 20,000 randomly selected television households in the United States. The **set meter**, first used in 2,000 homes in 1987, attaches to a television set and records when it is turned on and off and to what broadcast, cable, and satellite programs and channels the set is tuned. The set meter requires an accompanying diary to track what programs household members are watching.

The **people meter** is a newer, more sophisticated version of a set meter. Instead of writing down the channels that are being watched, each household member is assigned an identification button on a remote control device that is supposed to be pushed whenever someone begins and ends watching television. The people meter records both what programs are being watched and who is watching them. Diaries are still used, but only to match viewers with their demographic characteristics.

FYI: To Meter or Not to Meter

In October 2002 when Knoxville, Tennessee, was the 59th-largest market in the United States with 552,000 television households it converted from diary reporting to the Nielsen meter system. Area television station executives expressed mixed feelings about the meters, especially as the $500,000 price tag for each station was about three times what they had paid for diaries. Beyond cost,

some were concerned that the meters would not make any difference in the stations' market positions and thus were not worth the expense. Yet others argued that the meters would give more accurate audience data than the diaries. Knoxville, home of the University of Tennessee, has a younger-than-average population that is less likely to fill out diaries, which could hurt the Fox-affiliated station that tends to draw younger audiences. Meters, which have a high participation rate among young viewers, could more accurately reflect larger audiences for Fox.

The big advantage of being a metered market is that such a market is more likely to draw national advertisers. The financial outlay for the metered ratings could be made up in increased revenue. One station's general manager said that meters "are the best thing for the market," but another sighed, "They are coming, and we are going to deal with it" (Flannagan, 2002; Local Television Markets, 2009; Morrow, 2002).

FIG. 8.3 A Nielsen people meter
Photo courtesy of Nielsen

The Portable People Meter

Hailed as a revolutionary audience-metering device, the **portable people meter (PPM)** is capable of capturing radio listening data through wireless signals. The PPM is a small, pager-sized device that picks up and decodes inaudible signals that radio broadcasters embed in their programs. An individual clips the PPM to his or her clothing or purse, like a pager or cell phone, and the PPM decodes the unique signal emitted by each radio station the individual hears, whether he or she is at home, in a shopping mall, in a car, or anywhere else. The PPM is the first electronic metering device that monitors in- and out-of-the-home exposure to over-the-air radio. The PPM stores each station's signals, and at the end of each day, survey participants place the device in an at-home base station that sends the codes to Nielsen for tabulation.

The PPM is used to measure radio listenership in 48 markets. PPM data suggest that radio listening habits may be different from what has been reported through diaries. For example, PPM numbers indicate that many more people listen to the radio but spend less time listening and they change stations more frequently than previously known.

Challenges of Gathering Radio and Television Audience Numbers

Ratings are based on *estimates* of numbers of viewers and listeners and are thus not absolutes. Each method of

FIG. 8.4 A Nielsen personal people meter

FIG. 8.5 The Nielsen people meter is attached to the viewer's television set.
Source: Photo courtesy of Nielsen

measurement is limited in some way. Although ratings companies do their best to ensure that the most accurate data possible are gathered, there are still many concerns and criticisms over how audience data are collected.

Samples

One of the most serious concerns about ratings data is the sample of viewers and listeners whose media habits are monitored. It is not possible to poll the entire U.S. population, so a subset, or *sample*, of listeners and viewers is selected. For a long time, Nielsen sampled 5,000 U.S. households, but many critics contended that the sample size was too small to produce accurate results. Others claim that the sample size was large enough as long as it represented all of the demographic groups within a population. Nielsen has since increased its sample to 20,000 homes, representing about 45,000 viewers. It has also changed some of its sampling methods to ensure demographic representation. In other words,

in a DMA survey area in which 30% of the people are elderly, diaries should be sent to a group of participants that includes that same percentage of elderly people to be sure that the data accurately reflect media use. Ratings companies make every attempt to ensure that samples accurately reflect the survey population and often mathematically weight samples to correct over- or underrepresentation of particular groups. Even so, critics insist that ethnic and other specific demographic groups are underrepresented in most samples, yielding an inaccurate picture of media use.

Location

Television and radio are not used only in the home but also in bars and restaurants, in hospitals, at work, in hotels, and so on. But out-of-home television and radio use is not recorded. Because most people listen to the radio at work or in cars, not recording such use is a huge detriment to radio stations, and the ratings may not accurately reflect the actual number of listeners. In comparison, television ratings are minimally affected, because people do most of their watching at home. The PPM should minimize this concern because it provides out-of-home radio listening estimates.

Accuracy

Ratings companies depend on participants to fill out diaries and use people meters correctly. Yet too many listeners and viewers fail to follow diary instructions, forget to record the stations they have listened to, and write down the wrong station or channel number. All of these mistakes result in inaccurate data and thus create an incomplete picture of media use and users.

Television-metering devices are also problematic. Although set meters record what is being watched, Nielsen researchers depend on participants pushing their ID buttons before and after each viewing session, but many forget to do so.

Data Collection Method

How data are collected also influences responses. People often overreport and underestimate their media use. For example, telephone survey participants may feel a bit shy, embarrassed, or hesitant to tell a researcher how much television they watch or how long they spend surfing the Internet. One study compared media usage data collected by telephone survey and diary to direct observation. They found that people tend to report they use the media much less than in reality. For example, participants reported via telephone survey they watched television an average of 121 minutes per day, but direct observation showed that they actually watched for 319 minutes per day, a reporting difference of 198 minutes. On the other hand, participants overestimated in a written diary that they devoted 26 minutes per day reading a newspaper, when indeed direct observations showed they only spent 17 minutes (34.6% fewer minutes). Because of these types of self-reported errors, researchers often conduct several studies using various protocols to research the same issue. When considered together, the various studies provide a more accurate estimate and increase the reliability of the findings.

THE INTERNET

Some of the data collection methods used by the television and radio industries are also used to measure Internet audiences, and new methods have been developed specifically for the medium as well. As with any medium, online advertisers are concerned with attracting the most eyeballs for their money. However, they face unique challenges in buying online space, calculating the number of users who saw their ads or clicked on a banner, and reaching their target audience. Web site operators struggle to get a clear picture of their consumers whose characteristics are critical to convincing advertisers to buy space on their sites.

Online Ratings Data and Measurement

A number of companies are in the business of supplying Web sites and advertisers with online ratings and customer profiles. Nielsen's Total Internet Audience panel of 200,000 online users across 30,000 sites measures digital media use including computers, smartphones, and tablets. There are many other companies, referred to as *third-party monitors*, that also employ various auditing techniques and audience-measuring methods, such as monitoring the number of times a banner ad is clicked on; providing site traffic reports by day, week, month; monitoring Web site activities; and developing customer profiles. Web site operators use this information to sell space to advertisers, and advertisers use this information to find the best sites on which to place their banner ads.

Challenges of Gathering Online Audience Numbers

The biggest challenge faced by Web sites and online advertisers is the lack of standardized data collection methods. **Third-party monitors** use different techniques and data collection methods to arrive at their figures, which often lead to contradictory and confusing reports. For example, they use different metering techniques, select their samples differently, and have different ways of counting Web site visitors. Some might count duplicated visitors (for example, one person who returns to a site five times will be counted as five visitors), and others

June 2001 Web Site Visitors per Jupiter Media Metrix (includes pop-up ads)		Web Site Visitors per Nielsen Media Metrix (excludes pop-up ads)	
1. AOL Time Warner	72.5 mil.	1. AOL Time Warner	76.9 mil.
2. Microsoft	61.5	2. Yahoo!	68.1
3. Yahoo!	59.9	3. MSN	62.7
4. X10.com	34.2	4. Microsoft	39.4
5. Terra Lycos	33.3	5. Terra Lycos	32.5
		116. X10.com	3.8

Sources: Hansell, 2001; Stelter, May 2009

might count only the number of unduplicated visitors (for example, one person who visits a site five times will be counted as one visitor). The unit of analysis may also be defined differently from one measurement service to another. For example, an *Internet user* could be defined as "someone who has used the Internet at least 10 times" or "someone who has been online in the past 6 months." These variations can yield very different results. For example, in 2009, Nielsen reported 8.9 million users for Hulu.com, yet ComScore gave the site 42 million users.

Several associations and organizations are taking the lead in establishing Web auditing and measurement standards. The Coalition for Innovative Media Measurement (CIMM) was formed in 2009 for pursuing new methods of monitoring media audiences. The council is a collaborative effort between media research companies and networks to come up with innovative ways to measure how consumers are using new media delivery systems. Further, the **Internet Advertising Bureau (IAB)** and the **Advertising Research Foundation (ARF)**, along with several other third-party monitors, are at the forefront of developing new and effective means of gathering audience data and standardizing definitions of terms. The IAB, along with prominent Web publishers and several advertising technology firms, has issued voluntary guidelines for online advertising measurement and many other aspects of Internet advertising.

CALCULATING AND REPORTING RADIO AND TELEVISION AUDIENCE NUMBERS

MEANS OF CALCULATING

Once ratings companies have gathered audience numbers and know how many people are watching television or listening to the radio, they conduct more meaningful analyses. For example, just knowing that 10,000 people listen to Station A does not have much meaning unless it is considered in the context of the total audience size. If 10,000 people listen to Station A, does that mean many people listen or just a few? The answer depends on the market size. If in a market of 100,000 people, 10,000 are listening to Station A, then that station has captured 10% of the population. However, if Station A is in a market

---

FYI: Measurement Differences

Back in 2001, when Jupiter Media Metrix was still in the business of measuring Web audiences, it included exposure to pop-up and pop-under ads as a Web site visit. However, Nielsen Media Metrix did not count either ad type as a visit. As a consequence of such measurement differences, the number of visitors to sites that posted a large number of pop-ups would be greater if pop-ups were included than if they were excluded. In the following example, Jupiter ranked X10.com as the 4th-most-visited site, but Nielsen ranked it as the 116th-most-popular online venue.

with a population of 200,000, it has drawn only 5% of the population. Translating the number of viewers or listeners into a percentage provides comparison across markets. These percentages are commonly referred to as ratings and shares.

RATINGS AND SHARES

Radio and television both rely on ratings and shares to assess their market positions. For both radio and television, the mathematical computations are identical. Ratings and shares are simply percentages. A *rating* is an estimate of the number of people who are listening to a radio station or watching a television program divided by the number of people in a population who have a radio or a television. Another way to think of a rating is that 1 rating point equals 1% of the households in the market area that have a television set or radio. A *share* is an estimate of the number of people who are listening to a radio station or watching a television program divided by the number of people who are listening to radio or watching television. Thus, the only difference between a rating and a share is the group that is being measured. Ratings measure everyone with a radio or television, whereas shares only consider those people who are actually listening to the radio or watching television at a given time.

FYI: How to Calculate Radio and Television Ratings and Shares

Radio ratings = Number of people listening to a station/Number of people in a population with radios

Radio shares = Number of people listening to a station/Number of people in a population listening to radio

TV ratings = Number of households watching a program/Number of households in a population with televisions

TV shares = Number of households watching a program/Number of households in a population watching TV

Radio Ratings (Example: In City Z, 5,000 *people have radios*. Compare Station A's rating to Station B's rating).

500 listeners tuned to Station A/5,000 people with radios = 0.1 or 10%, expressed as '10 rating.'

400 listeners tuned to Station B/5,000 people with radios = 0.08 or 8%, expressed as '8 rating.'

Radio Shares (Example: In City Z, 2,500 *people are listening to the radio*. Compare Station A's share to Station B's share).

500 listeners tuned to Station A/2,500 people listening to radio = 0.20 or 20.0%, expressed as '20 rating.'

400 listeners tuned to Station B/2,500 people listening to radio = 0.16 or 16.0%, expressed as '16 rating.'

Notice that when calculating ratings, everyone in the population with radio (5,000) is included in the denominator. But when calculating shares, only those who are listening to radio (2,500) are included in the denominator.

Television Ratings (Example: In City Z, 5,000 *people have television*. Compare Program A's rating to Program B's rating).

500 viewers tuned to Program A/5,000 people with TV = 0.1 or 10%, expressed as '10 rating.'

400 viewers tuned to Program B/5,000 people with TV = 0.08 or 8%, expressed as '8 rating.'

Television Shares (Example: In City Z, 2,500 *people are watching television*. Compare Program A's share to Program B's share).

500 viewers tuned to Program A/2,500 people watching TV = 0.20 or 20.0%, expressed as '20 rating.'

400 viewers tuned to Program B/2,500 people watching TV = 0.16 or 16.0%, expressed as '16 rating.'

Notice that when calculating ratings, everyone in the population with television (5,000) is included in the denominator. But when calculating share, only those who are watching television (2,500) are included in the denominator.

The examples show that ratings and shares for both television and radio are calculated in the same way but that radio measures individual listeners (**People Using Radio, or PUR**) and television measures households with televisions (**Households Using Television, or HUT**). Because the average household consists of 2.7 persons and contains 2.9 televisions, and because household members are increasingly watching television separately from others in the home, the HUT measure is giving way to PUT, a measure that may be more valuable to the media industry.

Notice that when ratings are calculated, all members of the population who have radios or televisions are included in the calculation, even if they're *not* listening or watching. When shares are figured, however, only members of the population who *are* listening or watching are included in the calculation. Shares, therefore, are always larger than ratings, because they include a smaller segment of listeners or viewers rather than the entire population. In the unlikely event that every person in the market watched or listened to the same show, then the rating and share would be equal.

One way to visualize the relationship of ratings to shares is to think of a chocolate birthday cake. If there are 10 people at a party, the cake must be cut into 10 slices (ratings). But if two partygoers are on a diet, another one is allergic to chocolate, and another one is too full to eat a piece of cake, then the cake will be 'shared' among 6 people, and thus each slice will be larger than if the cake was cut into 10 slices.

FYI: A Point to Remember

One rating point represents 1% of the population being measured. For example, one rating point represents 1,164,000 households (1.16 million) or 1% of the nation's estimated 116,400,000 (116.4 million) television homes.

Source: Nielsen, 2014

Average Quarter Hour

In almost all markets, there are more radio stations than television stations; therefore, radio station ratings are generally lower than television station ratings, because there are more radio stations competing for the same number of listeners. To compensate for lower rating points, radio measures audience in terms of **average quarter hour (AQH)**.

There are several AQH measures: average quarter hour *persons*, average quarter hour *rating*, and average quarter hour *share*. Nielsen defines AQH persons as "the average number of persons listening to a particular station for at least 5 minutes during a 15-minute period" (Terms for the Trade, 2014). AQH persons are **duplicated listeners** who can be counted up to four times an hour, one time for each quarter hour. For example, if Jim listens to Station A for at least 5 minutes during the 1:00 p.m. to 1:15 p.m. quarter hour, he is counted as one listener, and if Jim listens during the 1:15 p.m. to 1:30 p.m. quarter hour, he is counted again and thus is considered two listeners.

AQH rating and share are calculated as follows:

- AQH rating:
  (AQH persons/Survey area population) × 100 = AQH rating (%)
- AQH share:
  (AQH persons/Listeners in survey area population) × 100 = AQH share (%)

Cumulative Persons

Whereas AQH persons represent *duplicated* listeners, **cumulative persons** (or *cume*) represent *unduplicated* listeners. Nielsen defines cume as "the total number of different persons who tune to a radio station for at least 5 minutes during a 15-minute period" (Terms for the Trade, 2014). No matter how long the listening occurred, each person is counted only once. For example, if Sharon listens to Station A for at least 5 minutes during the 1:00 p.m. to 1:15 p.m. quarter hour, she is counted as one listener, but if she also listens during the 1:15 p.m. to 1:30 p.m. quarter hour, she is not counted again and is considered one listener.

Cume rating is calculated as follows:

- Cume rating:
  (Cume persons/Survey area population) × 100 = cume rating (%)

Although cumes are more commonly used to assess radio listenership, household television meters also produce cumes to measure program viewership. Because many prime-time television shows air once a week, 4-week cumes are sometimes reported to assess how many people watch a program over a month-long period.

ZOOM IN 8.3

Practice your media math with the online media math calculator at:

- www.srds.com/frontMatter/sup_serv/calculator/cpm/index.html
- www.wdfm.com/advertising.html

MEANS OF REPORTING

Nielsen data are reported in standardized ratings reports or are tailored to meet the specific needs of a particular client. The standardized radio market report provides radio audience estimates for each ratings period as well as market information (station and population profiles) and audience information. Rating, share, cume, and AQH numbers are broken out by age, gender, and time of day. Eyeing these figures gives station managers a good idea of how their station compares with other stations in the market. Whereas one station may be highly rated among men, another may draw more women. Or a station may have strong ratings during one part of the day but be weak in another. One station might draw the 25- to 54-year-old crowd, while another might dominate the teen listening group.

Television: Nielsen Services

Nielsen provides several standard television reports and customizes others for its clients. Below is a sampling of Nielsen's reports:

- **Nielsen Television Index (NTI)** provides metered audience estimates for all national broadcast television programs and national syndicated programs. Nielsen breaks out station and program rating and share figures by age and gender. These program ratings are used to compare and rank shows against one another. If a program is consistently rated poorly, there is a good chance it will not be on the air for very long. When newspapers or industry magazines like *Broadcasting & Cable* report the ratings for *The Simpsons*, *American Idol*, or *Dancing with the Stars*, it is more than likely the information has come from the NTI.
- **Nielsen Station Index (NSI)** provides local market television-viewing data through diaries and continuous metered overnight measurement. Metered information is processed overnight, and a report is sent to the station the next morning. Television executives rely on these overnights for immediate feedback on how their programs and stations rate compared to their rivals. Overnights are often used to make quick programming decisions to beat out the competition. NSI provides rating and share figures for programs that air on local affiliates, such as newscasts, as well as for network shows.
- **Nielsen Home Video Index (NHI)**: Nielsen reports cable ratings and shares in which, like its other reports, it breaks out viewing by age, gender, time of day, and cable network. Ratings for cable programs and networks are collected like those for broadcast ratings via diaries and meters, but cable ratings are reported separately and interpreted differently from broadcast ratings.

Television Ratings

Nielsen measures viewership of cable, pay cable, DVD, Web-linked television, and other program delivery systems with meters and diaries. Prior to the proliferation of cable networks and thus cable subscribers, ABC, NBC, and CBS vied for the largest share of the audience. Now hundreds of cable networks compete against each other and against the big three broadcast networks for the audience's attention.

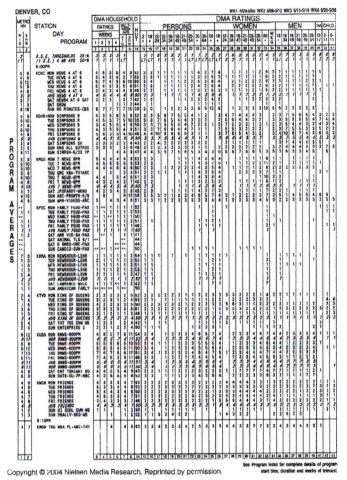

Copyright © 2004 Nielsen Media Research. Reprinted by permission.

FIG. 8.6 A page from an NSI ratings book

the week of December 8, 2014, was ESPN's regular-season football game (Atlanta vs. Green Bay) with an 8.5 rating (13.9 million viewers), whereas NBC's *Sunday Night Football* led the pack of broadcast shows with a 14.3 rating (24.2 million viewers), almost twice as high as ESPN's game. For that same week, the highest-rated nonsports cable program was FX's *Sons of Anarchy* that attracted 6.3 million viewers for a 3.9 rating, compared to the highest-rated nonsports broadcast show *The Big Bang Theory* with 15.4 million viewers (9.3 rating). In fairness, cable program and broadcast program ratings are interpreted differently. For example, a 3.9 rating for a cable network program is considered quite good, but it is considered very low for a broadcast network program. Other cable shows, such as AMC's *The Walking Dead*, often surpass many broadcast programs. For example, for the week of March 25, 2014, *The Walking Dead* was the fifth-most-watched program with 13.5 million viewers behind only *NCIS* (CBS, 17.1 million), *Dancing With the Stars* (CBS, 14.2 million), *NCIS: Los Angeles* (CBS 14.2 million), and *The Voice* (NBC, 13.5 million).

FYI: DVRs Boost Ratings (Sept. 2011—May 2012)

Program	Average 'Day of Broadcast Rating'	Average DVR Rating (Within 7 days of Broadcast)	Total Rating
Modern Family (ABC)	4.1	1.7	5.8
New Girl (Fox)	2.9	1.3	4.1
Grey's Anatomy (ABC)	3.1	1.2	4.3
The Big Bang Theory (CBS)	4.4	1.2	5.6
The Office (NBC)	2.4	1.2	3.6

Of the 100 top-rated programs during the 2008–2009 television season, 99 aired on the broadcast networks. The 80th-ranked National Football League's regular-season games shown on ESPN was the only cable program in the top 100. But by the 2012–2013 season, three cable shows had made it into the top 25—*The Walking Dead* (AMC, 10th), *Monday Night Football* (ESPN, 16th), and *Duck Dynasty* (A&E, 21st).

FYI: Viewer Trends

Rather than using rating points, another way of measuring who is watching a program is by audience numbers. For example, in 2014, an average of 6.31 million viewers watched *The Celebrity Apprentice* (NBC). The number of viewers represents a 50.9% gain from a previous years. The larger the audience gains, the more likely the show will remain on the air.

Source: Seidman, 2010

FYI: The Price of a Blackout

When the power blackout of August 14, 2003, plunged the eastern states into darkness, it also plunged media companies into millions of dollars of lost revenue. Broadcast stations collectively lost between $10 million and $20 million in commercial time for preempting regularly scheduled programming with continuous all-news coverage. Each of the major networks aired about 3 hours of news coverage, with no commercial breaks, starting shortly after the 4:11 p.m. blackout (Eastern Standard Time). By prime time, all the networks except ABC were carrying regularly scheduled programming, but NBC and Fox did not run their full commercial schedules.

The power failure also affected ratings, as about 13% of Nielsen's sample of households were without electricity to power their televisions. Nielsen expected viewership data to be lower than usual for this time period.

Broadcast and Cable Television Program Ratings

The large number of cable networks has diluted the viewing audience, and, when coupled with spotty coverage (as not all cable systems carry the same cable networks), cable network programs typically garner lower ratings than broadcast programs. For example, the highest-rated cable program for

Broadcast and Cable Network Ratings

The broadcast and cable networks duke it out for ratings both on the program level as well as on the overall network level. Although broadcast network ratings and shares are still far ahead of those for cable, cable is making serious inroads. For example, during the week of September 21, 2009, advertiser-supported cable networks combined surpassed the seven broadcast networks (ABC, CBS, NBC, Fox, ION, CW, MyNetworkTV [before it became a syndication service]) during prime time with a 56 share compared to broadcast's 42 share. Cable networks collectively threatened the broadcast networks for many years and eventually took over. By 2010, cable nets were hooking about 59% share compared to broadcast's 39%.

The newer broadcast networks—CW and the Spanish-language channels, Univision and Telemundo—face some of the same ratings challenges as the cable networks. They reach a smaller audience than the more established broadcast networks. For example, from September 2014 through the first week of January 2015, CW finished in fifth place with a 0.9 rating among 18- to 49-year-olds compared with NBC with an average 2.8 rating, CBS with a 2.4 rating, ABC 2.2, and Fox 2.0. Univision, however, is catching up with and has at times surpassed the big four networks. In July 2013 and 2014 sweeps, Univision was the number one rated network, and its telenovela *Lo Que La Vida Me Robo* (*What Life Took From Me*) attracted more 18- to 49-year-old viewers than many regularly scheduled prime-time programs on the broadcast networks.

In general, projections of television ratings show a slide of about 6% per year from 2015–2020, with the biggest decline among 18- to 24-year-olds, who prefer to watch programs through online services rather than through television networks.

FYI: Top 10 Prime-Time Regularly Scheduled TV Programs of 2015

Program	Rating	Viewers (millions)
NBC Sunday Night Football (NBC)	7.8	23.29
The Big Bang Theory (CBS)	7.1	21.06
NCIS (CBS)	7.1	20.91
The Walking Dead (AMC)	6.6	19.66
Empire (Fox)	6.0	17.74
NFL Thursday Night Football (CBS)	6.0	17.66
NCIS: New Orleans (CBS)	5.8	17.31
Sunday Night NFL Pre-Kick (NBC)	5.8	17.08
Blue Bloods (CBS)	5.1	14.97
Dancing With the Stars (ABC)	4.9	14.55

Source: Pennington, 2015

SEPARATING CABLE AND SATELLITE RATINGS

Cable networks such as Nick at Nite and ESPN are uplinked to satellite and then either downlinked by a cable service and then delivered to subscribing homes or downlinked directly to homes with satellite dishes. Thus, when cable ratings are calculated, they include both the number of viewers who watched programs via cable and those who also watched programs via satellite. This double counting increases the cable networks' ratings by making it seem as though they are drawing larger audiences. Nielsen has proposed reporting cable- and satellite-delivered program ratings separately, but doing so could lower the price of cable advertising, because the audience numbers will be lower without satellite viewers. Currently, cable network audience estimates include those viewers who see the network via cable and thus receive local and regional commercials *plus* those who view the network via satellite but do not receive the commercials. Thus, cable ratings might be inflated to the detriment of advertisers, who pay for more viewers than who are actually exposed to their commercials.

WHEN AUDIENCE NUMBERS ARE NOT ENOUGH

Numbers cannot reveal personal characteristics that determine media choice. For the radio, television, and online industries to really know their audiences, they need to look at them in ways that numbers cannot provide. For example, ratings and shares might reveal which programs people watch or listen to most frequently, but they do not tell why viewers prefer one program over another, whether people prefer certain characters on a program, or why and when they like to listen to certain types of music.

Nielsen and other companies thus study the personal characteristics of media consumers. Psychographics categorize listeners based on their overall lifestyle, including hobbies, travel, interests, attitudes, and values. For example, a company advertising snowboards probably wants to reach young, athletic, daring individuals, and diaper manufacturers aim for parents. Another segmentation scheme is based on technographics—an individual's use of and attitudes toward technology. For example, how many television sets someone owns or how often someone watches a video on his or her smartphone is a technographic measure. From all this information, ratings companies obtain a clearer profile of the types of people who watch or listen to particular programs. Because advertisers depend on the media to reach their target consumers, it is crucial to match up with a medium, station, or program that is used by their customers. Psychographic and technographic information guides advertisers to programs that draw viewers who are most likely to use their products.

The media industry also uses specialized types of research such as music preference, pilot and episode testing, television quotient data, and online consumer profiling to further understand its users and their media preferences.

MUSIC PREFERENCE RESEARCH

Used by radio stations, music preference research assesses radio listeners' musical likes and dislikes to determine station playlists. Music testing is an expensive venture but often well worth the effort. Music research is conducted either over the telephone, over the Internet, or in an auditorium-like setting. Researchers either call a sample of listeners or bring them together (usually hundreds) in one location to play song hooks (5 to 20 seconds each of various songs). Participants are asked to evaluate how much and why they like or dislike each song. Sometimes, smaller focus groups of listeners, usually less than 20, are convened to listen and discuss their preferences in depth.

FYI: Michael Jackson's Funeral

Michael Jackson was always good television, and even in death in 2009, he still had the power to draw a large audience. An estimated 31 million (20.6 rating) U.S. viewers tuned into his funeral to say goodbye to one of the most popular singers of all time.

Here is what 31 million viewers means. It is more than the number of viewers who tuned into the funerals of former presidents Ronald Reagan (20.8 million, 15.7 rating, 2004) and Gerald Ford (15 million, 11.3 rating, 2007) and Pope John Paul's funeral (8.8 million, 7.2 rating, 2005), but less than Princess Diana's funeral (33 million, 26.6 rating, 1997).

It's also less than the number of viewers who watched President Bill Clinton's apology address (67.6 million, 46.9 rating, 1998) and the O. J. Simpson verdict (53.9 million, 42.9 rating, 1995).

Although 31 million viewers is quite impressive, Michael Jackson's funeral drew about as many viewers as the first 2009 *American Idol* episode (30 million, 18 rating) but lagged behind the 2009 Academy Awards show (36.3 million, 23.3 rating) and was surpassed by the Super Bowl in 2010 (98.7 million, 42 rating).

TELEVISION PROGRAM TESTING

Television programs are subject to rigorous concept testing before they make it to the air—if they ever do make it that far. Even after programs are on the air, the testing is not over.

New programs and episodes of current programs are tested to gauge audience reactions to the plot, characters, humor, and other program elements and to the overall program itself. The test audience watches perhaps the pilot episode of a new program or the season finale of an existing show, and each person indicates his or her level of like or dislike on the test meter at any time during the show. The meter records the viewer's reaction every time he or she changes the dial. Executives and writers rely on these screening data for content decisions. If, for example, most viewers move their dials down after hearing a joke's punch line, the program's writers might delete it.

On a broader scale, test meter information is matched up with demographic characteristics—such as age, gender, education, and income—for information about what types of viewers prefer what types of programs. For instance, the data may show that males over the age of 40 who have high incomes and are sports minded prefer action-oriented programs, that women between the ages of 18 and 34 prefer situation comedies over police dramas, or that children prefer shows with teen characters. Given this information, the producers might add a character or a storyline that appeals to a certain type of viewer.

FYI: Television City

Perhaps one of the most well-known testing centers opened in 2001 in Las Vegas. CBS's Television City Research Center is housed within the MGM Grand Hotel and attracts about 70,000 program rating volunteers a year from all over the United States. CBS and Viacom partnered with A. C. Nielsen Entertainment to provide the testing equipment and audience feedback reports. Basically, groups of about 20 to 25 volunteers are ushered into one of six studios to watch and rate commercials and new program pilots as well as episodes of existing shows from CBS, MTV, Nickelodeon, and other Viacom networks. Each screening lasts for about an hour (about 45 minutes to watch a commercial-free program and a 15-minute follow-up survey). Learn how to participate in program testing at www.vegas.com/attractions/on_the_strip/televisioncity.html

TELEVISION QUOTIENT DATA

Marketing Evaluations/TVQ, Inc., is an independent audience research company that assesses viewers' feelings about programs and performers. Television quotient data (TVQs) focus on viewers' perceptions of programs and stars and are often used to supplement Nielsen ratings and other data. TVQ findings are collected through mail and telephone surveys, diaries, and panel discussions that examine the familiarity and popularity of programs and actors. A low TVQ score could mean the death of a character, and a high TVQ score could shift a secondary character to a more prominent role.

PHYSIOLOGICAL TESTING

Advanced laboratories are commissioned to measure human response to advertising and programs. Participants are wired to instruments that measure heart rate, eye movement, skin temperature, facial muscle movement, and other physiologic changes that are induced by exposure to television and online content. For example, researchers measure whether participants attend to or ignore advertising, how long their eyes rest on an ad, and whether a scene evokes an emotional response.

Researchers also assess viewers' opinions of television celebrities. For example, researchers may want to know whether Anderson Cooper, Jon Stewart, and Steven

Colbert are 'cool,' 'friendly,' 'fun,' 'intelligent,' 'trustworthy,' 'witty,' and so on. Presumably, viewers are drawn to celebrities they like. Researchers may also ask viewers how much they like/love a particular program and about their 'emotional connection' to a show. Many users are happy to fill out surveys or participate in focus groups, especially if they get a discounted price or some other reward for doing so.

TRANSLATING AUDIENCE INFORMATION TO ADVERTISING SALES

Media outlets develop audience profiles based on audience numbers (such as ratings and shares), audience information (such as age, ethnicity, gender, and other personal characteristics), and lifestyle preferences (such as program and music likes and dislikes). Once an audience profile has been established and the media outlet knows who listens to its station or watches its programs, how it rates in the market, and what its most popular programs are, it can establish commercial rates for its airtime and begin selling its audience to advertisers. Again, airtime is nothing but empty seconds unless someone is listening to or watching the spot; thus, it is really the media *audience* that is being sold. Viewers and listeners are not the consumers but the product being sold to advertisers. For example, if a radio station knows that it has a strong rating among women between the ages of 25 and 49 who enjoy sports and especially golf, it has a compelling case to make to a golf or general sports store or fitness center. The station can demonstrate to the retailer that it reaches the retailer's customers.

Not only are advertisers interested in buying media that reach their target consumers, but they also want to know how they can reach the largest number of target consumers for the lowest cost. Although the price of commercial time is largely based on program and station rating points, advertisers often need more information when planning a media buy. Advertisers compare the costs of reaching their target audience using various stations and types of media. For example, perhaps teens primarily listen to three radio stations in their town and also watch a television affiliate's Saturday-morning lineup. Advertisers need to have some way to compare the price of airtime on each of the three radio stations and the television program. To do so, they often rely on figures such as cost per thousand, gross ratings points, and cost per point:

- **Cost per thousand (CPM)** is generally used to sell print and broadcast media. CPM is one of the most widely used means of comparing the costs of advertising across different media, such as television and radio. Many people often wonder why *cost per thousand* is abbreviated *CPM* instead of *CPT*. It is because *thousand* is derived from the Latin word *mil*, which means 'one thousandth of an inch.'

Web publishers also use CPM to sell online space, but with a slight difference. With nononline media, CPM reflects the cost of reaching one thousand consumers, but for online ads the measurement is called CPI (cost

per impression)—the cost of reaching an individual consumer.

Formula for cost per thousand:

(Cost of a spot or schedule/population × 1000)

$10,000 cost for ads/200,000 viewers × 1000 = $50 per thousand.

Formula for cost per impression:

(Cost of a spot or schedule/number of impressions)

$10,000 cost for ads/200,000 impressions = $.05 per impression

- **Gross rating points (GRPs)** are the total of all ratings achieved by a commercial schedule. Put more simply, it is the *reach* (average rating) multiplied by the *frequency* (the number of spots).

Gross Rating Points:
Reach (ratings) × Frequency (number of spots)

- **Cost per point (CPP)** is the cost of reaching 1% (1 rating point) of a specified market. The CPP gives advertisers a way to compare the costs of rating points in various markets. The CPP can be determined using this formula:

CPP = Cost of schedule/Gross rating points

FYI: Cost per Thousand—Prime-time Broadcast Network Television

	Cost per 30 seconds	Cost per thousand
1965	$19,700	$1.98
1970	$24,000	$2.10
1975	$32,300	$2.39
1980	$57,900	$3.79
1985	$94,700	$6.52
1990	$122,200	$9.74
1995	$95,500	$8.79
2000	$82,300	$13.42
2005	$129,000	$21.45
2010	$103,600	$19.74
2014	$112,100	$24.76

Source: TV costs and CPM trends, 2015

Advertisers are concerned with cost per thousand viewers or listeners and with cost per point, the cost of reaching 1% of the market population. The advertisers' goal is to pay the least amount of money for the largest possible audience. They want to know how many viewers or listeners will see or hear their ads (number of viewers) and how many times on average they are exposed to the ads (gross ratings points). With that knowledge, they calculate the CPP and CPM to compare the cost of various outlets.

PRICING AIRTIME FOR RADIO AND TELEVISION

Radio and television stations set base prices for their commercial spots. Some stations, especially television, do not give out their base rates to their advertisers but rather use them as a starting point for negotiation. With the availability of new computer software, many television and radio stations have maximized their revenues through **yield management**, in which commercial prices change each week or even each day, depending on the availability of airtime. If there is high demand to advertise on a particular program, the last advertiser to commit may have to pay a higher price than the first advertiser. The station may also sell a high-demand time slot to whichever advertiser is willing to pay the highest price.

Dayparts

Radio is usually sold by **dayparts**, which are designated parts of a programming day. Radio breaks out the 24-hour clock into these five segments (Eastern Standard Time):

Morning drive:	6:00 a.m. to 10:00 a.m.
Midday:	10:00 a.m. to 3:00 p.m.
Afternoon drive:	3:00 p.m. to 7:00 p.m.
Evening:	7:00 p.m. to midnight
Overnight:	Midnight to 6:00 a.m.

Television is also sold by dayparts but more commonly by specific program. Dayparts and program ratings determine commercial costs. For example, a commercial spot aired during daytime will cost less than the same spot aired during prime time. Further, the cost of commercial time during a daypart may vary with the popularity of the program in which it is inserted. The cost of running a 30-second spot during a top-rated prime-time program will be higher than for airing the same spot during a lower-rated prime-time show.

Here are common television dayparts (Eastern Standard Time), though there is no universal agreement among stations.

Early morning:	6:00 a.m. to 9:00 a.m.
Morning (Daytime*):	9:00 a.m. to noon
Afternoon:	Noon to 4:00 p.m.
Early fringe:	4:00 p.m. to 6:00 p.m.
Early evening:	6:00 p.m. to 7:00 p.m.
Prime access:	7:00 p.m. to 8:00 p.m.
Prime time:	8:00 p.m. to 11:00 p.m.
Late fringe:	11:00 p.m. to 11:30 p.m.
Late night:	11:30 p.m. to 2:00 a.m.
Overnight:	2:00 a.m. to 6:00 a.m.

*9:00 a.m. to 4:00 p.m. is also known as 'daytime.'

Run of Schedule

Radio stations further classify dayparts by advertiser demand and the number of listeners. The most desired and thus most expensive time is categorized as Class AAA, followed by Class AA, Class A, Class B, and down to Class C, the lowest priced. Because most advertisers do not want to buy what they perceive as second-rate time, some stations do not even designate a Class B or Class C time; some stations even prefer to classify their rates by daypart instead of by class.

One of the most common discounted buys is called **run of schedule** (*ROS*). In this type of buy, the advertiser and the salesperson agree on the number of times a commercial is going to air, but the station decides when to broadcast the spot, depending on available time. For example, on radio, instead of purchasing 5 spots in Class

AAA time, 10 in Class AA time, and 10 in Class B time, the advertiser is offered a discounted rate for the package of 25 spots, and the station will air them during whatever time classes are available. In many cases, ROS benefits the advertiser, which may end up getting more spots on the air during prime dayparts than it actually paid for or otherwise could afford. Television ROS works similarly to radio, except that spots are rotated by daypart rather than by time classification.

Fixed Buys

Radio and television stations charge higher rates for **fixed buys**, in which advertisers specify what time they want their commercials to run. Perhaps a fast-food restaurant wants to dominate the 10:30 a.m. to 1:00 p.m. period and thus wants its commercials broadcast every 15 minutes. Most stations, especially radio stations, can accommodate these specific needs, but they charge a premium price for doing so.

Careful planning and buying do not guarantee that commercials will run during agreed-upon times or that the audience will be as large as promised. Suppose a radio station inadvertently runs a spot at the wrong time or a television station overestimates the number of viewers it anticipated would watch a particular program, but the station has already charged the advertiser for the commercial time. In these and other situations in which the terms of advertising agreements are not met, advertisers are offered *make-goods* as compensation, which usually consist of free commercial time, future discounts, or return payment.

Frequency Discounts

Radio and television stations are willing to offer frequency discounts when advertisers agree to air their commercials many times during a given period, such as 6 months. Advertisers who buy a large number of spots reduce their cost per commercial, and the more commercial time they buy, the bigger the discount per spot. For example, suppose an advertiser could buy 10 30-second spots at $200 each for a total cost of $2,000. But if the advertiser agrees to run the spot 20 times instead of 10 times, the station may lower the cost per spot from $200 to $125 for a total of $2,500. So, for an extra $500, the advertiser has doubled the number of commercial spots. Other frequency discounts might include 6-month or yearly deals or other longer-term advertising agreements.

Local Discounts

Radio and television stations often offer discounts to local businesses. National chains, for example, benefit from reaching a station's entire market area. McDonald's may have 20 or more restaurants in a market, any of which might be visited by the station's listeners. In contrast, a local business, such as a produce market with one store on the east side of town, will probably draw only shoppers who live in close proximity, and thus the business will not want to pay to broadcast to the west side of town. Because a large part of the station's audience may live too far from the produce market, the station may offer it discounted airtime.

Bartering

Some radio and television commercial time is **bartered** rather than sold. Also known as **trade** or **trade-out**, bartering basically involves trading airtime in exchange for goods or services. For example, instead of charging a promoter to advertise an upcoming concert, a station may exchange airtime for tickets of equal value. Or a station in need of office furniture might air a local office furniture store's spots in exchange for desks and chairs.

Cooperative Advertising

Radio and television cooperative (co-op) advertising is a special arrangement between a product manufacturer and a retailer; namely, a manufacturer will reimburse a retailer for a portion of the cost of advertising the manufacturer's product. For example, Sony might reimburse a local retailer for promoting Sony televisions during the retailer's commercials. Co-op agreements vary, depending on the manufacturer, time of year, new-product rollouts, previous-year sales, and other factors. For example, Sony may have a 3% of sales co-op agreement with a retailer. If the local retailer sells $100,000 worth of Sony television sets, then Sony will reimburse the retailer $3,000 of its advertising expenditure that was spent promoting Sony televisions.

Co-op agreements are often very specific and require close attention to detail to get sufficiently reimbursed. Many stations hire a co-op coordinator, whose primary responsibility is to help the station's advertisers track co-op opportunities. The more reimbursement advertisers receive from manufacturers, the more commercial time they can afford to purchase.

CABLE TELEVISION

Selling commercial time on cable television follows a slightly different process from selling network or affiliate time. Cable network programming is transmitted through a local cable company, such as Comcast.

There are hundreds of cable program networks, and most cable companies provide subscribers with hundreds of cable networks plus the major broadcast networks. Because there are so many more cable networks than broadcast networks, the audience for each cable network is much smaller; thus, the price for commercial time is lower on cable than on broadcast networks.

Interconnect

Just over two-thirds of cable buys take place at the network level such that advertisers buy time on a cable network, such as ESPN, and the spot is shown in selected locations or throughout the country. The remaining time is then sold locally by the cable service. For example, Comcast sales representatives will sell time on cable channels to local advertisers. Local cable companies sell advertising spots within and between cable network programs, similar to the way broadcast

Career Tracks: John Montuori. Digital Sales Manager–Radio Division. Scripps Digital

FIG. 8.7 John Montuori

What led to your present job?

After graduating with a degree in broadcasting from the University of Tennessee in 2003, I dove into a career in media sales. As a young account executive for a local Fox affiliate, I spent most of my time developing new local direct business. Most of my day consisted of prospecting, door knocking, and pitching packages to decision makers.

As I gained experience and consistently hit my goals, I earned opportunities to pick up agency accounts. I quickly excelled due to my ability to connect with key media buyers and offer them finely detailed account management and high-level customer service. My good work led to a very attractive offer from the competing CBS affiliate. My initial position was as an account executive, but I was quickly promoted to the national sales manager desk, where I led multiple external sales teams and managed relationships with major advertising agencies that placed national campaigns for major brands. Other responsibilities included managing and pricing station inventory and leading our local team through agency negotiations.

After 10 years in television, I had a strong reputation in the market and contacts at all the major networks in town as well as stations and agencies across the country. Despite a crippling economic decline in 2008, I was able to sustain a significant income and managed to maintain consistent growth across my accounts. However, an obstacle that I could not overcome was the rapidly changing media landscape. Networks starting streaming their programs and platforms like YouTube, Netflix, and Hulu offered users the ability to catch clips, if not full episodes of programs online. These changes led to the decline of the ever-powerful rating point. Because revenue follows ratings, TV account executives began earning less and less each year because audiences were moving away from viewing programs on TV to viewing them online.

What are your primary responsibilities?

In 2013, I answered a call from a recruiter who asked me join a brand-new company selling customized digital strategies. It was the best decision I ever made. The position came with a base salary and a rewarding commission plan but *zero* accounts. The role was 100% acquisition based, and for me, it meant completely starting over. I knew that for me to stay relevant in the media field, I would need to take this leap of faith, learn a whole new industry, and build a new book of business from scratch. The digital world is unlike anything I had previously known. The language was foreign, and the accountability to data was overwhelming. I spent days and nights learning the technology. Through hard work, dedication, and probably some luck, I became one of the top-producing reps in the company.

By my 1-year anniversary, I was promoted to digital sales manager. I oversaw our corporate remote digital team as well as 30 account executives and managers on the newspaper side of our company. I was responsible for 100% of the digital revenue associated with our local property. I coached our teams to create, articulate, and execute effective digital strategies for our clients and partner agencies. We conducted thorough client needs analysis, identified and analyzed their profit centers, revenue streams, and ideal prospects and customers, and offered customized executions to help clients reach their goals.

I was recently offered a new position with Scripps Digital. As the Divisional District Sales Manager, I travel to all of our radio properties and facilitate our digital revenue. I conduct trainings and workshops to help our sales teams articulate our product line in language that resonates with our clients.

What advice would you have for students who might want a job like yours?

If you feel that you belong in sales and you have the slightest doubt, you need to develop a tough skin, a poker face, and a self-confidence that radiates through the unexpected. People will beat you down in this industry. You have to expect it. Your goal is not necessarily to always fight back but to be able to stand back up when it's over, shake the dirt off, and move forward. You must also be observant. You will come across a wide variety of coworkers and clients. Listen, watch, and absorb everything—good and bad. Every lesson is a lesson well learned.

affiliates sell time. Where broadcast and cable differ, however, is with multiple cable system buys. Large markets are often served by several different cable companies, each assigned to a specific geographic region. For example, one cable company might serve the east side of the New York market, another the west side, another the north side, and yet a fourth the south side. If a local advertiser has stores located all over the market, then it will want to reach the entire market. The advertiser can do so through an *interconnect buy*, in which it buys time on all the area cable companies through a one-buy pricing agreement. In some cases, arrangements are made for commercials to appear at the same time across the city.

THE INTERNET—COUNTING ONLINE AD EXPOSURES

The Internet's struggles with audience measurement have resulted in imprecise ratings and inconsistent pricing schemes. It is often difficult to measure how many people see an online ad and thus how much it should cost. Most Web sites and auditors rely on the number of hits, number of click-throughs, and other means of counting their visitors.

Hits

In the Web's early years, the number of Web site visitors was largely determined by the number of hits received on a page. A **hit** is typically defined as the number of times a page is accessed. Hits' is a very poor measure of the number of site visitors, however. Fifteen hits on one site could mean 15 visitors, 1 user visiting 15 times, or a small number of users visiting several times. It is virtually impossible to say that the number of hits equals the number of visitors.

Caching presents another challenge in determining the number of visitors to a site. Computers often store a copy of a visited Web page as a **cache file** (or temporary storage area) on the hard drive rather than on the server, so when a user returns to a Web page he or she previously visited during the same Internet session, the browser retrieves the information from the cache file rather than from the server. Consequently, the site does not record the second hit, even though it is a repeat visit or second exposure.

Adding to the confusion are newer **push technologies**, which rely on automated searches (also called **robots** and **spiders**) that comb the Web looking for information. Each time an automated agent scans a Web page, it is counted as a page view, although a human never saw any of the content. Even if automated searches account for only a small percentage of all Web site traffic, a small overestimation of visitors could cause an advertiser to spend millions of dollars more than warranted for a banner ad. Conversely, deflating the number of human page views could cost Web site publishers millions of dollars in lower ad rates.

Cost Per Click

Rather than count people (or hits) who visit a site and might not even notice a particular banner ad, many sites and advertisers rely on *click-through* counts, which identify the percentage of visitors who actually click through a banner ad.

Click-through rates (CTRs), also called **cost-per-click (CPC)** or **pay-per-click (PPC)**, are based on the percentage of Web site visitors who click through a banner ad. Advertisers are charged only for the people who actually expressed an interest in the ad, not for all the others who landed on the site but did not click on the banner.

Advertisers hoping to reach the click-happy consumer are often disappointed with numbers that show few click-throughs. In 1994 when the Internet was still new and a curiosity, about 10% of users clicked on banners. By 2009, only about 3% of Web users clicked on a banner ad, and by 2015, that percentage dropped to about .86%.

Cost-per-click is calculated by multiplying the number of people who click on an ad by an agreed-on rate. For example, if a Web site charges $0.10 per click and 1,000 users click on the ad, the site would charge the advertiser $100 ($0.10 × 1,000 users).

Cost Per Transaction (CPT)

An online advertiser who is charged by cost per transaction pays only for the number of users who respond to an ad. Web sites and advertisers negotiate a fee or a percentage of the advertiser's net or gross sales based on the number of inquiries that can be directly attributed to a banner ad. Online advertisers are assessed a minimal charge or, in some cases, no charge at all for ad placement, but they are charged for the number of people who ask for more information or buy the product as a result of seeing the banner. CPT is a good pricing system for online merchants who are skeptical of investing advertising dollars that probably will not lead directly to sales and for high-traffic Web sites that are eager to earn a commission on top of a small fee for delivering an audience.

Cost Per Impression (CPI)/Cost Per Thousand Impressions (CPM)

Dollar for dollar, Web advertising can be more expensive than television advertising, though recently online CPMs seem to be trending downward. But in fact, lower CPMs may actually translate into higher costs when advertisers pay for viewers who are not within their target audience. About one-third of online revenues are priced on a cost-per-thousand basis.

Time Spent Online

Online users who spend longer periods of time reading a page and traveling within a site are more valuable to advertisers than those who just land on a page for a few seconds and then move on to another site. The longer a user spends on a page, the more likely he or she is to click on a banner; thus, some online pricing schemes are based on how long the ad stays on the screen and how long the average user stays on the page.

Size-Based Pricing

Borrowing from the conventional newspaper advertising pricing structure, the cost of a banner ad is sometimes based on the amount of screen space it occupies. The charge for a newspaper ad is assessed according to a specific dollar amount per standard column inch area, whereas the rate for a banner ad is measured by pixel area. *Pixels* are tiny dots of color that form images on the screen. Size-based pricing is calculated by multiplying an ad's width by its height in pixels. The price is determined per pixel or by the total number of pixels.

Discount Online Advertising Sales and Buys

Online advertising can be a risky venture, especially for smaller businesses with thin wallets that are hesitant to spend advertising dollars on a new and unproven medium. But there are several online buying strategies that advertisers can employ without risking their finances. **Advertising exchanges**, co-op deals, discounts, and auctions all offer discounted rates for unsold spots. Advertisers benefit from low fees, and providers of Web ad space benefit by filling spots that might otherwise go unsold.

Cooperative Advertising/Product Placement

Cooperative advertising is a strategy in which a marketer promotes a product on a Web site, usually in the form of a review or endorsement, and shares sales profits with the hosting site. For example, Amazon and other online booksellers charge book publishers to promote their titles on their sites. The publisher writes a promotional review of its own book, and Amazon places the review next to an image of the book either for a set fee or for a percentage of sales made from the site. This type of co-op advertising, also known as **product placement**, has been sharply criticized for failing to make it clear to customers that the reviews are promotional pieces, not unbiased editorials written by Amazon's staff. Amazon now posts a page that explains its publisher-supported placement policy and lets users know which ads are part of the co-op program.

Discounts

Some Web sites offer discounted rates if online ads come **online ready**, or fully coded in HTML and complete with all graphic, audio, and video files. In addition, sliding fees are formulated based on how much formatting the Web site manager has to do to an ad before posting it online. Frequency discounts are also available on many Web sites.

Ad Exchanges/Ad Auctions

Although many small businesses have terrific Web sites, they typically cannot afford aggressive online campaigns. One way to get the word out is through an *advertising exchange*, in which advertisers place banners on each other's Web sites free of charge. For example, a company selling beauty products could place its banner on a site that sells women's shoes, and in turn, the shoe company could put a banner on the beauty product site. Neither company charges the other; they simply exchange ad space. Advertising exchanges are gaining in popularity, especially among marketers who do not have much money and who do not have a large sales team. By trading space, advertisers find new outlets that reach their target audiences that they would not otherwise be able to afford.

Ad exchanges more commonly connect advertisers and Web sites through an auctioning system. Ad space is available on millions of Web pages. Web sites with unsold banner spaces put them up for auction at reduced rates through specialized third-party sites. The thinking behind this arrangement is that getting discounted rates for unsold space is better than not getting any money at all. Plus, advertisers are more willing to take chances on less popular sites if they are paying a discounted price for them. Just like the name implies, ad auctioneers put the space up for bid, and the advertiser with the highest bid buys the space, sometimes at less than half price.

Sophisticated target marketing technologies have moved the auction model beyond selling discounted space to matching advertisers with targeted purchasers. Ad auctions, such as those conducted by Google and Yahoo! Ad Exchange, help advertisers identify Web sites that attract their target market. The auctions are set up so that advertisers determine how much they are willing to pay for targeted clients.

The key players involved are as follows:

Audience: Web site users who will see the ads

Sellers: Web site publishers with space to sell

Bidders: Advertisers who bid on the ad space

Ad Exchange: Operates the auction by accepting and processing bids.

Typically, the seller auctions space based on impressions—the expected number of viewers who will see the ad. Advertisers bid on the space/impressions. How much money they bid depends on how likely they think the audience is to buy their products. **Ad auctions** are high-stakes, high-energy bidding not unlike stock trading, where prices fluctuate in an instant and buys are made at the moment visitors land on a page. In real-time bidding, advertisers adjust the price-per-click they are willing to pay based on the competition and the value of impressions. A consumer is no longer just a consumer; some are more valuable than others. A wealthy, 40-year-old risk taker who likes luxury cars is worth more to BMW than an unemployed 25-year-old who drives an older-model sedan. At one point after learning that Mac users spend more on hotel rooms than PC users, the travel site Orbitz started sending Mac users ads for hotels that were 11% more expensive than the ads it sent to PC users.

Google offers the following example of how its ad auctions work. An ad is given a quality score based on its past click-through rates, how well the ad matches the auctioning site's audience, and the quality of the advertiser's landing page (the page a user is connected to after clicking on the ad).

Example: Three advertisers with the same quality score

Advertiser	Price-Per-Click Bid	Quality Score
Company A	$5	10
Company B	$3	10
Company C	$1	10

In this case, Company A wins the auction to place its ad on the Web site. But because the actual amount of money Company A needed to bid to beat the competitors was $3.01, that is the amount Company A will actually pay (About the ad auction, 2015).

ZOOM IN 8.5

To learn how an ad auction works, watch www.youtube.com/watch?v=PjOHTFRaBWA

Read about how ad exchanges work at https://support.google.com/adxbuyer/answer/6077702?hl=en

Mobile Marketing/Advertising Pricing

Similar to traditional media and the Internet, the price for mobile advertising is based on audience numbers. Mobile advertisement is most commonly priced by cost-per-impression (the cost of reaching each mobile consumer) and by cost-per-click (advertisers pay per consumer who clicks on their ad). Advertisers might also be charged on conversion (price per number of consumers who buy the product) or other interactive measure. Because less attention is paid to mobile ads, online companies must price them at a lower rate to attract advertisers. The average price per mobile viewer is about 5 times lower than those per desktop view and about 10 times lower than per print magazine reader. Google earns only about 51 cents per click on mobile search ad but more than $1 per click on computer ads.

Ads placed on social media sites are charged according to a site's model. For example, LinkedIn and Facebook commonly charge by cost-per-click or cost-per-impression, but Facebook also has special prices for embedded sponsor stories, wall photos, or for placing an app on a page, and Twitter commonly charges per click. But like other Web sites, social media sites might also charge by cost per transaction, size, or other method. But audience numbers are not enough to gauge a successful mobile or social media campaign. More important is the level of audience activity.

Career Tracks: Trey Fabacher, Television General Sales Manager

FIG. 8.8 Trey Fabacher

What are your primary responsibilities?
In the role of general sales manager, I am responsible for all revenue aspects of the television station operation. These responsibilities include managing a team of sales managers and sellers that work with clients and develop revenue, television commercial revenue as well as digital revenue, a growing part of our business.

What was your first job in electronic media?
In my 27 years since graduation from the University of Tennessee, I have taken a progressive sales track from account executive to national sales manager, local sales manager, general sales manager, station manager, VP/general manager, and now general sales manager again. We are always focusing on servicing our clients and the overall sales performance. My first job in broadcasting was as an overnight radio DJ in Lafayette, LA, so I can really say that I started my career in the graveyard. I was able to create spec spots for salespeople, which gave me my first taste of sales. From that point, I knew I wanted to sell media, and the broadcasting department at the University of Tennessee exposed me to the connections to make that happen. Through various scholarships and intern programs, I was able to make the proper connections to get a foot in the door and demonstrate my abilities.

What led you to your present job?
By the time I had graduated in 1989, I had worked 2 years in radio as a DJ, 1 year in a sales research internship at the local ABC station, and one summer at NBC New York as a sales pricing analyst for the NBC owned and operated stations (O&Os). That aggressive experience allowed me an opportunity to enter a sales training program with Blair Television and get my first sales job 6 months later in Minneapolis. While there, I moved to CBS, where I managed three TV stations in three different markets. I left CBS to join Meredith to manage the CBS affiliate in Atlanta. I moved from that role after 2 years, and we decided as a family to prioritize our location, with our kids attending college in Georgia, which led me to Columbus, Georgia, and Raycom Media.

SEE IT LATER

THE NEVER-ENDING QUEST FOR RATINGS

The radio, television, and Internet industries are all on a quest for accurate and reliable ways to measure their audiences. Their biggest help may come from new technologies and new ways of reporting data. There are several new ways to measure the audience and to interpret numbers.

MOVING TARGET AUDIENCE

One of the biggest challenges to marketers is how to reach a moving target audience. Think of a target audience—those an advertiser wants to reach—but one that resides in a mobile world. **Global positioning system (GPS)** technology is being developed to track consumers' exposure to billboard and other out-of-home advertising, such as an ad on a gas pump or one in a bathroom stall. Motorists and pedestrians wear small battery-operated meters that track their movements every so many seconds. The GPS data, along with travel diaries, are then matched up to a map of billboard and other outdoor ads to determine the 'opportunity to see' an outdoor advertisement. GPS systems hold promise to deliver "reliable reach, frequency, and site-specific ratings data on real people, passing real sites, in real time" (Wi-Fi & GPS Combined, 2008).

Years of research has honed methods of reaching consumers through television and radio and other media, but catching the attention of those who flit among mobile phones, tablets, and laptops is a formidable task. The old ways of targeting must be redesigned for a new, increasingly mobile world. For example, using GPS on a smartphone to locate a restaurant or opening a recipe app in a sense exposes users to advertising. Smart marketers are tapping into methods of measuring mobile users who see their products and brand names. For example, a Dockers campaign encouraged consumers to shake their iPhone to get a model wearing a pair of khakis to dance. Exposure to the ad was measured by tracking how long users kept shaking their phones.

The proliferation of smart televisions that hook into the Internet has initiated major changes in the way Nielsen counts television viewers and commercial exposure. Nielsen's 'TVandPC' meters include 'cord cutters,' viewers who spurn cable delivery for online streaming, as television viewers. Although about 15% of U.S. viewers and 19% of Millennials are considered true cord cutters or have never had cable, industry executives believe the trend will accelerate. Typically the commercials shown on streamed programs through services such as Hulu are not the same as those shown on the original broadcast, but that has changed with the advent of new services that stream shows with the original commercials, like with the live-streamed Super Bowl in 2016. If, for example, the same commercials shown on an original broadcast of an episode of *The Simpsons* are also shown when the episode is streamed on the Fox Web site or through its program app or on Hulu, Nielsen would capture ad exposure to both the broadcast and online views and count them equally. But the technology for counting the streaming audience has not caught up to streaming technology, and thus billions of hours of viewing on Netflix, Amazon, and Hulu are not being captured accurately. Nielsen is perhaps best positioned to come up with an effective way to measure audience viewing across television and digital platforms, but the pressure is on its executives to perfect a technique before an upstart measurement company beats them to it.

Smart television and online streaming have redefined the meaning of 'television viewer' and 'television program,' especially when considering that a television set is a screen and that computers and tablets are also screens that one day could be considered 'television sets.'

FYI: Who Is Tweeting About TV

	Reality Shows	Drama	Comedy	Sports
Male	35%	39%	53%	79%
Female	65%	61%	47%	21%
35 and over	37%	34%	28%	25%
Under 35	63%	66%	72%	75%

Source: Nielsen Social TV, 2014

Television ratings and program popularity are no longer about just the number of viewers but also about what they are talking about. Nielsen Social tracks television-related conversation on Twitter, and there is a lot of it. In 2013, 36 million people sent 990 million tweets about television and television content. Nielsen estimates that the size of the audience that reads the tweets is 50 times the number of those who write the tweets. Research shows that the more viewers talk about a program, the more engaged and involved they are, and thus the more they watch it. Tweeting about television shows, however, seems confined to a small percentage of viewers who are also extreme tweeters. One study found that only about 16% of respondents use social media while watching prime-time television, and of them, only about one-half tweeted about the show they were watching. Moreover, slightly less than 7% were drawn to a show because of something they read on Twitter. The degree of influence Twitter has on television ratings is still unclear, but it seems that it is powerful among heavy users but less so for those who are not as dependent on the social medium.

FYI: Number of Tweets—Top Shows (2015)

Average Tweets Per Episode

Empire	559,000
The Walking Dead (AMC)	424,000
Pretty Little Liars (ABC Family)	285,000
American Horror Story: Hotel (FX)	239,000
Scandal (ABC)	222,000
The Bachelor (ABC)	156,000
Game of Thrones (HBO)	152,000
Grey's Anatomy (ABC)	120,000
The Bachelorette (ABC)	114,000
Parks and Recreation (NBC)	89,000

Source: Tops of 2015, 2015

The music industry is also finding ways to monitor listenership. In early 2013, *Billboard* teamed with Nielsen to include YouTube video streams with authorized audio in its calculations of most popular songs. Nielsen's methodology calculates digital download sales, on-demand audio streaming, and online radio streaming. The hip-hop single "The Harlem Shake" was one of the first songs to benefit from the new measurement system. On sales totals alone, it would have placed within the top 15 on the Hot 100 list, but with downloads, it ranked number one on both the Hot 100 and Streaming Songs charts.

SUMMARY

Researchers started measuring radio audiences in 1929 to give stations a base on which to price time for commercial spots and to give advertisers a way to compare stations by number of listeners and demographic characteristics. Many of the techniques used for measuring radio audiences were later used to measure audience numbers for broadcast stations and cable and broadcast networks. Today, audience ratings figure prominently in the media world and often determine the success of a program or a station. The new ways of measuring audio in partnership with YouTube and of monitoring streamed television program viewing give a clearer picture of changing media habits.

Nielsen is the most prominent media audience measurement company. To get a sense of the number of listeners and viewers, Nielsen relies on media use diaries and metering devices. Nielsen has updated its methodology and measuring techniques to capture uses of digital media. The Internet poses new challenges to traditional audience measurement techniques. Metering computer use is more complex than metering television or radio use, and the definition of an *Internet user* has not been standardized. Researchers are grappling with how users should be counted, such as by number of hits or number of click-throughs. Although several companies monitor traffic at Web sites, there are many discrepancies regarding how the audience is measured.

Audience numbers are not the only way to evaluate a media audience. Online and traditional media and research companies often use methods that delve into users' likes and dislikes as well as their motivations for using a particular medium or program. Twitter conversation reveals what viewers really think about television shows and can forecast a program's success or failure. Online marketers use customer profiling to deliver banner ads to promote products similar to those on other sites users have visited.

It is essential for media outlets to understand their audiences, because after all, the media are in the business of selling audiences to their advertisers. Television and radio stations and networks and online sources depend on audience numbers and profiles to establish how much they should charge for advertising. Calculations such as cost-per-thousand, cost-per-point, gross rating points, average-quarter-hour, and cume figures are used to set ad costs. Advertisers use these figures when making media buys to compare costs among various media.

Radio and television offer a variety of pricing structures, depending on availability, daypart, program, and other factors. Most stations offer fixed buys, run-of-schedule, frequency discounts, and barter arrangements. Online advertising pricing structures also include CPM and discount buys. Unique to the medium, banners are charged by click-through rates, size-based pricing, cost-per-transaction, and hybrid deals.

Audience measurement tools yield, at best, gross estimates of audience characteristics and media use, but they are only as accurate as the technology allows. Technological innovations may one day produce audience numbers that are far more accurate and reliable than what is currently available.

New communication technologies have forced marketers to rethink tried-and-true methods of reaching an audience. With so many new ways of communicating product information and for persuading an audience to make a purchase, marketers need to expand their media choices. But it is difficult to measure a mobile media audience and to know how to capture its attention yet not annoy customers with ads that are pesky and intrusive.

BIBLIOGRAPHY

52.7 million watched President Bush's economic crises address. (2008, September 26). *Nielsen Wire*. Retrieved from: blog.nielsenwired/media_entertainment/bush-economic-address/ [September 25, 2009]

2009/10 season has begun and cable is in the lead. (2009, September 28). *Cable Television Advertising Bureau*. Retrieved from: www.thecab.tv/main/whyCable/ratingstory/09ratingsnapshots/200910-season-has-begun-and-cable-is-in-the-lead.shtml

ABC 2009 May sweep. (2009, May 21). *The Futon Critic*. Retrieved from: www.thefutoncritic.com/news.aspx?id520090521abc01 [February 5, 2010]

About the ad auction. (2015). *Google.com*. Retrieved from: https://support.google.com/adsense/answer/160525?hl=en

AC Nielsen Entertainment partners with CBS for real-time audience research [press release] (2001, April 18). *Nielsen*. Retrieved from: www.acnielsen.com/news/corp/2001/20010418.htm

Academy awards rise in ratings. (2009, February 23). *The Live Feed*. Retrieved from: www.thrfeed.com/2009/02/academy-awards-oscar-ratings-1.html [September 25, 2009]

Adalian, J. (2009, February 23). Oscar TV ratings: Early returns show gain. *Crains New York Business.com*. Retrieved 9/25/09 from: www.crainsnewyork.com/article/20090223/FREE/902239975.

Albiniak, P. (2003, February 10). Cold weather, hot ratings. *Broadcasting & Cable*, p. 13.

Angel, S., & Walfish, M. (2013, August). Verifiable auctions for online ad exchanges. In *Proceedings of the ACM SIGCOMM 2013 conference on SIGCOMM* (pp. 195–206). ACM.

Atwell, L. (2013). Social media advertising: The total cost breakdown of a 'like'. *Adpearance*. Retrieved from: http://adpearance.com/blog/social-media-advertising-the-total-cost-breakdown-of-a-like

Bachman, K. (2010, January 22). Nielsen preps TV and PC report. *Adweek*. Retrieved from: www.adweek.com/news/television/nielsen-preps-tvandpc-report-101308

Barnes, B. (2009, July 27). Lab watches Web surfers to see which ads work. *The New York Times*, pp. B1, B6.

Bauder, D. (2009, December 29). Anchor changes at ABC's 2 biggest newscasts make no immediate difference in Nielsen ratings. *Business News*. Retrieved from: blog.taragana.com/business/2009/12/29/anchor-changes-at-abcs-2-biggest-newscasts-make-no-immediate-difference-in-nielsen-ratings-16060/ [February 6, 2010]

Berr, J. (2014, November 20). Has the cord-cutter threat to pay TV been exaggerated? *CBS Money Watch*. Retrieved from: www.cbsnews.com/news/has-the-cord-cutters-threat-to-pay-tv-been-exaggerated/

Beville, H. M., Jr. (1988). *Audience ratings*. Hillsdale, NJ: Erlbaum.

Bibel, S. (2015). 2014 2014 season: NBC leads among adults 18–49 & CBS tops total viewers through week 15 ending January 4, 2015. *ZapToIt.com*. Retrieved from: http://tvbythenumbers.zap2it.com/2015/01/07/2014–2015-season-nbc-leads-among-adults-18–49-cbs-tops-total-viewers-through-week-15-ending-january-4–2015/346929/

Boyce, R. (1998, February 2). Exploding the Web CPM myth. *Online Media Strategies for Advertising* [Supplement to *Advertising Age*], p. A16.

Carter, B. (2009, August 26). Comedy Central tries to gauge passion of its viewers. *The New York Times*, p. B3.

Clifford, S. (2009, March 11). Advertisers get a trove of clues in smartphones. *The New York Times*, pp. A1, A14.

DiPasquale, C. B. (2002, October 8). Nielsen to test outdoor ratings system. *Advertising Age*. Retrieved from: www.adage.com/news.cms?Newsld536254#

Dreze, X., & Zufryden, F. (1998). Is Internet advertising ready for prime time? *Journal of Advertising Research*, 38(3), 7–18.

Eastman, S. T., & Ferguson, D. E. (2002). *Broadcast/cable/web programming*. Belmont, CA: Wadsworth.

Flannagan, M. (2002). Local TV will pay big bucks for Nielsens. *Knoxville News-Sentinel*, pp. A1, A13.

Friedman, W., & Fine, J. (2003). Media sellers pay the price for blackout. *Advertising Age*, pp. 1, 31.

Fung, B. (2015, December 22). Survey: Many cord-cutters don't have broadband. *The Washington Post*, p. A12.

Glaser, M. (2007a, July 25). The problem with web measurement, part 1. *PBS*. Retrieved from: www.pbs.org/mediashift/2007/07/the-problem-with-web-measurement-part-1206.html [February 5, 2010]

Glaser, M. (2007b, August 1). The problem with web measurement, part 2. *PBS*. Retrieved from: www.pbs.org/mediashift/2007/08/the-problem-with-web-measurement-part-2213.html [February 5, 2010]

Gorman, B. (2009, January 18). Super Bowl TV ratings. *TVbytheNumbers*. Retrieved from: tvbythenumbers.com/2009/01/18/historical-super-bowl-tv-ratings/11044 [September 25, 2009]

Graft, K. (2009, August 7). 40 percent of U.S. homes have a gaming console, as HDTV adoption rises. *Gamasutra*. Retrieved from: www.gamasutra.com/php-bin/news_index.php?story524757 [September 28, 2009]

Hall, R. W. (1991). *Media math*. Lincolnwood, IL: NTC Business Books.

Hansell, S. (2001, July 23). Pop-up ads pose a measurement puzzle. *The New York Times*, pp. C1, C5.

Hot 100 news: Billboard and Nielsen add YouTube video streaming to platforms. *Billboard*. Retrieved from: www.billboard.com/articles/news/1549399/hot-100-news-billboard-and-nielsen-add-youtube-video-streaming-to-platforms

Hutheesing, N. (1996). An online gamble. *Forbes*, p. 288.

IAB Internet advertising revenue report: 2013 full year results. (2014, April). *Interactive Advertising Bureau*. Retrieved from: www.iab.net/media/file/IAB_Internet_Advertising_Revenue_Report_FY_2013.pdf

Ibarra, S. (2009, February 2). Super Bowl TV ratings change: Last minute victory. *TV Week*. Retrieved from: www.tvweek.com/news/2009/02/super_bowl_tv_ratings_change_l.php [September 25, 2009]

Improving their swing. (1998, February 2). *Online Media Strategies for Advertising* [supplement to *Advertising Age*], p. A45.

James, M. (2012, August 31). Nielsen 'people meter' changed the ratings game 25 years ago. *Los Angeles Times*. Retrieved from: http://articles.latimes.com/2012/aug/31/entertainment/la-et-ct-nielsen-people-meter-anniversary-20120830

Kang, C. (2015, November 10). Online advertising predicted to top TV by 2017 as viewing habits evolve. *The Washington Post*, p. A16.

Kaye, B. K., & Medoff, N. J. (2001). *The World Wide Web: A mass communication perspective*. Mountain View, CA: Mayfield.

Levin, G. (2014, March 25). Nielsen ratings: 'Dancing' drops; '100' opens big. *USA Today*. Retrieved from: www.usatoday.com/story/life/tv/2014/03/25/nielsen-tv-weekly-ratings-highlights/6866439/

Local television market universe estimates. (2016). *Nielsen*. Retrieved from: www.tvb.org/media/file/2015-2016-dma-ranks.pdf

Mandese, J. (2004, March 1). Observers rock research. *TelevisionWeek*, p. 35.

Market survey rankings, frequency, type and population. (2016). *Nielsen*. Retrieved from: www.nielsen.com/content/dam/corporate/us/en/docs/nielsen-audio/populations-rankings-fall-2015.pdf

Mitchell, R. L. (2014, January 15). Ad blockers: A solution or a problem? *Computerworld*. Retrieved from: www.computerworld.com/article/2487367/e-commerce/ad-blockers—a-solution-or-a-problem-.html

Mobile banners continue to boast high click rates. (2012, August 27). *eMarketer*. Retrieved from: www.emarketer.com/Article/Mobile-Banners-Continue-Boast-High-Click-Rates/1009299

Moreno, C. (2014, August 1). Univision trumps English-language network giants again. *Huffington Post*. Retrieved from: www.huffingtonpost.com/2014/08/01/unvision-number-1-network_n_5642872.html

Morrow, T. (2002, May 14). Knox stations replacing Nielsen system. *Knoxville News-Sentinel*, p. D1.

Most watched series finales. (2010). *Wikipedia*. Retrieved from: en.wikipedia.org/wiki/List_of_most-watched_television_episodes#Most-watched_series_finales [February 10, 2010]

Nastic, G. (2013, July 1). Nielsen launches Twitter TV ratings. *CSI Magazine*. Retrieved from: www.csimagazine.com/csi/Nielsen-launches-Twitter-TV-Ratings.php

Nielsen and Twitter establish social TV rating. (2012, December 17). *Nielsen*. Retrieved from: www.nielsen.com/us/en/press-room/2012/Nielsen-and-twitter-extablis-social-tv-rating

Nielsen estimates more than 116 million TV homes in the U.S. (2014, August 29). *Nielsen*. Retrieved from: www.nielsen.com/us/en/insights/news/2014/nielsen-estimates-more-than-116-million-tv-homes-in-the-us.html

Nielsen Media. (n.d.). Online. Retrieved from: www.nielsen.com/us/en/solutions/measurement/online.html

Nielsen Media (n.d.). Glossary of terms. Retrieved from: http://www.nielsenmedia.com/glossary/terms/R/R.html

Nielsen Social TV. (n.d.). Retrieved from: www.nielsen.com/us/en/solutions/measurement/social-tv.html

Nielsen Social TV. (2014). *Who Is Tweeting About TV*. Retrieved from: www.nielsensocial.com/whos-tweeting-about-tv/

Nielsen to test electronic ratings service for outdoor advertising [press release] (2002, October 8). *Nielsen*. Retrieved from: www.nielsenmedia.com/newsreleases/2002/Nielsen_Outdoor.htm

Nielsen Twitter TV ratings. (n.d.). *Nielsen Social TV*. Retrieved from: www.nielsensocial.com/product/nielsen-twitter-tv-ratings/

Pennington, G. (2015, December 8). Here are 2015's top rated TV shows, according to Nielsen. *St. Louis Post Dispatch*. Retrieved from: www.stltoday.com/entertainment/television/gail-pennington/here-are-s-top-rated-tv-shows-according-to-nielsen/article_ab1f8852–0808–51f0–9be7–0615ef7ca1ce.html

The portable people meter. (2009). *Arbitron*. Retrieved from: www.arbitron.com/portable_people_meters/ppm_service.htm [September 23, 2009]

Price of a 30 second ad in network TV. (2009). *Media Literacy Clearinghouse.* Retrieved from: www.frankwbaker.com/mediause.htm [September 9, 2009]

Programs most watched using DVRs. (2012, June 24). *The New York Times.* Retrieved from: www.nytimes.com/interactive/2012/06/24/business/media/top-programs-watched-using-dvrs.html

Ratings/Nielsen Week. (April 12–18, 2010). *Broadcasting & Cable.* Retrieved from: www.broadcastingcable.com/channel/TV_Ratings.php

Rosen, J. (2012, December 2). Who do they think they are? *The New York Times Magazine*, pp. 40–45.

Schumann, D. W., & Thorsen, E. (1999). *Advertising and the World Wide Web.* Mahwah, NJ: Erlbaum.

Schneider, M. (2013, June 10). America's most watched: The top 25 shows of the 2012–2013 TV season. *TV Guide.* Retrieved from: www.tvguide.com/News/Most-Watched-TV-Shows-Top-25-2012-2013-1066503.aspx

Scott, N. (2015, January 28). 10 Best Super Bowl commercials of all time. *USA Today.* Retrieved from: http://ftw.usatoday.com/2015/01/10-best-super-bowl-commercials-of-all-time

Seals, T. (2013, July 1). TV cord-cutters, cord-nevers total 19 percent of US adults. (2013, July 1). *Cable Spotlight.* Retrieved from: http://cable.tmcnet.com/topics/cable/articles/2013/07/01/344134-tv-cord-cutters-cord-nevers-total-19-percent.htm

Seidman, R. (2010, February 4). Bravo's "Shear Genius" makes waves with highest-rated season premiere ever. TVBytheNumbers.com. Retrieved from: tvbythenumbers.com/2010/02/04/bravos-shear-genius-makes-waves-with-highest-rated-season-premiere-ever/41130 [February 5, 2010].

Shane, E. (1999). *Selling electronic media.* Boston: Focal Press.

Shaw, R. (1998). At least there are no Web sweeps (yet). *Electronic Media*, p. 17.

Singer, N. (2012, November 18). Your attention, bought in an instant. *The New York Times*, Business pp. 1, 5.

Small crowd, young and old watch Super Bowl on Web. (2016, February 4). *The New York Times.* Retrieved from: www.nytimes.com/reuters/2016/02/04/arts/04reuters-nfl-superbowl-streaming.html

Snyder, H., & Rosenbaum, H. (1996). Advertising on the World Wide Web: Issues and policies for not-for-profit organization. *Proceedings of the American Association for Information Science, 33*, 186–191.

So many ads, so few clicks. (2007, November 12). *BusinessWeek.* Retrieved from: www.businessweek.com/magazine/content/07_46/b4058053.htm [September 23, 2009]

The state of the news media. (2012). *The Pew Research Center's Project for Excellence in Journalism.* Retrieved from: www.stateofthemedia.org/2012/network-news-the-pace-of-change-accelerates/network-by-the-numbers/

The state of the news media. (2013). *The Pew Research Center's Project for Excellence in Journalism.* Retrieved from: www.stateofthemedia.org/2013/network-news-the-pace-of-change-accelerates/network-by-the-numbers/

Steel, E. (2016, February 2). Nielsen plays catch-up as streaming era wrecks havoc on TV raters. *The New York Times.* Retrieved from: www.nytimes.com/2016.02/03/business/media/Nielsen-playing-catch-up-as-tv-viewing-habits-change-and-digital-rivals-spring-up.html

Stelter, B. (2009, December 2). Nielsen to add online views to its ratings. *The New York Times*, p. B4.

Stelter, B. (2009, May 14). Hulu questions count of its audience. *The New York Times.* Retrieved from: www.nytimes.com/2009/05/15/business/media/15nielsen.html?_r51&ref5media [February 5, 2010]

Stelter, B. (2009, September 11). Media group to research new methods for ratings. *The New York Times.* Retrieved from: www.nytimes.com/2009/09/11/business/media/11ratings.html [September 25, 2009]

Stelter, B. (2013, February 21). Nielsen adjusts its ratings to add Web-linked TVs. *The New York Times*, B1, B2.

Stross. R. (2009, February 8). Why television still shines in a world of success? *The New York Times.* Retrieved from: www.nytimes.com/2009/02/08/business/media/08digi.html?_r51&scp51&sq5%22Why1television1still1shines1in%22&st5nyt

Sutel, S. (2007, September 30). Electronic ratings give new reality to radio. *The Tennessean*, pp. 1E–2E.

Terms for the trade. (2014). *Nielsen.* Retrieved from: www.arbitron.com/downloads/terms_brochure.pdf

Thompson, D. (2013, March). The incredible shrinking ad. *The Atlantic*, pp. 24, 26, 28.

Top 10 online advertisers in the U.S. (2012, March 3). *Digital Strategy Consulting.* Retrieved from: www.digitalstrategyconsulting.com/intelligence/2012/03/top_10_online_advertisers_in_u.php

Top-rated programs of 2008–09 in households. (2009). *Television Bureau of Advertising.* Retrieved from: www.tvb.org/rcentral/ViewerTrack/FullSeason/08–09-season-hh.asp [October 2, 2009]

Tops of 2015: TV and Social Media. (2015, December 8). *Nielsen.* Retrieved from: www.nielsen.com/us/en/insights/news/2015/tops-of-2015-tv-and-social-media.html

TV audience of 31 million1 for Michael Jackson memorial. (2009, July 8). *Radio Business Report.* Retrieved from: www.rbr.com/tv-cable_ratings/15685.html [September 25, 2009]

TV costs and CPM trends—Network TV primetime. (2015). *Television Advertising Bureau.* Retrieved from: www.tvb.org/trends/4718/4709

TV Ratings. (2010, April 19). *The Nielsen Company.* Retrieved from: http://en-us.nielsen.com/rankings/insights/rankings/television

TV shows, Top-ten list. (2014, December 8). *Nielsen.* Retrieved from: www.nielsen.com/us/en/top10s.html

Voight, J. (December, 1996). Beyond the banner. *Wired*, p. 196.

Vonder Haar, S. (June 14, 1999). Web retailing goes Madison Avenue route. *Inter@ctive Week*, p. 18.

Walmsley, D. (March 29, 1999). Online ad auctions offer sites more than bargains. *Advertising Age*, p. 43.

Warren, C. (Fall, 1999). Tools of the trade. *Critical Mass*, p. 22.

Webster, J. G., & Lichty, L. W. (1991). *Ratings analysis.* Hillsdale, NJ: Erlbaum.

Webster, J. G., Phalen, P. F., & Lichty, L. W. (2000). *Ratings analysis.* Hillsdale, NJ: Lawrence Erlbaum.

Who we are and what we do. (2002). *Nielsen.* Retrieved from: www.nielsenmedia.com/who_we_are.html

Wi-Fi and GPS combined move outdoor audience measurement indoors. (2008, April 10). *Business Wire.* Retrieved from: www.businesswire.com/portal/site/google/?ndmViewId5news_view&newsId520080410006247&newsLang5en [September 23, 2009]

Zipern, A. (2002, January 15). Effort to measure online ad campaigns. *The New York Times.* Retrieved from: www.nytimes.com/2002/01/15/business/media/15ADCO.html

Social Media: Private Conversations in Public Places 9

Contents

Most people these days cannot imagine life without social media. Whether on blogs, Facebook, Instagram, LinkedIn, Vine, or some other site, today's social world revolves around online connections. Millennials often ask Boomers how they set up dates when they were young or knew what their friends were doing or just got through the day without social media. Instant and constant communication with friends and family is the new normal and has created a new world of social expectations and anxieties.

This chapter explores the rise of social media. It begins with a historical overview of the first social media sites and progresses to the online social world of today. This chapter is not so much about how to use social media as it is about the cultural and social shifts brought on by the ability to always be digitally connected.

SEE IT THEN

Historically, the communication process has been somewhat linear. There is a sender with a message designed to reach an audience or person. The rise of the information age has created a paradigm shift in communication. Instead of a linear process, communication is now a circular feedback loop. In other words, communication is no longer just one way from a communicator to a person or mass audience but instead is a two-way conversation that empowers the receiver to provide feedback. Social media enhance this new communication landscape by giving consumers the opportunity to connect to information and post their own content on their own schedule 24/7.

PREWEB SOCIAL NETWORKING

It is not easy to pinpoint what was the first social media network. Several early Internet venues could be loosely considered types of social networks. **Multiuser dungeons (MUDs)** were preweb Internet games set in a fantasy world in which players took on fake identities, and using various keyboard commands could slay a dragon or complete a quest. MUDs were played online as early as 1978 and could be considered the origin of interactive, social gaming.

The **bulletin board system (BBS)** created in the late 1970s was a way for users to exchange messages. Users could post a message or read what others had posted. Early BBS allowed only one person to log on at a time. LISTSERV went online in 1986 as an automatic mailing list. Listservs are still used today to automatically email messages to everyone signed up to a group.

BBS and listserv ushered in the era of consumers as content creators. Internet Relay Chat (IRC) is the precursor to instant messaging as known today. Created in 1988, it was the first form of one-to-one real-time chatting.

The Whole Earth 'Lectronic Link (WELL) went online in 1985. It networked writers and readers of *The Whole Earth*

Review, which was a magazine dedicated to technological innovation and making the world a better place through science.

POSTWEB SOCIAL NETWORKING

The creation of browser technology (HTTP-hypertext mark-up language) changed the face of the Internet. Instead of being text-based, browsers enhanced the Internet with graphics and photos and later with audio and video. Pages of information were now linked through hypertext so users could bounce among them instead of having to read them in order like pages of a book.

New user-friendly and intuitive Web pages were prime for linking users together. *Six Degrees* is commonly thought of as the first Web-based social network. Named for the concept of 'six degrees of separation,' the site was conceptualized as a place for "meeting people you don't know through people you do know" (Adams, 2011). The site contained user profiles, friend lists, and an area for private messaging. At its peak, the site boasted close to 1 million users but it could not sustain itself and closed up in 2001.

The year 1999 is known for the emergence of several key Web sites. *Move-On* connects politically interested Web users, *LiveJournal* provides a place to follow blogs and other writings, *AsianAvenue* is an online community for Asian Americans, and *BlackPlanet* connects African Americans. Although these sites are not generally considered 'social networks' by today's definitions, they were all set up as sites to connect users based on common interests and backgrounds.

Of all the new Web sites that popped up in the late 1990s and early 2000s, *Friendster* is considered the groundbreaking social network site that led to the ones of today. It was also the first network to attain more than 1 million users—within the first few months of its existence, 3 million users had signed on. Anyone could join Friendster and create a personal profile that consisted of such information as occupation, favorite music and television shows, location, and relationship status. Users would invite their friends into their networked circle and thus become friends of their friends, plus their friends and so on. For its first 2 years, Friendster was the most popular social network, but by 2004, the 1-year-old *MySpace* had overtaken it. Friendster never regained its status as a social network, and in 2011 it was repositioned as a social gaming site.

FIG. 9.2 Friendster toolbar

As Friendster declined, *MySpace* soared. By the end of 2007, *MySpace* claimed 100 million users worldwide, mostly teenagers and young adults. MySpace built its community around music and entertainment. After several high-profile events, MySpace quickly found itself embroiled in scandal and having to fend off skeptics who claimed that it was a place where child molesters and murderers lurked. A 16-year-old Michigan girl made international headlines when she ran off to the Middle East to meet a young man she met on the site, and a 13-year-old Missouri girl committed suicide after being bullied on the site. When the Connecticut Attorney General claimed that children were being exposed to pornography on MySpace, the site began providing the names of registered sex offenders who used the site. With all eyes turned on MySpace as a "vortex of perversion" (Gillette, 2011), parents started making their children close out their accounts and shifted them to what they deemed safer online havens, such as the new *Facebook*. MySpace limped along despite the gradual decline of users. In February 2011 alone, traffic fell 44% from 95 million to 63 million users. In 2013, MySpace rebranded itself as "the new MySpace," aimed at 17- to 25-year-olds. The new site focuses more on music, games, movies, and the entertainment industry and less on social connections. MySpace is alive and well. It attracts about 50 million people who generate more than 300 million video views per month. Facebook, however, is the 'now' social media site.

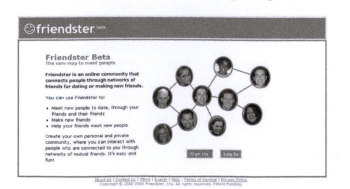

FIG. 9.1 Friendster homepage

FYI: Who Uses Social Networking Sites

All Internet users	74%
Men	72%
Women	76%
18–29	89%
30–49	82%
50–64	65%
65+	49%
High School Grad or Less	72%
Some College	78%
College+	73%
Less than $30,000 per year	79%
$30,000–$49,999	73%
$50,000–$74,999	70%
$75,000+	78%

Source: Social networking fact sheet, 2014

FYI: Top 13 Most Engaged Social Media Users by Country (2015)

Country	Average Hours Per Day
Argentina	4.3
Philippines	4.3
Mexico	3.9
Brazil	3.8
Thailand	3.8
United Arab Emirates	3.6
Malaysia	3.5
South Africa	3.2
Vietnam	3.1
Saudi Arabia	3.0
Turkey	2.9
Indonesia	2.9
USA	2.7
Average of 30 top countries	2.4

Source: Social Media Today, 2015

SEE IT NOW

SOCIAL MEDIA VERSUS SOCIAL NETWORKING SITES

It is ironic that social media sites have become the hubs of the real-life world—users go online to find out what is going on offline. It is not just teenagers and young adults but people of all ages who flock to social media. The use of social media has surged in recent years. The percentage of online adults with a social media profile has soared from 8% in 2005 to 35% by the end of 2008 to 43% by mid-2009 to 74% in 2015 to 78% as of early 2016. But more astounding than these percentages is the sheer number of social media users. Out of about 7.3 billion people on earth, just over 2 billion use some form of social media.

The number of users tells only one part of the story; the amount of time spent with social media reveals a true obsession. U.S. users alone spend about 121 billion minutes on social media sites per month. Collectively, those minutes equal 230,060 years spent socializing in front of a screen instead of face to face. On an individual level, U.S. Internet users spend about 2.7 hours per day with social media. Those 18 to 34 years of age socialize online about 3.8 hours per day compared to 50- to 64-year-olds, who do so for about 2.4 hours per day.

But what are social media? The term 'social media' is often used loosely as a catchall to describe any type of user-generated content (UGC). But for user-generated content to be considered social media, it must meet three primary standards:

1) UGC must be published either on a publicly accessible web site or on a social networking site accessible to a selected-group of people; 2) UGC needs to show a certain amount of creative

effort, 3) and it needs to have been created outside of professional routines and practices. (Kaplan & Haenlein, 2010, p. 61)

Subcategorization of social media further complicates the understanding of it. Some social media are also classified as **social network sites**, which are

Web-based services that allow individuals to 1) construct a public or semi-public profile within a bounded system, 2) articulate a list of other users with whom they share a connection, and 3) view and traverse their list of connections and those made by others within the system. (boyd & Ellison, 2007, p. 211)

For simplicity, the remainder of this chapter uses the term **'social media'** as a descriptor of any type of online site that is driven primarily by user-generated content. The term 'social network sites' refers specifically to sites that are social networks by design, such as Facebook.

SOCIAL NETWORKING SITES

The number of social networking sites is anyone's guess. Estimates range from 300 well-known ones to thousands of obscure specialty sites. What is known, however, is that the number of SNS is growing faster than they can be counted. Several social networking sites, however, far outrank others in terms of numbers of global users. The exact number of monthly users is elusive, so the figures that follow are best approximations as of mid-2016.

Facebook	1.5 billion active monthly visitors
YouTube	1.0 billion
Tumblr	555 million
Google+	540 million
Instagram	400 million
Twitter	320 million
Vine	200 million
Pinterest	100 million
LinkedIn	100 million
Flickr	92 million

Source: Leading social networks worldwide, 2016; Moreau, 2016; Social media active users, 2016

FACEBOOK (2004)

Facebook has good reason to brag about its 1.5 billion monthly users. It is really quite astounding to think that one site draws so many people from around the world. Facebook has grown into the largest social media portal on the planet. Yet several countries ban Facebook: North Korea, Iran, China (except in the trade-free zone of Shanghai), and the tiny island nation of Nauru. Other countries, such as Pakistan, Syria, and Egypt, have at various times enacted temporary bans.

In the U.S., Facebook is the center of attention. It is the most popular social networking site and is used by about 71% of online adults. U.S. users even spend more time with Facebook each day than they do with their pets—39 minutes per day taking care of and feeding furry and feathered friends, 40 minutes on Facebook. Recent trends indicate that the over-55 crowd is the fastest-growing age group, with an 80.4% increase

FIG. 9.3 History of social media
Photo courtesy of Cendrine Marrouat (research), http://socialmediaslant.com; Karim Benyagoub (design)

five new 'reactions'; love, haha, wow, sad, angry. Now when someone posts news about Grandma's heart attack, a friend can click on 'sad' rather than 'like,' which could have been interpreted as being happy that Grandma is sick.

Mark Zuckerberg along with several other partners founded 'The Facebook' (as it was originally known) in 2004 while he was studying psychology at Harvard University. A keen computer programmer, Zuckerberg had already developed a number of social-networking Web sites for fellow students, including *Coursematch*, which connected Harvard students based on the courses they were taking, and *Facemash*, where students could rate their fellow students' attractiveness. Within 24 hours of opening Facebook online, 1,200 Harvard students had signed up, and after 1 month, more than half of the undergraduate population had a profile.

The name 'Facebook' was taken from the printed booklet that was distributed to freshman to familiarize them with Harvard students and staff. Many other universities also distributed to potential employers printed 'facebooks' that contained profiles and resumes of graduating seniors.

Facebook first started as a way to connect college students. Only subscribers with an '.edu' email address could join. But registrants soon wanted to include nonstudent friends and family members and leaned on Facebook to open the site to the general public. Facebook's early business model of restricting access to students is today considered social suicide. The entire point of social media is to connect to as many people as possible.

There are so many people in the world with a Facebook account that dead users could one day outnumber the living. As creepy as it sounds, 30 million Facebook users died in the first 8 years of its existence. Because it is so difficult to delete someone else's profile, it stays online in perpetuity. In some ways, knowing that a loved one's profile is still accessible is comforting. Faced with the possibility that by the year 2065, there could be more dead people than live ones on the site, the company now allows a profile to be turned into a memorial page on which friends can post comments and memories.

Facebook dominates global social media in terms of number of users and time spent on the site. In the U.S., it also captures the largest market share in terms of number of visits.

YOUTUBE (2005)

YouTube is the world's most popular online video site, with users watching hundreds of millions of hours of video each day and uploading 300 hours of video every minute. YouTube lets anyone upload short videos for private or public viewing. Although YouTube was not created as a social medium per se, but as a way to showcase amateur and professional videos, it has quickly become a place to network. Social connections are created through friend requests, channel subscriptions, and commenting on videos.

On YouTube, anyone can be a filmmaker, a teacher, an artist, or a performer. Featuring videos it considers entertaining, YouTube has become a destination for ambitious videographers, as well as amateurs who fancy making a

in users between 2011 and 2014. In turn, 32.8% of 13- to 24-year-olds who are turned off by having to 'friend' their parents and other relatives have abandoned the site. Facebook photos of young adults playing beer pong and drinking shots of tequila are being quickly replaced with ones of proud parents with their little soccer stars.

Users are attracted to Facebook's interface, which makes it easy to quickly scroll through and scan photos, updates, and news and to keep up with friends just by clicking the 'like' button. But because the 'like' option often fell short of conveying the appropriate response, in 2016 Facebook added

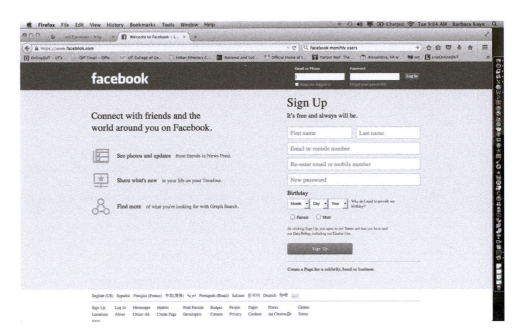

FIG. 9.4 Facebook homepage

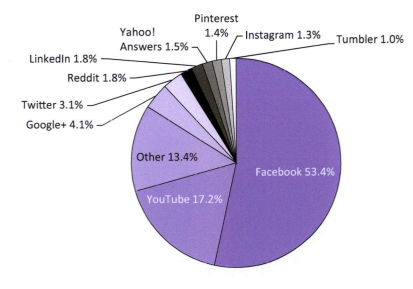

FIG. 9.5 Most popular social media by market share
Created by author

statement of some kind. In addition, YouTube has emerged as a major venue for videos of political speeches. In the 2008 presidential campaign, videos of Barack Obama and John McCain were viewed more than 3 billion times, and 39% of voters watched at least one campaign-related video on YouTube. The 2012 Obama-versus-Romney contest yielded just about as many video views. YouTube has become a formidable force in setting and changing public opinion.

But YouTube is not all political. Videos run the gamut from professional instruction to amateur 'how-to' to humorous clips of silly pets and silly people to recordings of live events and even of a tornado as it bears down on a town. Viewers turn to YouTube for a variety of reasons. YouTube's comments section keeps users connected socially. YouTube is often used as a substitute for television—someone without cable or satellite can watch what he or she missed. Users think the videos are informative, funny, and inspiring, and they like to find out what other people think of the same videos. Some people connect to

YouTube just because they are curious or because a friend has sent them a link.

YouTube is a global medium—88% of its traffic comes from outside the U.S., and it is available in 61 countries and 61 languages. Its visibility and strength and its ability to bring nations together compelled Google to acquire YouTube in 2006 for $1.65 billion.

TUMBLR (2007)

Tumblr is a cross between a social media site, live streaming application, and microblog, on which users post short-form blog entries, music, and visuals (photos, graphics, video). Although there are no length restrictions, Tumblr encourages short updates. Tumblr's interface is easy to use. It categorizes posts by 'text,' 'photo,' 'quote,' 'link,' 'chat,' 'audio,' and 'video.'

Yahoo!-owned Tumblr hosts about 238 million blogs on which 81 million posts are created each day. The site is most popular among teens and college-aged online users—more than half of its visitors are under the age of 25.

GOOGLE+ (2011)

Introduced in 2011 as the social media arm of the formidable Google, Plus is the "social layer across all of Google's services" (Bosker, 2012). Google+ offers many of the same features as other social media but calls them by different names. On Google+, users can 'circle' their friends by category, such as family, college buddies, and office colleagues. 'Hangouts' is the name for video chat and video conferencing, 'Hangouts-On-Air' lets users create and distribute Webcasts on YouTube, 'Huddles' is for group messaging, 'Stream' displays 'Circle' updates, and 'Sparks' pushes content tailored to each user's interest.

The biggest criticism of Google+ is that once a user signs up for the service, it becomes the portal to all Google services. In other words, Gmail accounts, searches, YouTube, maps, and any other Google activity are all linked through Google+. On the surface, this aggregation may seem like a convenience for users, but it gives Google unprecedented access to users' personal and shopping information and Internet habits, and it taps into their friends' data. Privacy activists claim that providing a social network is a mask for the primary purpose of Google+—selling in-depth consumer information to advertisers.

Although Google+ attracts millions of active users per month, it has a way to go to catch up to Facebook. And the numbers are a bit misleading. Because Google+ becomes a registrant's entry to all Google services, the figures include users who are signed in but may be using a different product like YouTube or Google search. Google+ has not caught on in the same way as Facebook, and those who do go on are not very engaged in the site. For example, Facebook users spend an average of 6 hours and 15 minutes on Facebook per month, while Google+ users are on for a paltry 7 minutes, with smartphone access lifting that time by 4 minutes.

INSTAGRAM (2010)

Instagram is the largest photo-sharing social media site. Since its inception in 2010, Instagram has added *direct messaging, tagging, hashtags,* and *video sharing.* From the very beginning, Instagram was a success. It amassed 1 million users within 2 months after its launch. Its wild success led Facebook to snap it up for a cool $1 billion in 2012.

Instagram is not just a place to store visuals, but it is a rich social network as well. The point is to share experiences by uploading photos and videos in real time as an event is occurring. Similar to other social networks, account holders have a profile, news feed, and a network of 'followers.' Instagram is a mobile platform, which means that though visuals may be viewed, commented on, and liked on the Web, uploading must be done through the mobile app.

Instagram is used by just over one-half of all young adults ages 18 to 29. About 68% of all users are female, and nine of ten users are under the age of 35. Instagram users are very engaged with the site, spending 257 minutes per month, as compared to 170 minutes on Twitter. The number of Instagram photos is astounding. By late 2015, 18.7 billion photos had been shared, and 58 million photos were being posted daily. Although Instagram is for all types of visuals, over 35 million selfies have been posted.

TWITTER (2006)

Twitter is a microblogging social networking site (combination of a blog and a social network site) that distinguishes itself from other social media by limiting messages to 140 characters. Twitter was first created in 2006 and was a bit slow to catch on, but by 2009, it was ranked as one of the top 50 Web sites worldwide, with about 106 million registered users. By 2015, Twitter was claiming about 320 million monthly active users who fire off 500 million tweets per day, 10 times more than 5 years earlier.

About 23% of online adults, 19% of the entire U.S. adult population, use Twitter, but it is more popular with younger users. Almost one in four 18- to 29-year-olds tweet regularly, whereas only about one-quarter of 30- to 49-year-olds do so. In fact, Twitter is one of the top venues where the Millennial generation goes for news. The percentage is much lower for older individuals. Only 10% to 12% of those over 50 use Twitter. Early on, parents were concerned that it was unsafe for their children to tweet their locations or what they are doing, so only about 4% of Twitter users were under the age of 18, but these days one-quarter flock to Twitter.

Different from social network sites where users approve who gets to become a friend and only friends can see postings, Twitter is an open-access network that anyone can read. Whereas connections on Facebook and other such social network sites are bidirectional, meaning that users have to accept friends into their network before they see each other's updates, email, and chat, Twitter does not require the same reciprocation—users can follow whoever they want to follow. Connections on social networking sites are typically among real-life friends, while the majority of connections on Twitter are among strangers who do not need to reveal their true identities.

Twitter is best used for delivering instant news and engaging in conversation with *trending* topics. Users tweet and *retweet* or *favorite* to populate their timeline. A Twitter *feed* is in real time, and the timeline is generated based on the number of followers and what they saying at any given snapshot in time. By adding a **hashtag** to a message, others can find out everything being said about the particular subject. Other social media like Pinterest, Instagram, and Facebook also use hashtags to make it easy to search for topics.

The Twitter network was aflutter when in early 2016 the company announced that it would allow tweets up to 10,000 characters in length. Only the first 140 characters would be displayed, but users could tap on a link to see full-length text. Reactions to longer tweets were mixed. Some users worried that longer Tweets would interrupt Twitter's sleek flow and would make it more like any ordinary blog with long babblings. Others, though, welcomed the idea of clicking on a short post for more information. But after 2 months of anticipation, Twitter's chief executive made a brief, 81-character announcement on the *Today* program referring to the original character limit, "It's staying. It's a good constraint for us. It allows for of-the-moment brevity."

So how did Twitter get its name? The definition of twitter is 'a short burst of inconsequential information' and 'a series of chirps from birds.' The name was fitting, and so the new platform became Twitter. Soon the 'chirps' of many 'twitterers' would be heard/seen throughout the Twitterverse as the microblogging platform caught on with Internet users. As with texting, users have developed Twitter shorthand to stay below the maximum number of words. For example, 143 means 'I love you,' and SYL means 'see you later,' W2F 'way too funny,' TY 'thank you,' and chillax 'chill and relax.'

Twitter is a relatively new source that has built a strong brand name and is used for a variety of purposes. For example, Twitter is used as a way to teach students to write concisely. Expressing an idea in fewer words is often more difficult than rambling for pages on end. Professors and teachers also encourage students to tweet questions and share information. On the other hand, it is difficult to understand how writing in tweet-speak contributes to intellectual discourse and good writing skills.

Many professionals use Twitter to stay in contact with associates and clients and for public relations and advertising purposes. For example, a mobile vendor in San Francisco tweets his daily street location. A sushi restaurant owner tweets about his fresh fish. A spa manager tweets about daily massage specials. Twitter is an easy, low-cost, and efficient way to create a digital word-of-mouth professional network.

But Twitter has established itself most notably as a way for ordinary citizens to break big stories such as the dramatic emergency landing of the US Airways flight into the Hudson River and the terrorist attacks in Mumbai, India. Activists used Twitter to mobilize citizens and set up protests during the Arab Spring, a series of anti-government uprisings and protests throughout the Middle East in early 2011.

FYI: Barack Obama on Twitter

On May 18, 2015, then-President Barack Obama surprised the Twitterverse with his first Tweet: "Hello Twitter! It's Barack. Really! Six years in, they're finally giving me my own account." Obama's @POTUS account is notable for being the fastest to achieve 1 million followers—in 4 hours and 52 minutes.

Source: Crunched, 2015

VINE (2012)

Twitter owned Vine is a social media site built around looping video clips of ordinary life. It is basically a collection of moments linked together in a 6-second video that is shared among followers. Immensely popular among Millennials (71% use), Vine videos show others what is going on at events around the world.

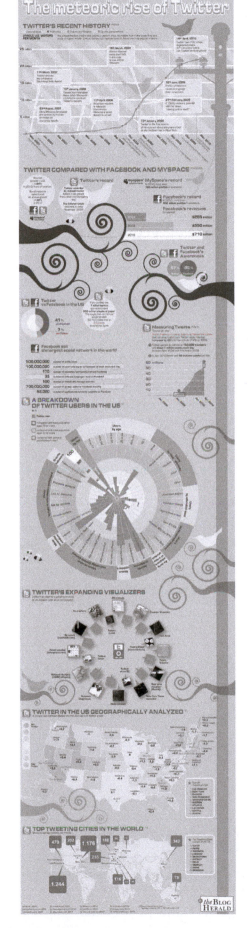

FIG. 9.6 Rise of Twitter
Photo courtesy of Splashpress Media

Vine videos are posted for social reasons as well as professional ones, such as clips of musical performances, stand-up comedy, stage plays, and athletic events. Even companies like Dunkin' Donuts 'vine' television commercials. (In October 2016, Twitter announced plans to shut down Vine sometime in the next few months, giving users a chance to move their videos to other platforms.)

PINTEREST (2009)

Many proud parents keep a pinboard in the kitchen or family room on which they display vacation photos and their children's hand-drawn artwork or leave notes for family members. Pinterest is basically the online version of a pinboard that is used for sharing 'interesting' images, known as 'pins'; hence the name Pinterest. And it is also a social medium but one that is centered on photos and graphics.

Just like pinning a montage of photos on the kitchen pinboard, Pinterest users create their own boards that group pins of similar topics. For example, a user could have a vacation board, a recipe board, or a movie board. But Pinterest is not just a static pinboard. It is a social medium because users can share someone else's pin—known as 'repinning.' 'Pins' can also be shared on Facebook and Twitter, and like these social media, Pinterest users add friends to their accounts and everyone follows each other, and a user's main page displays a 'pin feed' that chronologically lists friends' Pinterest activities.

In the first year or so of its existence, Pinterest was largely the domain of female users. Males, however, have started gravitating to the site but still account for only about 30% of its audience. Pinterest's typical user is female, between the ages of 18 and 49, and has attended some college or earned a 4-year degree.

Several theories have been offered to help explain why females are attracted to Pinterest. One explanation is that while men enjoy hunting and seeking, females enjoy collecting, and Pinterest is the perfect online venue for gathering images of one's life. Pinterest is more than a platform for storing visuals; it is also a 'self-expression engine,' on which users create montages of images that reflect a mood or vision. The site's platform is ripe for creating pinboards that reflect female interests. Well-known brands, such as Kate Spade and Martha Stewart Living, are pinning all over the site, thus keeping females tuned into the latest trends. Moreover, 81% of females trust recommendations they read on Pinterest compared to Twitter (73%) and Facebook (67%).

LINKEDIN (2003)

The digital age has changed the way people look for jobs. The old ways of reading newspaper employment listings and knocking on doors are long gone. Simply finding an appealing job description and submitting a resume is not enough to land employment—now companies are looking for personalized approaches and recommendations from networks, and LinkedIn is the best way to make business connections. LinkedIn brings to life the adage "It's not what you know but who you know."

LinkedIn is the largest online professional profiling platform. It boasts 414 million registrants across 200 countries and territories. It is designed for *posting* professional resumes, *connecting* with industry leaders, and *endorsing* others to enhance users' *profiles*. LinkedIn is an effective way for those early in their careers to introduce themselves to industry leaders. About 28% U.S. adult Internet users have a profile on LinkedIn. The typical LinkedIn user completed some college or has a college degree, is between the ages of 30 and 64, and makes more than $50,000 per year.

Once reserved only for those who had a foot in the door, LinkedIn aspires to give everyone an equal chance of being seen and heard and showing what they do best. Users' profiles are enhanced when others in their network 'endorse' them for their skills and knowledge. Most important—about 93% of professional recruiters access LinkedIn for potential employees.

> ### ZOOM IN 9.1
>
> Learn more about LinkedIn and see the infographic of World Wide Web membership at https://press.linkedin.com/about-linkedin

FLICKR (2004)

Flickr is an image-storing and -hosting site that is used for sharing photos and video among an online community of users. By sheer volume, Flickr's 1.4 million photo uploads pale in comparison to Facebook's 350 million. But Yahoo!-owned Flickr is not so much about the act of sharing a photo as about the photo itself. On Facebook, once a photo slips down the news feed scroll, it is largely forgotten, whereas Flickr showcases photos on "endlessly scrolling pages" (Roush, 2013), and photos can easily be sorted and categorized into albums or 'sets.' One of the coolest features is that geotagged photos are displayed on a world map.

Flickr's photo editing tools attract a community of professional and serious amateur photographers and those who could use a terabyte of free storage. Like on any other social media site, Flickr users form social networks and comment on each other's photos and videos.

OTHER SOCIAL MEDIA

The list of 10 social networking sites highlighted here does not include Reddit or SnapChat, which are not always classified as a social networking site per se. Regardless, they are both very influential and popular social media that deserve recognition in this chapter.

Reddit (2005)

Reddit is yet another student success story. Thought up by two University of Virginia roommates in 2005, Reddit's success comes both from its content and from its fun-to-use, interactive interface. Reddit is a combination news and entertainment Web site, a social networking platform, and online bulletin board system. Registered users, known as 'redditors,' post links and content on the board. Community members then vote each submission 'up' if they like it or find it interesting or 'down' if they are not in favor of it. The goal is to send the best stories or submissions to the top of the main page. The homepage lists the 'hot topics,' but other tabs show 'new,' 'rising,' 'controversial,' 'top,' 'gilded,' 'wiki,' and 'promoted' content lists. Thousands of subreddits, which are topic-specific subcommunities, such as science, gaming, music, fitness, and gadgets and technology, make the site search friendly. The site gets its social media flair from users' comments. In 2015, users made 525 million comments, which were up-voted or down-voted more than 2 billion times.

The name 'Reddit' is a play on the words 'read it,' like "I 'read it' on the Internet yesterday." Reddit's spiffy name plus the way the intuitive way it is organized give it enough online cachet to attract about 172 million unique monthly users from 215 countries. About 6% of online adults in the U.S. are regular users, and the largest user segment is 18- to 29-year-old males.

FIG. 9.7 Screen shot of Reddit

ZOOM IN 9.2

Visit this Web site and click on the short 'about Reddit' video: www.theatlantic.com/technology/archive/2013/09/what-is-reddit/279579/

SnapChat (2011)

SnapChat is a photo-messaging app and mobile-device sharing service, but the photos disappear a few seconds after they are posted. SnapChat is a welcome antidote to a cyberworld where every movement, activity, and emotion is photographed, video recorded, or written and forever archived. SnapChat is a here-and-now app that builds social bonds through photos and messages, but users do not have to worry that an unflattering picture or a silly post will come back to haunt them years later. SnapChat's impermanence is its strongest draw.

SnapChat is for sharing a special moment among friends. Once a photo is shared, the receiver has a set number of seconds to view it before it disappears. If someone tries to defeat the purpose of SnapChat by taking a screen shot of the photo, the sender is informed. Senders can personalize a photo by drawing on it or adding a caption. SnapChat is especially attractive to Millennials, who account for 70% of its users. SnapChat users send out about 9,000 snaps per second.

Blogs

Blogs are not social network sites per se. Early diary-style blogs such as LiveJournal were the precursor to today's social network sites, but they evolved in a different direction and became more issue oriented and less personal. Blogs stand out from Facebook, Twitter, and other social network sites because conversation occurs between the blogger and fellow blog users who might be friends, acquaintances, or strangers, but users do not create networks or circles of 'friends.' Blogs are both conversational and informational but are not used primarily for social purposes. Although blog users express their opinions, on the major blogs of today, there is very little exchange of photos or conversation about personal matters as on Facebook. Instead, most blogs tackle political and social issues. Blogs, then, are a social medium in that conversation occurs, but they are not used primarily for social networking but for the exchange of ideas and opinions.

Blogging has exploded on the cyberscene. A new blogger jumps on the bandwagon every 40 seconds. Although estimates vary widely on the number of blogs (depending on how 'blog' is defined), there were between 20,000 and 30,000 in the late 1990s, between 0.5 million and almost 3 million by the end of 2002, and 70 million blogs worldwide in 2007. By 2009, an astounding 133 million blogs inhabited cyberspace, and by 2015, that number had reached about 250 million.

Today's **blogosphere** (the collective world of blogs) is inhabited mostly by White, highly educated, high-income, conservative and libertarian males. But the blogosphere is becoming more diverse as new bloggers tackle multicultural and feminist issues and meet the needs of minority users.

The basic or general-topic blog is an open forum, in which anyone can read and participate in discussions about a myriad of topics. Some general-information blogs tackle many issues and topics, whereas others focus on specific topics such as dog breeding or gardening.

Media/journalism blogs encompass those that post news stories and opinions and often focus on issues concerning journalism and the media. These bloggers are usually, but not always, journalists, and the blogs are often hosted by media organizations. A warblog centers on terrorism, war, and conflict, often with a promilitary stance. A military blog (milblog) is a blog written by members or veterans of any branch of the U.S. armed services, posting directly from the front lines. A political blog primarily comments on politics and often takes a clearly stated political bias. A corporate blog is published and used by an organization to reach its organizational goals. These examples are just some of the major types of blogs. There is a blog out there for everyone, though social networks such as Facebook seem to have taken over blogs as places to share personal and social information.

FYI: Millennials and Media Use (Hours per Day)

Browsing the Internet	3:34
Social Networking	**3:12**
Watch live TV	2:19
Watch time-shifted TV	1:47
Play videogames	1:47
Listen to the radio	1:15
Use email, text, texting apps	1:04
Read print newspapers and magazines	0:32

Source: Statista, 2014

SOCIAL AND CULTURAL BENEFITS AND CONSEQUENCES OF SOCIAL MEDIA

Social media have drastically changed the way we live, love, and work. Whether for good or for ill, social media are here to stay for a very long time. Eventually, some other new social technology might replace the social networking of today, but even so, we will never go back to the old ways of making friends and maintaining relationships. Social media have even altered the way we do our jobs and our bosses' expectations, and it has affected the way we learn about current events. This chapter will next explore several key ways that social media have altered our reality.

FYI: Importance of a Mobile Device

- Just over one-half (53%) of Millennials would rather lose their sense of smell than their favorite electronic communication device.
- 40% of Millennials say that giving up their cell phone would be a bigger hardship than giving up their car.
- Adolescents would rather lose a pinky finger than their cell phone.

Sources: Dickinson, 2013; Elmore, 2014; Our Social Times

NEWS AND INFORMATION

Social media have ushered in the age of immediate information. News from across the globe shows up on our mobile feeds in an instant, 6-second Vine videos grace our screens a few seconds after they are recorded, and the latest rumors are ours to judge and spread long before they are verified. Social media has created the need to know, now. Waiting until the next day to read a printed edition or even a couple of hours for a television newscast is passé; after all, by then, something more interesting might have happened.

The intersection of social media and news is the number-one online place for keeping up with the world. Twitter users get 36% of news from family and friends and 27% from journalists. Today, Facebook users receive 70% of news links from family and friends and only 13% from news organizations and journalists. This trend is driven by immediate access to the Internet and plethora of sources. One news story or one perspective is no longer satisfying; news-hungry consumers rely on many sources, both professional and amateur. Social media have become our eyes on the world.

Citizen Journalism

Social media empower the ordinary person on the street to act as an observer and as a reporter. A mobile device is a tool that captures an event, and a social medium is the tool that spreads the word of the event. Anyone with both tools can become a roving reporter. Amateur Vine videos were among the first to show the aftermath of the 2013 Boston Marathon bombings and the protests in Ferguson, Missouri, in 2014.

While there have been a countless number of false bits of information circulating in cyberspace, at other times, online news scooped the media. For example, Keith Urbahn, chief of staff to former Defense Secretary Donald H. Rumsfeld, tweeted the news of Osama bin Laden's killing about 20 minutes before the news media made the official announcement. While the news media were verifying the account and putting their broadcasts together, Urbahn tweeted, "So I'm told by a reputable person, they have killed Osama bin Laden. Hot damn." That tweet became the first credible report of what happened on May 2, 2011, in Abbottabad, Pakistan.

Blogs were probably the first 'social medium' on which the ordinary person could become a reporter or a columnist. Since the first blogs appeared in the late 1990s, they have evolved from personal diaries meant for a few friends and family members to formidable agents of public opinion. Blogging redistributes traditional media's power "into the hands of many" (Reynolds, 2002). Bloggers often scoop the traditional media and bring stories into the limelight that the media might otherwise have overlooked or buried. One of blogging's first coups came in 2002 when then-Senator Trent Lott glorified Senator Strom Thurmond's 1948 desegregationist campaign. The story was printed on page 6 of the *Washington Post* and omitted entirely by the *New York Times*. Only when bloggers homed in on the remark did the mainstream press run with the story. But it was pressure from the bloggers that eventually led to Lott's resignation.

The media consider themselves the watchdogs of the government, and now blogs have taken on the role of watchdogs of both the government and the media. Like vultures, bloggers hover over media reports, often criticizing and commenting on news stories and looking for errors even before the printed versions have hit the newsstands or the electronic versions have zipped through the airwaves. Bloggers immediately swoop into action when they read or hear of a controversial report or a story they deem biased or factually incorrect. As one blogger commented, "This is the Internet and we can fact check your ass" (Reynolds, 2002). Bloggers are an "endless parade of experts . . . with Internet-style megaphones ready to pounce on errors." Conversely, traditional journalists claim that bloggers are really just "wannabe amateurs badly in need of some skills and some editors" (Rosenberg, 2002).

Some media executives and journalists concerned that blogs spread misinformation and blur fact and opinion have set up their own blogs as a counterpoint. Almost all major news sites contain several blogs that are hosted by their own journalists and pundits. But because media-hosted blog content is edited and fact-checked and is sometimes merely expanded print or on-air stories that may reflect the views of the organization, blog purists do not consider them true blogs. Blog purists claim that real blogs are those that provide a space for uninhibited public deliberation and an open marketplace of ideas where everyone's views are considered seriously. The downside to independent blogs is that they are not always subject to the strict editorial standards and source- and fact-checking that characterize the traditional media blogs. A good deal of the information found on independent blogs is, in fact, not carefully scrutinized and might often be in error.

Perceptions of blog credibility vary. Some contend that information found on non–media-hosted blogs is not as credible as that found on media blogs or those posted by credentialed journalists, which are usually fact-checked. Others claim that blogs are credible precisely because the online world scrutinizes the information found there. Even mainstream journalists access blogs to look for tips and story ideas. Many online users consider blogs more credible than the traditional media, which they deem biased, either to the left or to the right.

Political Efficacy and Grassroots Advocacy

When blogs, and later Facebook, Twitter, and other social media, first started appearing in cyberspace, they were hailed as catalysts for democratic discussion and political understanding and unity. It was hoped that these sites would not only bring people together socially but would also strengthened political self-efficacy, the confidence to bring about governmental change. It does seem that 'get out the vote' messages are effective on social media, especially on Facebook. Those 'I voted' messages strongly influence others to go to the polls. When Facebook users see their friends voting or engaging in some other political activity, they are likely to do so as well, a phenomenon known as the 'social contagion effect.'

Facebook, Twitter, and other social media are credited with fueling revolutions and protests around the globe. From the Arab Spring uprisings to the occupation of Cairo's Tahrir Square that toppled President Hosni Mubarak

to Hong Kong's Umbrella Revolution to the Occupy Wall Street sit-ins in the U.S., social media amplified the voices of the discontented and empowered ordinary citizens to initiate change.

But despite the marches and demonstrations, the online push for democratic deliberation, and worldwide attention, once the initial exhilaration wore off, protestors gradually returned to their everyday lives with little social or political change. Although there were some successes, social media have a way to go before becoming true instruments of social and cultural change. Rumors, lack of deep and thoughtful conversation, a mob mentality, 'echo chambers' in which discussion takes place only among like-minded individuals, and a general atmosphere of rudeness hinder the power of social media.

When Traditional Media Rely on Social Media

When Andy Carvin was a social media strategist for NPR, he was tweeting hundreds of messages per day chronicling the latest news from items picked up on the Internet and turning himself into a one-man wire service. This type of news tweeting is still evolving and looking for its newsroom niche. And people who are doing it are looking for a job title. Maybe they could be called a "tweet curator," "social media news aggregator," or "interactive digital journalist" (Farhi, 2011a).

But Twitter's successes mask factors that keep it from emerging as a credible source of news. First, it is still used by a minority of the population; 23% have used it, but only about 10% do so in a typical day. Although Millennials account for the largest group of Twitter users, they are not big consumers of news—only about 4 in 10 access news every day. Further, only about 2 in 10 Tweets can be considered news; the rest are celebrity or personal in nature, with television, film, sports, and music as the most common topics. Also, the 140-character count does not allow for in-depth exploration of hard-news issues. Anyone can tweet anything, true or false, so it is not surprising that Twitter news is regarded as fairly low in credibility.

Opinions about Twitter run the gamut from idiotic to its being a historical 'godsend.' When the Library of Congress announced in May 2010 that it had acquired the Twitter archive, historians were elated. The spontaneous nature of tweets contributes to their value as an authentic cultural record. But try convincing those who think tweets are 'culturally vacant.' The 'snarkosphere' lit up with complaints and cynical comments about why anyone would want an archive of tweets (Hesse, 2010, p. 1).

News and Information Consumption

In his book *The Shallows: What the Internet Is Doing to Our Brain*, author Nicholas Carr contends that the Internet is a system of interruption that constantly pings users with morsels of information. The flashing of sights and sounds stimulates our brains like a drug: the more we get the more we need. Online hyperactivity has changed the way we process information. It has diluted our ability to read deeply and concentrate for long periods of time. We are in a constant state of distraction: looking here, looking there;

reading this sentence, reading that sentence; watching this part of this video, watching that part of that video; writing this text, writing that text; viewing this news clip, viewing that news clip. Although we believe that hyper information intake is making us smarter and better informed, it actually has the opposite effect. We know less, we understand less, and we think less.

Social media have added to the information frenzy. Social media information is delivered in rapid fragments rather than in thought-out pieces. As a result, we know less about more rather than more about less. But many believe that is an acceptable trade-off. In his book *Smarter Than You Think*, author Clive Thompson claims that because the Internet is a collection of specks of information about a wide variety of subjects, it is making us smarter and better informed.

Social media are also blamed for dumbing down the news and precipitating a decline in journalistic substance. After all, a 140-character story is not in-depth news. And much of what is out there is unverified and written by anonymous sources. In sorting through the changes brought on by social media, scientists have discovered that Millennials consume less news and information than Baby Boomers, and they are more interested in entertainment and cultural news than 'hard' news or international events. Further, half as many Millennials as Boomers keep up with problems facing the country. In just a few generations, our brains have changed, and so has the way we learn and consume information.

RELATIONSHIPS

Many sociologists, psychologists, and other experts question whether social media have enriched our relationships or diluted them. There is little debate that the ways we socialize and interact with one another has changed dramatically, but the broader consequences and benefits of social media are just beginning to emerge.

> ### FYI: Social Media Trivia
>
> - One in five divorces involves social media.
> - Every second, two new members join LinkedIn.
> - The fastest-growing demographic on Twitter is seniors.
> - 53% of Twitter users recommend products in their tweets.
> - 93% of shopping decisions are influenced by social media recommendations.
>
> *Source: Our Social Times*

Loneliness

The typical Facebook user has 338 friends, and 15% of account holders have more than 500 friends. Not all friends are 'known'; some are strangers and some are friends of friends. In a world of increasing personal isolation, social media friends have in some ways taken the place of in-person contact. The elderly living in small apartments on fixed incomes, Boomers busy with travel,

work and family, and Millennials absorbed with college and careers have little time, money, and opportunity to get together with friends and family. Although it may seem contradictory, the modern ways of online socializing might actually be making us lonelier.

Americans live more solitary lives than ever before. In 1950, less than 10% of households contained only one person, but by 2010, that percentage had jumped to 27%. Although the quality, not the quantity, of relationships determines loneliness, proximity is a major determinant of with whom and how often we socialize.

Loneliness and being alone are not the same— loneliness is a psychological state, being alone a physical one, and sometimes a blissful one. A study conducted in 2010 discovered that 35% of adults over the age of 45 were chronically lonely, as compared to 20% 10 years earlier. Moreover, in 1985, only 10% of Americans said they had no one to confide in, and by 2004, that figure had risen to 25%.

Now here comes Facebook, well timed as a panacea for loneliness. But at question is whether Facebook and other social media alleviate loneliness, or do they heighten it by making users keenly aware of their physical isolation from others?

The relationship between social media and loneliness is complex. On one hand, social media widen social circles, increase feelings of belonging, and keep people in close contact with distant friends and family. On the other hand, they make it easy to be physically isolated without feeling friendless. Social media create 'ambient awareness.' Photos, videos, and messages make people feel physically close though miles away.

Ambient Awareness

'Sharing' is the buzz word of social media, but just how much is 'good sharing' and how much is 'bad sharing'? Confessions of extramarital affairs, binge drinking, failing grades and drug use, revelations of hook-ups, car accidents, health problems, relationship break-ups, whinings about mean bosses, college course assignments, bad hair days and bad days in general, and brags about new jobs, smart pets, marathon runs, expensive cars, highly accomplished children, and the best marriage in the world are the fabric of social media. Viewing social network sites is like being pummeled with annoying holiday letters every day of the year. We roll our eyes and mutter, "oh, brother," yet we cannot tear ourselves away. Social scientists call this type of incessant online contact 'ambient awareness,' because it is so much like being with someone. We have become accustomed to being what sociologist Sherry Turkle calls 'alone together.' We can be in the same classroom, at the same dining table, or in the same living room as friends and family, yet we are each in our own digital world and thus together but not together. We mistake a 140-character conversation for intimacy and Facebook banter as close friendship. It may seem ironic that we use digital technology to forge closeness, yet in the long run it keeps people apart and makes us feel lonely. We feel uncomfortable being alone; thus we are lonely. Further, the constant need for ambient intimacy is a form of narcissism in which people think that every one of their thoughts or actions is of profound interest to someone else. The need to share every little moment in life and to know about everyone else's moments can be a bit too much ambient awareness. Almost 4 in 10 Facebook users dislike it when their friends share too much information, and just as many resent it when their photos and personal information are posted without permission.

Like it or not, ambient awareness can be a satisfying substitute for getting together with someone. As we would in real life listen to a friend sharing his or her latest accomplishments and disappointments, through social media we listen online. In real life, though, we can control the conversation somewhat, but in social media life, we take what we are given.

Social Anxiety

Social networking, whether through social media or texting, is ripe for misinterpretation and misunderstanding. When people speak face-to-face or even on the telephone, body language, facial expressions, and vocal intonations reveal what they are feeling. We know when someone is feeling happy, angry, sad, frustrated, or calm, but emotional cues are missing from social media, often resulting in barren conversation. Matter-of-fact messaging might be fine for business, commercial, or other formal relationships, but it is not so good for interpersonal ones.

Emoticons were created as a way to let others know what we are feeling. Emoticons represent moods and cues recipients on how they should respond. The word 'emoticon' is a cross between 'emotion' and 'icon.' Although emoticons, such as smiley faces, have long been used in casual and humorous writing, the earliest Internet versions

began with an online bulletin board message written by Carnegie Mellon University professor Scott Fahlman:

19-Sep-82   11:44   Scott E Fahlman :-)

From: Scott E Fahlman <Fahlman at Cmu-20c>
I propose that the following character sequence for joke markers:
:-)
Read it sideways. Actually, it is probably more economical to mark things that are NOT jokes, given current trends. For this, use
:-(

Although character sequences are still used to represent emotion, most software automatically replaces them with a corresponding icon. Thus typing :-) automatically becomes ☺.

Emoticons are different from **emoji**, which are characters that represent words. The word 'emoji' is a combination of the Japanese words, 'picture' and 'character.' Although some emoji represent moods and emotions, emoji are standardized icons that are used as shortcuts to writing out a word.

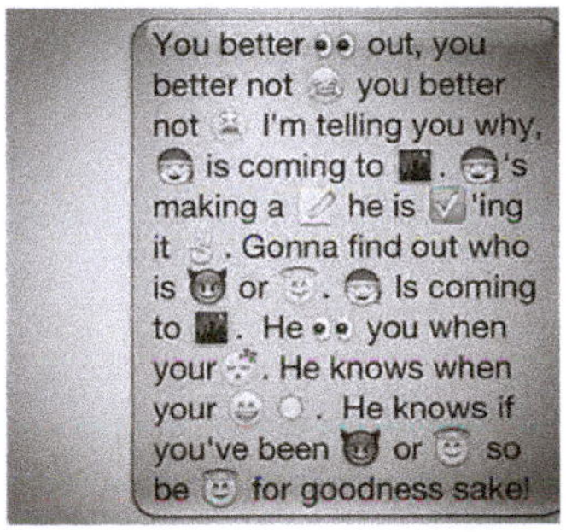

FIG. 9.9 Examples of emojis

Until there is a set of standardized emoji or software capable of translating them as users intend across all electronic devices, users are relying on emoticons to set the mood. Even grammatical punctuation has come to have meaning beyond merely signifying the end of a sentence or a clause. Even a seemingly innocuous text confirming dinner plans can turn into a nightmare if the proper punctuation is not used. For example, a text that reads, "Dinner still on. What time" could be interpreted as meaning that the sender does not want to go because the question marks were left off. But if more than one question mark was used, "Dinner still on?? What time??" then it could be thought that the sender is not sure that a dinner date was set in the first place. Inserting an exclamation mark, "Dinner still on! What time??" makes the sender seem excited, but two exclamation marks, "Dinner still on!! What time??" could make the sender seem overeager. In the dating scene, the inclusion or exclusion of a punctuation mark could make or break a relationship. Ending a text sentence with a period or using just the letter 'K' instead of 'OK' could signify that someone is angry or exasperated. Texting "I caaaannn't wait for toooonight" might make someone seem clingy. Knowing the new and ever-changing rules of punctuation is a must when texting. Improper or unacceptable punctuation causes much relationship angst and anxiety.

FIG. 9.8 Examples of emoticons
Photo courtesy of Shutterstock

FYI: Social Media and Love

- People who use Twitter daily or very often are more likely than less frequent users to cheat, break up, or divorce.
- 81% think it is acceptable to respond to a text or email during a date as long as there is an explanation.
- 67% of committed couples share online passwords with each other.
- 17% of single people 21 to 34 have used their phones during sex.
- 7% of couples married 2005–2012 met through a social networking site, the same percentage that met offline in school.

Source: Shelasky, 2014

FYI: Email Etiquette

How should you sign your email messages? Best? Fondly? Yours truly? Sincerely? Cheers? Love? XOXO? Kisses? Hugs? Apparently, the sign-off can take on a bigger meaning than you think. For instance, the wrong sign-off to a boyfriend or girlfriend could signify an emotional intensity or coolness. Or a too-casual or too-familiar sign-off could be a show of disrespect.

One survey found that the most frequently used professional signoff is 'sincerely,' followed by a thank you of some kind, followed by no close at all. With personal emails, 'love' is the most popular sign-off.

Source: Best for Last, 2009

WORK

Savvy employers are using social media as a way to tap into 'collective intelligence.' Coming up with new ideas and solutions to problems are no longer confined to brainstorming sessions in windowless conference rooms; the process is now open on collaborative platforms. Employees now share what they are working on, with whom they are meeting, how they are feeling, and what they are accomplishing. Thanks to social media, the work environment is much more transparent and less hierarchical.

Wikipedia, though not a social medium per se, is the ultimate collaborative product. Since its creation in 2001, the online encyclopedia, built entirely from user-generated content, has grown to 17 million articles, 3.5 million written in English. About 1,000 new articles are added each day and read over by 100,000 volunteer editors. Wikipedia has shown employers that collaboration often results in a better product than otherwise would have materialized from one person or from assigning several people rigid tasks.

While social media have in many ways improved the workplace, they have also created more work and have changed expectations. For example, although email is not a 'social medium' in the strict sense of the term, it is the most common way colleagues connect one-on-one.

Just over 6 of 10 workers say that email is 'very important' to doing their job, 54% feel the same way about the Internet, and only 35% think so about their telephones. Many employees feel overburdened and overwhelmed with emails. Business users send and receive an average of 141 emails per day, accounting for about 70% of all email traffic. Email boxes are clogged with tens of thousands of messages. Some employers require that all messages be saved, while others leave it to the employees' discretion.

Slack is one of several new applications that aim to replace interoffice email and foster collaboration. Slack is customizable and easy for departments and companies to maintain. Slack, for instance, defaults to open communication so anyone in the company can see what anyone has written, which could be unsettling to some employees. Whether such applications actually cut back on the relentless number of emails or just increase the barrage of messages remains to be seen.

Email has become the primary cause of workplace stress, and social networking sites are becoming a big distraction as well. Although only 4% of workers say that social networking sites are important to their job, many workers take a break to spend a few minutes checking Facebook or some other social medium. All too often, however, a few minutes turns into 30 minutes and more, leading some employers to block access to social media sites altogether.

Other pitfalls of social media result from who owns the data. Lawsuits are being heard in courtrooms across the country over who owns a Twitter account and thus the messages—the employee who tweets as part of his or her job requirement or the company. At question is whether an employee is required to cede the Twitter account and its followers to the company when he or she finds another job.

Other lawsuits and much furor have popped up concerning employees being fired for inappropriate remarks made on Twitter. Most notably, a senior director of communication at IAC was vilified for a tweet she made to her 170 followers just before boarding a flight from London to South Africa. "Going to Africa. Hope I don't get AIDS. Just kidding. I'm white!" Unbeknownst to her, while she was on her flight, her tweet went viral. Some eleven hours later as she walked off the plane she was met with a maelstrom of thousands of hate messages and calls for her firing, which happened promptly. Even workers at the hotel she had booked threatened to strike if she showed up. Many others have also been fired and publicly attacked for their tweets: a woman who dressed as a Boston Marathon bombing victim; a California Pizza Kitchen employee who complained about the new uniforms; a waiter who told about getting stiffed by actress Jane Adams; a guy at a conference whose stupid joke to a friend was overheard by a woman sitting nearby who took a photo of him and tweeted his comment. Getting fired does not seem like the worst part of these stories—the horrible, stinging comments, death threats, and public ostracism sent some victims into hiding and emotional depression.

PRIVACY

Once any bit of information—whether true or false, trivial or important—goes online, it will stay there, forever. Drunk dialing is a dumb thing to do, but now drunk posting is the ultimate in stupidity. As the saying goes, "if you don't want the world to know, don't put it on the Internet," and hope that your friends do not do so either. Many Baby Boomers rambling around in their parents' attic or garage have stumbled across their dusty old diary. Reading the entries in private is embarrassing enough, but the thought of such information being online can make even the thick skinned cringe. Boomers were probably the last generation to not have their formative years plastered all over cyberspace.

Online users are very concerned about the surveillance of their digital communications, especially about the invasion of privacy and the ability to retain and control confidentiality. Privacy is a hot button that has different meanings to different people. Some online users are fearful of government surveillance; others are more worried about commercial spying. Some say privacy breaches are acceptable as long as they are used to monitor terrorist activity; others claim their civil liberties are being violated when data are collected for any reason.

Apple Computer has taken a stand against breaches of privacy by refusing to unlock an encrypted cell phone belonging to Syed Farook, one of the killers of 14 people in San Bernardino, California, in December 2015. Suspecting that the phone held crucial evidence about the shooting, the FBI attempted to unlock the phone but did not know the password. The FBI asked Apple to disable the feature that erases data on the phone after entering an incorrect password 10 times. Apple refused to do so, stating that it would set a dangerous precedent if asked to decrypt phones at the request of law enforcement or anyone. Apple maintains that only the person who owns the phone or who knows the password should be able to unlock it. If phones can be broken into, then no one is ensured privacy, even of personal conversation or information. The Justice Department stepped in and ordered Apple to create a way into the seized phone, but Apple still refused, stating that creating such software for one phone would make all phones vulnerable and violate customers' civil liberties. Apple challenged the order, and a federal magistrate judge ruled for the company and stated that the government had overstepped its authority. While there have been at least 70 instances in which the government has compelled Apple to break into a phone to help gather evidence of criminal activity, this is the first time that Apple has fought back. The Justice Department has appealed the latest ruling, and so the dispute continues. Dozens of technology companies have presented a united front against the government by filing legal briefs in support of Apple's position. As of spring 2016, Apple remains steadfast in its resolve to protect its customers' privacy, even if it means going to Congress or the Supreme Court.

Trying to control personal information is like trying to swim against a tsunami. It is impossible to control the wave; online users are swept in whatever direction privacy policies take them. Nine of 10 online adults say they have no control over how personal information is collected or used. While almost 6 of 10 online users are willing to share personal information with companies in return for free services, about the same number would like the government to do more to regulate the types of information advertisers can collect.

Users feel especially vulnerable on social media sites; only 15% of online adults believe social media are secure. Texting and emailing are deemed more secure than social media, but only about 40% of online users trust that personal information will not be released. While some users like to track their online reputation and profile, only about 60% have ever 'googled' themselves, and 8 of 10 acknowledge that it would be very difficult to remove or correct false information.

ZOOM IN 9.3

Privacy Settings for Social Media Platforms

Facebook: www.facebook.com/help/193677450678703

Twitter: https://support.twitter.com/articles/14016-about-public-and-protected-tweets#

Instagram: https://help.instagram.com/116024195217477/

Pinterest: https://help.pinterest.com/en/articles/change-your-privacy-settings

YouTube: https://support.google.com/youtube/answer/157177?hl=en

LinkedIn: https://help.linkedin.com/app/answers/detail/a_id/66/~/managing-account-settings

FYI: Social Media Best Practices

- Start with a clear plan. Do not send messages without knowing exactly why, when, and where they are going.
- Set goals. Know in advance what you want to achieve.
- Positively promote your cause or organization.
- Put your networking power into play with a smile.
- Create a user experience and dialogue. Listen and respond to your audience.
- Build relationships and a community.
- Be open and honest.
- Be respectful, professional, and nonargumentative.
- Post timely and up-to-date information.
- Prepare to lose control of your message. Others will edit, interpret, change, and repost it.
- Think before you act. An inappropriate or emotional outburst cannot be undone.
- Manage your reputation.

SEE IT LATER

The proliferation of social media has outpaced even the most generous projections. Some thought it was a fad. Facebook's beginning as a site for collegiate socializers was not the best predictor of what was to come. Today, how we communicate has changed. In fact, most scholars agree that this paradigm shift is cemented as 'normal.' We are hard pressed to find a situation in today's media world that is not affected by the proliferation of social media.

SOCIAL MEDIA AND SOCIETY

Companies, schools, civic groups, and individuals must take into account social media ethics, posting protocol, standards, and best practices. The legal landscape is changing rapidly. The push to 'share' has created a wave of litigation that will take years to sort out and settle. Slowly, we are beginning to understand that our social media profiles never go away.

Companies are now using social media to hire and fire employees. A traditional resume has turned into a *Klout* score and a profile search. Klout is a Web site and app that uses analytics to compute an individual's popularity on social media. The Klout score algorithm measures a person's online influence on a scale of 1 to 100. The Klout formula takes into account several variables such as frequency of updates and the number of followers, retweets, likes, and shares. Individuals with high Klout scores are eligible for 'Klout Perks,' free goodies from advertisers hoping to have their products touted by online movers and shakers. In today's competitive media and employment environment, a high Klout score (average = 40, top 5%= 63+) could secure a promotion or a job.

Individuals invest in social media monitoring and management tools as well. Whether one is a chief marketing officer managing a global brand or an editor in a local newsroom, properly managing and curating a social profile is essential. Growing up as digital natives, the Millennial demographic does not know a world without social media. However, even the most prolific posters, pinners, and tweeters need to know what they are doing and how it will stay with them forever.

As social media change the way we think and consume information, the news media and educators will need to adapt to the next generation's learning style. Established professionals might balk at what they consider the dumbing down of the nation's intellect. But perhaps the swing between extremes will balance somewhere in the middle and social media will foster in-depth conversation and bolster knowledge of ourselves and of our world.

FYI: Attention Span

The average attention span for humans was 12 seconds in 2000 but had dropped to 7 seconds by 2015. For a goldfish, it is 9 seconds.

Source: Attention Span Statistics, 2015

Blogs are also places to share information and opinion. With so much information available over blogs, it is understandable that people may prefer connecting to a source on which they are encouraged to talk to rather than wait for television or radio to talk at them. If bloggers continue to scoop traditional media, provide in-depth commentary and diverse viewpoints, and point out media errors, they may become even more influential in shaping cultural ideology, especially among blog readers who dislike or distrust traditional media.

Although some social media users complain about the invasion of privacy, they are still willing to share their lives with the world. Social media has brought on the belief that the need to know trumps the need for privacy, and this collective cry for transparency is unlikely to change in the near future. Ambient awareness is likely to intensify as new online tools make it easier to keep in touch and know what others are doing 24/7. Mobile devices and social media have given us a new sense of security we are unlikely to cede to independence.

A record breaking 200 million new social media sign-ups occurred in 2015, and it is expected that by early 2017, there will be more than 2 billion social media users. Mashable's prediction for the digital world of 2020 includes virtual avatars on screen and in holographic projections, the online crowd will increasingly connect through the cloud, social and mobile media will continue their explosive growth while other media decline, 25 to 50 billion devices will contribute to the avalanche of data collection, and data mining and new algorithms will provide unprecedented insight and forecasts.

SUMMARY

For most people, deleting their Facebook or Instagram account would be like banishing themselves to a desert island. For as much as we hate social media, we love them just as much. This chapter covered the rise of social media, which began in the late 1970s when the Internet was limited to text and traversed by using complicated commands. Early online socializing meant sending an email, reading a bulletin board, or joining a MUD. Social media as known today came about after the emergence of the Web and hypertext connections. The earliest social network sites, *Six Degrees* and *Friendster*, were slow to catch on and never really had much of an online impact. *MySpace* is credited for starting the social media movement. Teenagers and young adults went wild over the site. Media coverage, both positive and negative, fed the frenzy. But several high-profile incidents involving bullying, kidnapping, and other crimes turned users from *MySpace* to the newer and seemingly safer *Facebook*, which is currently the most popular social network site in the world.

Social media have changed the way news and information are delivered and consumed. Journalists and ordinary citizens share snippets of events with their friends and followers, who share the news items within their circles. The news cycle is now 24/7 with not day-to-day or hour-to hour updates but second-to-second news flashes.

The word of an event goes out even before it is verified. Facts are checked after the news breaks, not before.

Critics complain that in-depth professional coverage has been taken over by shallow amateur accounts, and thanks to social media, Millennials are less knowledgeable than previous generations about national and world affairs and turn their attention instead to frivolous celebrity gossip.

This chapter examined privacy issues related to social media. Although about 85% of online users believe their personal information is at high risk, they are willing to create a personal profile, post photos, and reveal intimate details of their lives. Peer pressure, work requirements, and Klout scores compel even the most reluctant to jump into the social media world. Further, social media users are willing to trade personal autonomy and independence for the feeling of security, to feel like they belong, and for the ability to peek in on their friends.

Social media have also changed the workplace. Employees are expected to have social media profiles on which they promote their employers' products and services. Yet employers are hypervigilant about what is posted about them and are merciless when it comes to an employee who inadvertently crosses the line with an unacceptable or offensive post.

This chapter wraps up with a listing of social media best practices. Protocols, ethics, and standards are being developed to maximize outcomes and keep social media users out of trouble. The new world of social media is complex. It is enlightening yet uninformative, serious yet frivolous, and egalitarian yet repressive. Despite the contradictions, social media are omnipresent and here to stay.

BIBLIOGRAPHY

1 in 10 adults has microblogged. (2009, February 19). *Marketing Vox*. Retrieved from: www.marketingvox.com/1-10-adults-has-microblogged-on-twitte-or-else where-043248 [July 8, 2009]

9 bold predictions for the digital world of 2020. (2012, April 12). *Mashable*. Retrieved from: http://mashable.com/2012/04/04/predictions-digital-future/

24 Twitter acronyms and abbreviations. (2015). *AllAcronyms.com*. Retrieved from: www.allacronyms.com/twitter/topic

About Reddit. (2015). *Reddit*. Retrieved from: www.reddit.com/about/

Adams, D. (2011). The history of social media. *InstantShift.com*. Retrieved from: www.instantshift.com/2011/10/20/the-history-of-social-media/

Afshar, V. (2013, March 24). Five ways social media has forever changed the way we work. *Huffington Post*. Retrieved from: www.huffingtonpost.com/vala-afshar/social-media_b_2944407.html

Agarwal, A. (2015, May 27). 5 predictions for the future of social media. *INC*. Retrieved from: www.inc.com/aj-agrawal/5-predictions-of-the-future-of-social-media.html

Amis, D. (2002, September 21). Weblogs: Online navel gazing? *NetFreedom*. Retrieved from: www.netfreedom.org

Arrington, M. (2009, January 22). Facebook now nearly twice the size of My Space worldwide. *TechCrunch*. Retrieved from: http://techcrunch.com/2009/01/22/facebook-now-nearly-twice-the-size-of-myspace-worldwide/

Attention span statistics. (2015). *Statistics Brain Research Institute*. Retrieved from: www.statisticbrain.com/attention-span-statistics/

Baer, J. (2013). 11 shocking new social media statistics in America. *Convince & Convert*. Retrieved from: www.convinceandconvert.com/social-media-research/11-shocking-new-social-media-statistics-in-America

Bartlett, J. (2015, June 7). Why VK is beating Facebook in Russia: It lets you search for pirated movies—and sex. *The Telegraph*. Retrieved from: http://blogs.telegraph.co.uk/technology/jamiebartlett/100011938/why-vk-is-beating-facebook-in-russia-it-lets-you-search-for-pirated-movies-and-sex/

Bates, D. (2012, July 31). You've got (more) mail: The average worker now spends over a quarter of their day dealing with email. *DailyMail.com*. Retrieved from: www.dailymail.co.uk/sciencetech/article-2181680/Youve-got-mail-The-average-office-worker-spend-half-hours-writing-emails.html

Benner, K., & Goldstein, J. (2016, February 29). Apple wins ruling in New York iPhone hacking order. *The New York Times*. Retrieved from: www.nytimes.com/2016/03/01/technology/apple-wins-ruling-in-new-york-iphone-hacking-order.html

Benner, K., Lichtblau, E., & Wingfield, N. (2016, February 26). Apple goes to court, and the F.B.I. presses Congress to settle iPhone privacy fight. *The New York Times*, pp. B1, B7.

Bertrand, N. (2015, May 6). One of the world's smallest countries just banned Facebook. *Business Insider*. Retrieved from: www.businessinsider.com/one-of-the-worlds-smallest-countries-just-banned-Facebook

Best for last? (2009, August 3). *The New York Times*, pp. C1, C8.

Biggs, J. (2011, December 25). A dispute over who owns a Twitter account goes to court. *The New York Times*, pp. B1, B2.

Bosker, B. (2012, March 10). Vic Gundotra explains what Google+ Is (but not why to use it). *Huffington Post*. Retrieved from: www.huffingtonpost.com/2012/03/10/vic-gundotra-google-plus_n_1336601.html

boyd, d. m., & Ellison, N. B. (2007). Social network sites: Definition, history, and scholarship. *Journal of Computer-Mediated Communication*, 13(1), 210–230.

Bradford, K. T. (2010, December 8). What Tumblr is and how to use it: A practical guide. *Laptop*. Retrieved from: http://blog.laptopmag.com/tumblr-tips

Brustein, J. (2014, July 23). Americans now spend more time on Facebook than they do their pets. *Bloomberg Business*, Retrieved from: www.bloomberg.com/bw/articles/2014-07-23/heres-how-much-time-people-spend-on-facebook-daily

Brusilovsky, D. (2009, July 13). Why teens aren't using Twitter: It doesn't feel safe. *The Washington Post*. Retrieved from: www.washingtonpost.com [July 14, 2009]

By the numbers: 25 amazing Vine statistics. (2016, February 26). *DMR*. Retrieved from: http://expandedramblings.com/index.php/vine-statistics/

Carr, N. (2011). *The shallows: What the internet is doing to our Brains*. London: W.W. Norton & Co.

Company info. (2015, March 15). *Facebook*. Retrieved from: https://newsroom.fb.com/company-info

Constine, J. (2016, January 5). Twitter may increase tweets to 10,000 characters, but hide all past 140. *TechCrunch*. Retrieved from: http://techcrunch.com/2016/01/05/information-density/

Costill, A. (2014, January 16). 30 things you absolutely need to know about Instagram. *Search Engine Journal*. Retrieved from: www.pewinternet.org/2015/01/09/demographics-of-key-social-networking-platforms-2/

Corcoran, M. (2009, July 12). Death by cliff plunge, with a push from Twitter. *The New York Times*. Retrieved from: www.nytimes.com/2009/07/12/fashion/12hoax.html [July 29, 2009]

Coyle, J. (2009, July 1). Is Twitter the news outlet for the 21st century? *The Washington Post*. Retrieved from: washingtonpost.com/wp-dyn/content/article/2009/07/01 [July 14, 2009]

Crunched. (2015, August 23). *The Washington Post Magazine*, p. 8.

Dickinson, B. (2013, November 26). What's in a name? *Brand Marketing*. Retrieved from: www.paceco.com/whats-in-a-name-defining-the-millennial-generation/

Doyne, S. (2015, March 2). Does punctuation in text messages matter? *The New York Times*. Retrieved from: http://learning.blogs.nytimes.com/2015/03/02/does-punctuation-matter-in-text-messages/

Dugan, M., Ellison, N. B., Lampe, C., Lenhart, A., & Madden, M. (2015, January 9). Demographics of key social networking platforms. *Pew Research Center*. Retrieved from: www.pewinternet.org/2015/01/09/demographics-of-key-social-networking-platforms-2/

Eddy, N. (2010, February 23). Twitter logging 50 million tweets per day. *eWeek.com*. Retrieved from: www.eweek.com/c/a/Midmarket/Twitter-Logging-50-Million-Tweets-Per-Day-234947/

Elmore, T. (2014, March 4). I'd rather lose my Ford than my phone. *Huffington Post*. Retrieved from: www.huffingtonpost.com/tim-elmore/id-rather-lose-my-ford-or-my-finger-than-my-phone_b_4896134.html

Eltantawy, N., & Wiest, J. B. (2011). Social media in the Egyptian Revolution: Reconsidering resource mobilization theory. *International Journal of Communication*, 5, 1207–1224.

Fallows, J. (2011, April). Learning to love the (shallow, divisive, unreliable) new media. *The Atlantic*, p. 34–49.

Falman, S. E. (n.d.). Smiley lore. Carnegie School of Computer Science, Retrieved from: http://www.cs.cmu.edu/~sef/Orig-Smiley.htm

Farhi, P. (2011a, April 13). In a tweet spot. *The Washington Post*, pp. C1, C3.

Farhi, P. (2011b, May 3). Beat by a tweet: How the bin Laden story broke. *The Washington Post*, pp. C1, C7.

Friedman, T. L. (2016, February 3). Social media: Destroyer or creator? *The New York Times*. Retrieved from: www.nytimes.com2016/02/03/opinion/social-media-destroyer-or-creator.html

Gillette, F. (2011, June 21). The rise and inglorious fall of my space. *Bloomberg Business*. Retrieved from: www.bloomberg.com/bw/stories/2011–06–21/the-rise-and-inglorious-fall-of-MySpace

Guadin, S. (2009, July 15). Americans spend most online time on Facebook. *Computer World*. Retrieved from: www.networkworld.com/news/2009/071509-americans-spend-most-online-time.html?hpg15bn

Hamilton, A. (2003, April 7). Best of warblogs. *Time*, p. 91.

Harlow, S., & Johnson, T. J. (2011). Overthrowing the protest paradigm? How the New York Times, Global Voices and Twitter covered the Egyptian Revolution. *International Journal of Communication*, 5(feature), 1359–1374.

Hermida, A. (2010). Twittering the news: The emergence of ambient journalism. *Journalism Practice*, 4(3), 297–308.

Hesse, M. (2010, May 6). Twitter archive at Library of Congress could help redefine history's scope. *The Washington Post*. Retrieved from: www.washingtonpost.com/wp-dyn/content/article/2010/05/05/AR2010050505309.html

Hesse, M. (2011, January 13). Celebrating a decade of the age of Wikipedia. *The Washington Post*, pp. C1, C4.

Heussner, K. M. (2010, June 8). 'The Facebook effect': Inside Zuckerberg's coups, controversies. *ABC news*. Retrieved from: abcnews.go.com/Technology/Media/facebook-effect-inside-zuckerbergs-coups-controversies/story?id=10853306

Hiscock, M. (2014, June 26). Dead Facebook users will soon outnumber the living. *The Loop*. Retrieved from: www.theloop.ca/dead-facebook-users-will-soon-outnumber-the-living/

The history of Instagram. (2015, February 3). *Social Experiment*. Retrieved from: http://socialexperiment.net/uncategorized/the-history-of-instagram/

Hughes, D. J., Rowe, M., Batey, M., & Lee, A. (2012). A tale of two sites: Twitter vs. Facebook and the personality predictors of social media usage. *Computers in Human Behavior*, 28(2), 561–569.

Is social networking changing childhood? (2009). *Common Sense Media*. Retrieved from: www.commonsensemedia.org/teen-social-media [February 10, 2010]

Isaac, M. (2016, March 19). Twitter opts to keep its characters. *The New York Times*, p. B2.

Johnson, T. J., & Kaye, B. K. (2016). *Some like it lots: The influence of interactivity and reliance on credibility. Computers in Human Behavior*, 61, 136–145. doi: 10.1016/j.chb.2016.03.012.

Joinson, A. N. (2008). "Looking at," "looking up" or "keeping up with" people? Motives and uses of Facebook. *Paper presented to the conference on Human Factors in Computing Systems. Proceedings of the 26th Annual SIGCHI Conference*. Florence, Italy. April 5–10.

Kang, C. (2014, August 23). Famous in a flash. *The Washington Post*, pp. A1, A4.

Kaplan, A. M., & Haenlein, M. (2010). Users of the world, unite! The challenges and opportunities of social media. *Business Horizons*, 53(1), 59–68.

Kaye, B. K., & Johnson, T. J. (2015). I only have eyes for YouTube: Motives for political use. *Journal of Social Media Studies*, 1(2), 91–104. doi:10.15340/2147336612841.

Kavulla, K. (2012, January 19). Pinterest: What it is, how you use it, and why you'll be addicted. *SheKnows.com*. Retrieved from: www.sheknows.com/living/articles/852875/pinterest-what-it-is-how-to-use-it-and-why-youll-be-addicted

Kemp, S. (2015, January). Time spent on social media. Social Media Today. Retrieved from: www.socialmediatoday.com/content/global-digital-social-media-stats-2015

Kinzie, S. (2009, June 26). Some professors' jitters over Twitter are easing; Discussions expand in and out. *The Washington Post*, p. B1.

Kopytoff, V. (2009, July 8). Most Facebook users are older, study finds. *SFGate.com*. Retrieved from: www.fgate.com [July 8, 2009]

Korkki, P. (2009). An outlet for creating and socializing. *The New York Times*, p. B2.

Kovach, S. (2011, June 28). Everything you need to know about Google+ (including what the heck it is). *Business Insider*. Retrieved from: www.businessinsider.com/what-is-google-plus-2011–6?op=1

Krashinsky, S. (2012, November 2). How YouTube has transformed the 2012 presidential election. *The Globe and Mail*. Retrieved from: www.theglobeandmail.com/news/world/us-election/how-youtube-has-transformed-the-2012-presidential-election/article4871244/

Kushin, M. J., & Yamamoto, M. (2010). Did social media really matter? College students' use of online media and political decision making in the 2008 election. *Mass Communication & Society*, 13(5), 608–630.

Lampe, C., Ellison, N., & Steinfield, C. (2007). A Face(book) in the crowd: Social searching vs. social browsing. *Paper presented to the proceedings of the 2006 20th Anniversary Conference of the Association for Computing Machinery. Banff, Vancouver, Canada*.

Landry, T. (2014, September 8). How social media has changed us: The good and the bad. *Social Media Today*. Retrieved from: www.socialmediatoday.com/content/how-social-media-has-changed-us-good-and-bad

Lange, P. G. (2007). Publicly private and privately public: Social networking on YouTube. *Journal of Computer-Mediated Communication*, 13(1), 361–380.

Leading social networks worldwide as of January 2016, ranked by active users. (2016, January). *Statista*. Retrieved from: www.statista.com/statistics/272014/global-social-networks-ranked-by-number-of-users/

Lenhart, A. (2009a, January 14). Pew research center. Retrieved from: pewresearch.org/pubs/1079/social-networks-grow [September 28, 2009]

Lenhart, A. (2009b, January 14). Social networks grow: Friending Mom and Dad. *Pew Internet & American Life Project*. Retrieved from: pewresearch.org/pub/1079/social-networks-grow [March 20, 2009]

Lever, R. (2016, January 28). Apple-FBI case has wide implication. *Phys.org*. Retrieved from: http://phys.org/news/2016-02-apple-fbi-case-wide-implications.html

Levy, S. (2002, August 26). Living in the blogosphere. *Newsweek*, pp. 42–45.

Lichtblau, E., & Benner, K. (2016, March 10). Apple and U.S. bitterly turn up volume in iPhone privacy fight. *The New York Times*. Retrieved from: www.nytimes.com/2016/03/11/technology/apple-iphone-fight-justice-department.html

Lohr, S. (2010, April 14). Library of Congress will save tweets. *The New York Times*. Retrieved from: www.nytimes.com/2010/04/15/technology/15twitter.html?_r=0

Madden, M. (2014, November 12). Public perceptions of privacy and security in the post-Snowden era. *Pew Research Center*. Retrieved from: www.pewinternet.org/2014/11/12/public-privacy-perceptions/

Manjoo, F. (2015, March 12). An office messaging app that may finally sink email. *The New York Times*, pp. B1, B6.

Manjoo, F. (2016, February 25). Maintain privacy in an always-watch future. *The New York Times*, p. B1.

Marche, S. (2012, May). Is Facebook making us lonely? *The Atlantic*. Retrieved from: www.theatlantic.com/magazine/archive/2012/05/is-facebook-making-us-lonely/308930/

Markoff, J. (2012, September 13). Social networks can affect voter turnout, study says. *The New York Times*, p. A17.

McAlone, N. (2016, October 28). *Business Insider*, Retrieved from: http://www.businessinsider.com/twitter-shutting-vine-down-2016-10

McCarthy, N. (2014, March 13). Millennials rack up 18 hours of media use per day. *Statista*. Retrieved from: www.statista.com/chart/2002/time-millennials-spend-interacting-with-media/

Millennials get news on Twitter, Facebook. (2014, March 25). *Yahoo! Finance*. Retrieved from: http://finance.yahoo.com/news/facebook-front-page-millennial-news-194700637.html

Miller, C. C. (2009, August 26). Who's driving Twitter's popularity? Not teens. *The New York Times*, pp. B1, B2.

Miller, C. C. (2009, July 23). Marketing small businesses with Twitter. *The New York Times*, p. B6.

Miller, C. C. (2014, February 14). The plus in Google plus? It's mostly for Google. *The New York Times*. Retrieved from: www.nytimes.com/2014/02/15/technology/the-plus-in-google-plus-its-mostly-for-google.html?_r=0

More people own a mobile device than a toothbrush, really? (2013). *OurSocialTimes*, Retrieved from: http://oursocialtimes.com/more-people-own-a-mobile-device-than-a-toothbrush-really/

Moreau, E. (2016, February 13). The top 25 social networking sites people are using. *About Tech*. Retrieved from: http://webtrends.about.com/od/socialnetworkingreviews/tp/Social-Networking-Sites.htm

Musgrove, M. (2009, June 17). Twitter is a player in Iran's drama. *The Washington Post*, p. A10.

Nakashima, E. (2016, February 17). Apple vows to resist FBI demands to crack iPhone linked to San Bernardino attacks. *The Washington Post*. Retrieved from: www.washingtonpost.com/world/national-security/us-wants-apple-to-help-unlock-iphone-used-by-san-bernardino-shooter/2016/02/16/69b903ee-d4d9-11e5-9823-02b905009f99_story.html

Neal, R. W. (2014, January 16). Facebook gets older: Demographic report shows 3 million teens left social networks in three years. *International Business Times*. Retrieved from: www.ibtimes.com/facebook-gets-older-demographic-report-shows-3-million-teens-left-social-network-3-years-1543092

O'Reilly, L. (2015, December, 28). Why Snapchat is 'the one to watch in 2016'—at the expense of Twitter. *Yahoo! Finance*. Retrieved from: http://finance.yahoo.com/news/why-snapchat-one-watch-2016-103905717.html

Original bboard thread in which :-) was proposed. (1982, September 19). Retrieved from: www.cs.cmu.edu/~sef/Orig-Smiley.htm

Percentage of U.S. population with a social network profile 2008 to 2016. (n.d.). *Statista* Retrieved from: www.statista.com/statistics/273476/percentage-of-us-population-with-a-social-network-profile/

Pfanner, E. (2009, June 24). MySpace to cut two-thirds of staff outside of U.S. *The New York Times*.

Phillips, S. (2007, July 25). A brief history of Facebook. *The Guardian*. Retrieved from: www.theguardian.com/technology/2007/jul/25/media.new media

Plambeck, J. (2016, March 3). An industry lines up behind Apple. *The New York Times*. Retrieved from: www.nytimes.com/2016/03/04/technology/an-industry-lines-up-behind-apple.html

Poindexter, P. (2012). Why Millennials aren't into news. In *Millennials, news, and social media: Is news engagement a thing of the past?* (pp. 16–34). New York, NY: Peter Lang Publishing.

Popkin, H. A. S. (2012, December 4). We spend 230,060 years on social media in one month. *CNBC*. Retrieved from: www.cnbc.com/id/100275798

Purcell, K., & Ranie, L. (2014, December 30). Technology's impact on workers. *Pew Research Center*. Retrieved from: www.pewinternet.org/2014/12/30/technologys-impact-on-workers/

Ramirez, J. (2008, November 9). The YouTube election. *Newsweek*. Retrieved from: www.newsweek.com/youtube-election-85069

Reynolds, G. (2002, January 9). A technological reformation. *Tech Central Station*. Retrieved from: www.techcentralstation.com/1051

Ronson, J. (2015, February 12). How one stupid tweet blew up Justine Sacco's life. *The New York Time Magazine*. Retrieved from: www.nytimes.com/2015/02/15/magazine/how-one-stupid-tweet-ruined-justine-saccos-life.html?_r=0

Rosen, R. J. (2013, September 11). What is Reddit? *The Atlantic*. Retrieved from: www.theatlantic.com/technology/archive/2013/09/what-is-reddit/279579/

Rosenberg, S. (2002). Much ado about blogging. *Salon*. Retrieved from: www.salon.com/tech/col/rose/2002/05/10/blogs

Roush, W. (2013, May 31). 11 reasons why Flickr, not Facebook, is the place to put your photos. *Xconomy*. Retrieved from: www.xconomy.com/national/2013/05/31/11-reasons-why-flickr-not-facebook-is-the-place-to-put-your-photos/2/

Rusli, E. M. (2012, April 9). Facebook buys Instagram for $1 billion. *The New York Times*, Retrieve from: http://dealbook.nytimes.com/2012/04/09/facebook-buys-instagram-for-1-billion/

Shapira, I. (2009). In a generation that friends and tweets, they don't. *The Washington Post*, pp. A11, A1.

Sharabi, A. (2007, November 19). Facebook applications trends report. *No Man's Blog*. Retrieved from: no-mans-blog.com/2007/11/19/facebook-applications-trends-report-1/ [January 12, 2009]

Shelasky, A. (2014, September). The love-tech connection. *Self*, p. 60.

Shields, M. (2015, January 14). MySpace still reaches 50 million people each month. *The Wall Street Journal/CMO Today*. Retrieved from: blogs.wsj.com/cmo/2015/01/14/my-space-still-reaches-50-million-each-month/

Smith, A. (2014, February 3). 6 new facts about Facebook. *Pew Research Center*. Retrieved from: www.pewresearch.org/fact-tank/2014/02/03/6-new-facts-about-facebook/

Smith, A., & Brenner, J. (2012). Twitter use 2012. *Pew Research Center*. Retrieved from: www.pewinternet.org/Reports/2012/Twitter-Use-2012.aspx

Smith, C. (2015, May 23). By the numbers: 40+ amazing Reddit statistics. *Digital Marketing*. Retrieved from: http://expandedramblings.com/index.php/reddit-stats/3/

Smith, C. (2016, February 19). By the Numbers: 60 amazing Reddit statistics. *DMR*. Retrieved from: http://expandedramblings.com/index.php/reddit-stats/3/

Smith, C. (2016, March 6). By the Numbers: 70 amazing SnapChat statistics. *DMR*. Retrieved from: http://expandedramblings.com/index.php/snapchat-statistics/

Social media active users. (2016, January 28). *The Social Media Hat*. Retrieved from: www.thesocialmediahat.com/active-users

Social networking eats up 3+ hours per day for the average American user. (2013, January 9). *MarketingCharts.com*. Retrieved from: http://marketingcharts.com/online/social-networking-eats-up-3-hours-per-day-for-the-average-American-user-26049

Social networking fact sheet. (2014). *Pew Research Center*. Retrieved from: www.pewinternet.org/fact-sheets/social-networking-fact-sheet

Social networking statistics. (2015, December 15). *Statistics Brain*. Retrieved from: www.statisticbrain.com/social-networking-statistics/

Stanley, A. (2009, February 28). What are you doing? Media Twitters can't stop typing. *The New York Times*, pp. C1, C6.

State of the media: The social media report. (2012). *Nielsen*. Retrieved from: https://postmediavancouversun.files.wordpress.com/2012/12/nielsen-social-media-report-20122.pdf

Statistics. (2015). You Tube. Retrieved from: www.youtube.com/yt/press/statistics.html

Steinmetz, K. (2014, September 19). In praise of emoticons. *Time*. Retrieved from: http://time.com/3341244/emoticon-birthday/

Stenovec, T. (2011, June 29). My Space history: A timeline of the social network's biggest moments. *Huffington Post*. Retrieved from: www.huffingtonpost.com/2011/06/29/myspace-histroy-timeline_n_887059.html

Sterling, G. (2013, May 21). Pew: 94% of teenagers user Facebook, Have 425 friends, but Twitter and Instagram adoption way up. *Marketing Land*. Retrieved from: http://marketingland.com/pew-the-average-teenager-has-425-4-facebook-friends-44847

Stevenson, S. (2012, April 24). What your Klout score really means. *Wired*. Retrieved from: www.wired.com/2012/04/ff_klout

Sullivan, W. (2009, April 16). Twitter grows 131 percent in one month. *PoynterOnline*. Retrieved from: www.poynter.org [October 26, 2009]

Swartz, J. (2009, October 15). For social networks, it's game on. *USA Today*. Retrieved from: www.usatoday.com/money/media/2009-10-15-games-hit-social-networks_N.htm [January 16, 2009]

Sydell, L. (2009, October 21). Facebook, MySpace dived along social lines. *National Public Radio*. Retrieved from: www.npr.org/templates/story/story.php.?storyID5113974893 [October 21, 2009]

Tate, R. (2012, December 12). Social media is eating our lives (and Pinterest is chewing fastest). *Wired*. Retrieved from: www.wired.com/2012/12/social-spike/

Thompson, C. (2008, September 5). Brave new world of digital intimacy. *The New York Times*. Retrieved from: www.nytimes.com/2008/09/07/magazine/07awareness-t.html?pagewanted=1

Thompson, C. (2013). *Smarter than you think*. New York: Penguin Press.

Teens fact sheet. (2012). *Pew Research Center*. Retrieved from: www.pewinternet.org/fact-sheets/teens-fact-sheet/

Top 15 most popular social networking sites. (2015, June). *eBiz/MBA Guide*. Retrieved from: http://ebizmba.com/articles/socal-networking-web sites

Tsukayama, H. (2016, March 19). As it turns 10, Twitter to stick to 140 characters. *The Washington Post*, p. A16.

Tsukayama, H. (2016, February 24). Facebook officially expands beyond the 'like'. *The Washington Post*. Retrieved from: http:www.washingtonpost.com/news/the switch/wp/2016/02/24/facebook-officially-expands-beyond-the like/

The Twitter landscape. (2013). *Brandwatch*. Retrieved from: www.brandwatch.com/ twitter-landscape/

Twitter usage. Company facts. (2015, March 31). *Twitter*. Retrieved from: https://about.twitter.com/company

Wagner, K. (2014, December 15). Instagram hits 300 million users, now larger that Twitter. *Recode.net*. Retrieved from: http://recode.net/2014/12/10/instagram-hits-300-million-users-now-larger-than-twitter/

World's largest photography libraries. (2013, December). *Smithsonian*, p. 24.

Worthan, J. (2013, February 8). A growing app lets you see it, then you don't. *The New York Times*. Retrieved from: www.nytimes.com/2013/02/09/technology/snapchat-a-growing-app-lets-you-see-it-then-you-dont.html

Yeung, K. (2013, May 5). LinkedIn is 10 years old today: Here's the story of how it changed the way we work. *TNW News*. Retrieved from: http://thenextweb.com/insider/2013/05/05/linkedin-10-years-social-network/

The Business of Entertainment and Media Ownership

10

Contents

Thousands of college students across the United States study electronic media to prepare for careers in radio, television, cable, satellite, and the film industry. Most of them have selected electronic media as a major because they believe that a career in this field will be creative and exciting. Others enter the field in the hope of achieving fame, success, and perhaps wealth. Some students study electronic media simply because they have been entertained by it most of their lives, and still others feel that they can create programming that is at least as entertaining as what is available now.

These are all good reasons to study electronic media and pursue a career in this field. However, many students miss the big picture, because they do not understand that the various fields of electronic media are, first and foremost, *businesses*. Although the programming of media is a creative or artistic endeavor, the driving force behind the industry is business. (Maybe that is why they call it show *business*, not show *art*.)

SEE IT THEN

FINDING A BUSINESS PLAN THAT WORKED

In the early part of the 20th century, wireless radio was used to send information from one point to another. Ship-to-ship, ship-to-shore, and shore-to-shore radio communications were frequently used to transmit all sorts of information. Radio was still new, and a revenue-building strategy had not yet been devised. The economics of the industry revolved around manufacturing the equipment and licensing the patents used to build the equipment. Early experimenters and broadcasters used wireless radio for noncommercial purposes. They had no expectations of making money but enjoyed sending messages and entertainment via wireless to other experimenters and anyone with a wireless receiving set.

This economic situation changed when early broadcasters incurred expenses from purchasing equipment, building studios, and hiring people to manage the stations. Broadcasting shifted from the transmission of Morse code signals between ships at sea or amateur radio enthusiasts to the transmission of the human voice and live music and other entertainment programming. In the early 1920s, broadcasters looked for a way not only to reimburse themselves for their expenses but also to provide enough income to sustain a business.

BUSINESS MODELS

The owners and operators of radio stations considered a number of financial models in the 1920s to help them pay for expenses and perhaps even make a profit. In fact, other forms of electronic communication offered models for making radio pay for itself. These models are discussed in the following sections.

The Telegraph Model

Telegraph companies bought their own equipment and network (telegraph lines) and also had to pay for professional telegraph operators. To recoup their expenses and generate additional revenue, the companies charged for each word in a message sent by customers. Shorter messages were less expensive than longer ones. Similar to today's tweets, telegraphed messages were often cryptic and written using abbreviations to keep the messages as short as possible. Both the sender and the receiver were charged for the cost of the message.

The Telephone Model

The telephone system was and still is a closed system, meaning that messages travel only from telephone-to-telephone. Up until the 1980s, telephones were owned by one of the telephone companies and users simply 'rented' them. When someone moved out of their home and disconnected their phone service they were required to return the telephone to the company. After the breakup of the Bell system in the 1980s, extension telephones were sold on the consumer market, and purchasers could simply hook one up to existing wiring in their homes. Telephones are easy to use, and unlike the telegraph, they do not require a professional sender and receiver to decode the signals. Telephone companies, nevertheless, have many other expenses and therefore need to generate revenue. Telephone companies charge either a flat fee per month for calls or a combination of a flat fee for local calls and a per-use (or per-minute) charge for long-distance calls.

MODELS FOR GENERATING REVENUE FOR RADIO

Radio looked at how telegraph and telephone companies charged for services. Telegraphs and telephones are both one-to-one models of communication in which both the sender and receiver are known, whereas radio is a one-to-many model in which the receivers are not known. Radio companies knew they had to find a different way to charge for their services or, if possible, adapt the methods used by the telegraph and telephone companies.

The Per-Set Tax Model

David Sarnoff proposed a 2% tax on each radio a consumer purchased. The money collected was to be sent to broadcasters to pay for programming. This idea never got the approval from legislators it needed to become part of sales tax law.

The Voluntary Audience Contribution Model

Another idea for generating revenue was to get listeners to pay for radio programming. In 1922, a station in New York tried this plan but collected only $1,000 from listeners instead of the $20,000 that it wanted to obtain. The station ended up returning the collected money. The contribution model, however, did not go away. Public broadcast stations conduct on-air pledge drives asking their listeners and viewers to donate money, and noncommercial stations routinely raise money by selling printed program schedules and other promotional items.

The Government Subsidy or Ownership Model

Just after World War I, the U.S. government debated whether to keep control of the broadcast operations that the Navy had taken over for security reasons. The government observed the fierce competition between competing telegraph and telephone companies and preferred to avoid that situation among broadcast companies. Despite some strong lobbying for the Navy to maintain control, the idea of government control over electronic media was not acceptable to the radio industry or to the public, and the idea was dropped.

The Toll Broadcasting Model

Toll broadcasting started in 1922 at WEAF, the AT&T station in New York. The station borrowed from telephone's revenue model by charging an advertiser based on minutes of airtime used. A Long Island real estate firm was the first company to pay for radio time. It bought 10 minutes of airtime to promote real estate it was selling. The broadcast is often thought of as the first infomercial. The toll model also included the concept of network buying. Because WEAF was part of a 13-station group owned by AT&T, toll advertising was sold for all stations at one purchase price, which was less expensive than buying time on each station individually.

The toll model is still used today in the form of the info-mercial, in which program-length time is bought from a media outlet. The station airing the program does not produce it but simply schedules and airs it in exchange for a fee.

The Sponsorship Model

In the early 1920s, advertisers were discouraged from including direct sales pitches in their messages. Instead, they were encouraged to sponsor entire programs and performers by paying for the costs involved. Often, this practice led to naming the program (*The Kraft Music Hall*) or the performers (Astor Coffee Dance Orchestra) after the sponsor. Mixing advertising with entertainment programming was not initially acceptable to many people, but by 1928, advertising sponsorship had established itself as the primary model for providing financial support to the radio broadcasting industry.

The Spot Advertising Model

In the late 1950s, the networks started moving away from single-advertiser sponsorship of a program. The main reason for this shift was that sponsors wanted too much control over program production, stars, and content. Program sponsors had quite a bit of power, including the ability to rig the outcomes of the television quiz shows. In short, pressure from sponsors to gain huge audiences led to the television quiz show scandal of the late 1950s.

The move away from sponsorship also came about because the advertising marketplace was experiencing maturation. Advertisers were willing to pass up 60-second spots in favor of less expensive 30-second spots. The spot advertising model gave the networks more commercial inventory to sell to sponsors. Rather than feature a single sponsor for a program, multiple sponsors, buying shorter 30-second ads, could be featured in the programming. Shorter, less expensive spots gave smaller advertisers the opportunity to reach a network audience and also gave the networks the opportunity to generate more revenue for selling the same amount of airtime.

The Subscription Model

The subscription model for radio borrowed from the print media, which have used product subscriptions as a revenue stream for years. Readers pay a monthly or annual fee to subscribe to a magazine or newspaper.

The print media receive a greater portion of their income from audience subscriptions than from single-issue rack sales. Because the vast majority of newspapers and magazines also receive advertising revenue, subscription income is part of a dual revenue stream, with the subscription fee usually supplementing the overall revenue picture.

OWNERSHIP BY BROADCAST NETWORKS

THE REPORT ON CHAIN BROADCASTING

While the early radio networks were becoming stronger and more profitable, the affiliates were losing control over their programming. For instance, radio network affiliation contracts required local stations to broadcast network programming even if they had to cancel local programming. Local stations complained, and in 1938, the FCC challenged the power structure.

The FCC took almost 3 years to write *The Report on Chain Broadcasting*. The report found that NBC and CBS controlled the vast majority of prime-time radio programming across the country, and the FCC deemed this monopolistic and counter to the idea of localism. The report mandated that affiliation contracts be limited to 3 years, that affiliates could reject network programs, that networks could own only one affiliate per market, and that networks had the right to offer to nonaffiliate stations the programs that affiliates do not want to air.

The networks were unhappy with the FCC's new rules and began a legal challenge that went to the U.S. Supreme Court (*NBC v. United States*, 1943). The FCC won, which meant the networks had to change the way they conducted business with their affiliates. The most notable result of the ruling was that in 1943, NBC was forced to divest the smaller of its two networks, the Blue network, which it sold to LifeSavers candy mogul Edward J. Noble. The Blue Network was subsequently renamed the *American Broadcasting Company* (ABC). The three major broadcast networks were now in place, ready to dominate broadcasting for the next 50 years.

NETWORK OWNERSHIP SINCE 1945

At the end of World War II, NBC, CBS, and ABC were still all radio networks, but that changed rapidly as the former radio networks began new interconnections for television stations to form television networks. Affiliate stations were contractually committed to airing the networks' programs in exchange for **network compensation**, or money paid to them for their airtime. In return, the networks gained an audience large enough to sell national advertising time. This arrangement helped the television networks and the affiliates prosper and grow.

The television networks also owned and operated their own stations in some of the largest markets. This arrangement was very profitable because network programs drew large audiences that the network sold for high prices to advertisers. Because the networks owned the stations, they did not have to compensate the stations for clearing the airtime to run the network's programs. The only network that did not own stations was the Mutual Broadcasting System (MBS), which was a programming cooperative.

In the early 1950s, another television network emerged. Allen DuMont, the owner of a television manufacturing business, started the DuMont network. Because Allan DuMont owned a company that manufactured televisions, he presumably started the television network to help sell more of his television sets. But because the stronger television stations were already affiliated with the top three established networks, the DuMont network affiliated with weaker stations and was never able to capture much of an audience. The network was never higher than fourth place behind ABC, NBC, and CBS. In fact, many markets only had three television stations, so DuMont did not even have the opportunity to distribute programs in those markets. The DuMont network ceased operating in 1955.

ABC, born from the divested NBC Blue network, did not have the power of either NBC or CBS. It affiliated with the leftover stations that did not already affiliate with the two bigger networks. Because of the lower audience share of its affiliate stations, it could not generate enough revenue to stay profitable. By 1951, ABC began negotiating with companies that could serve as potential partners, and in 1953, it merged with United Paramount Theaters. Although the merger brought ABC the cash it needed to continue network operations, it remained the weakest of the three networks for more than 20 years.

The big three networks continued television operations relatively unchanged from 1953 until 1985. Then, by contrast, big changes occurred. A television group owner, Capital Cities Communications, took control of ABC in 1985. Known for its ability to control expenses and show a profit, Capital Cities brought some stability to the network. Just 1 year later, in 1986, RCA, the parent company of NBC, was sold to GE, the company that had helped start RCA almost 60 years earlier. GE then sold NBC's radio network to a radio program syndicator, Westwood One. In 1995, Westinghouse Electric Corporation, a company rooted in broadcasting history, acquired CBS. Westinghouse sold its electronics business (appliances) and renamed itself CBS Corporation. In 1996, ABC was sold again; the Walt Disney Company acquired Capital Cities/ABC for $18.5 billion in the hope of building a synergistic media company consisting of television, movies, and entertainment brands—including the Disney theme parks. CBS was sold again in 1999, this time to conglomerate Viacom, a company with roots in the drive-in theater business that was created years earlier to syndicate old CBS television series.

In 1986, media mogul Rupert Murdoch created the Fox television network. Named after Murdoch-owned Twentieth Century Fox film studio, the new network at first lost money—$80 million in 1988 alone. Fox eventually caught on, and 5 years later, its operation became similar to that of the big three networks when it began scheduling prime-time programming 7 nights per week.

RADIO NETWORKS

Although there were radio networks after the rise of television, those networks were not seen as being essential to local radio. National radio programming was most noticeable with National Public Radio.

Many radio stations had preproduced material, but the majority of that material was supplied by syndicators rather than by networks. Audiences did not care about a radio station's network affiliation, but the audience did care about a station's format. Radio station ownership included some small groups and many stations owned by small companies, often local to a market or region. 'Mom-and-pop' radio stations were common for many years.

DEREGULATION'S INFLUENCE ON BROADCAST MEDIA

After some 40 years of increasing its control over broadcasting, the FCC began to relax its control in the late 1970s. Two factions in Congress struggled over FCC oversight of electronic media. Liberal legislators maintained that the FCC needed to keep strict regulatory control over broadcasting because the phrase "public interest, convenience, and necessity" mandated doing so. Conservatives, on the other hand, felt that FCC regulation came at a high cost and that the marketplace of the media, with stations and other program suppliers (e.g., cable television) vying for audience attention, should be allowed to guide station behavior, not the government.

The conservatives and their marketplace concept won out, and in 1981, after many years of complaining about increased station workloads and costs to comply with government regulations, the radio industry was somewhat deregulated by the FCC. As a result, radio stations were no longer beholden to the public-interest aspects of programming. The television industry received some deregulation relief in 1984.

Radio stations were no longer required to consider the public interest when making programming decisions, freeing them from mandated community involvement and simplifying license renewal. Stations no longer had to demonstrate that they operated in the public interest, and thus renewal became little more than filling out the postcard renewal form.

The move to deregulation put the fairness doctrine in the spotlight. The fairness doctrine was put in place in 1949 by the FCC to require broadcasters to present both sides of controversial issues important to the public. The doctrine stipulated that the presentation of the issues was honest and balanced. This was not necessarily an 'equal time' requirement, but it intended that opposing views on important issues were represented.

The call to eliminate the fairness doctrine began in the early 1980s, but the FCC did not vote to stop enforcing it until August 1987. The spirit of deregulation stems from both the First Amendment and the marketplace concept. Broadcasters wanted the same freedom from regulation that newspapers enjoyed and felt that programming should be determined by the marketplace (the audience and stations), not bureaucrats.

The deregulation march continued forward when in 1983 the courts ruled that the National Association of Broadcasters' NAB Code of Ethics, which had set voluntary programming and advertising standards, was illegal. The FCC had used these voluntary standards as a quasi-measuring stick for judging whether a station was meeting the 'public interest' requirements. The courts ruled that the code set an 'arbitrary limit' on the supply of commercial minutes available to advertisers, thereby allowing the television networks and radio and television stations to continue to raise their prices on the limited number of commercials available to advertisers over the air.

In sum, the FCC agreed to some extent with broadcasters and legislators about the marketplace concept but perhaps took the notion further than was expected by Congress. The FCC developed more opportunities for programming and advertising to reach audiences. Looking to the newspaper industry, which had no licensing barriers to entry, the FCC sought to reduce the barriers for entry into the broadcast and electronic media businesses. The result was the addition of more stations and increased competition from alternative delivery systems like cable, satellite, and multipoint multichannel delivery systems.

In the late 1980s, the push for broadcast deregulation slowed considerably, as Congress decided that the FCC had moved beyond the loosening of control that legislators had intended. Congress also decided it should play "a more direct role in the policy sandbox" (Sterling & Kittross, 2002, p. 560).

The broadcast deregulation trend that began in the late 1970s continued throughout the 1980s. Some have even said that the *de*regulation of the 1970s became the *un*regulation of the 1980s during the Reagan Republican years. There were fewer public service programs, no limit on the number of commercials per hour, a much easier license renewal process, and no trafficking (the selling of stations) restrictions. Previously, an owner had to hold a broadcast station at least three years before it could be sold. In 1984, the FCC upped ownership limits from the **rule of sevens** (which allowed owning only seven stations in any service, AM, FM, or television) to 12 stations in either radio service and up to 25% of the total television households nationwide. In 1992, the limits were raised again, allowing up to 40 radio stations per owner and some relaxation of the duopoly rules.

CABLE GROWTH AND OWNERSHIP

In the 1960s and 1970s, cable ownership occurred mostly in small and medium-sized communities that did not have a television station or only had one or two local television broadcast stations. Big name companies like TelePrompTer (maker of teleprompters for TV studios), Westinghouse, and Cox began investing in cable. The number of cable subscribers nationwide increased from 1 million in 1963 to 4.5 million in 1970. In the 1970s, the number of cable operators grew steadily as cities adopted franchising agreements so cities could be wired by cable companies.

In 1975, HBO unveiled its plans to use satellite distribution to send its programming to cable systems nationally. HBO triggered an explosion of interest and investment in cable television ownership nationwide. Larger companies with deeper pockets bought small companies that could not afford the satellite equipment. By the end of the 1980s, more than 50 million households had cable subscriptions. The number of cable-only (i.e., not a broadcast station or network) channels grew from 28 in 1980 to 79 by 1989. By the end of the 1990s, 7 in 10 households (more than 65 million) subscribed to cable.

By the late 1990s, the bulk of cable subscribers were held by the top 10 **multiple system operators** (MSOs). Although not a true oligopoly, the industry was consolidating. A typical example of the ownership consolidation is the company that was the leading cable company in the late 1990s, Tele-Communications, Inc. (TCI). TCI had almost 14 million subscribers. In 1999 it was purchased by AT&T. Charter Communications then acquired some of the cable systems owned by AT&T.

SEE IT NOW

THE BROADCAST STAR MODEL

Radio first used toll broadcasting, and then it changed to the sponsorship model, and finally it adopted the spot advertising model. Spot advertising has been used by both radio and television networks and stations since the late 1950s and is still the dominant model today.

For a local station to be financially successful, other components of the industry must be involved. Using a network-affiliated television station as an example shows how some related groups, organizations, and businesses function to make the broadcast television system work. Together, these elements form the shape of a star, with a network-affiliated television station at the center. If the center of the star is replaced (see Diagram 10.1) with a network-affiliated radio station, it could also be used to represent the radio industry as it was from the late 1920s until the mid-1950s.

At the center of the star model are television stations, consisting of three different types: network owned and operated (O&Os), network affiliates, and independent stations. These types of stations differ by their relationship to the network. A network O&O is owned and operated by a network. A network affiliate is by far the most common type of station and has a long-term agreement with a network to run its programs and commercials. An independent station does not have an agreement with a network and thus must find other sources for programs.

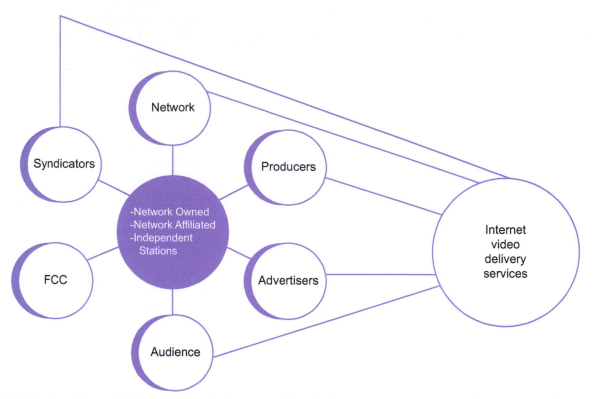

DIAGRAM 10.1 Broadcast star model. This model shows the basic interrelationships between stations, program suppliers, regulators, advertisers, and the audience. Although the model is still descriptive of these relationships, the business is changing. As a larger percentage of the audience gets its programming from sources other than standard broadcast television, cable, and DBS, the networks are not depending upon the network-affiliated stations as the only way to reach the audience. There have been network discussions about severing relationships with local affiliates and forming relationships with the cable or satellite companies. There is now a new 'planet' in the star system. This new planet consists of the Internet video delivery services (Hulu, Amazon Video, and Netflix). These relatively new services have become powerhouses in the industry by attracting huge audiences. These new players do not have a direct relationship to broadcast stations and are not directly regulated by the FCC. Also, program providers like the networks are going 'over the top' to reach audiences directly through the Internet and may leave affiliated stations with smaller audiences and less leverage with advertisers.

FYI: Network, Syndication, and Off-Net Programs

In the past, network programs were produced by independent producers with some support from the networks, but the producers maintained control of the program. After a program accumulated a few years of episodes (or about 100 episodes), the producer could employ a syndicator, who would license (or rent) the program directly to stations or groups of stations at whatever price the market would bear. The program would continue to air on the network until no new episodes were being made. At that point, the program would be considered off-network.

Networks are now allowed to own a controlling portion of the shows they air, and that has changed the life cycle of a television program. The network can decide where the program will be shown after (or even during) its network run. The networks often have 'sister channels' (such as Fox's FX) that get shows from the networks at a lower price than what a producer would get from other stations in the syndication market. The relationship between the network and its sister network can seriously reduce the value of a program, since the network can have an agreement to outbid any other network or station for the show at any time. Because of that type of agreement, the other networks or stations may not even attempt to get the show, thus reducing the market value of the show.

An example of a show following this path is *The X-Files*, which ran on Fox for a number of years. Fox made a deal with FX, its cable channel, to run the show 5 days per week. The producer and the star (Chris Carter and David Duchovny, respectively) sued Fox, claiming that the show would have a reduced value when it did go into syndication because of the many airings it was to get on FX. Clearly a reduction in the value of the show results in a reduction of the royalties paid to the show's producer and star.

The star model shows the interdependence that exists among the groups/organizations, businesses, and the broadcast television stations. In the past, all of the entities depended on the television stations to stay in business, although that is changing. In the illustration of the model, each line connecting two entities represents a two-way relationship. For example, the network supplies its affiliate station with programs, and the station provides the network with an audience for its programs. These relationships have proven good for all involved, as the model has endured in television for more than 65 years. Some businesses play an important role in the television industry but are not shown in the star model, including the delivery systems that provide the majority of the U. S. audience with television programs. Cable television, satellite delivery services, and other television multichannel delivery systems are used by about 90% of the households in this country.

FYI: Recording Industry: A Business Plan That Has Failed

The music industry has seen its revenue drop consistently since 1999. Album sales have decreased by about 8% per year. The industry has been changing from an ownership model, where individuals own a version of the album or songs that they like, to an access model, where individuals have the ability to listen to any music that appeals to them at any time and practically any place. Beginning in 1999 with Napster, the first peer-to-peer file-sharing site, the idea of getting music from the Internet became a very popular trend. At first, huge numbers of people were downloading copies of albums and songs at peer-to-peer sites. However, this practice was illegal, since it violated the copyrights of many songwriters. In fact, some people were prosecuted for facilitating the illegal downloads. Some audience members who downloaded songs for their own use were also prosecuted. A more recent trend has been to stream music instead of downloading the songs. Downloading music takes up memory space and, unless paid for, is still illegal. Streaming music takes up almost no memory space and is usually done through a legal service. Instead of owning a large library of music, more and more people are simply listening to one of the various audio services available on the Internet, such as Pandora or iTunes. These services are free with some restrictions or can be subscribed to for a low monthly fee and provide practically limitless listening to a huge library of music 24/7.

THE TELECOMMUNICATIONS ACT OF 1996

The major piece of legislation that governed media for most of the 20th century was the Communication Act of 1934. It was written at a time when the only electronic medium was radio. Television was experimental and not commercially viable. But television and cable television became viable delivery systems shortly after World War II, and the Communication Act of 1934 became somewhat obsolete. After decades of attempts and at least 1 full year of political negotiating and compromise, Congress passed the Telecommunications Act of 1996 by an overwhelming majority vote. This act, signed into law on February 8, 1996, became effective immediately and was the first comprehensive rewrite of the Communications Act of 1934. This Act of 1996 paved the way for dramatic changes in station ownership.

Starting in 1950 with the 'rule of sevens,' the FCC continued to prevent broadcast ownership groups from owning large numbers of stations. But with the Telecommunications Act of 1996, ownership restrictions in radio were removed. A group could own as many stations around the country as it wanted, as long as it did not own too many in any one market.

In the decade following the passage of the act, radio stations were bought and sold at a dizzying pace, with many small owners succumbing to the tremendous pressure and big money offered for stations by large group owners. Several groups owned hundreds of stations, including Clear Channel (1,216 stations in 190 markets), Cumulus (310 stations in 61 markets), Citadel (now part of ABC, with 223 stations in 56 markets), CBS (185 stations in 40 markets), and Entercom (110 stations in 23 markets). As those owners tried to minimize their operational costs by sharing personnel and other resources among their stations, the radio industry became more corporate and formulaic.

Programming began to sound similar from market to market as groups that owned similarly formatted stations in different markets used similar (and sometimes identical) music playlists and shared DJs.

Technology has enhanced the trend toward format similarity, as automation allows many programming and recording tasks to be done on a computer. **Voice tracking**, the prerecording of DJ talk and announcements for use later at one or many stations, can be accomplished from anywhere in the world with a microphone, a computer, and an Internet connection. Some DJs have never even visited the markets in which their voices are broadcast. Thus, radio has lost much of its localism.

The business plan for most of the larger groups did not work out as expected. The cost savings were less than expected, and station groups were unable to use their size to leverage significant increases in advertising. In the decade following 1996, radio's share of the total advertising market grew only from 7% to 8%. A combination of factors thus forced most station groups to alter their approach to the business and, in some cases, to sell off stations or find other means to recapitalize their company. This pattern of aggressive station acquisition, declining audiences due to listening habit changes, and declining advertising revenue worsened during the recession in 2008 and 2009. In addition, advertising priorities shifted money from radio to competing media.

The Telecommunications Act of 1996 called for changes in five major areas: telephone service, telecommunications equipment manufacturing, cable television, radio and television broadcasting, and obscenity and violence in programming.

TELEPHONE SERVICE

The 1996 Act overruled state restrictions on competition in providing local and long-distance service, which allowed the regional Bell (Telephone) operating companies (also known as Baby Bells and **RBOCs**, pronounced 'ree-boks') to provide long-distance service outside their regions and required them to remove barriers to competition inside their regions. New universal service rules guaranteed that rural and low-income users would be subsidized and that equal access to long-distance carriers would be maintained.

TELECOMMUNICATIONS EQUIPMENT MANUFACTURING

The RBOCs were allowed to manufacture telephone equipment. The FCC was empowered to enforce nondiscrimination requirements and restrictions on joint manufacturing ventures and monitor the setting of technical standards and accessibility of equipment and service to people with disabilities.

RADIO AND TELEVISION BROADCASTING

The 1996 act relaxed media concentration rules for television by allowing any one company to own stations that could reach 35% of the nation's television households, an increase from the previous limit of 25%. The major broadcast television networks (ABC, CBS, NBC, and Fox) were allowed to own cable systems but not other major networks (excluding the smaller networks: WB, UPN, and PAX). Media concentration rules for TV changed again in 2003, when television groups were allowed to reach 39% of the national audience.

Limits on radio station ownership were also repealed, but some restrictions remained on the number of local stations a company could own in any one market. In markets with 45 or more stations, one company could own 8 stations but no more than 5 in either AM or FM. In markets with 30 to 44 stations, a company could own 7 or fewer with only 4 in any one service. In markets with 15 to 29 stations, one company could own only 6 stations (4 in any or either service). In a market with less than 15 stations, only 5 could be owned by one company—3 maximum in either AM or FM and only 50% or less of the total number of stations in that market. In addition, the U.S. Department of Justice's Antitrust Division was assigned to look closely at large transactions to prevent any group that already had 50% of the local radio advertising revenue or was about to obtain 70% of the local advertising revenue in any one market from acquiring more stations in that market.

CABLE TELEVISION

The Telecommunications Act of 1996 relaxed the rules set down in the 1992 Cable Act, essentially removing rate restrictions on all cable services except basic service. In addition, telephone companies were permitted to offer cable television services and to carry video programming. Other provisions were that cable companies in small communities were to receive immediate rate deregulation, that cable systems had to scramble sexually explicit adult programming, and that cable set-top converters could be sold in retail stores (instead of being available only as rental units from the cable companies).

FYI: Cable Subscriptions 2010–2015

2010	99.0 million households
2011	100.9
2012	100.8
2013	99.3
2014	98.0
2015	97.2

The number of cable subscriptions peaked in 2012 and has been slowly decreasing since then. The projection for 2017 is that there will be fewer than 95 million subscribers to cable television.

Source: Based on information from TDG Research and Stiles, 2016.

FYI: Top 10 Video Subscription Services

Netflix	41.4 million
Comcast	22.4
DirecTV	20.4
Dish	14.0
Time Warner Cable	11.0
Hulu	9.0
AT&T U-Verse	5.9
Verizon FiOS	5.6
Charter	4.3
Cox	4.1

Netflix and Hulu are Internet video subscription services, DirecTV and Dish are satellite subscription services, Comcast, Time Warner, Charter, and Cox are cable subscription services, and AT&T and Verizon are telecommunication companies that also offer multichannel subscription service.

Source: National Cable Television Association, July 2015

Despite all of the fervent merger activity, the number of cable subscribers is declining. In 2012, there were slightly more than 100 million subscribers. By 2015, that number had dropped to just over 97 million. For the most part, cable companies are compensating for the loss in revenue due to fewer subscribers by raising prices. This practice, combined with the dissatisfaction among consumers with the bundling of channels and the availability of television programs and movies online, has led to the 'cutting the cord' trend.

Taxation

Taxation has generally not been an issue for broadcasters and other electronic media. Recently, however, two issues have concerned the industry: general taxation of the Internet and the proposed performance tax.

The **Internet Tax Freedom Act**, initially signed into law on October 21, 1998 (H.R. 4328), placed a moratorium on Internet taxes and continues to be the law that demonstrates the federal government's stance on Internet taxation. The U.S. government pays attention to the market to help create policy, and the result is that the government allows the Internet to continue to develop before placing tax restrictions on it. In 2014, Congress passed the Permanent Internet Tax Freedom Act, which amends the Internet Tax Freedom Act (H.R. 3086) by making permanent the ban on state and local taxation of Internet access and on multiple or discriminatory taxes on electronic commerce.

The *Performance Tax* is a proposed tax that is a result of pressure from record labels to impose on radio stations for airing free music. Radio stations have traditionally paid licensing fees to the organizations that represent composers and songwriters (e.g., BMI, ASCAP) for airing their music. Internet radio stations pay a fee for streaming music. The record companies are pushing Congress to impose a tax on radio stations to compensate performers when the stations air their music. Not surprisingly, the National Association of Broadcasters, representing radio stations, has vehemently opposed this new tax. The broadcast station logic is that they already pay a licensing fee for the music, and radio is the best promotional vehicle for the record label and artists. Adding another cost to the operation of local radio stations could have serious implications for the viability of many stations, especially smaller ones, whose profitability has eroded since 2008.

THE ROLE OF THE FCC

A number of legal issues concern both radio and television professionals and audiences. Traditionally, the electronic media have received special treatment under the law because they use the electromagnetic spectrum, and their programs can reach young children. This treatment has not necessarily meant better or more privileged treatment by the government or the courts but rather different treatment than is given to newspapers, magazines, and books.

The print media and Web sites, because of their ease of entry into the marketplace and that they do not use spectrum space, do not require a license to begin business. Moreover, the print media enjoy the full protection of the First Amendment. The print media and the Internet are not directly overseen by a regulatory agency that controls the technological and business aspects of their operations.

On the other hand, radio and television receive only partial protection from the First Amendment and have the FCC regulating most aspects of their technology as well as some of their business practices.

Although the FCC originally had seven commissioners, that number was reduced to five in 1982 as part of federal cost cutting. The President of the United States nominates the commissioners, and the Senate must approve them. Only three commissioners can come from any one political party. The FCC now has seven bureaus or divisions, each with a specific area of administration: the Consumer and Governmental Affairs Bureau, the Enforcement Bureau, the International Bureau, the Media Bureau, the Public Safety and Homeland Security Bureau, the Wireless Telecommunications Bureau, and the Wireline Competition Bureau.

various licensing terms and conditions. This bureau investigates and responds to potential unlawful conduct to ensure (1) consumer protection, (2) a level playing field to promote vigorous and fair competition, (3) efficient and responsible use of the public airwaves, and (4) strict compliance with public safety-related rules.

The International Bureau administers international telecommunications and satellite programs and policies, including licensing and regulatory functions. The bureau also promotes fair competition abroad, coordinates global electromagnetic spectrum activities, and advocates U.S. interests in international communications and competition.

The Media Bureau oversees broadcast radio and television, as well as cable and satellite services, on behalf of consumers. It also administers licensing for broadcast services and cable and handles postlicensing matters for satellite services.

The FCC's Public Safety & Homeland Security Bureau (PSHSB) is responsible for public safety communications issues, including 9-1-1 and E9-1-1; coordination of public safety communications; communications infrastructure protection and disaster response; and network security and reliability.

The Wireless Telecommunications Bureau oversees fair licensing of all wireless services, from fixed microwave links to amateur radio to mobile broadband services. This includes nearly 2 million licenses. The bureau also conducts auctions to award services licenses and manages the communication tower registration process. The bureau produces an annual assessment of the wireless industry—the Mobile Wireless Competition Report—and manages interactive Web tools such as the Spectrum Dashboard, which deliver key information on wireless services to the public.

The Wireline Competition Bureau works to ensure that all Americans have access to affordable broadband and voice services. Its programs help ensure access to affordable communications for schools, libraries, health care providers, and rural and low-income consumers. The bureau provides the public with accurate and comprehensive data about communications services, including broadband.

ZOOM IN 10.1

The Bureaus of the FCC

The Consumer and Governmental Affairs Bureau (CGB) is the bureau of the FCC that develops and implements the commission's consumer policies, including disability access. It serves as the public face of the commission through outreach and education, as well as through the Consumer Center, which is responsible for responding to consumer inquiries and complaints. The bureau also partners with state, local, and Tribal governments to provide support for emergency preparedness and implementation of new technologies.

The Enforcement Bureau (EB) is responsible for enforcing the Communications Act and the Commission's rules, orders, and

RULE MAKING AND ENFORCEMENT

In addition to the existing rules that have come from congressional legislation, the FCC creates its own guidelines for electronic media. To do so, the agency must announce its intent to create a new regulation and then publish a draft of it. After publishing the proposed rule, the FCC must wait for comments from both audiences and professionals in electronic media. Only after these comments have been considered can the FCC officially enact the regulation.

Through its Office of Strategic Planning and Policy Analysis, the FCC also reviews the overall past economic performance of the electronic media and makes projections for the future of the industry. The FCC often asks media owners for input on developing rules or policies to help support the future success of the broadcast medium.

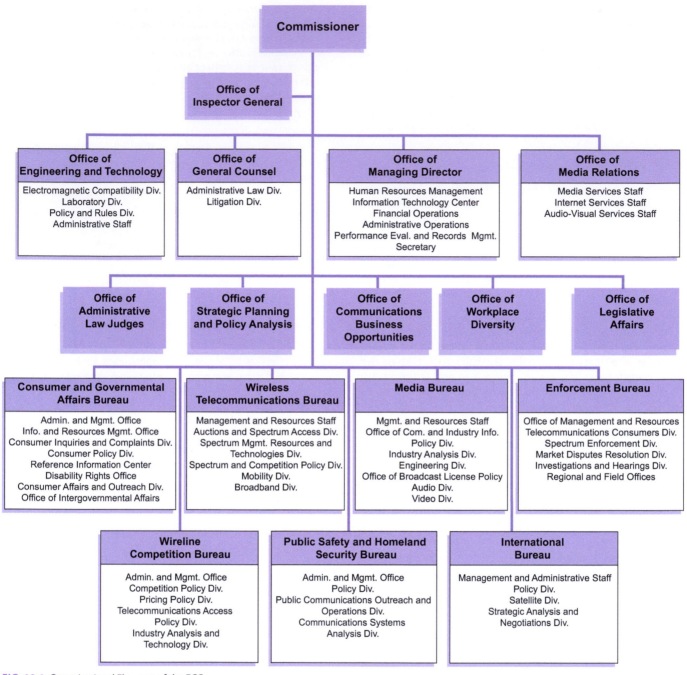

FIG. 10.1 Organizational Flowcart of the FCC
Source: Based on www.fcc.gov/fccorgchart.html

The Enforcement Bureau of the FCC has the responsibility of making sure that broadcasting and other electronic media follow the rules. If the Enforcement Bureau finds that a rule has been broken, it usually sends a letter of reprimand to alert the station that it allegedly violated a regulation. The station then is required to take corrective action. If the station does not, then the next step is for the FCC to issue a cease-and-desist order. If the station does not comply, it can be fined (often referred to as a *forfeiture*). Fines range from $2,000 to as much as $250,000 for each violation. Other actions that can be taken against a station include a short-term renewal (i.e., 6 months to 2 years instead of 8 years) or, in the case of repeated violations and a failure to correct problems, a nonrenewal or a revocation of the station's license. Although these last two drastic actions are rare, they do occur.

Jawboning is another, less punitive tactic that the FCC uses to urge regulatory compliance. Rather than waiting for a station to violate a rule and then punishing it, the FCC alerts broadcasters to new regulations and suggests ways to avoid fines. FCC staffers keep in close contact with broadcasters by attending state, regional, and national conventions, where they discuss best practices for the broadcast industry.

OWNERSHIP OF BROADCAST STATIONS

One of the most important and newsworthy functions of the FCC has been and continues to be the licensing of broadcast stations, including the process of renewing licenses and overseeing the transfer of licenses when stations are sold. Obtaining a new broadcast license is a daunting process that requires time, money, and energy. Consulting engineers and communications attorneys are often needed to complete the necessary paperwork and to respond to questions and challenges by other entities competing for the same license.

The FCC promotes diversity of ownership in electronic media properties. Its goal is to encourage competition in advertising, programming, and news and political viewpoints. In addition, the FCC encourages station ownership by women and minorities. Some of the recent ownership issues considered by the FCC are discussed below.

In June 2003, the FCC commissioners voted to raise the existing television ownership cap to 45% of the national audience, opening the door to more consolidation in the broadcast industry. The court received a strong response from audience groups against further consolidation and halted the FCC's new rules. Instead, it brokered a compromise that allows any company or individual to own television stations that reach 39% or less of the national audience. Despite the compromise, this remains a hotly contested issue, however, and serious debate in the marketplace continues about the need for more diversity of ownership in the television industry. Simply put, the FCC wants to avoid having broadcasting become an oligopoly dominated by just a few major media companies.

The consolidation in the television industry remains a topic of concern, not only for the FCC but also for the Department of Justice. This department of the U.S. government monitors mergers and acquisitions to prevent monopolies and unfair competition in the electronic media industry.

OWNER QUALIFICATIONS

Broadcast station ownership requires licensing by the FCC. Originally, licensing meant that a prospective owner would apply for the right to build and operate a licensed station. Today, there are few available frequencies for new radio stations or television stations in any given market. The scarcity of spectrum space is testament to the ubiquity of broadcast stations, to their importance as a source of entertainment, information, and news, and to their role as economic stimulators through advertising. Prospective owners must still qualify in four areas: citizenship, character, ownership structure, and programming.

CITIZENSHIP

Owners of broadcast stations must be citizens of the United States, a requirement that has roots in the past. Just before World War I, foreign-owned companies like British Marconi had taken an aggressive interest in radio. However, the U.S. government was disinclined to let foreigners or foreign companies control the powerful new medium, feeling that radio transmitter ownership should be in the hands of citizens. Further, the U.S. government took control of all radio transmitters in the country for the duration of World War I. Since then, the government's policy has been that owners of broadcast stations must be U.S. citizens, though non–U.S. citizens (individuals or companies) may hold up to 20% ownership.

CHARACTER

Although this is a very broad criterion, the law expects broadcast station owners to be people of good character. This is difficult to show in an application, but one obvious rule is that convicted felons are not allowed to own broadcast stations. Another standard requires applicants to certify that they have not been convicted of distribution or possession of controlled substances. The FCC would also prefer to see owners who will be directly involved in the day-to-day operation of a station.

OWNERSHIP STRUCTURE

Knowing who actually owns a new station may seem to be a simple issue, but the FCC places great scrutiny on ownership structure. The FCC wants to know if the applicant owns or operates other stations and whether that applicant intends to build a station from scratch or buy an existing station.

PROGRAMMING

The FCC has maintained the policy that broadcasters can program to the public in whatever way suits the marketplace. Prospective licensees are expected to familiarize themselves with the obligation to provide programming responsive to the needs and interests of the community of license.

Because the FCC cannot censor a program or get directly involved in either radio formats or television programs, an applicant can be granted a license despite having a programming plan that the FCC does not like. For instance, the FCC cannot withhold a license from an otherwise qualified owner because he or she intends to program country music in a small market that already has three other country music stations. However, the FCC can withhold a license from a potential owner that intends to conduct faith healing all day long because it would not be in the public interest.

Much of the licensing and ownership process involving the FCC has taken the role of marketplace observer. The FCC's presumption is that reasonable people will operate a station in the public's best interest. The FCC expects owners will have a programming, financial, and technical plan that will ensure community service and economic viability. When a station gets into trouble, the FCC may step in to enforce penalties for technical or other violations or to facilitate the transfer of the license to a new owner.

COMPETING FOR A LICENSE

In the past, the FCC screened all applicants who applied for a broadcast frequency to determine which one would be the best future broadcast station owner. This procedure was often very slow and agonizing, though, for both the applicants and the FCC. In recent years, the FCC has adopted two new procedures designed to streamline the process. The first procedure is the channel or frequency lottery, in which the FCC selects an applicant at random. Once a winner has been selected, he or she is thoroughly screened to make sure that the prospective owner is fully qualified and that the application is complete.

The lottery procedure is being used less often now that the FCC has realized that another procedure, the auction, can bring money to the government for the awarding of the license. Once the FCC has deemed a frequency appropriate for a broadcast station license, it may conduct an auction for use of the frequency. As in any auction, bidders compete by submitting their offers for the license. Once the highest bidder has been identified, the FCC closely scrutinizes that applicant to ensure that the criteria for ownership are met.

CONSTRUCTION PERMITS

Once an applicant for a new broadcast station has been selected, reviewed, and approved (a process that could take 3 years or longer), the applicant receives a *construction permit* (*CP*), which gives authorization to build and test a broadcast station. Once the facility has been built, the applicant must receive a license to begin broadcasting. During this period, the station is physically constructed and the equipment is purchased and put in place. After that, the station begins testing the equipment to make sure that it works properly and that the signal is in compliance with all FCC technical regulations. If the FCC determines that the testing is successful, the station is granted a full-service license to begin broadcasting a regular schedule of programming.

KEEPING THE BROADCAST STATION LICENSE

The FCC seldom revokes a station license. A station would have to blatantly disobey FCC rules to lose its license. Being convicted of a felony or being convicted of drug possession could be grounds for a loss—though even then, the convicted individual would have to be a direct part of the ownership and not just an employee of a parent corporate licensee. It is hard to lose a license in part because of the large number of broadcast stations on the air; the FCC simply cannot keep a close watch on all of them. Moreover, the FCC has a graduated schedule of notifications and warnings that are sent to troublesome stations to get the attention of their owners. A station would have to ignore the so-called raised-eyebrow letter (stating that the FCC is aware of a possible violation of rules), the cease-and-desist order, the forfeitures (i.e., fines), and the threats of short-term or conditional renewal before its license would be in danger of being revoked.

One area of operation that the FCC *does* monitor closely is employment practices. For the past 30 years, the FCC has been monitoring broadcast station and multichannel program providers' employment practices. The FCC has intended to make sure that they are hiring more women and minorities. Every station must now file a yearly report of its hiring activities with the FCC. Failure to comply with hiring guidelines and other FCC rules results in fines.

In 1998, the FCC's Equal Employment Opportunity (EEO) hiring practice rules were found to be unconstitutional. The court ruled that the FCC could not show that these rules were in the public interest. New rules were put in place in early 2000 to guide broadcasters in using the methods of recruitment expected from the FCC. The EEO recruitment rules have three prongs. Stations that employ five or more full-time employees (defined as those regularly working 30 hours a week or more) must:

- Widely distribute information concerning each full-time job vacancy, except for vacancies that need to be filled under demanding or other special circumstances
- Send notices of openings to organizations in the community that are involved in employment if the organizations request such notices
- Engage in general outreach activities every 2 years, such as job fairs, internships, and other community events

A second area of FCC scrutiny is the station's public file. Every station is required to keep a file of important documents that can be inspected by the public—for instance, permits and license renewal applications, change requests, the FCC pamphlet *The Public and Broadcasting—How to Get the Most Service from Your Local Station*, employment information, letters from the public, information about requests for political advertising time, station programming that addresses community needs, time brokerage agreements, and information about children's programming and advertising. A public file is required for all stations regardless of their commercial or noncommercial status. A radio station must keep these documents for 7 years and a television station for 5 years. A cable company need keep only those documents relating to employment practices. This file must be kept at the station or at some accessible place to allow the public to inspect it during regular business hours.

RENEWING THE BROADCAST STATION LICENSE

The vast majority of stations applying for license renewal are renewed with little or no trouble from the FCC. In some cases, an individual or group will challenge the right of an existing station to keep its license by filing a petition to deny renewal. If the FCC feels that an existing station is not serving the public, it can schedule a hearing about the renewal. Serious challenges are not very common, but they can be both time consuming and expensive.

In recent years, the process of license renewal has become streamlined and easier than in times past. The renewal form is 9 pages but includes another 30 pages of instructions and must be submitted electronically. The form asks the renewing station to affirm past actions as an upstanding member of its community of license.

BROADCAST STATION FINANCES

Broadcasting is a business. Just because a station can keep its license does not mean that it can stay in business. Programming is important, but advertising sales and financial management keep the signal on the air and prevent the station from going 'dark.' The station must generate enough money to pay its bills. This does not just mean a monthly electric bill or the employee payroll. Station liabilities include payroll and utility costs as well as technical equipment and repair and certainly news and programming expenses. Often overlooked is debt service. Debt service, repaying money borrowed from investors, is a primary concern for most broadcasters. Prior to the economic collapse of 2008/2009, the sale of broadcast stations was brisk. New owners paid high prices for stations at a time when the U.S. economy was very strong. As the economy collapsed, advertising revenue decreased, but the debts accrued from buying a station remained. In poor economic times, people watch even more television—it is inexpensive entertainment. But the lack of consumer spending has a deleterious effect on advertising spending and station profitability.

In good times and bad, a station's business department keeps track of all money coming into and out of the station using various accounting procedures. A profit and loss statement, referred to as a P&L, is generated that accounts for all revenue (i.e., money received) and all expenses (i.e., money spent) during a given time period. The balance sheet shows a station's net worth. The profit and loss statement shows the station's profit or income for a given time.

OWNING VERSUS OPERATING BROADCAST STATIONS

Some stations are owned by one company, but they are operated by another in a relationship known as an *LMA*: a **local marketing agreement**. An LMA lets one station take over programming and advertising sales for another station in the same market. LMAs are used by both television and radio stations. Much of the movement to adopt local marketing agreements stems from the need to control costs. With an LMA, the station owner holds the license while putting another broadcaster in charge running the station. The managing station benefits from cost savings

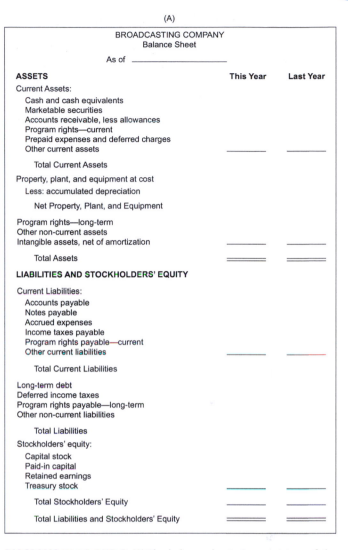

(A)

DIAGRAMS 10.2A AND B (A) The balance sheet gives a picture of the station's overall financial health at one point in time. It shows what the station owns (assets) and what the station owes (liabilities and stockholders' equity). (B) A statement of operations, or profit and loss statement, tracks a station's financial performance over a set period of time. The difference between station revenue and station expenses yields the profit or income of the station for that period of time.

through personnel and resource sharing with the second station. An example of an LMA pairing is an ABC affiliate partnered with a Fox affiliate. With Fox Network programming currently ending at 10:00 p.m. EST, an ABC affiliate could repurpose news content by airing an early newscast on the Fox station, at 10:00 p.m., prior to the conclusion of ABC Network programming at 11:00 p.m. EST and the start of the ABC station's newscast at 11:00 p.m. The Fox station could air the newscast itself, but the local market economy might not produce enough advertising revenue to support a full news operation. The LMA allows the ABC station to program the newscast with personnel from the ABC station and to sell the advertising on the Fox affiliate with minimal increases in expenses.

Another type of sharing is the the Joint sales agreement, which is similar to an LMA but limited to sales. That is, the sales representatives for one station also sell time for another station in exchange for a percentage of the

(B)

BROADCASTING COMPANY
Statement of Operations

As of _____
(Date)

MONTH: _____   YEAR-TO-DATE

Actual	Budget	Last Year		Actual	Budget	Last Year
			OPERATING REVENUE			
			Local			
			National			
			Network			
			Production			
			Other Broadcast			
___	___	___	Misc. Revenue	___	___	___
			Gross Revenue			
			Less:			
___	___	___	Commissions	___	___	___
			Net Revenue			
			OPERATING EXPENSES			
			Technical			
			Programming			
			News			
			Sales and Traffic			
			Research			
			Advertising			
			General and Admin.			
___	___	___	Depr. and Amort.	___	___	___
___	___	___	Total Op. Expenses	___	___	___
___	___	___	Op. Profit (Loss) Before Taxes	___	___	___
___	___	___	Provision for Taxes	___	___	___
===	===	===	Net Income	===	===	===

DIAGRAMS 10.2A AND B (Continued).

revenue. Some stations also produce newscasts for independent stations that want to air an evening newscast but cannot afford to staff a newsroom. Depending on the relationship, the station producing the newscast may buy the airtime on the independent station or reach an agreement for revenue sharing.

DUOPOLY OWNERSHIP

Changes in public viewing and listening patterns have led the FCC to review ownership practices. At one time, ownership of television and radio stations had a one-per-market limit; owning more than one FM station or television station in the same service area was prohibited. A *duopoly* is an ownership structure in which one licensee owns two or more stations in a local market area. Allowing a duopoly or common ownership of two or more media stations or services produces economic efficiency for the owners through controlling expenses and repurposing programming.

Radio ownership in a market is determined on a sliding scale, according to market size. As an example, one entity may own up to 5 commercial radio stations, not more than 3 of which are in the same band (i.e., AM or FM), in a market with 14 or fewer radio stations and up to 7 commercial radio stations, not more than 7 of which are in the same service (AM or FM), in a radio market with between 30 and 44 radio stations.

Under the local television ownership rule, a single entity may own two television stations in the same local market if (1) the so-called Grade B or 'acceptable' signals of the stations do not overlap or (2) at least one of the stations in the combination is not ranked among the top four stations in terms of audience share and at least eight independently owned and operating commercial or noncommercial full-power broadcast television stations would remain in the market after the combination.

Four factors are part of the FCC's analysis of any proposed combination: (1) the extent to which cross-ownership will serve to increase the amount of local news disseminated through the affected media outlets in the combination; (2) whether each affected media outlet in the combination will exercise its own independent news judgment; (3) the level of ownership concentration in the DMA; and (4) the financial condition of the newspaper or broadcast station and, if the newspaper or broadcast station is in financial distress, the owner makes a commitment to invest significantly in newsroom operations. Public interest groups argue that relaxing the ownership rules is detrimental to the public because it limits the number of voices or opinions that reach the audience. Economic circumstances are cyclical; rule changes during an economic slump might not be appropriate during robust economic times. What is known, however, is that public tastes and expectations are changing. Rules enacted prior to the advent of cable and the Internet may not be appropriate for the changing viewer marketplace.

CROSS-OWNERSHIP

One of the biggest issues in broadcast station ownership involves cross-ownership. Mergers between large media companies often result in one company owning both a print publishing business and a broadcast business.

Cross-ownership refers to an individual or company owning a newspaper and a broadcast station in the same market. The federal government has generally sought to maintain diversity in broadcasting and to keep as many voices and viewpoints on the air as possible, but diversity becomes more difficult to maintain as more cable and online content competes with newspapers and broadcast stations. The cross-ownership investigations began in 1941, when Congress noted that newspaper companies owned many of the early broadcast stations. This pattern of ownership of two media outlets was of particular concern when the newspaper and the broadcast station were located in the same market. (Interestingly, newspaper ownership of broadcast stations first came as a result of government encouragement; see Chapter 2.) The cross-ownership rules were formalized in 1970. Until a rule change in 1992, the FCC would not allow one company to own more than one AM and FM station or more than one television station in the same market. Cross-ownership of broadcast stations is now permitted;

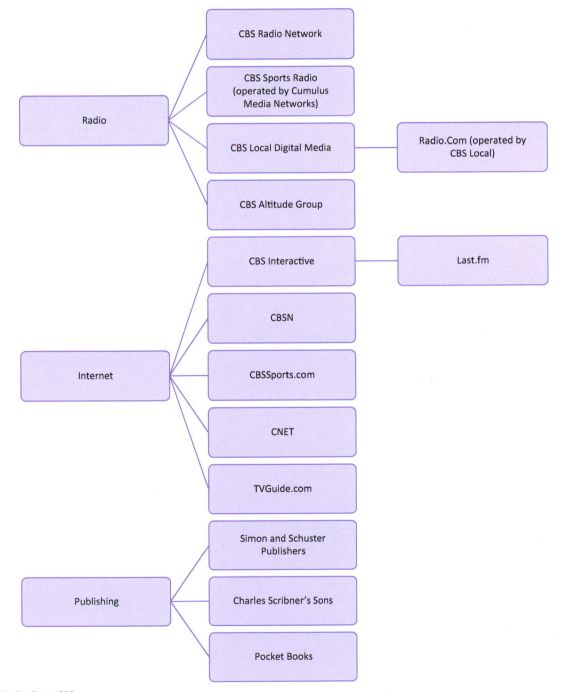

FIG. 10.2 A Media Giant: CBS
Source: CBS Corporation, www.cbscorporation.com/portfolio.php?division=96CBS

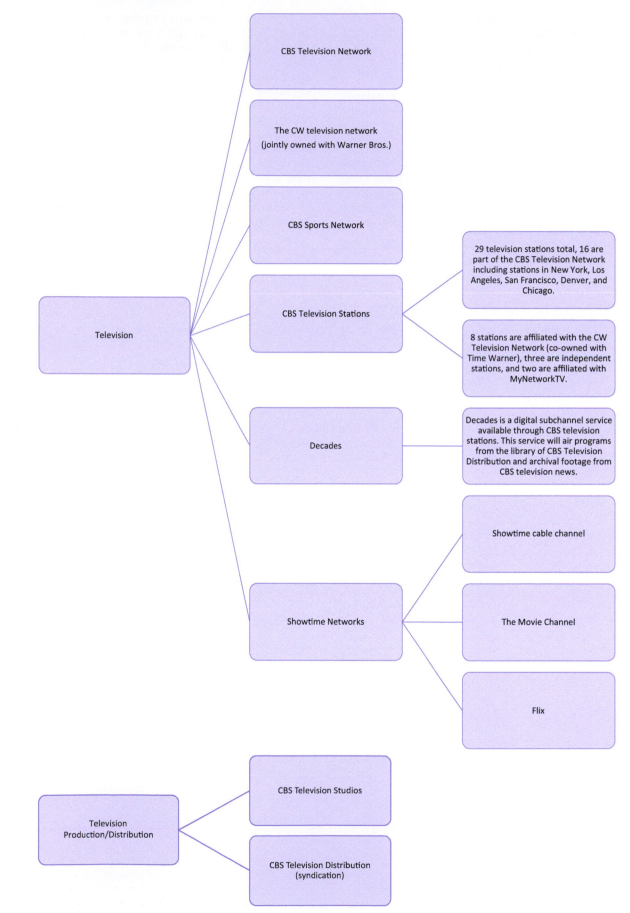

FIG. 10.2 (Continued)

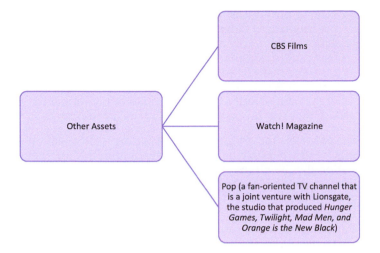

CBS Films

Other Assets

Watch! Magazine

Pop (a fan-oriented TV channel that is a joint venture with Lionsgate, the studio that produced *Hunger Games, Twilight, Mad Men, and Orange is the New Black*)

FIG. 10.2 (Continued)

even newspaper and broadcast cross-ownership is permitted when conditions of financial distress are presented. It remains to be seen whether financial distress will take on a new definition as 'old media' compete with 'new media.'

The FCC rules prohibit cross ownership of a daily newspaper and a full-power broadcast station (AM, FM, or TV) if the station's main broadcast signal encompasses the newspaper's city of publication. This rule has prevented some large mergers and has forced some media companies to split their newspaper business from their broadcast business. For example, Gannett owned both the major daily newspaper and an NBC television affiliate station in Phoenix, Arizona. Despite the attempt by Gannett and many media companies to have the cross-ownership rule relaxed (or abolished), the FCC has maintained the rule. In order to comply, Gannett has split off its newspaper business from its broadcast business.

ETHICAL GUIDELINES

Organizations such as professional associations often create formal ethical codes or guidelines for their members. These guidelines are provided to help individuals in the group make good decisions and to enhance the reputation of the group in general. Sometimes referred to as *applied ethics*, such guidelines can give group members specific information about what is considered acceptable or ethical behavior. When a group publishes a code of ethics, it is endorsed by the membership as a whole.

FCC regulations are meant to guide the operations of broadcast television and radio stations and networks. For instance, FCC rules cover technical and ownership issues, hiring practices, programming content, and ethics to ensure that stations and networks operate in the interest of the public. For example, broadcast stations may not conduct lotteries (although stations may accept advertising for government-run lotteries), operate with more power than their license permits, or refuse to sell advertising to a political candidate if his or her opponent has already purchased advertising on that station. Even though most of the regulations are fairly straightforward, situations arise that might not violate a rule per se but may be unethical. For example, is it okay for a station

account representative to sell advertising to a client who will not benefit from the spot? Should a disc jockey play a song performed by his friend's band and tout it as the number-one hit in the region, even if it is not? Is it okay for a news producer to decide not to air a negative story about an advertiser?

These are not legal issues, but they are questions of ethics. Ethics are the set of moral principles or values that guide behavior. In a situation in which there are no specific legal guidelines, ethics are often relied on to help make the best decision.

In electronic media, an example of an ethical code is the one from the **Radio Television Digital News Association (RTDNA)**. Issued to all members of the organization, this code recommends how electronic media news personnel should collect and report the news. The **National Association of Broadcasters (NAB)** established a code of ethics in 1929 for radio and another one for television in 1952. Both covered issues in programming and advertising and were amended over time to keep pace with changes in the business and society.

ZOOM IN 10.2

Go to www.rtdna.org/content/rtdna_code_of_ethics to see the RTDNA Code of Ethics.

In general, ethical codes are good resources for personnel. Station managers, advertising salespeople, and programmers benefit from having a standard set of ethical guidelines to help them make everyday decisions in the workplace. However, in the case of the NAB guidelines, some members of the organization felt that the guidelines were too restrictive and did not translate well in providing guidance in specific situations. Specifically, the NAB code created some concern among members who felt that its standard of limiting advertising minutes per hour of programming hampered their ability to make money for the station. In fact, this advertising limitation led to an antitrust suit alleging that limiting advertising minutes per hour was forcing higher advertising prices. As a result of this litigation and other complaints, the NAB revoked its code of ethics in 1983.

Another issue regarding ethical guidelines is whose perspective is being followed. Deciding what perspective should guide business behavior puts ethics more in the category of art than science. Resolutions of ethical dilemmas are not exact but rather ongoing and evolving just as standards in society change with the times. Moreover, individuals will differ as to what constitutes ethical behavior at any given point in time. Only when there is widespread agreement about what is ethical does an ethical standard become a law.

Broadcast news operations frequently face ethical dilemmas. Though it is not possible to list every possible situation, some of the most common are discussed below.

- Reporting: Reporters are often pressured to cover or not cover a certain story. For example, suppose an advertising client of a local broadcast television station is opening a new store in the mall and insists that the grand opening is a newsworthy event that should be covered by several reporters and a videographer. Generally, broadcast stations do not consider this type of event important news and would choose not to cover it. But what if the client insists, saying that if the station does not cover the grand opening on the news, the store will withdraw all advertising from the station?
- Paying for interviews: Most reputable broadcast and electronic media news operations in the United States will not pay interviewees. Some will pay, however, if it gets them an exclusive interview or one that could not have been set up through any other means. The shows that practice so-called **checkbook journalism** are not typical network hard news shows but rather shows that lean toward a tabloid style of news.
- Recreating news events: In November 1992, *Dateline NBC*—a prime-time news magazine—produced a story on General Motors (GM) trucks and their tendency to explode into flames upon impact. The producers of the story tried to show how a pickup truck would ignite when hit by another vehicle but could not recreate it successfully. So to save the story and create some exciting video footage, the producers hired someone to place a small incendiary device inside the GM truck and then trigger it when the truck was struck by another vehicle.

    The plan worked perfectly, and the resulting video was dramatic. However, when GM found out that NBC had staged the scene, it slapped the network with a lawsuit. NBC publicly apologized in a *Dateline NBC* program and paid almost $2 million to GM to settle the suit. In addition, the president of NBC's news division was fired, as were others associated with the story, and the network's reputation was considerably damaged.
- Using unnamed sources: Using unnamed or anonymous sources is another practice that could be considered unethical and that could undermine the credibility of a news program or journalist. The problem with anonymous sources is that their validity usually cannot be checked using external sources. In other words, the audience cannot verify the information and facts contained in the story.
- Playing dirty tricks: Broadcasting is a highly competitive industry, to the extent that the audience is measured continuously—for instance, in markets that have overnight audience ratings that show a station's performance from the night before. Given this level of competition, broadcasters may do everything they can to gain an edge in the ratings. And when they run out of traditional tactics, they will on occasion resort to unethical ones.

For example, the programming department of a station may be so interested in the future programming of another station that it will resort to any means of finding out what that station is planning to air. Besides hiring away programming personnel from a competitor, stations have actually paid people to go through the garbage of another station to look for discarded internal memos, drafts of contracts, old research, and other bits of information that might give insight into what the competition is doing.

Competition among local newscasts is often quite intense, which sometimes prompts news operations to look for any opportunity to gain an advantage. For instance, in a market in which an attractive woman anchored the number-one news show, the number-two station tried to find her a good job in another market. In fact, the competing station recorded her newscast and, without her knowledge, sent copies to stations in distant markets that were looking for a new anchorperson.

Where should the line be drawn between aggressive competitive behavior and unethical behavior? It is difficult and sometimes even vexing, as many issues in electronic media fall in the gray area between what is and is not appropriate.

ZOOM IN 10.3

Ethical Dilemmas: What Would You Do?

- A salesperson from a radio station has sold airtime to a local restaurant, whose manager wants to see if advertising will bring in new customers for the 'two for the price of one' special. The restaurant is running spots on the station for only one day, and it has not placed ads for the special with other media. Obviously, the salesperson wants the commercial to prove successful and feels pressure to get a large crowd out to the restaurant. One option is to call his or her friends and relatives and tell them to go to the restaurant that night, mentioning that they heard the spot on the radio. Another option is for the salesperson to add bonus spots for the restaurant to that day's commercial announcements but not tell the client. Both options involve questionable ethics on the part of the salesperson.
- The musical director of a college radio station has the responsibility for calling record companies to request CDs for airplay but knows that some companies will not provide free service to small stations, especially college stations. The director is under a great deal of pressure to get new music for the station. Should he or she simply avoid mentioning to the record companies that the station is a college station to get the free records?
- During a sweeps month, a television station has a contest to win $50 in free gas during a time when gas prices have

jumped significantly. Viewers can win if they know a secret code word that will be given during a newscast. The station has its news anchors deliver numerous teases, and reporters do live shots in front of gas pumps during the newscast. In a 1-hour newscast, the station devotes 10 minutes to contest information and talk. Is this information worthy of being in the newscast, or is it strictly promotional?

In addition to professional ethics, there are also personal ethical issues that involve electronic media, especially Internet usage. Here's another example to think about:

- Since Napster gained popularity in 1999, Internet users have downloaded songs without paying for them. As a consequence, the songwriters, copyright holders, performers, and record companies that produced the songs are not paid for their work. Movies can also be downloaded from numerous sites. Should individuals continue to download music, movies, and other copyrighted material without paying for them?

DEFAMATION

Defamation is a legal term, not just an issue regarding ethics. Defamation refers to a false attack on a person's reputation that causes humiliation, ridicule, or loss of good name or makes him or her the target of hatred. Defamation is the general term for the offense, and it is further classified as either **libel** or **slander**. Slander and libel both require publication but they are distinguished based on the form of publication. *Slander* involves oral publication, such as words, sounds, or gestures that injure a subject's character or reputation. Libel, on the other hand, is the printed publication of offending words or images. Libel is considered more serious because print is traditionally regarded as being more permanent than the spoken word. The terms *libel* and *slander* are sometimes interchangeable, and a person can be both libeled and slandered in the same program. For instance, if a written script that causes harm to a person is published on a broadcaster's Web site, it is libel. If the broadcaster also airs the offending script, it becomes slander as well.

The constitutional standard for defamation comes from the U.S. Supreme Court case *New York Times v. Sullivan* (1964). In this case, a public official (Sullivan) sued *The New York Times*, accusing it of defamation of his character. In a defamation case, the burden of proof is on the person who alleges the defamation or libel. In other words, the person who is attacked must prove that the media source was in error or even negligent in publishing the story. The finding was that the *Times* was not guilty of defamation and did not show a reckless disregard for the truth. This finding established the legal standard that unless there was actual malice in the false reporting of a story, the news outlet was not liable. *New York Times v. Sullivan* made it more difficult for individuals to win libel suits. This case is one that is considered to be a landmark in protecting a free press. The result of the case was to discourage lawsuits by public officials who do not like the stories that are published about them.

Although the First Amendment and the standard established by the *New York Times v. Sullivan* case protects

journalists in their pursuit of the truth, there are some safeguards in place to compensate people who have been publicly wronged by the media.

Three criteria are applied to determine if material is libelous or slanderous: Was the material published? Does the material defame a person (or persons)? Was the person identified clearly? If the material meets all three criteria, then the person mentioned may have been libeled and/or slandered and may take legal action against the publisher of the material. A medium, however, cannot be held accountable for libel or slander if the defamatory statement is true; when charges are made after the statute of limitations has run out (i.e., 2 years in most states); when the comments have resulted from coverage of statements made by a government official or political candidate or a normally reliable source; or when the statements are known to be commentary rather than news, in other words, stated as an opinion rather than as a fact.

BROADCAST GROUP OWNERSHIP

Consolidation (more stations owned by fewer owners) increased dramatically after the passage of the Telecommunications Act of 1996. There are limits on the ownership of stations in an individual market or service area, but national ownership limits have continued to expand. Subject only to market and signal issues, there is no national limit on the number of radio stations an individual or company can own. For the television industry, no company or individual may own stations reaching more than 39% of all U.S. households. As of 2016, there are more than 118 million TV households in the U.S.

FYI: Affiliation Switching

When the Fox network helped New World Communications buy a group of television stations in 1994, it did so because it had a plan to get its own strong affiliates in major markets: The deal specified that all stations in the New World purchase would become Fox affiliates.

This touched off a mad scramble for network affiliations among stations in some major markets.

The stations in Phoenix, Arizona, underwent these interesting changes:

- KTSP (channel 10), the CBS affiliate, became KSAZ, a new Fox affiliate.
- KNXV (channel 15), the Fox affiliate, became the ABC affiliate.
- KTVK (channel 3), the ABC affiliate, became an independent station.
- KPHO (channel 5), an independent station, became the CBS affiliate.

Fox wanted to have a VHF affiliate in the Phoenix market and achieved this goal when KTSP was purchased and switched its affiliation from CBS to Fox.

Consolidation has led to a market that is increasingly **oligopolistic**. Some very large companies, including all four major networks, own 10 or more television stations. The big four are interested primarily in owning stations in the largest U.S. television markets. The five largest markets are New York, Los Angeles, Chicago, Philadelphia, and Dallas–Ft. Worth. Outside of the top 20 markets, there are station groups that own a large number of stations (e.g., group owner Nexstar owns or operates many stations through local **management service agreements** [MSAs]), because the markets are small, the national reach is considerably less than the eight stations owned by ABC. In 2014, the FCC began to increase its scrutiny of SMAs to prevent their use as a loophole to circumvent FCC ownership rules.

The number of stations owned by large media companies changes rapidly as new deals are made to strengthen overall market position. Large media companies own groups of stations and are often referred to as group owners. Some of these groups are so large that they are referred to as **supergroups**. Supergroups often own many other properties in both broadcasting and related industries. Most non–network group owners have stations affiliated with several different networks. Having a diversity of network affiliations lowers the group's business risk, because if one network performs especially poorly with viewers, the stations in that group will suffer financially. But one network's downturn usually means that other networks are doing better, and other stations in the group that are affiliated with other networks will probably benefit financially. All of the big networks own stations, and therefore the networks are group owners as well as being content providers to all of their network affiliates.

Although owners claim that group ownership becomes more efficient and thus profitable as the number of stations in a group increases, the general population fears that large groups will decrease diversity in both programming choices and opinions. Group ownership is not something that will be eliminated by regulation. The real question is how ownership prohibitions will be maintained, especially in a world where viewers increasingly do not distinguish between broadcast and cable channels and where programming can be watched online through a local station's Web site, a network Web site, or a provider such as Hulu.

Another reason that companies are buying local television stations is retransmission consent. Local stations receive money from cable companies for the privilege of having their station's programming 'retransmitted' by the cable companies. In some stations, retransmission consent is an important financial factor because it brings significant revenue to the station without additional work or expense.

FYI: Advantages and Disadvantages of Consolidation of Broadcast Stations

Advantages:

1. More resources are available for program production and creation.
2. There is a guaranteed distribution of programs to other stations in the same group.
3. Programming can be repurposed easily among stations.

4. Advertising packages can be sold for more than one station or for a combination of radio, television, or other media.
5. Radio station groups can offer a wide variety of program formats.
6. Economies of scale and efficiency help groups and their stations stay economically healthy.
7. Career opportunities available to employees can be favorable. Employees of a group are often be hired internally for similar or better jobs at other stations in the group. There are both horizontal (similar job at another station) and vertical (moving up to more responsibilities) job opportunities in a large group.
8. A group of stations has more leverage than individual stations when negotiating with syndicators and cable companies.

Disadvantages:

1. Fewer voices are heard, and therefore there is less diversity of opinion and programming.
2. Formulaic programming is often used, even though many groups have the resources to produce unique and innovative programming.
3. Large groups can undercut the advertising prices of smaller groups and thus create financial hardship for competition.
4. Career opportunities are reduced because the stations can function with fewer people running them; this downsizing also creates an overall reduction in the number of jobs available to media professionals.
5. There is less local programming, especially in news.
6. In some companies, there is anti-union sentiment, lower wages, and less job security.

The recession that began in 2007 eventually reached the nation's top media companies in 2008. In that year, the top 100 media companies showed a paltry 0.8% revenue growth. It got even worse in 2009, with many companies actually showing a decrease in revenue compared to the previous year. About 10% of the top 100 firms went into bankruptcy reorganization as a result of shrinking revenue and huge debts. Print media firms were the hardest hit, but many of the large media companies such as the Tribune Company owned both print and electronic media operations. Stock prices dropped precipitously, and one of the results of the downturn was that many stations reduced their staff sizes, and thus full-time jobs in the industry remain harder to find. In the years following the big recession, a number of large media companies (e.g., the Tribune Company) split off their print business from their other media businesses.

BROADCAST TV NETWORKS

Several factors can be identified that led to emergence, convergence, and disappearance (1946–1948) of the small networks. First, there was a growth spurt in over-the-air television stations. Television channels that were allocated to local communities saw investors lining up to apply for these new channels. The addition of new stations caused further splintering of the audience. Second, cable television's growth in systems and subscribers plus the must-carry regulation ensured that the stations would receive carriage on the local cable system providing a picture quality equal to that of existing stations. UHF and VHF stations, despite huge disparity in signal strength and quality, appeared the same when watched on cable. Third, the arrival of Fox and cable channels showed that the big three networks were vulnerable to competition. Fourth, television was becoming big business; the network buying/selling station binge among the big four led investors to consider the possibility of investing in new television networks. However, the audience share of the existing networks got smaller as the popularity of cable television grew. The availability of new cable networks gave viewers an unprecedented choice of programs. Fifth, the audience was generally shifting away from programs intended for a mass audience to niche cable programs that appealed to specific demographic groups or those with similar lifestyle interests, such as those shown on the Golf Channel or HGTV. The multitude of new cable television channels, VCRs, and availability of low-priced color receivers shifted viewership away from a single television set receiving only four or five channels to multiple-set households that could receive more than 200 channels.

With the reduction in audience share, the cost of advertising on the big three networks became more competitive, reducing the profitability of the networks. This decrease in profits made shareholders nervous and eager to consider selling the networks.

NETWORKS AND STATION COMPENSATION AND RETRANSMISSION FEES

The income for network-affiliated television stations traditionally has been comprised primarily of two revenue streams. One is selling advertising, both to local merchants or services and to national advertisers. The second has been the **station compensation** given to the local station in exchange for airing network programs and commercials. A third component, in practice since the 1990s, is **retransmission consent**.

The compensation model of the networks paying their affiliates for airtime is complicated, depending on which network, which market, and which stations as well as the economic situation at the time.

Relationships between the networks and local affiliates continue to evolve. Local stations want to make more money from the networks through station compensation for airing network shows, and they want more advertising time available to them during prime time. Networks are reluctant to pay station compensation and now demand **reverse compensation** from affiliates who receive network programs.

For example, in 1994, after the Fox network had negotiated a contract with the National Football League to air NFC games, Fox faced having to pay its affiliated stations for clearing airtime to show the games. Fox's affiliates agreed to lower their compensation to help the network. The NFL contract helped establish Fox as a viable fourth network, and the affiliates benefited from carrying the games because of the high viewership numbers.

The higher ratings translated to higher advertising revenue. This network-affiliate arrangement opened the door to future shifts in this station–network relationship. Since that arrangement, network compensation to its affiliates has changed, with stations getting little or no compensation.

Another important revenue source for television stations is the retransmission consent payments made by cable companies to television stations for the privilege of carrying the station on their cable systems. Local cable providers are required by the FCC to include all over-the-air television channels on the cable company's basic programming lineup, the **must-carry rule**. This rule dates to a time when local television stations worried about competition from other affiliates in distant markets. Must-carry also ensured that small networks, like The CW or ION, would have an easily accessible cable channel with consistently high quality to all cable subscribers. As the number of cable channels increased, they began to demand payment from the cable and satellite companies (ultimately from the subscribers) for the exclusive programming created by the channel, such as ESPN. This payment model led local television stations to demand the right for similar payment from the cable companies, known as *retransmission consent*. Broadcast stations contend their programming is equally valuable to consumers as cable programming. Threatening to remove their signal from local cable systems gave broadcast groups the power to negotiate successfully for retransmission compensation. Retransmission fees were a small portion of station revenue until recently. In 2008, retransmission fees accounted for about 4.5% of the revenue for Texas-based television station group owner's stations. Just 4 years later, the fees accounted for 14.5% of its station's revenues. Retransmission fees diversify station revenues that formerly relied upon advertising and network compensation to pay the bills. This additional revenue source has allowed Nexstar to increase the number of stations it owns and have more leverage in dealing with the cable companies. Negotiating these fees has been contentious, sometimes resulting in standoffs between broadcasters and cable companies that led to blackouts of the stations on local cable systems. In 2015, retransmission fees paid to broadcasters amounted to approximately $5 billion.

Retransmission fees have provided needed revenue for station groups, but it complicated the relationship between affiliates and networks. Networks claim they are due a share of the retransmission revenue because their prime-time programs keep local stations in business. Local stations contend that the unique character of local news, weather, and event coverage is what keeps viewers tuning in. In 2015, retransmission fees sent from local stations and groups to the networks amounted to more than $1.5 billion. It is projected that this amount will rise to more than $3 billion by 2020.

RECENT CHANGES IN TV NETWORK OWNERSHIP

Comcast, the largest U.S. cable company, attempted to purchase the Disney company (owner of ABC and ESPN) in 2004 but was unsuccessful. It was successful in purchasing NBC from GE in early 2011. In 2014, Comcast attempted to purchase Time Warner, a deal that would have combined the two largest cable companies in the U.S., but final approval for the sale never occurred because of the concern by the FCC and the public about the creation of one media company controlling a huge portion of the cable subscription audience. Since the collapse of that proposed merger, Charter Communications, another cable company, has purchased Time Warner Cable.

MULTICHANNEL TELEVISION DELIVERY

In 2007, the FCC extended the rule requiring multichannel delivery system owners (e.g., cable and satellite television) who own or provide programming for their own delivery systems to make their channels available to the rest of the multichannel industry at a reasonable price. Without this rule, a company with large interests in broadcast and cable television, could avoid making its programming available to the direct broadcast satellite (DBS) industry. Without access to a full array of programming, satellite services cannot compete with cable.

CABLE TELEVISION

The Telecommunications Act of 1996 relaxed the rules set down in the 1992 Cable Act, essentially removing rate restrictions on all cable services except basic service. In addition, telephone companies were permitted to offer cable television services and to carry video programming. Other provisions were that cable companies in small communities were to receive immediate rate deregulation, that cable systems had to scramble sexually explicit adult programming, and that cable set-top converters could be sold in retail stores (instead of being available only as rental units from the cable companies).

The cable industry has experienced considerable consolidation. In 1970, about 20% of U.S. households in the top five markets were served by a multiple-system operator (MSO). In the top 50 markets, 63.5% were served by MSOs. By 1999, almost 60% of households in the top five markets were served by MSOs, and almost 92% of households in the top 50 markets were served by MSOs. Though television ownership is capped at 39% of household reach, efforts to regulate cable company growth have been tossed out by federal courts. An FCC–imposed limit of 30% video subscriber reach was declared "arbitrary and capricious" in 2009 by a U.S. court of appeals. Unlike broadcast stations, which involve FCC licensing of the electromagnetic spectrum, cable systems do not use the public airwaves and thus are not under the same control and scrutiny by the FCC.

Over the past 25 years, cable's market share of television subscribers has dropped significantly. In the mid 1990s, cable had about 95% of all television subscribers. Now, because of competition from satellite services and telcos, cable's market share is about 50%.

FYI: Time Warner, Time, and Time Warner Cable

Time Warner was formed after a merger between Time and Warner Communications in 1990. In 2009, it spun off its cable operations (service to homes, not cable channels) to form Time Warner Cable. In 2014, it spun off its publishing/print business to form Time Inc. Time Inc. publishes 90 magazines and owns Life.com, a Web site for news and photography.

Time Warner is now one of the world's largest producers of entertainment, including three divisions: Home Box Office, Turner Broadcast Systems, and Warner Brothers Entertainment. In late 2016, AT&T made an $86 billion bid to buy Time Warner. If this deal goes through it will create an AT&T/Time Warner mega giant in the media industry.

Recently, Comcast attempted to purchase Time Warner Cable to add subscribers to its cable business. Various groups opposed that deal, and it fell through in 2015. Another bidder, Charter Communications, bought the company shortly after the collapse of the Comcast deal.

Time Warner Entertainment

Home Box Office

HBO

CINEMAX

Turner Broadcast Systems

Cable Networks
  TBS
  TNT
  CNN
  Cartoon Network
  Turner Classic Movies
  truTV

Warner Bros. Entertainment

Warner Bros. Television

Network TV programs: *The Big Bang Theory*, *The Mentalist*, *Mike and Molly*, *Two and a Half Men*
  Cable programs: *Longmire*, *Rizzoli & Isles*
Reality programs: *The Bachelor*, *The Voice*
Warner Bros. Pictures (movie studio)

  Movies (produces 18–24 films per year): *American Sniper*, *Mad Max: Fury Road*, *Entourage*

DC Comics (largest comic book publisher in the world)

  Comic books: Superman, Batman, Green Lantern

SATELLITE TELEVISION

Since the late 1990s, two companies have been providing satellite television directly to the audience. These direct broadcast satellite companies (DBS), DirectTV and Dish-Network, began business when the size of the satellite dish required to receive a high-quality television signal was reduced to a small size (36 in. × 22 in. for DirecTV; 20-in. round for Dish Network) that could be installed without raising the ire of neighbors who objected to the large and unsightly satellite dishes of the 1970s and 1980s.

DirecTV was launched in 1994 after joining with the United States Satellite Broadcast company (USSB). At the time, the company was a subsidiary of General Motors and owned by Hughes Electronics. DirectTV bought out USSB in 1998 and acquired PrimeStar, a competitor, a year later. Numerous changes occurred in the partial ownership of the company, including the Liberty Media company and News Corp. (owned by Fox Network's Rupert Murdoch). DirectTV has about 20 million subscribers in the U.S. and an additional 18 million subscribers in Latin America. AT&T acquired the company in 2015 for the purpose of expanding its multichannel television delivery in the U.S. and in Latin America, a fast-growing market. The deal had a value of approximately $48.5 billion.

Dish Network began as part of EchoStar, a company formed in 1980 to distribute satellite dishes. Dish Network began signing up subscribers shortly after the launch of Echo-Star I in late 1995 and began sending signals to customers in 1996. After numerous acquisitions, Dish Network was spun off from parent company EchoStar in 2008. Dish Network controls programming, customer service, and marketing, while EchoStar manages the satellites and other technology aspects of the service. In 2012, the company rebranded itself as simply 'Dish.' In 2015, the company announced the start of the Sling TV service, an over-the-top (Internet-based) service that allows customers to get multichannel television service on many different devices for only $20 per month. The company markets the service as a no-commitment, no-installation, no-extra-fee service that can be cancelled anytime online without penalty.

THE INTERNET

The barriers to entry into the broadcasting, cable, or satellite business are many. For example, obtaining a broadcast license is a difficult, time-consuming, expensive endeavor, as applicants are scrutinized by the FCC and then face the prospect of a costly spectrum auction. The cable business requires a municipal franchise agreement. The satellite business requires licenses for spectrum space and communication satellites in orbit around the globe. By contrast, the Internet offers an opportunity for large corporations, with deep pockets and available technical support, down to individuals of modest means and limited technical knowledge, to send content to a potentially large audience. Mainstream media firms are using the Internet to reach audiences that otherwise might be reached only through broadcast affiliates. Television networks and cable channels use the Internet to cross-platform-promote other programming and to allow program viewing. In a broadband world, the question becomes not whether we need traditional television but how many years it will take for all programming to shift to online availability.

Sling television is a live television service that allows subscribers to watch a collection of cable television (note: not broadcast television) channels on their television sets, computers, or mobile devices. These are live television channels complete with commercials, just like those offered by a cable or satellite television service, except they are delivered directly to subscribers via the Internet. The service is offered for $20 per month and is aimed at young people who are 'cable nevers.'

SEE IT LATER

In a May 2003 speech, former FCC chairman Michael Powell stated that technology, more than politics and lobbyists, will soon drive communications public policy. His comments are still relevant today. The FCC is rooted in broadcast technology, but interest in newer technologies has been a consistent part of its new vision. Information and entertainment delivery has been moving from the traditional phone line, coaxial cable, and airwaves to wireless phone, satellite, and the Internet. It would be fair to say that technology also drives ownership changes and mergers.

Although the role of the FCC has been to guide and protect broadcasting and related electronic media, the future of the FCC may involve more emphasis on developing a broadband strategy and building a broadband infrastructure that is an "enduring engine for job creation, economic growth, innovation and investment" (Puzzanghera, 2009). In late 2014, FCC chairman Tom Wheeler stated that the underpinning of FCC broadband policy is that competition is the most effective tool for driving innovation, investment, and consumer and economic benefits.

THE FCC LOOKS TO THE FUTURE

The FCC strives to be a highly productive, adaptive, and innovative organization that maximizes the benefit to stakeholders, staff, and management from effective systems, processes, resources, and organizational culture. But it continually has the struggle to stay relevant in a media world in which technology and distribution change rapidly, and the role of the FCC to control and regulate the changes is often in question. Below are some of the FCC's goals for the future:

— Americans should have affordable access to robust and reliable broadband products and services. Regulatory policies must promote technological neutrality, competition, investment, and innovation to ensure that broadband service providers have sufficient incentive to develop and offer such products and services. The FCC is looking for ways to increase the availability of electromagnetic spectrum space to provide more broadband access. One attempt that will be put forward is a 'voluntary giveback' of spectrum space from broadcasters. Although it seems unlikely that broadcasters will give back any of their frequencies willingly, the FCC is pursuing this and other methods to offer more of the electromagnetic spectrum to broadband providers.

— Competition in providing communications services, both domestically and overseas, supports the nation's economy. The competitive framework for communications services should foster innovation and offer consumers reliable, meaningful choice in affordable services.

— Efficient and effective use of the electromagnetic spectrum domestically and internationally promotes the growth and rapid deployment of innovative and efficient communications technologies and services.

— The nation's media regulations must promote competition and diversity and facilitate the transition to digital modes of delivery.

— Communications during emergencies and crises must be available for public safety, health, defense, and emergency personnel, as well as all consumers in need. The nation's critical communications infrastructure must be reliable, interoperable, redundant, and rapidly restorable.

BROADCAST STATIONS

RADIO

The radio industry will continue to consolidate, creating an oligopolistic industry in which a few gigantic group owners dominate the major markets. For most of its history, individuals or small groups owned radio stations. Much of the focus of the industry, at least since the 1950s, has been local. This focus has changed as the industry became consolidated and large radio groups were formed. Consolidation has resulted in a move away from localism and local entrepreneurship to a more corporate style of business that emphasizes cost cutting to pay for the enormous debt that financed radio's rapid expansion after 1996. In many cases, the debt repayment has had to be deferred or refinanced, often at a higher interest rate. This debt essentially makes radio stations less attractive to investors, especially small investors interested in owning a small radio station or even a small group of stations. The furious pace of station buying and selling over the past 20 years has eliminated most potential station buyers except those with deep pockets capable of purchasing stations.

The audience may be changing as well. The future of radio may well be in the mobile audience. Autos still have dashboard radios, and that means a captive audience may still be available for many years. Instead of transistor radios, popular since the 1960s, and then Walkman-style (first tape then CD) players, the audience of the future will probably best be reached through their smartphones. HD radio, though a provider of multiple signals from one channel and excellent sound quality, has not caught on to a great extent. Smartphones could easily become radio receivers, but it is more likely that streaming technology will be the best access to radio. Generally, radio stations have adhered to the old model of over-the-air reception, which will most likely become less relevant to the audience of the future.

TELEVISION

The finances of television stations will be strained for years to come because of the transition from analog to digital broadcasting coupled with a weak advertising economy, decreasing audience shares and a variety of competing methods of accessing video programming. The FCC–mandated conversion to digital transmission in 2009 cost stations millions of dollars for the equipment and engineering. Digital transmission does allow television stations to broadcast more than one program channel over their digital signal. Most consumers watch local television signals provided by cable or satellite providers. Some cable systems do provide the digital subchannels on their systems, but these channels have not flourished or provided large influxes of revenue for broadcasters. Stations will continue to change hands, especially as economic weakness chips away at previously flush profits. Changes in the national television ownership cap, television duopoly rules, or television–newspaper cross-ownership rules could lead to additional consolidation. Erosion of the television audience by cable, satellite, and the Internet presents a serious threat to those stations without signature local news broadcasts.

Networks own the vast majority of their programming, and the flow of programming is changing. Although programs do progress from the networks to services like Hulu and to syndication, some are now going from the networks to other network-owned outlets.

Cable company mergers and broadcast group owner consolidation could change retransmission consent deals, and the network–affiliate relationship could also change. It is predicted that in 2020, cable companies will pay $9 billion to broadcasters in retransmission fees. This amount of money is an incentive for television station groups to continue to consolidate. Big station groups have bargaining leverage in retransmission negotiations. Retransmission fees give a financial boost to stations that have seen decreases in local ad sales.

It is difficult to predict, but one possible approach could be for a cable company like Comcast that owns NBC to bypass local affiliates altogether and deliver NBC programming directly to its enormous number of cable customers. Local, over-the-air television can survive, but there is likely to be a similar upheaval in television as has taken place among newspapers. As dire as circumstances have been for newspapers, television could be worse; there are more television stations in a local service area than there are newspapers. Few of these stations could survive without a national network partner or strong program provider. Some stations have relied upon increasing local news to compensate for the lack of network programming or the high cost of network programming.

If broadcast networks reach audiences directly through the Internet or mobile TV without the help of their local affiliates, local stations will have a serious problem. The broadcast network model traditionally has been that networks supply programs to local stations in exchange for reaching the local audiences. Because the local station gives up program and advertising time to the network, the networks previously paid station compensation to their local affiliates to reach the audience. In the future, the networks may decide to reach their audiences directly and stop paying station compensation. Currently, networks pay station compensation in major markets only. Networks may choose to affiliate with cable or satellite channels or simply expect audiences to watch programs at their online sites. For example, if viewers in Phoenix want to watch *Saturday Night Live* on NBC, they might go directly to NBC.com or Hulu.com instead of tuning to the local NBC affiliate, KPNX-TV, channel 12. If the networks drop local affiliation agreements, the local stations will be working hard to find programming and advertising to fill the void left by network programming. This situation would be somewhat similar to the dilemma that local radio stations faced when the radio networks morphed into television networks in the late 1940s and early 1950s. If this does happen, television stations will have a very difficult time filling in for the missing network programs, especially in prime time.

ZOOM IN 10.4

CBS Goes 'Over the Top'
In October, 2014, just a day after HBO announced a direct-to-audience streaming service, CBS announced its own digital service that allows the audience to stream its service directly without a cable or satellite subscription.

Full seasons of its prime-time shows, as well as daytime and late-night programming, are available, along with a library of some older shows. New shows are made available the day after the show airs on its broadcast network.

The service initially cost $5.99 per month. Live streaming of CBS shows is available in some large markets where there is a CBS television station.

This All Access service does not include CBS's NFL games. The All Access service is set to appeal to loyal CBS television viewers and 'cable nevers,' young people who have never subscribed to cable or satellite television but have access to Internet connections either at home or through their smartphones.

CABLE AND SATELLITE TELEVISION

The companies in the cable and satellite businesses that now sell subscriptions to audiences are trying to devise a strategy that will avoid the problems experienced by the music and newspaper industries. Free music-streaming sites and numerous news and news-blogging sites have thoroughly undermined music and newspapers, formerly big moneymakers. Now that streaming television is easy, inexpensive, and sometimes free, satellite and cable companies are looking for ways to protect their business model. Essentially, they are trying to make sure that their content is available to paying audiences, but out of the reach of nonpaying audiences. Subscribers to any cable or satellite service get an identification number or password that allows them to access online programs that are owned by the companies. In other words, the companies that have the rights to programming channels will collect fees from subscribers regardless of how the subscribers access the programs. A paywall means that subscribers will have more reason to bypass their cable or satellite service and go straight to the Internet for television programs. As more and more audience members rely solely on the Internet for television programs, paywalls might save the companies from a battle with the Internet that they cannot win.

The number of cable television subscriptions will continue to drop because of a variety of factors. First is the simple competition that comes from satellite delivery and free over-the-air broadcast signals. Second is the trend of cable companies raising prices to compensate for subscriber losses. Third, cable companies are some of the most mistrusted and maligned companies in the country. Fourth, cable most young people today have no fear of cutting the cord. Their experience with watching television on computers, game consoles, tablets, and smartphones gives them more opportunities to get television programs than older generations. The trend in going over the top to send programs from content providers to audiences has been supported by some of the biggest technology companies—Apple (Apple TV), Google (Chromecast), and Amazon (FireTV). Cable will be the most common Internet provider to homes and will therefore maintain many customers who just want broadband Internet service.

ZOOM IN 10.5

Least-Liked Companies: Time Warner Cable and Comcast
Each year, the University of Michigan conducts a study called the American Consumer Satisfaction Index. This study polls Americans to find out which household brands are liked or disliked the most. In 2015, Time Warner Cable was ranked lowest among 230 brands. In fact, Time Warner Cable ranked in the two lowest positions, one for its Internet service and one for its cable television service. Comcast was also rated poorly. The study found that high prices, slow data speeds, and unreliable service caused these companies to be rated low by consumers (Hess & McIntyre, 2015). Unless these and similar companies pay close attention, they may find that consumers look elsewhere for television delivery and Internet provider services.

Cable companies may maintain their business, or lose it more slowly, if they decide to adopt an 'a la carte' approach to their business. Consumers dislike paying for many channels that they rarely watch, like the shopping channels, religious channels, or foreign-language channels. Most consumers would opt to pay less for fewer channels, especially if those channels were the ones they watched most often.

Dish Network will probably initiate a multiple-channel service for a much lower price than their satellite service but with fewer channels. DirecTV, the third-largest video subscription service after Netflix and Comcast, has yet to announce a subscription service to compensate for a decrease in its subscriber base. Unlike cable, satellite is not inherently a two-way service; satellite subscribers do not have a simple uplink to the Internet as they do through cable modems. Recently purchased by AT&T, DirecTV will have more options for offering Internet service to customers and bundling the service with multichannel video delivery and telephone services.

SATELLITE RADIO

There is only one company in the satellite radio business, SiriusXM, but there are many companies in the business of providing audio services. Satellite radio provides many channels of music and audio, some with DJs, others with hosts, multiple news sources, and sports. SiriusXM faces increases in royalty payments and decreases in subscribers due to competition from Internet radio. Since newer vehicles are being manufactured with Internet connections, satellite radio's competitive advantage may be diminished. SiriusXM provides commercial-free music and a variety of exclusive nonmusic content including news, sports, talk, and comedy. However, even with an eroding subscriber base, SiriusXM supplements its revenue with advertising on many but not all channels. SiriusXM will stay in business, but Internet-connected cars provides easy connections to online audio services (Pandora, Spotify). Internet-connected cars may lead to a strong decline in SiriusXM's subscriber base.

TELCOS

Telephone companies like Verizon are providing fiber-optic connections to homes. Fiber-optics positions them to compete directly in the one-stop telecommunication business. Not content to just provide landline telephone service, these companies pursue multichannel television, telephone, and Internet service in direct competition to the cable companies. Their ability to package many services into one bill, which can duplicate the cable companies but also provide cell phone service and mobile TV, makes the telcos strong players in the electronic media future.

For example, Verizon has a multichannel programming service called FiOS that recently began to offer packages of channels to its customers. Instead of the tiered services packages that required a subscriber to pay for hundreds of bundled channels, subscribers can customize their own smaller channel packages for lower cost. Although subscribers have anxiously awaited the idea of 'a la carte' channel packages, cable, satellite, and telco companies have been reluctant to adopt newer, smaller packages of channels. In 2016, Verizon sold its FiOS business in California, Florida, and Texas to Frontier Communications for more than $10 billion. Verizon sold this part of its business to allow the company to focus on its businesses in the northeast part of the country.

THE INTERNET AND WIRELESS CONNECTIVITY

Perhaps more than the *content* of the Internet, *access* to the Internet is what is driving changes in viewing and listening habits. Dial-up connections have been abandoned in favor of broadband connections via DSL and cable modems. On college campuses and at corporate offices, hotels, and even commercial enterprises like coffee shops, grocery stores, and restaurants, wireless access is commonly available free of charge. The future of the Internet as well as the consumption of information, entertainment, and news likely belongs to wireless Internet connectivity. While wireless access is found in more and more places, the standard IEEE 802.11 wireless and 4G, **long-term evolution (LTE)** technology is available by most of the wireless industry. Connectivity puts listeners and viewers in charge of their media consumption. Faster connection speeds and the accompanying devices that work on the new networks are creating demand for user-programmed entertainment, information, and news. The networks continue to embrace the Internet as a means of reaching the audience directly. The advent of the 'full-episode players' on network sites like abc.go.com enables the networks to reach their audience without the involvement of a local broadcast affiliate.

Hulu.com (a joint venture of ABC, NBC, and Fox) has experienced huge growth since its inception on the Internet. Viewers watch full episodes of some network shows with fewer commercials without having to adhere to the linear schedule of the networks. Hulu offers paying subscribers current-season episodes of network shows with some commercial interruption. There is now an option for subscribers to pay more and receive programs with even fewer commercials than the regular Hulu service. This option, getting more programming with fewer commercials, is a growing trend. The networks are letting go of the linear programming paradigm. They are beginning to recognize that the audience expects to view programs when and where they are ready to watch and with fewer commercials.

It is perhaps not a question of whether connectivity will replace passive over-the-air broadcasting but how long it will take for the changes to be meaningful. While traditional radio and television broadcasts reach large daily audiences, user-programmed 'radio' in the form of MP3 song files and podcasts, DVR time-shifted programming, and Internet/online content viewing are increasingly becoming regular patterns of media consumption.

Some programming services are now entering a new era. HBO is offering its programming directly to audiences without going through a cable company. This over-the-top plan is simply a subscription model that depends upon the Internet as the carrier rather than cable or DBS satellite. Other popular programming services (e.g., ESPN and CBS) are trying this business model as well. This practice could be extremely troublesome for cable and satellite systems. If the audience can get its programming directly from the source, the business model for cable and satellite systems may suffer in the future. Cable and satellite systems might become extraneous middlemen.

SUMMARY

An undeniable fact about broadcasting and other electronic media is that they are businesses. They exist to generate enough revenue to cover their expenses and perhaps bring a profit to their owners.

Finding a business plan that would ensure success was the task of the early broadcasters. A number of business models were considered. The one that worked best for broadcasting was first based on toll broadcasting and then spot advertising, or the practice of charging advertisers for using airtime to reach potential customers. In other words, the broadcasters provided free programming to the audience, and the advertisers paid the broadcasters to reach that audience.

The federal government has traditionally encouraged diversity of opinion and ownership and discouraged the formation of monopolies. Since the late 1970s, the government has been relaxing electronic media regulation, thus permitting the marketplace to decide how electronic media should function in U.S. society.

A license must be obtained from the FCC to own a broadcast station. The FCC expects the prospective owner to be a citizen of this country, a person of good character, someone with enough money to build and operate the station, and someone who will program the station with an eye toward public interest, convenience, and necessity. When a new frequency becomes available, it is now commonly auctioned to the highest bidder, who must then apply and meet the requirements for licensing. To keep a license, the owner must continue to operate with the local market in mind and maintain a public file, containing documentation of employment information, letters from the public, information about requests for political advertising, and other information about the station's advertising and programming for children.

The three major networks—ABC, CBS, and NBC—dominated broadcasting for many years. Their ownership remained stable until the mid-1980s, when ABC and NBC were both sold and a new television network, Fox, began. In the mid-1990s, three new networks were formed by Warner Bros. (The WB), United Paramount (UPN, later to become The CW), and Paxson Communications (PAX)—now ION TV.

The cable industry has also undergone considerable consolidation. Small systems have been bought by MSOs, which can afford the technology, forming even larger companies. At present, there are no ownership limits on the size of cable MSOs.

The Internet presents many possibilities for audio and video program delivery—generally for relatively little cost and readily available technology. Both radio and television stations can simulcast their signals on the Internet using streaming technology or park programming online for access at other times.

Cross-ownership occurs when one company or individual owns a broadcast station and a newspaper in the same market. The FCC has traditionally forbidden these combinations but attempted to drop this restriction in 2003. A broad rule change has been delayed, but newspaper–broadcast cross-ownership is permitted in cases of financial hardship.

Technological changes will continue to influence ownership patterns. Traditional media owners will keep buying companies that provide new delivery systems. Television ownership will see further consolidation, but the audience cap will most likely remain at 39% of the national audience, preventing the largest owners gaining too large of the audience share. Broadcast viewing hours are expected to drop, while other delivery systems will enjoy increased viewing hours. Methods of accessing the Internet will change from wired broadband to increased wireless access as media consumption becomes user directed.

Business models may change as program providers are going over the top to reach audiences directly without involving cable or satellite television companies for distribution. As consumers cut the cord and cancel cable or satellite television, this could cause big changes in the ownership of these services.

ACKNOWLEDGMENT

The authors would like to acknowledge Gregory Pitts for his contribution to an earlier edition of this chapter.

BIBLIOGRAPHY

Albarran, A. (2002). *Media economics* (2nd ed.). Ames, IA: Iowa State Press.

Albarran, A. (2010). *Management of electronic media* (4th ed.). Boston: Wadsworth.

Albiniak, P. (April 19, 2009). NAB 2009: Top 25 station groups. *Broadcasting & Cable*, www.broadcastingcable.com/article/209430-NAB_2009_Top_25_Station_Groups.php

Baig, E. (2015, January 11). *Cord-cutting TV fans can rejoice.* New York: USA Today.

Brodkin, J. (2015, September 4). Verizon sale of Fios and DSL network in three states clears FCC hurdle. *ARS Technica*. Retrieved from: http://arstechnica.com/business/2015/09/verizon-gets-approval-to-sell-part-of-fios-and-dsl-territory-to-frontier/

Eile, E. (2014, August 6). Gannett to spin off publishing business. *USA Today*. Retrieved from: www.usatoday.com/story/money/business/2014/08/05/gannett-carscom-deal/13611915/

Epstein, A. (2015, April 3). The really important policy affecting the future of TV that no one is talking about. *TV Wonk*. Retrieved from: http://qz.com/375566/the-really-important-policy-affecting-the-future-of-tv-that-no-one-is-talking-about/

Erb, K. (2014, June 20). Will congress keep the Internet tax free? *Forbes*. Retrieved from: www.forbes.com/sites/kellyphillipserb/2013/05/06/theres-a-lot-wrong-with-the-internet-sales-tax-proposal-so-why-arent-more-people-talking-about-it/

FCC's review of the broadcast ownership rules. *Federal Communications Commission*. Retrieved from: www.fcc.gov

Future is bright for powerline broadband [press release] (2004, July 14). *FCC*. Retrieved from: www.fcc.gov [July 29, 2004]

Goldman, D. (February 2, 2010). Music's lost decade: Sales cut in half. *CNN.money.com*. Retrieved from www.money.cnn.com/2010/02/02/news/companies/napster_music_industry/

Greenspan, R. (2003, September 5). Home is where the network is. *Cyberatlas*. Retrieved from: cyberatlas.internet.com/big_picture/applications/article/0, 1301_3073431,00.html [July, 2004]

Greppi, M. (2001, November 19). NBC, Young and Granite roll the dice. *Electronic Media*. Retrieved from: www.craini2i.com/em/archive.mv?count53&story5em178785427096220069 [February 12, 2004]

Hess, A. & McIntyre, D. (2015, Jan. 17). 10 Most hated companies in America. *24/7 Wall St.* Retrieved from: http://247wallst.com/special-report/2015/01/14/americas-most-hated-companies/#ixzz3P7NDElCY

Higgins, J. (2004, January 5). Cable cash flow flows. *Broadcasting & Cable*. Retrieved from: www.broadcastingcable.com/archives

History of cable. (2016). *California Cable & Telecommunications Association*. Retrieved from: www.calcable.org/learn/history-of-cable/

Jessell, H. (2004, January 5). Dotcom redux. *Broadcasting & Cable*. Retrieved from: www.broadcastingcable.com/archives

Kerschbaumer, K., & Higgins, J. (2004, January 12). Utah's uncable surprise. *Broadcasting & Cable*. Retrieved from: www.broadcastingcable.com/index.asp5articlePrint&articleID5CA374134 [February 25, 2004]

Knee, A., Greenwald, B., & Seave, A. (2009). *The Curse of the Mogul: what's wrong with the world's leading media companies.* London: Portfolio.

Kuchinskas, S. (2003). Research firm says Wi-Fi will go bye-bye. *Internetnews.com*. Retrieved from: www.internetnews.com/wireless/print.php/2233951

Lind., R., & Medoff, N. (1999). Radio stations and the World Wide Web. *Journal of Radio Studies, 6*(2), 203–221.

Maglio, T. (2014, October 27). U.S. TV station retransmission fees will hit $9.3 billion by 2020, *SNL* Kagan projects. *The Wrap*. Retrieved from: www.thewrap.com/u-s-tv-station-retransmission-fees-will-hit-9–3-billion-by-2020-snl-kagan-projects/

Malone, M. (2009, February 18). Countdown begins on the great network affiliate divorce. Retrieved from: www.broadcastingcable.com/blog/Station_to_Station/11136-Countdown_Begins_on_the_Great_Network_Affiliate_Divorce.html

March, E. (2014, March 19). *TV subscriptions fall for first time as viewers cut the cord.* New York: Edmund Lee.

Naftalin, C. (2001). Strike two: FCC's EEO rules repealed for the second time in three years. *Telecommunications Law Update, 3*, 1. Retrieved from: www.hklaw.com/content/newsletters/telecom/3telecom01.pdf [February, 2004]

National Cable Television Association retrieved from this web site: https://www.ncta.com/industry-data

NBC v. United States. (1943). 319 U.S. 190.

Neiger, C. (2014, October 22). The future of cable: Bleak in the face of more cord cutters. *The Motley Fool*. Retrieved from: www.fool.com/investing/general/2014/10/22/the-future-of-cable-bleak-in-the-face-of-more-cord.aspx

Owen, B., & Wildman, S. (1992). *Video economics*. Boston: Harvard University Press.

Ownership limits split stations and networks. (July 27, 1998). *Television Digest*, p. 16.

Parsons, P., & Frieden, R. (1998). *The cable and satellite television industries*. Boston: Allyn & Bacon.

Performance royalties: The state of play. (2014, July 30). *Intellectual Property*. Retrieved from: www.commlawblog.com/2014/07/articles/intellectual-property/performance-royalties-the-state-of-play/

Puzzanghera, J. (2009, July 20). FCC chairman has broad approach to Net access. *Chicago Tribune*. *Retrieved from:* www.chicagotribune.com/business/la-fi-genachowski20-2009jul20,0,3591928.story

Retransmission fee race poses questions for TV viewers. (2013, August 2). *USA: USA Today*. Retrieved from: www.usatoday.com/story/money/business/2013/07/14/tv

Romano, A. (2002, December 20). Cable's big piece of the pie: In 2002, it had larger share of audience than broadcast networks. *Broadcasting and Cable*, p. 37.

Schechner, S., & Dana, R. (2009). Local TV stations face a fuzzy future. *Wall Street Journal*. p. A1.

Seward, Z. (2014, January 21). The race is on to launch an internet TV service in the US. *Over the Top*. Retrieved from: http://qz.com/169249/the-race-is-on-to-launch-an-internet-tv-service-in-the-us/

Smith, F., Wright, J., & Ostroff, D. (1998). *Perspectives on radio and television* (4th ed.). Mahwah, NJ: Erlbaum.

Sterling, C., & Kittross, J. (February 10, 2002). *Stay tuned: A history of American broadcasting*. Mahwah, NJ: Erlbaum.

Stiles, G. (2016, July 25). Meet the cable-cutters. *Mail Tribune*. Retrieved from: http://www.mailtribune.com/news/20160725/meet-cable cutters

Television isn't really dying—but there is a war over it. (2014, July 2). *Business Insider*. Retrieved from: www.businessinsider.com/what-people-get-wrong-about-the-death-of-tv-2014–7

Thorsberg, F. (April 20, 2001). Web radio goes silent in legal crossfire. *PC World*. Retrieved from: www.itworld.com/Tech/2427/PCW010420aid47983/pfindex.html [March 2002]

Trick, R. (May 2003). Powell: Newspapers to "fare well' in cross-ownership decision. *Presstime*, p. 8.

Veronis, S., & Stevenson Media Merchant Bank. (2003). Communication industry forecast, 2003. Retrieved from: www.vss.com [February 4, 2004]

Walker, J., & Ferguson, D. (1998). *The broadcast television industry*. Boston: Allyn & Bacon.

Weidner, J. (2015, April 21). HBO now and the future of cable TV, Retrieved from: www.investopedia.com/articles/investing/042115/hbo-now-and-future-cable-tv.asp

Media Operations 11

Contents

About 70 years ago, television won the public's hearts and eyes, much to the chagrin of the radio industry. But through station formatting and focusing on music, radio found a way to live side by side with television. As new types of music have hit the airwaves, formatting options have expanded, which has led to more competition in many radio markets. And of course, digital technology vastly improved sound quality. But otherwise radio station operations have remained basically the same.

The same can be said for television. Though technology has changed and improved picture and sound quality and new types of programs have been created and old types circle in and out of favor, television stations still operate in relatively the same ways they have for decades. Major improvements came about as improved technology led to new production techniques and digital delivery became possible.

Internet delivery of broadcast radio and television fare further complicates the role of traditional media and how they operate, especially as what were once strictly considered radio or television companies have merged with Internet, film, newspaper and other types of media corporations.

This chapter looks at how the operation, production, and distribution of electronic media programming has changed through the years, especially with the advent of cable, satellite, and Internet as program delivery systems.

SEE IT THEN AND SEE IT NOW

RADIO STATIONS

From radio's origin in the early 1920s, six basic functions have kept broadcast stations in business: general management, engineering, production, programming, sales, and promotions. These operations and business functions were carried out in much the same way in all stations. General management oversaw all business needs including personnel; engineering was in charge of recording machines, microphones, cameras, lighting, and the studio; programming decided what programs and music should air; production created and recorded the shows; sales generated revenue by filling airtime with commercials; and promotions got the word out about the station.

From the early 1950s until the mid-1990s, the radio audience grew along with revenues. The number of stations increased dramatically, and most stations realized a profit and stayed in business. Although the audience had shifted from AM to FM, most stations had a management structure that stood the test of time and worked well in both the small, independently owned stations and the stations that belonged to corporate groups.

The Telecommunications Act of 1996 kick started ownership consolidation and convergence in both radio and television. Radio stations changed their organizational chart and operations, and consolidated station jobs among group personnel became quite common.

Before 1996, a radio station often had a full array of managers and staff in each of its departments. But after 1996, some of the managers working for radio station groups performed managerial tasks for more than one station. This allocation of duties was especially common after an acquisition. For example, suppose an owner of a group of stations like Clear Channel acquires four stations in a market. Before the acquisition, each of those four stations might have had its own program director. After the acquisition, one *regional* or *market program director* would make the program decisions for all four stations. This trend also extends to other departments at the station. One sales director and one assistant director would replace the four sales directors and manage sales for all four stations. It is also common practice for salespeople to sell for more than one station and possibly for all stations in a group. Although selling for more than one station presents some conflicts, like competition among stations within the same group, these issues are small when compared to the value of reducing expenses by decreasing personnel.

In some cases, station ownership groups save money by combining the studios of several stations into one facility. Operating one large facility with several studios and on-air control rooms is often less expensive than operating several fully equipped studios at different locations—consolidation reduces overhead costs like bookkeeping, billing, subscriptions, professional membership fees, and insurance.

GENERAL MANAGEMENT

The general management of a radio station is responsible for policy planning, hiring and firing, payroll and accounting, purchasing, contract administration and fulfillment, and the maintenance of offices, studios, and workplaces. In addition, general management provides leadership, promotes communication among station departments, and makes all final decisions that affect the station and its employees.

In many cases, the general manager supervises more than one station; often the same general manager is in charge of the AM and FM operations or clusters of stations owned by the same company.

The station manager typically oversees one station and often answers to the general manager. Overall, the station manager has three important roles: to make the station function efficiently, to provide the maximum return to the owners or shareholders, and to adequately serve the commitments of the station's license. He or she is involved in day-to-day operations of the station and supervises all department managers. The station manager selects key people to manage and operate the station and is in charge of hiring and firing station personnel. The station manager must be familiar with the various labor unions active in broadcasting and must be knowledgeable about legal issues that may affect the station. Knowledge of FCC rules and regulations, as well as knowledge of local, state, and federal laws that impact the station and its operation, is crucial.

Career Tracks: Dave Kessel, General Manager, Yavapai Broadcasting

FIG. 11.1 Dave Kessel

What is your job? What do you do?
I am the general manager of Yavapai Broadcasting, which owns and operates six radio stations and a cable television facility in Northern Arizona. As GM it is my responsibility to oversee all departments within our organization: sales, news, engineering, traffic, on-air, accounting, promotions, etc. The GM is also responsible to make sure the stations are abiding by FCC guidelines, obeying all federal, state and local laws, and operating in the public interest necessity and convenience.

How long have you been doing this job?
I have been the GM of Yavapai Broadcasting since 1995. I have also held administration positions at radio stations in Flagstaff, Sedona, and Phoenix.

What was your first job in electronic media?
While attending Northern Arizona University in 1972, I took a part-time position as an evening announcer on a country-and-western station in Flagstaff. That evolved into a position as program director.

What led you to the job?
Radio broadcasting has always been my love due to the intimate relationship between announcer and the audience. It is the ultimate word-of-mouth medium. Radio entertains, informs and enlightens in a more personal way than any other entertainment

form. The closeness between the listener and their favorite station really makes radio the ultimate 'social media.'

What advice would you have for students who might want a job like yours?
The two greatest assets an aspiring broadcaster can have are a well-rounded education and a positive attitude.

Be versed in all of the liberal arts: history, art, literature, political science, philosophy, etc. Having the ability to talk on a wide range of topics and interests is a huge asset in any profession but especially in the communications industry. Don't focus so myopically on your chosen field that you lose sight of the things that are important to the people you want to communicate with. Loyalty and trust are most easily found when there is some common ground. That is true for business relationships as well as between communicator and audience.

A positive, upbeat, and confident attitude can go a long way in compensating for a lack of experience. Demonstrating a willingness to learn and putting in the time and effort to gather experience can open doors. Attitudes are contagious—both positive and negative. Make sure what you are projecting is positive.

Important aspects of an effective broadcast manager are flexibility, conviction tempered with compassion, people and motivational skills, financial understanding, communication skills, community involvement, engineering fundamentals, awareness of laws and regulations, and the ability to set goals and work toward achieving them.

ENGINEERING

Broadcasting is the result of scientific and engineering discoveries and applications. Since the beginning of broadcasting, the technical functions of a station included the construction, maintenance, and supervision of a station's broadcast equipment. The chief engineer—who maintains the transmitter and antenna, manages all recordings, supports operations, is the liaison with the FCC and labor unions and installs and maintains all broadcast equipment—oversees the engineering department of the station.

Every station has a chief engineer (either full or part time) who is guided by station managers and the FCC. The chief engineer's basic responsibility is to provide a strong and clear signal. To do so, the engineer must make sure that station's transmissions complies with government rules regarding the station's power, frequency, antenna location and height, and hours of operation.

The engineering department (which often includes information technology, or IT) is responsible for the operation, care, and installation of technical broadcast equipment. In some stations, this department is now split into an engineering department, which maintains and installs the transmitter and all technical equipment, and an operations department, which runs the equipment and makes sure that there is appropriate workflow from one area of the station to another. *Workflow* refers to a process like making sure that the new commercial gets to where it needs to be for airing at the right time.

The chief engineer hires personnel to install and maintain the equipment and also makes recommendations to the station manager or general manager about new technologies and equipment needed by the station. Knowledge about computers and computer networks has become much more important recently, because all broadcast stations now rely heavily on computer technology for daily operation.

PRODUCTION

A basic radio production system consisted of microphones to pick up sounds or phonograph records and magnetic recordings of prerecorded shows. The audio signal then travels to an audio console or audio board, which modifies sound by increasing or decreasing the volume, changing the frequency, removing parts of the sound, mixing and combining the audio with other audio, and then sends the combined signal to the station transmitter. The radio station transmitter combines the audio signal with an electromagnetic wave signal, which is then sent to the transmitting antenna. The combined signal is broadcast from the transmitting antenna in the form of invisible electromagnetic waves on the frequency assigned to the station.

Television production consists of studio television that may be live, as in television news, or recorded for later playback, as in a public-service program. Television also includes recording in the field in two primary areas: news-gathering and commercial production.

PROGRAMMING

The programming functions of early radio included securing programming from the network if the station was owned and operated by the network or a network affiliate and creating enough programming to fill the rest of the broadcast transmission schedule. The programming department, led by the program director, was responsible for planning ahead and implementing (or executing) the programming plan. Programming personnel selected and scheduled programming based on the decisions and guidelines set by general management.

In addition to network programs, local radio stations produced their own shows. Some programs originated in the station's studio. Sometimes musicians and singers would play live from the station's studio. Talk shows, news shows, and live readings were also broadcast from the station. Later, as broadcast technology improved, a station could air music from a remote location, such as a

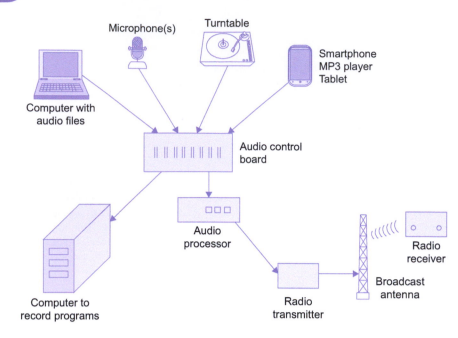

DIAGRAM 11.1 Audio sources like a computer, microphones(s), turntable, and audio player (smartphone, MP3 player, or tablet) supply signals to the audio control board. The control board or audio console, mixes, and controls the audio signals and controls the level of each signal. Then audio control board (or audio console) combines the signal(s) and sends it out to a recording device (computer) and to an audio processor, which then sends the signal on to the radio transmitter. The transmitter creates a radio wave on a specific frequency (e.g., 107.1 mHz) and sends the signal to an antenna. The signal is then broadcast out from the antenna to radio receivers in homes and cars.

concert hall. The term **live remote** was coined to describe these types of programs. Live music was sent via telephone wires to the radio station studio, and from there to the station's transmitter, which broadcast the signal over the airwaves. Later live remotes covered sports matches and other events. Live remotes were often sponsored by the concert hall, hotel, ballroom, arena, or wherever the action was taking place. Some live remotes were aired daily. Once all the microphones and telephone lines were in place, the station engineer merely had to flip a switch to get the remote on the air. Some stations relied heavily on live remotes. For example, in the 1920s, station WBT in Charlotte, North Carolina, played live band remotes from the Charlotte Hotel every day from 12:30 p.m. to 2:00 p.m.

Musicians were often on a station's payroll. It was not unusual for some radio stations to have 10 or more musicians on the full-time staff. The studio musicians performed live numbers, announcers' background music, or even a sponsor's jingle. The musicians performed in the station's sound studio, which was equipped with microphones that were connected to an audio console that controlled the sound level and combined or mixed the sound from the various microphones in the studio.

Before the advent of audio magnetic tape, live programs could not be stored for later use, nor could programs be preproduced for play at a later time. If, for example, a performer was caught in a traffic jam, suddenly got sick, or otherwise did not get to the station by airtime, no tape was available to fill time until a substitute performer could be found. When these kinds of problems came up, the station often relied on the studio musicians to perform on short notice.

Beginning in the 1920s, local station audio programs were sometimes recorded on phonograph records, referred to as **electrical transcriptions (ETs)**, using the 78 RPM (revolutions per minute) speed for music or the slower 33 1/3 RPM speed for voice. At either speed the sound quality

was crackly and poor. By the late 1930s, recording equipment used steel tape or wire, and though the sound was better, it still was not very clear.

Radio networks wanted only the best-quality shows on the air and did not allow their affiliate stations to air recorded programing. Thus, big stars like crooner Bing Crosby were forced to perform two live shows a night so that each could be played in different time zones. Wanting to make life easier on himself and other radio stars, Bing Crosby started the Crosby Research Foundation to seek patents on equipment that would improve audio recording. By the early 1950s, plastic-based magnetic tape was being used to produce audio recordings that were practical, affordable, and of high quality. As a result, performers could tape a live show that could be aired later in different time zones, and the high quality of the sound was intact.

The **program director (PD)** is responsible for everything that goes out over the air. He or she usually works closely with a **music director** if the station has a music format, and a **news director** if the station produces its own newscasts. Next down in the chain of command is a **chief announcer**, who supervises all the disc jockeys, newscasters, and other announcers. The PD often supervises a **director of production**, who in turn oversees the creation of commercials, promotions, and other prerecorded messages.

The PD has always been responsible for music selection and scheduling. In addition, in many stations, the news operation is part of the programming department—especially stations that are almost all music with some news. Stations with a large news component, such as those that are news–talk or sports, often have a separate news department.

The news director is responsible for writing, scheduling, and delivering newscasts throughout the day. If the station has a syndication service or a network that supplies some or all of the news, then the news director is in charge of the contractual agreements with those businesses. In

ZOOM IN 11.1

FIG. 11.2 Bing Crosby ad

Bing Crosby

During the Golden Age of Radio, the 1930s and 1940s, performers often had to perform their live shows twice: once for the east coast and once for the west coast. Bing Crosby, who worked for NBC, disliked having to do each show twice and sought a method to reduce the repetition. Until the late 1940s, the radio networks did not have much interest in recording methods because at the time, the quality of recorded programs was noticeably worse than live programs. Crosby performed in Europe during World War II and observed audiotape recordings and felt that high-quality audio recordings could simplify the lives of performers and the networks. Crosby left NBC, where there was no interest in recorded shows, and started working for ABC, which was accepting both of Crosby himself as a star performer but also of his ideas about audio recording. Crosby became the first national performer to prerecord his shows onto audio magnetic tape.

Through the medium of recording, Bing Crosby produced his radio programs with the same tools and procedures (rehearsing, doing a 'take' with high-quality recording, and editing) used in motion picture production. This style of producing radio became the industry standard.

Crosby invested money in the Ampex Corporation that created this country's first commercial reel-to-reel audio recorder. Crosby gave one of the recorders to a musician friend, Les Paul, a famous guitar player, who pioneered multitrack audio recording and is often credited as being the 'father' of the electric guitar.

---

addition, the news director is responsible for the entire staff, which may include the following:

- **News producers**, who put together news stories and news programs
- **Reporters**, who go out of the studio to gather news stories
- **Newscasters**, who deliver the news to the audience in front of the camera or microphone
- **Specialized news personnel**, like play-by-play and color announcers for sports events
- **Weather specialists and meteorologists**, who are educated in meteorology and deliver the current weather and forecasts
- **Special reporters**, who work as business analysts, environmental reporters, and other special topic journalists.

SALES

Since 1922 when WEAF in New York broadcast the first 'toll broadcasting' message, commercials have provided broadcast stations' most significant revenue stream. Businesses paid broadcasters for the privilege of reaching consumers with messages about products and services. Commercials became part of the basic structure of commercial radio and television in this country. Stations had two divisions in the sales department, national sales and local sales. Often the sales manager handled the national accounts (e.g., Coca-Cola) and the sales staff handled local advertising sales (e.g., Ralph's Barber Shop).

As the principal contact with businesspeople, salespeople have to be good communicators and knowledgeable about local businesses. In addition, a salesperson must like to sell and understand broadcasting, as well as the best ways to interpret business clients' needs into commercial campaigns. Most stations have a sales manager, usually an experienced sales person who also is very familiar with broadcasting. Most of the sales work is done by the sales staff; the sales people who are responsible for direct contact with businesses. As stated in *Broadcast Management* by Quaal and Brown (1976, p. 245), sales people should have "an extroverted personality, a high degree of intelligence, a gregarious nature, signs of perseverance, an ability to get along with people, a good appearance, correct manners, sincerity, a true interest in sales, and, frankly, a strong desire to make money."

In most large markets, the salespeople are specialists in selling. In smaller markets, announcers and other station personnel often sell commercials in addition to their regular station responsibilities. In small markets, the on-air personalities often became household names and are typically welcomed to visit businesses in the market, even though the visits are sales calls.

Preparing for a sales call requires a salesperson to study the past and current advertising of potential businesses. Information like the type of business, seasonal fluctuations, product quality, and general business needs of the client are important variables for salespeople to study before a sales call. Salespeople make a presentation and ask for the sale. If successful, the salesperson has to consider follow-up after the sale, which involves getting commercials written and produced, scheduled, and then accurately billed.

Program sponsorship was the standard practice for advertising on radio in the 1930s and 1940s. Advertisers enjoyed the local notoriety of being associated with a program and its host. Sponsorships were sold for a variety of programs and program lengths. Some programs were only 15 minutes in length, while others, such as baseball games, would last for several hours. The title of the show may have included the sponsor's name, such as the Eveready Hour (sponsored by Eveready batteries), or a musical show would feature a band named after the product, such as the Lucky Strike Orchestra.

Spot advertising was also sold, but was not very popular until the 1950s because it was believed that the direct association of the sponsor with a show was more effective than buying one or more of many spots included in or between shows.

Since the 1950s, spot advertising has dominated the local radio sales activities. Sales has become an increasingly sophisticated endeavor.

The sales manager hires and trains the sales staff and sets sales goals. In larger stations, the sales manager's position is often split into two positions: **local sales manager** and **national** (or **regional**) **sales manager**. In large markets, salespeople must be equipped with the latest audience rating information, the availability of spots on the station, and ability to provide advertisers with both demographic (e.g., age, income, and size) and psychographic (e.g., lifestyle) information about the audience. In larger stations, one or more employees also help with research and ratings, sometimes even conducting audience surveys to inform the sales effort.

In some stations, the sales department is responsible for scheduling commercials as well as selling them. These stations will have a **traffic manager** and staff to help with scheduling commercials and other program elements.

PROMOTIONS

Before the introduction of television, the need for radio promotion was less than it is now. Radio was unique and drew an audience as soon as radio receivers became available. It was the only electronic medium before World War II, and the audience could choose only from those radio stations they could receive in their homes. The audience did not have television, cable, satellite, or the Internet fighting for its attention. Nonetheless, stations did promote themselves to stand out from competing stations but to nowhere near the degree that they do now.

Before the 1970s, a station usually inserted a promotion only in unsold commercial time. But station managers realized the benefits of regularly scheduled self-promotion and started setting aside time for promotions when they were most likely to be seen. On-air promotions often plug programs that air later in the day or week and giveaways, contests, and special events initiated by the station. Promotions are very effective if they are aired at the same time or within the same show each day.

Radio promotes itself internally and externally for several main reasons:

1. **Audience acquisition:** To entice people who do not listen to sample the station
2. **Audience maintenance:** To urge listeners to continue to listen to the station
3. **Audience recycling:** To give listeners a reason to listen to the station on various days and at different times of the day (vertical recycling gets the audience to return later in the same day, and horizontal recycling gets the audience to listen during the same time period each day of the week)
4. **Sales promotion:** Radio stations promote themselves to advertisers so they will buy advertising on the station.
5. **Morale building:** Stations also set up in-station promotions to generate and maintain the sales staff's energy and self-motivation.

Promotion has become an integral part of the marketing strategy for broadcast radio stations. Building an audience for programs or formats in a competitive environment requires a careful and energetic plan, and the promotion must reach the potential audience. In addition, effective promotions are interesting enough to catch the attention of the intended audience.

Promotions are expected to generate ratings, revenue, and goodwill (i.e., to enhance the image of the station). These goals have to be accomplished while making sure that the promotional materials are in good taste, congruent with the overall station image, and realistic and factual enough not to create false expectations in the minds of the audience or advertisers.

The most effective promotional tool of a station is its own airwaves. On-air promotion includes simple promotional announcements about upcoming programs, station call letters and slogans ("KRCK radio: RoCK Radio at its best!") and special promotions like contests. Radio stations also use a variety of other media such as local newspapers, magazines, and social media to promote the station.

TELEVISION STATIONS

Television adopted the network radio model for business and operations and became immediately successful. Early television stations looked to radio operations to set up their enterprises. They adopted the same six basic functions as radio: general management, engineering, production, programming, sales, and promotions.

Some of the basic ways television stations operated in the past have carried over into the present. Technological improvement and innovation, however, have brought about changes in management structure and the types of jobs performed in modern stations. Production techniques are very different than they were ten years ago.

Operating a television station involves many different organizational schemes. Each group or station in a group might have a slightly different organization, but the departments discussed in the following sections are the ones that are most common to television stations today.

(A)

Typical Structure of a Small Radio Station

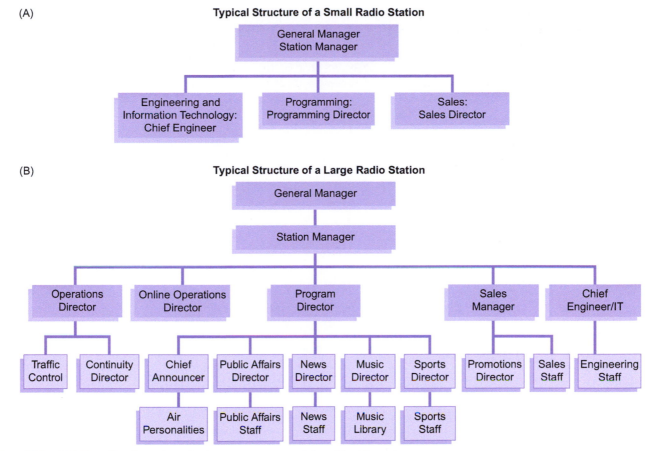

(B)

Typical Structure of a Large Radio Station

FIG. 11.3A & 11.3B These flowcharts show the departments commonly found in radio stations.
Source: Sherman, 1995

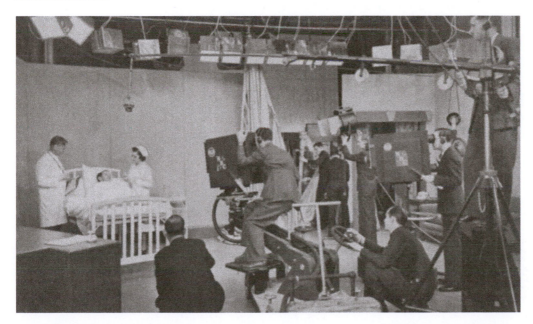

FIG. 11.4 In the early days, a television studio camera often required many people to operate it.
Photo courtesy of MZTV Museum

GENERAL MANAGEMENT

Operating a television station before the 1950s required many more people than were needed to operate a radio station. Moreover, television personnel had to perform a variety of tasks, many of which were different from those in radio. In particular, the visual element of television created the need for makeup, hairstyling, lighting, and costumes plus a wide range of sets and props.

The FYI on the next page shows the many departments and jobs that were needed to run a large television station in the 1950s and the modern-day equivalents. Note the similarities and differences between the jobs and personnel of 1950s television production and contemporary studio functions.

The general manager of a television station is responsible for all of its departments. The title of this person varies by

FYI: Television Station Departments and Job Titles: 1950s versus Now

1950s Jobs		Modern-Day Jobs
Department	Personnel	General manager
Executive Offices	Station manager	Sales manager
	Sales manager	• Sales associates
	Program manager	News director
	Engineering	• Executive producers
Program Production	Writers	• Producers
	Directors	• Writers
Engineering	Engineers	• Editors
	Operating maintenance	• Graphic designers
		• Assignment editors
Sales and Service	Salespersons	• Photographers
		• Reporters/Anchors
		• Sports
Scenic	Designers	Director of engineering
	Artists	• Managers
Carpentry	Carpenters	• Maintenance engineers
Property Shop	Property workers	• Operating engineers
Electrical	Electricians	• Building maintenance engineer/Carpenter
Visual and Sound Effects	Effects specialists	
Paint Shop	Artists	
	Painters	Program development manager
Wardrobe	Costumers	• Producers
Control Room	Operating engineers	• Directors
Studio	Operating crew	• Editors
	Actors	• Promotions editors/Photographers
Dressing Rooms	Makeup artists	
	Hairdressers	Programming director
Film Studio	Projectionist	Traffic director
	Operating engineer	Marketing/Community relations director
	Librarian	• Station promoters
Master Control	Operating engineers	IT administrator (computer network/telephone people)
Transmitter	Operating engineers	Human resource director
	Maintenance engineers	Accounting director
		• Payroll coordinator
		• Business manager
		Interactive director
		• Web site designers/editors

company and depends on whether the person is responsible for one station or more than one. If more than one station is involved, the title of the person in charge is either regional vice president or general manager. The manager of a single station is usually called the station manager.

Whatever their title, managers typically have assistants and staff to help with planning and setting goals, evaluating employee performance, hiring and firing, leading and motivating, and representing the station to the public and business community.

The general manager also oversees the business or finance department, which controls the flow of money both into and out of the station. The sales department makes the sales; the business department takes care of billing accounts receivable, such as payments that advertisers owe the station, and then collecting the money. When the station purchases equipment, like new cameras, the business department is usually consulted beforehand to see if the station can afford it. The business department also produces reports required by the government and by general management to determine the station's financial situation.

Business Department

The head of the business department is known by a number of titles, such as business manager, chief financial officer (CFO), or controller. Other personnel are accountants and bookkeepers, who record transactions and debit or credit the transactions to the appropriate station accounts. CFOs also create budgets and track expenses.

Human Resources

Stations with a large number of employees typically have a human resource department. Human resource managers recruit and hire new employees and provide and explain the benefits and services available to employees. They also conduct exit interviews and sometimes offer placement help to employees who are leaving the station.

Career Tracks: Kyle Majors, Director of Technology, Fox5 San Diego

FIG. 11.5 Kyle Majors

What is your job? What do you do?
Director of Technology. I oversee the technical operations and engineering of a Fox affiliate in San Diego, California. In this role, I am responsible for the station's technical operations, including studio, news, commercial, syndicated, network, and remote productions.

I am also responsible for general building maintenance like HVAC systems, electrical power, plumbing, and all safety and emergency systems for the building and staff.

In other words, my team makes sure anything that plugs into an outlet or takes batteries at a TV station works properly.

My role includes making yearly operating plans and budgets and accomplishing capital projects.

I keep the station in compliance with policies and regulations set by the Federal Communications Commission in order to be able to broadcast on public airways. I make sure we adhere to all local,

state, and federal laws, including the Sarbanes-Oxley Act and Occupational Safety & Health Administration or OSHA regulations.

I report to the station's general manager, who is the head of the TV station.

How long have you been doing this job?
Since 2011. Before this I was the station's operations manager and oversaw the personnel associated with the production of the news and operations of the station.

What was your first job in electronic media?
My first job was at KNAZ in Flagstaff, AZ, where I ran cameras, VTRs, audio mixers, and video switchers in the studio. I also worked as an announcer at an FM radio station.

What led you to the job?
Starting as a camera operator at a small station, I worked through many different positions at different television stations until I understood how a television station operates. Television production requires a team of people who work in unison using different systems to put on live television and cover local news. At one point in my career, the general manager of the station thought I understood how the people and systems worked together and promoted me to this position.

What advice would you have for students who might want a job like yours?
While jobs in television are collective and specialized, computer skills are necessary in all roles. Most people know how to use a computer, but we are always looking for people who understand how a computer works.

Television is part craft, part technical. It is important to know what makes good television as well as how the specialized equipment used to create television works. Those good at the what and how go far in this industry.

Engineering

The engineering department has the responsibility of installing and maintaining all broadcast equipment used by the station, which includes not only the audio and video production equipment but also the signal transmitting equipment. Engineers are often classified as maintenance engineers, who keep production equipment working; operating engineers, who keep the station on the air; building/maintenance engineers, who take care of the facility; and information technology engineers, who deal with computers and telephone networking.

PRODUCTION

Television production in the 1940s and 1950s was quite a bit different than it is now. Cameras were huge, expensive, and limited in terms of technology. Because the early broadcasts were all in black and white, little consideration was given to color. Clothing or objects in the colors of pink and blue often appeared as identical tones of gray on a black-and-white screen, and colors like brown, purple, and even dark green yielded the same shade of dark gray. Moreover, to ensure that the performers' facial features would be apparent onscreen,

white face makeup and black lipstick were used. Lighting equipment was heavy and lights had to be bright enough for the cameras to capture images. The lights generated quite a bit of heat, often raising the temperature of a small television studio set to an uncomfortable level. Visual effects were primitive in these early days. Miniature sets made of cardboard were common. Titles were drawn or printed on cards and then placed in front of the camera.

Shooting outside the studio meant taking a heavy studio camera into the field and supplying it with power from an electrical outlet. (Keep in mind that cameras were not battery operated and portable until the late 1970s.) These factors limited the types of programs that could be offered live on television to those that justified the huge outlay of equipment, personnel, and vehicles to transport equipment, sets, props, and so on. On-location shooting was thus limited to sporting events, concerts, parades, and special events like political party conventions, which were scheduled far ahead of their broadcast dates.

Prior to the 1970s, editing was done by linear editing. The system had two or more videotape recorders and a device called a **controller** that gave commands to each machine. These systems had one playback machine and one record machine and were only able to perform simple edits, or 'cuts.' Using linear editing, once a sequence of scenes was edited together, new scenes could only be added to the end of the program. If one or more shots had to be added in the middle of the program, the whole program had to be rerecorded.

Rehearsal time

The production of a television drama or comedy requires the actors to know their lines. Such memorization is not needed in radio, where actors can voice their lines with a script in hand. As such, television is more like live theatre, requiring quite a bit of rehearsing for actors to learn their lines as well as their movements.

(A)

(B)

(C)

FIG.S 11.6A, 11.6B, & 11.6C Early television production included the use of cardboard cards for prompting the talent with their lines (A), cardboard miniature sets like this harbor scene (B), and other simple devices like a drum showing program credits (C).

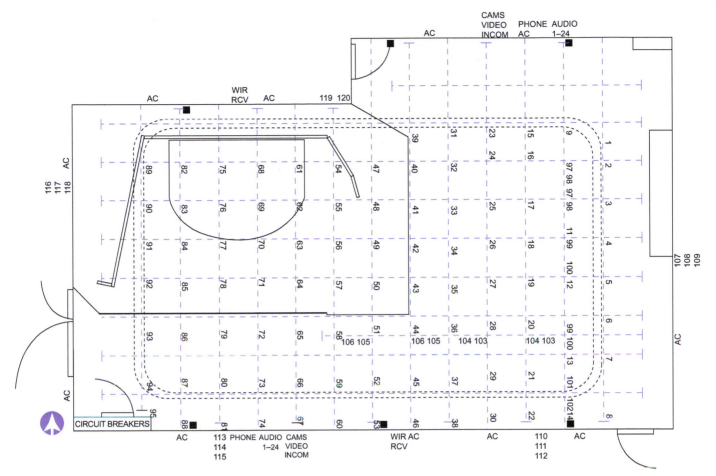

FIG. 11.7 Typical television studio floor plan. A news set is in the upper-left corner. The numbers along the dotted lines indicate the locations of lights. The connections for cameras, audio, telephone, and power are shown along the perimeter of the drawing.

A rehearsal requires everything that is needed for the final taping: lighting, props/scenery, cameras, camera operators, engineers and other personnel, and studio space. Also, since one program is rehearsing while another program is airing, the station needs to have at least two studios. For this reason, the typical broadcast schedule includes out-of-studio productions, such as sporting events (e.g., boxing, baseball), as well as in-studio productions.

Television production personnel work in a variety of departments in the station. For example, those who operate television production equipment often work in the programming department or sometimes in the engineering department. Production personnel create the **station promotional announcements (SPAs)**. Sales managers sometimes oversee the production of commercials. In a station that produces some of its own programs, production personnel work in the program development division of the programming department, and staff members who shoot news video on location, known as **news photographers**, **video journalists** (and now sometimes referred to as **multimedia journalists**), are usually part of the news department.

Camcorders

The dominant trend in camcorders is for small, lightweight cameras that store images on a flash drive. Two styles of cameras are common; the traditional video camera, which is often used with a shoulder mount, and the **DSLR (digital single-lens reflex)** camera, which is similar in style and shape to the 35mm still film cameras that have been around since the 1960s but shoot both excellent video and still pictures and record audio as well.

Studio Production

Television programs are produced either in a studio or outside the studio in the field. Studio television production requires large, specially designed spaces, lights, many electrical power outlets and vast amounts of electrical cable, portable cameras, and a crew able to take on many different functions.

A television studio is a large, open space designed to control several aspects of the production environment. The lighting and sound, for instance, are under complete control of the television crew and manipulated to

FIG. 11.8B This camera resembles the 35mm film cameras used for many years for still photography. This digital single lens reflex or DSLR camera is small and lightweight but can shoot very high-quality video.

FIG. 11.8A This portable video camera can be used in the field or in a television studio. The quality of the video it produces approaches the quality of a cinema (film) camera.

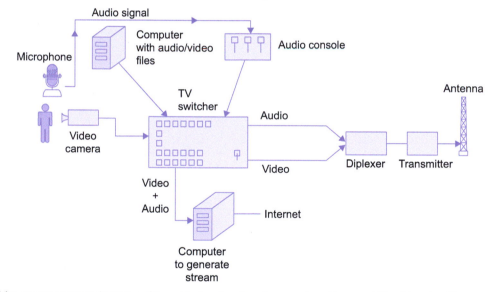

How a Television Signal Is Produced

DIAGRAM 11.2 A video camera captures the image of the subject and a microphone captures the sound. The video signal is sent to the TV switcher, while the audio signal is sent to the audio console. The video is then combined with other video signals that are generated in computers to add titles or graphic material to the signal and then sent to the diplexer. The audio can be combined with other audio sources at the audio console and then sent to the switcher or directly to the diplexer. The diplexer combines the audio and video signals, and the combined signal is then sent to the transmitter and the broadcast antenna. For nonbroadcast program sources for cable, satellite, or the Internet, instead of an antenna, the signal is sent to a satellite uplink and on to a satellite that either sends directly to the audience or to cable head ends to be distributed via cable systems. Video programs for streaming are sent to the Internet via a streaming computer that is connected to the Internet (or cloud).

fit the production. The television studio is temperature controlled so performers are comfortable under the hot lights. The studio is soundproof as well; that is, no noise from the outside world can be heard inside it. Microphones are placed in key locations around the set to maximize audio quality. Television studios are windowless to prevent unwanted light infiltrating a scene. Lighting is supplied by special video lights, most of which are attached to a system of pipes, called the lighting grid, that is attached to the ceiling. The lights are controlled from a centralized lighting board, which allows a member of the lighting crew to connect numerous lights and control their intensities for use in a production.

Studio cameras are mounted on large, heavy, roll-around devices called **pedestals**. Using a pedestal, the camera operator can make smooth and easy camera movements in any direction. The pictures from the cameras are fed through cables to a small room near the studio called the **control room**. Inside the control room, the camera cables are connected to a **camera control unit (CCU)**, which is used to control the color and brightness of each camera. The video signal travels to a device called a **switcher**, which is similar to the audio board that processes the audio signal. The switcher combines signals from other cameras, videotape, or other video storage devices.

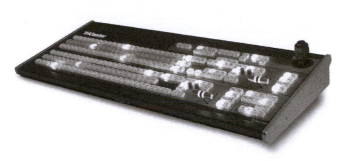

FIG. 11.9 A Tricaster TV switcher

Character generators are like sophisticated word processors. They add titles to pictures to create a **credit roll** at the end of a program, showing the names of the people who made the program, or a **crawl** across the bottom of the screen, as in the case of a severe weather watch or special bulletin. **Special effects generators (SEGs)** can make a multitude of creative changes to the video, such as slow motion, color variations, or a strobe effect.

The switcher also produces the changes between cameras or video sources that keep the program moving. The change can be a **dissolve**, in which one shot slowly changes to another; a **fade**, or a slow change from black to a picture or from a picture to black; a **cut**, or an instantaneous change from one camera to another; or a **wipe**, as when one video image pushes another off the screen.

After leaving the **switcher**, the video signal goes to one or more monitors so the director, the person in charge of the actual production, can see what each camera sees, what pictures are available via other sources (e.g., digital video recorder or special effects generators), and which picture is actually going on the air. The director can make changes as the production is happening. For example, he or she might suggest that one of the camera operators get a **close-up shot (CU)** instead of a **wide shot (WS)**. The director might also tell the **technical director (TD)**, the person who runs the switcher, to change from one camera to another to get a **medium shot (MS)** of the host of the program.

Shot changes are accomplished by using a set of commands that tell the technical director what video picture to use next. The director might tell the TD, "Dissolve to camera 2," which means that the TD should use a dissolve transition from the current video source to camera 2. The TD will push the appropriate button or move the appropriate lever on the switcher to accomplish the change requested by the director.

The video signal is then either stored on some type of storage device, like a DVR, DVD, or solid-state memory unit, or the signal can be sent directly to the transmitter for broadcasting. In some productions, like when ESPN is covering a sports event, the signal is sent to a satellite uplink so the program can be transmitted by satellite to the cable system. Some productions use the Internet to bring a remote signal to the studio.

In many larger markets, the operation of studio television has changed because of technology and budgets. Studios often had three or more cameras for news shows or studio shows; each camera had its own operator. In addition, a crew member would act as a **floor director** and direct the talent's attention to the correct camera. Since budgets have been tight and camera-control technology has become more sophisticated and cost effective, some studios are now automated. In other words, one person controls three or more studio cameras and performs all necessary camera shots and camera movements via one remote-control unit.

Numerous types of programs are produced in television studios: news, interview, talk, game, and quiz shows, along with dramas (mostly soap operas). Local news shows are almost always shot live from the local station's studio. Most other programs are prerecorded and broadcast at a later time.

Portable/Field Production

After battery-powered cameras became available for news and general production in the late 1970s, news stories and entertainment programs could readily be shot outside the studio. Field production added a sense of realism to television that was missing in the earlier days of studio television. Instead of constructing a set for each scene or program, a portable video crew could go to a location appropriate for the scene or program.

In addition, portable video has allowed news photographers to shoot breaking news and have it aired almost immediately at the station in a process known as *ENG*, or **electronic news gathering**. Stations regularly use field crews to do live shots during newscasts, which are sent to the stations via microwave. After the video is shot, a videotape editor prepares it for airing. Due to advances in technology and decreases in budgets, reporters are now sometimes expected to shoot and edit their own stories.

Editing

All editing systems are now **nonlinear** computer-based systems that allow random access to any video shot or scene without having to fast forward or fast reverse to find it. Nonlinear systems can create an array of special effects, such as slow motion, wipes and dissolves. Another highlight of a digital nonlinear system is its random access process that makes it easy for an editor to find desired shots or scenes without having to spend time fast forwarding or rewinding videotape. With nonlinear editing, shots or scenes can be easily added or removed anywhere in the program, and the computer adjusts the program length automatically. Linear editing was like composing a paper on a typewriter. If a mistake was made or new information needed to be added the whole piece had to be retyped. Nonlinear editing, on the other hand, is like using a word processing program. If a mistake is made, it is easily deleted and fixed with a few keystrokes, and new information can be added easily.

Nonlinear editing is also conducive to experimenting with creative styles and different programs endings. Depending on test audience feedback or the director's preference, the best ending is easily inserted. Several versions of commercials are also shot, and the best parts are edited into the final product. Nonlinear editing also boosts **repurposing**, or adapting programming material for different uses. For example, a 90-second story shown on a network newscast can be edited down to a 30-second version for airing on a local newscast or placed on a Web site.

PROGRAMMING

Many network radio programs made a smooth transition to television, solving at least part of the question of what to show to audiences. The television networks delivered entertainment shows the same way that radio did: through stations interconnected by telephone wires. The networks provided the programming, and the stations merely broadcast it.

Television stations did not get all of their programs from the networks, however. To fill nonnetwork times, stations either produced their own programs or found other program sources. Television programming was more complicated and expensive to produce than radio programming, requiring more rehearsal time and involving more people and more equipment than radio programs.

The programming department in a television station is managed by the program director (PD) and deals primarily with program acquisition and scheduling rather than program production. Because most television stations are either network owned and operated or network affiliates, they get most of their programs from the networks. The program director must fill the remaining airtime, usually by obtaining programming that has been produced from an outside source, like a **syndicator**. Stations use outside sources for programming, primarily because the expense of producing local television entertainment shows is very high and requires time, equipment, and often studio space. However, the station almost always produces some local programs, such as local public affairs and news shows.

Most network-affiliated television stations and independent television stations in medium and large markets produce at least one local news program each day. Local news is often very profitable and draws an audience that may keep watching other programs on the same channel. Local television news departments often have large staffs and receive strong support from general management. Most stations air newscasts at several times during the day, requiring many hours of news gathering and production, both in the studio and in the field.

The **news director (ND)** is responsible for all newscasts and personnel in the department. In addition to the ND and staff, there are producers of newscasts and news stories, as well as writers, story and script editors, assignment editors, video photographers, reporters, and anchors. Most stations also employ several weathercasters and several sportscasters.

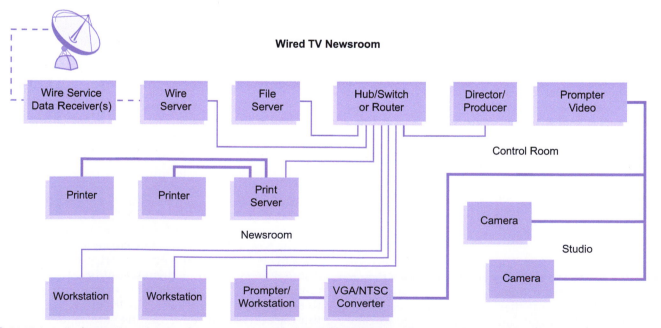

FIG. 11.10 In a wired newsroom, many people can access news information for editing and scriptwriting. The news information is sent from a wire receiver to a wire server. The files are then sent through a router to a file server and to a news director or producer. Other producers or writers can access the information from the file server at workstations in the newsroom. After the information is edited and put into script form, it can be sent directly to the prompters on the cameras in the news studio. The newscaster reads the story from the script in front of the camera.
Based on drawing from www.eznews.com

Career Tracks: Maria Hechanova, Reporter and Multimedia Journalist

FIG. 11.11 Maria Hechanova

What is your job? What do you do?

I'm a reporter/multimedia journalist for KOLD News 13, KMSB FOX 11, and TucsonNewsNow.com. I'm responsible for telling memorable stories while demonstrating strong news judgment, a keen sense of logistics, and the ability to work well under pressure. I gather, write, and present news for various newscasts including TucsonNewsNow.com, Facebook, and Twitter. I also operate my own live shots when needed.

How long have you been doing this job?

I've been an multimedia journalist since 2009. My 6-year anniversary in the business was September 2015. I've been with KOLD News 13/KMSB Fox 11/Tucson News Now for over three years.

What was your first job in electronic media?

Producer/multimedia journalist (MMJ)

What led you to the job?

I love to write. I love to talk. I love to ask questions. I took a journalism class in high school out of curiosity and discovered I could get paid for doing all three skills. Once I made up my mind that I wanted to be a broadcast journalist, I pushed to learn more about how I could make it a career.

What advice would you have for students who might want a job like yours?

Be a sponge. Watch the news. Study it. Apply for and earn a TV news internship. Ask questions. Find a mentor. Prove to your mentor you're worth his or her time. Don't get into the business just because you want to be on TV. Be persistent. Don't leave college without a resume tape.

Career Tracks: Mark Casey, Station Manager, KPNX Channel 12, Phoenix

FIG. 11.12 Mark Casey

What is your job? What do you do?

For the past 2 years, I've been station manager at KPNX-TV in Phoenix, AZ. However, for most of my career, I've been in newsroom management as a news director, executive producer, managing editor, and show producer. In each of these positions, including my current job, I've led coverage and presentation of television news reports, working with reporters, photojournalists, anchors, and the teams of people that come together to create a newscast. It's helpful that I've actually done just about every job in electronic news: reporting, photojournalism, studio production, etc.

Additionally, I've handled strategic planning, administration, recruiting, and government compliance for several newsrooms in several companies (Cox Broadcasting, Fox Broadcasting, The Walt Disney Company, and Gannett).

In recent years, our responsibilities as TV stations have expanded, and we've evolved into multiplatform news organizations. Our focus now is to meet the audience on their medium of choice with content that fits the native qualities of that platform to give them a quality user experience.

How long have you been doing this job?

I've been in journalism 41 years, almost all of it as a broadcaster.

What was your first job in electronic media?

In small-town commercial radio while at school (West Virginia University) in Morgantown, West Virginia, working days as a continuity writer (advertising copywriter) and nights as a reporter. While that experience laid the foundation, it was job #2 that stoked the fire: doing morning drive radio news in the small town of Americus, Georgia, in 1976. Ten miles down the road is Plains, Georgia, and at that time, a peanut farmer and former Georgia governor named Jimmy Carter lived there and used it as his base while running for

Career Tracks: Mark Casey, Station Manager, KPNX Channel 12, Phoenix (Continued)

president. Mornings were spent at WDEC-AM and afternoons and evenings covering the campaign and stringing for the Associated Press, United Press International, Mutual News (now Westwood One), ABC News, and the BBC. Real life, real news, and a tremendous real-time opportunity.

What led you to the job?
Luckily for me, there was never career confusion. Growing up in the 1960s and watching journalism come of age around the race to the moon, the Vietnam War, the Civil Rights movement, and the tragic political assassinations of that decade drove me into this business. Those historic, big-coverage moments captured by legends (Cronkite, *Life* Magazine, Huntley-Brinkley) were inspiring, and I just had to be in the middle of it. Fortunately, I had a support system of family and teachers at WVU who encouraged and nurtured that choice. When the job opened at the radio station, I took it even though I was already a full-time student and working another job.

What advice would you have for students who might want a job like yours?
Get out there and do it, NOW! Don't wait for an invitation, and don't worry about making mistakes. The hardware for creating journalism is in your smartphone, and the software is you serving witness to your world. Find a mentor—not a cheerleader, but a coach who will tell you what you do well and where you need work—and create journalism for that single person. Listen to the coaching and do it again and again and again until it's muscle memory.

It's also vitally important that you read, read, read. You learn about our world and its people by reading its history and absorbing the great daily deadline journalism that's produced in cities big and small. Backgrounding yourself—knowing a little bit about just about everything—is an ongoing, never-ending challenge. Embrace it! Finally, do this for the love of the craft and profession, not for money. Journalists haven't ever been high earners. If you want to make money, go to Wall Street. If you want to make a difference, be a journalist.

Career Tracks: Doug Drew, Executive Director, News Division, 602 Communications

FIG. 11.13 Doug Drew

What is your job? What do you do?
Executive director, News Division, 602 Communications. Conduct training for television stations and networks in reporting, producing, writing.

How long have you been doing this job?
More than ten years.

What was your first job in electronic media?
I was a reporter for the NBC affiliate in Flagstaff, Arizona.

What led you to the job?
I spent 4 years as a reporter for the television station at Ohio University, where I went to school. OU allows students to produce and present a half-hour newscast each night. A few of the students hold paid positions, and I was fortunate enough to obtain one of them. In addition, while I was at Ohio University, I also worked for the Associated Press Broadcast Division. With the opportunities to work at the television station and with AP while in school, in addition to my classwork in the radio-television program, I was more than ready to enter the real world of broadcast news.

What advice would you have for students who might want a job like yours?
Get involved ASAP as an intern, or work in a broadcast newsroom while still in school. The hands-on experience is invaluable. When you're applying for jobs, employers will look beyond your education to see what experience you have.

SALES

Television sales departments are similar to radio sales departments in that they are divided by national and regional/local categories. However, television stations often have more salespeople, more assistants, and more people involved in audience research than do radio stations. And because of television's larger audience, television advertising is usually more expensive and generates more dollars for the station than radio advertising does.

The sales department typically consists of a general sales manager, a national sales manager, and a local sales manager. It also employs account executives to sell spots; a traffic manager to schedule commercials; and researchers to collect, interpret, and prepare audience ratings information for use in sales. One of the easiest ways for young television professionals (especially those with management career objectives) to enter the business is through sales. The broadcast industry seeks energetic people to sell advertising time. The sales department is often the quickest path to upper-level management and has the side benefit of paying well to successful salespeople.

PROMOTIONS

A television station often has a community relations department that promotes the station and participates in community events. Another title for this department is **marketing**, as this department markets the station's product (i.e., its programs and personalities) to the audience and advertisers.

The promotions department has developed an increasingly important role since the 1970s due to industry changes—primarily the decreasing dominance of the networks, the rise in importance of local news, and the need to establish station identity among the numerous channels available from cable and satellite. The heavy competition for viewers (and, of course, advertisers) has created a promotions effort in stations that is based on consumer research, competitive positioning, long-range strategizing, and targeting specific audience segments.

CABLE SERVICES

OPERATIONS

Cable television was a small business from 1948 until the early 1970s. Most early ventures were operated by a handful of owners in a small town. Large cities had not yet been wired for cable, because franchise deals between cable companies and local governments were not in place, and free broadcast television signals were available. Early cable company owners took on many roles and often supervised many overlapping departments, including engineering, sales, and business. But since the 1980s, cable services have become well-organized businesses

with functions similar to those of radio and television stations as well as other functions unique to the cable business.

As the cable business became lucrative, it drew the attention of large media companies. They began buying up smaller cable systems in the 1970s, a trend that was later accelerated by the change brought on by satellite technology. To meet the demands of subscribers, cable systems had to become "satellite capable" to receive the new premium channels. HBO became very popular in the late 1970s, and all cable systems were pressured to carry it. Larger companies, especially those that owned other businesses, often had the capital required for the upgrade to satellite.

Larger companies also had an advantage in acquiring franchises. Small cable companies could not afford to bid on large-city franchises because the cost to build such a system was too high. Therefore, the larger franchises made deals with the bigger companies, who could raise the large amount of money needed for financing. Ultimately, many small systems had to sell out because they could not afford to upgrade the systems they owned or expand to larger, more profitable markets. Since the 1970s, most small systems have been acquired by large **MSOs (multiple-system operators)**.

GENERAL MANAGEMENT

Similar to broadcast groups that have stations in multiple markets, most cable systems have multiple systems. Therefore, general management must be able to manage distant systems with varying technological, personnel, and business needs. General management of a system is responsible to regional or national management to adhere to the principles of good business and the brand of the company that owns the systems. Most MSOs use the power and economies of scale to leverage deals with the various channels that are shown on their system. Management goals include getting the most desirable channels for the audience (e.g., ESPN, CNN, Fox News, HBO) at the best price from the channel provider and negotiate with local broadcasters to get an affordable retransmission fee plan in place.

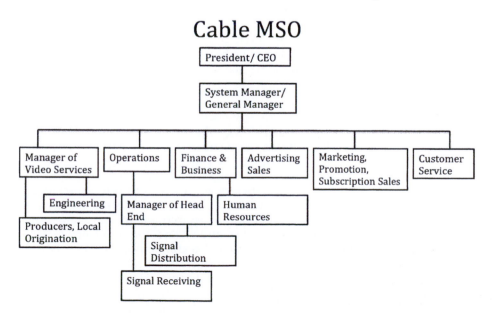

FIG. 11.14 Structure of a typical cable MSO (multiple cable system operator)

ENGINEERING

A cable service's primary function is to distribute programming to cable subscribers via cable wires, therefore engineering takes the driver's seat in cable operations. The engineering department receives television signals from both local and distant stations and sends them to the cable head end, the physical location where signals are assigned to particular channels.

Once the various television signals are sent to the head end and placed on channels, all of the channel signals are sent out through a system of shielded **coaxial cable wire** (i.e., a conductor with a metal sheath around a common axis or center). The system for distributing the signals to subscribers resembles a tree with a trunk and branches. Coaxial cable is used because it can carry many signals at once, and it provides shielding from electrical (AC power line) and signal interference from radio and television stations. Nonetheless, television signals sent through coaxial cable lose strength after traveling a distance on the wire, so amplifiers are set up at intervals along the path to enhance the signals. Newer systems are commonly wired with the use of fiber optic cable made of strands of optically pure glass as thin as a human hair that carries digital information over long distances. Compared to copper-based coaxial cable, fiber optic cable carries far more information.

Until the 1970s, most cable systems had 12 or fewer channels, and the television signals were placed on the 12 VHF channels (numbered 2 through 13), regardless of what channel each signal was originally assigned.

After HBO's satellite delivery debut in 1975 the number of cable networks available to local cable systems boomed, making it necessary for cable systems to provide more channels. Engineering's challenge was to work out a way to offer more channels when most television sets at the time were only capable of receiving 12 channels on the dial. Subscribers were offered set-top converters that basically expanded the number of channels that could be received on their television sets. Eventually, newer sets with built-in tuners capable of receiving hundreds of channels overtook the market, making channel-expanding converter boxes unnecessary.

Since the advent of HD television, most cable programs are typically received through a digital converter box or a DVR (digital video recorder). Although a direct cable wire to a television connection will work, the picture quality may not be up to digital standards nor as clear as the picture from an **HDMI (high-definition multimedia interface)** cable connection between the digital box and the television. Many of the channels distributed through a digital box are not available with a direct cable-to-TV connection.

PRODUCTION AND PROGRAMMING

Prior to the 1970s, cable companies transmitted broadcast programs to their subscribers but did not produce their own shows. Cable services began producing their own shows after the FCC mandated local-originated programming and **public-access channels**. Many systems built small television studios with sufficient equipment to make local talk or talent shows. Engineering personnel were available to keep equipment working properly. The channels provided to the local programs are referred to as **local-origination** channels. Although making local shows was an exciting idea to many, viewing the shows was not very popular. The idea that local audiences would want to watch local shows was just that . . . an idea. The reality was that local programs simply were too amateurish to garner loyal audiences large enough to justify the expense, space, and equipment needed to sustain a consistent local origination operation. Most cable systems abandoned the local-origination studio as soon as they were legally able to do so. The requirements changed from public access channels to **PEG channels**. PEG stands for public, educational, and government. The channels were designated to serve the local communities, providing channels for public information (e.g., libraries), education (local schools), and government (city council meetings, etc.).

ZOOM IN 11.3

The Federal Communications Commission issued its *Third Report and Order* in 1972, which launched hundreds of public-access television production facilities. The report contained a rule that required all cable systems in the top 100 U.S. television markets to offer three access channels, one each for public, educational, and local government use. The rule was amended in 1976 to require that cable systems in communities with 3,500 or more subscribers set aside up to four cable television channels and provide access to equipment and studios for use by the public.

Most local production in cable systems involves commercial and system promotional announcements rather than entertainment and information programs. Although some MSOs that have systems in very large markets have a local channel for news; budgets, audience size, and competition from local broadcasters make this a risky endeavor. In addition, some systems that do have a local news channel often partner with traditional noncable businesses, typically broadcast stations or newspapers, that provide some news content.

SALES

The task of a cable service's sales department was straightforward during the small-system era. The only product offered was a subscription to the cable system's television channels. A system did not insert its own locally sold commercials, and it did not have premium channels or pay-per-view programs to offer. Since only one cable system was franchised in a market, there were not any competitors for subscribers to sell against. As the small-system era ended and the MSO era began, local advertising sales became more important and more competitive with other local media outlets, such as the local television and radio stations and the local newspapers.

Today's cable world is much more competitive. Since direct broadcast satellite television became available, cable companies compete vigorously for subscribers. In addition, some markets have two or more cable companies that serve the same media markets and compete for the same subscribers. Probably the biggest change in cable sales, however, comes not from subscriptions but from advertising. There are now hundreds of advertising-supported cable networks. When a cable service picks up a cable network, for example, Comcast in Knoxville putting Food Network on its line-up, local sales reps sell commercial time on Food Network's programs to businesses in the Knoxville market. Comcast's sales reps, therefore, compete with sales reps from the local television affiliates who are also out trying to sell local businesses commercial time on television.

PROMOTIONS

Cable systems promote special television channel events or the addition of a new or desirable channel to their channel lineup to encourage subscriptions. Since competition for subscribers changed with the entry of satellite television in the mid 1990s, cable companies spend much more time and energy promoting their offerings and their 'bundles' to potential subscribers. Cable systems want customers to not only subscribe to cable television but also to their Internet service and their telephone service (VoIP). This requires cross-promotion involving other media, such as newspapers, radio, and direct mail, but also a nearly constant barrage of promotional announcements on their system's channels, filling unsold commercial time with the company's package deals for television, Internet, and telephone service. As cord cutting becomes more common, cable companies have stepped up their promotions to compensate for losing subscribers by increasing the number of people who sign on for the bundles and increasing prices.

SATELLITE DELIVERY

SATELLITE RADIO

The satellite radio industry paid the FCC more than $80 million in 1997 for the use of the frequencies for distributing its signals. The business began its operations in 2001 when XM satellite radio beamed its signal to subscribers. Sirius satellite radio began its service to subscribers in July of 2002. The two companies merged in 2008 and formed SiriusXM radio.

Although experiencing financial difficulties and slow growth in subscribers, SiriusXM continues to operate and adapt to the current media environment. The company is extending the service beyond satellite delivery by offering options for subscribers to personalize SiriusXM's commercial-free music channels using MySXM. Subscribers can also listen to thousands of hours of programming from the company's catalog using SiriusXM On Demand.

General Management

The satellite radio industry relies on an organization or structure similar to that of radio broadcasting. At the top of the organization chart is senior management, including a chief executive officer (CEO), an executive vice president, and a chief financial officer. This top layer of management has 5 areas of responsibility: engineering, production, programming, sales, and promotions. Each division or department has numerous subdivisions. For example, in the programming department, a satellite radio station will have separate subdivisions for music, news, and sports. Engineering has a satellite uplink subdivision and a studio subdivision. In satellite radio, the provider, SiriusXM, programs its music channels. Some news and sports channels are provided by other services like CNN, ESPN, and National Public Radio.

Career Tracks: Ryan Kloberdanz, National Imaging Director, News/Talk Radio, Salem Communications

FIG. 11.15 Ryan Kloberdanz

What is your job? What do you do?
I work with all of Salem's news/talk stations and help them develop and produce their on-air image.

How long have you been doing this job?
Eight years.

What was your first job in electronic media?
Nights/weekend board operator for a small station.

What led you to this job?
I always loved radio and was approached by a man that worked at the local radio station in my hometown who asked me if I wanted a part-time job. I took that part-time job, and here I am.

What advice would you have for a student who might want a job like yours?
Listen. Students have the ability to listen to stations all around the country and the world via the Internet. Listen to the production on the stations, learn different styles, and develop your own style based on what you have learned from the styles of others.

Engineering

The engineering department in satellite radio is responsible for maintaining the satellite uplink and downlink operations, construction and maintenance of studios (both on-air and production), and installation and maintenance of all production equipment. Engineering is also responsible for information technology, and owns and operates studios in New York City, Washington, DC, Los Angeles, Nashville and several other locations.

Engineering is also responsible for updating of the satellite radio receivers and the associated software. Unlike satellite television, satellite radio receivers often are part of existing radios in autos, and reception of the satellite radio signal does not require installation or special dish antennas. SiriusXM does not manufacture receivers but works closely with equipment manufacturers like Alpine, Audiovox, Cambridge Audio, Denon, JVC, and Clarion to make sure that equipment is available to the audience to receive a high-quality audio signal, in a home or in a car.

Production

Production work in satellite radio is similar to any audio production facility. Live recording takes place in an audio studio and involves a studio that does not have echoes and is isolated from sound generated outside the

FIG. 11.16 This satellite radio will receive SiriusXM satellite signals throughout the United States. On many newer model cars the standard AM/FM radio has become an AM/FM/XM radio with an additional "band" that receives the SiriusXM signal.

studio, similar to all production and broadcast station facilities. SiriusXM has many studios to supply programming for the hundreds of channels it programs. All of its programs are digital and sent to a central location for uplinking to the SiriusXM satellites. Production includes many talk shows, specialty music programs, news, and sports. Each type of show requires producers who specialize in that genre and individual producers who help produce each program. All audio sources such as prerecorded clips and microphone feeds are combined at an audio console that controls the sound level and quality.

Programming

A recent alignment of channels on SiriusXM included 72 channels of commercial-free music, 11 sports talk and play-by-play sports channels, 22 talk channels, 9 comedy channels, 15 news channels, 9 traffic and weather stations, 18 Latin stations, plus 14 'other' channels. The diversity of programming requires program directors for each of the various programming formats. In addition to the satellite channels that are similar to terrestrial radio station formats, there are numerous channels that carry play-by-play coverage of sports events. The number and specific format of the many channels in satellite radio is in constant flux. Channels change seasonally, for example, holiday music in December. As musical styles evolve, channels are added and deleted to suit subscribers' preferences.

Sales and Promotion

Satellite radio requires two kinds of selling. One type involves marketing to the general population to gain and keep subscribers. Satellite radio subscribers not only commit to a monthly subscription fee but also use special equipment to receive the programming. This is different than a decision to subscribe to video service like Netflix or audio service like Spotify that merely requires an account and an app to begin receiving the programming. Satellite radio requires the purchase of a satellite radio receiver for each listening location (e.g., home, car).

SiriusXM has more than 30 million subscribers and claims to have more than 51 million listeners in its audience. SiriusXM is a good way for national advertisers to reach a large audience with a single buy. However, the audience is splintered among the many commercial channels

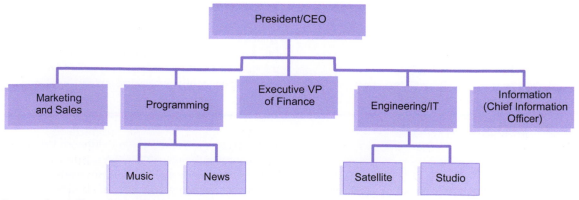

FIG. 11.17 Structure for satellite radio distribution company

offered. Advertising is placed by format (e.g., all of the comedy channels), individual channel, and region. The commercials are read live by on-air talent or produced by high-profile personalities and customized for either single spot placement or a customized segment sponsorship. Sales to advertisers requires a high level of sophistication because of the large number of channels available.

SATELLITE TV

Satellite television has existed since the 1970s, but the industry was very different than it is now. When HBO began distributing its programming to cable systems using satellites in 1975, enterprising companies began selling satellite dishes capable of receiving the HBO signal. Other cable channels followed HBO, and soon satellite **television receive only** (TVRO) dishes became somewhat common. These large dishes were popular in rural areas that were not being served by the local cable companies. At first the dishes were expensive, but by 1984, the price had dropped to about $1,000 per installation, and more than 600,000 TVRO dishes were installed in 1985. The installation charge (including the cost of the equipment) was the only charge to the customer. Once installed, the dishes received television signals that were sent to cable companies without the owners being required to subscribe to the pay channels. In other words, this practice was similar to the later practice of peer-to-peer sharing of copyrighted music and video—essentially pirating the signals.

The TVRO business changed dramatically beginning in January 1986 when HBO and other satellite-distributed channels began to scramble their signals. Once scrambled, the channels could not be viewed by TVRO owners; therefore, dish owners were no longer able to view the highly desirable pay channels. Within a year of HBO's scrambling, half of the TVRO companies went out of business. Although some channels were not scrambled, the lure of free premium channels was gone. The growth of cable into the suburbs and rural areas also contributed to the eventual demise of the TVRO business. The satellite television industry changed once again in the mid-1990s when direct broadcast satellite (DBS) companies began a subscription multichannel television delivery business. The dishes were small (18 inches across), the selection of channels was extensive, and the satellite television business was resurrected, but with a new business model.

At first the biggest disadvantage of direct satellite delivery was that the companies were restricted from delivering the local television stations' programs. The satellite industry lobbied Congress and eventually was given legal permission to deliver local station signals to satellite-subscribing homes through the Satellite Home Viewer Improvement Act of 1999. Cable and satellite still compete for subscribers, but cable has the edge because it can offer high-speed Internet service and other multichannel and audio services.

General Management

The general management of the satellite television companies has similarities with management of cable. The business of satellite television is to collect television channels from a variety of sources and bundle them into subscription viewing packages. The collection of television channels involves securing the rights to include the channels in the various bundles to market to individual homes and businesses. For example, gaining the right to send ESPN to the homes of their subscribers, management must negotiate with Disney, the owner of ESPN, regarding which ESPN channels (e.g., ESPN, ESPN2, ESPN Classic, ESPNews, ESPN Deportes, ESPN U) it will carry and for how long before renegotiations are necessary. General management oversees the acquisition of content (programming and legal departments), distribution of content and home installation and maintenance of equipment (engineering), marketing of subscriptions to the audience (promotions and sales), and business and finance (business).

Engineering

Although there was quite a bit of interest in satellite dish delivery, the early dishes were large and expensive. Viewers who bought the TVRO dishes were those who had space in their yards for the dish, could afford the dish, and could not or did not want to receive cable programming.

The two main responsibilities for satellite engineering have been to maintain the satellites in orbit that receive signals from the program channels and send them back down to earth and to provide and install satellite receiving equipment in subscribers' homes. Beginning now and in the future, satellite television channels will be expected to be online and available on all mobile devices. This requires extensive IT (information technology) work and supervision to route the outgoing television signals to servers that provide digital streams to subscribers on demand.

Production and Promotion

Satellite television companies are primarily multichannel program delivery services and currently do not produce their own entertainment programming. Their production is limited to producing informational programs about how to use the equipment, pay a bill, troubleshoot problems, or simply promotional programs about the channels offered, special free trials of channels, or information about video on demand.

ZOOM IN 11.4

Telecommunications company AT&T, a provider of multichannel video, Internet service, and telephone service, added satellite television to its services. In July of 2015, the FCC gave its approval for AT&T to merge with satellite television provider DirecTV. Amid fears of the lessening of competition in the multichannel video delivery market, the FCC and the Department of Justice felt that the combination of AT&T's land-based Internet and video business with DirecTV's satellite delivery of video does not pose a significant threat to competition and should not harm consumers.

Sales

When satellite television first started competing with cable, sales reps had the formidable task of selling an unfamiliar service. They had to explain satellite delivery, inform a potential client of its benefits, and get them to buy a huge, unsightly satellite dish. For a while, reps could use free premium cable programs as an incentive, but, as mentioned, premium cable channels soon caught on and began scrambling their signals in such a way that they could only be decoded by a special box. In general, television viewers did not think the expense of a TVRO dish was worth the additional programming.

The business changed when DBS satellite services began. Sales people were no longer selling equipment but instead were selling multichannel program delivery subscriptions. Also helping in their sales efforts was a growing distrust and dislike of cable companies' customer service. Thinking they had a lock on the market, cable representatives were often rude and surly, and subscribers they could not get help for problems with the service. Cable subscribers often changed to a satellite service out of their frustration with cable.

Satellite television has the technology to insert commercials into its programming. Satellite television offers 15-, 30-, and 60-second spots as well as longer commercial messages (such as infomercials) on unused or special channels. Dish offers advertising to tens of millions subscribers, but its advertising is addressable through the satellite receiving equipment. In other words, Dish knows who its subscribers are and can reach them through the receiving equipment. For example, if an advertiser wants to reach women, Dish can place ads in Lifetime Movie Network, MTV, E!, BBC, and other channels that have strong female viewership. Another aspect of satellite TV advertising is the ability to place a 'trigger' during an ad or program. A message is placed on the screen that tells the audience to 'press the select button now' to get more information about the product or service being advertised.

INTERNET

Most media outlets have been streaming video content to their viewers since about the 1990s. But streaming did not take hold with Internet users until major technological improvements in the late 2000s.

Streaming does not work through a middleman as cable and satellite delivery does. In other words, the audio and/or video signal information is not sent to a company that then sends it to the audience. All that is required with streaming media is a computer that accepts the audio and/or video signal and translates the information digitally. This digital information is then sent out from the 'server' computer to all 'client' computers that request the digital information via a stream. In this way, streaming functions as either a one-to-one media source or a one-to-many media source.

At first streaming media presented challenges to the audience. Because of slow download speeds, audience members often had to face many starts and stops in the stream while the computer displayed a 'buffering' message, during which the audio and/or video would stop and then start. Now, for most Internet users with broadband connections, buffering is much less common, and programs are usually streamed smoothly without interruption.

Streaming has become easy and dependable enough to allow millions to stream media from Netflix, Amazon Video, or HBO Go with few if any delays or problems. Now that streaming is nearly problem free, it can be said that streaming is replacing downloading of media. Streaming allows an audience member to 'rent' media—watch or listen in real time—as opposed to having to buy it. Streaming media is rapidly supplanting downloading media. Downloading takes time for the information to be stored on a computer and requires storage space on that device as well. Streaming takes up very little storage space, and that space is not needed once the listening or viewing of the streamed media ends.

Digital delivery online has some real advantages as compared to broadcast, cable, or satellite delivery:

— Streaming begins when the audience member requests the stream, anytime, day or night.
— Audience members can use interactive applications like creating a personal playlist.
— Provides a means for the content deliverer (e.g., the station) to monitor the size of the audience and the length of the audience's listening/viewing.
— Allows audience members to multitask while listening.
— While presenting an easy method for audience listening/viewing, streaming program content is discarded after being presented to the audience, protecting the copyright holder because the program is not stored on a computer/phone/tablet/game console.
— Broadcast stations and cable channels can stream live media as it is being broadcast or cablecast, but can also give access to a program library of recorded material.
— Once streaming media is set up for a station, it requires very little supervision or maintenance.

Streaming media technology has greatly improved in the 20 years it has been technologically feasible. The weak link in this delivery system is the user's connection speed.

SEE IT LATER

RADIO STATIONS

A recent study by the Pew Research Center showed that although online listening is a mainstream activity, radio still reaches an overwhelming majority of the American public. Radio reaches 91% of Americans age 12 and older each week, a reach that has not changed much in the past 10 years despite all of the new ways to listen to audio. By

2015 about 53% of people older than age 12 listened to online radio at least once a month. This percentage represents nearly twice the listenership from 5 years earlier (2010). Overall, AM/FM radio revenue for spot advertising declined slightly, but revenue from digital advertising showed strong gains.

The broadcast radio industry is working on new ways to improve audio quality and offerings as a way to compete with online and satellite delivery. **HD radio**, the digital service that provides digital signals and multiple channels for each station to broadcast, has seen slow adoption by audiences. Although some auto manufacturers offer HD radios as an option in new cars, the vast majority of vehicle owners are sticking with over-the-air analog stations. As of 2016, HD radios are still somewhat scarce in both cars and homes outside of the larger markets. Small-market radio stations have been very slow to adopt HD radio because of the steep cost involved and the lack of consumer awareness and demand.

Although radio stations battle with each other, MP3 players, satellite radio, and Internet audio services to capture an audience, the general radio model will not change much. Traditional radio will still maintain its advantage of being local. Local radio plays music that appeals to its market, runs commercial spots for local businesses, and airs local weather, traffic and event information. Radio's future depends not so much on new technology like HD radio but on staying competitive with its ability to localize and appeal to its local market, thus making it attractive to local audiences and advertisers.

Production styles are changing because of smaller production budgets, less advertising support, and increased competition from a variety of sources. In general, however, radio production will not change much, as so much of radio is programmed with prerecorded music.

Radio stations will continue to turn to consolidation to streamline operations to become more efficient and more profitable. For example, as a radio group increases the number of classic rock stations it owns, it can afford to produce high-quality programming in one location and send it to all of its stations. Or if a group owns 20 stations and 6 of them are of the classic rock format, buying programming for all 6 can be accomplished with one 'classic rock package' buy. Such group buys are also attractive to the program supplier, who can sell to six markets with one sale. Buying programming for many stations gives group owners buying power, which translates into a lower cost per station, because groups will often get a volume discount. Such economic leverage gives group owners a strong advantage over owners that buy programming for only one station.

Consolidation also streamlines sales operations. Fewer salespersons are needed as stations consolidate. One salesperson sells time for all the group-owned stations in the market or those with a similar format or audience demographic profile. Having fewer salespersons cuts down on training costs, travel costs, and costs in processing contracts and production. When a commercial spot is produced at one location and sent electronically via a broadband connection to many stations, it saves time and the costs of duplication, shipping, and handling.

Automation, consolidation, and mergers have helped streamline radio station operations so they are more efficient with fewer personnel. The consolidation and merger rage of the 1990s and early 2000s has slowed considerably, especially after many station groups incurred huge financial debt. Some of the behemoth radio groups like Clear Channel and Citadel (acquired by Cumulus in 2011) have been in serious financial trouble because they bought stations at the peak of radio station values. Many of the radio stations owned by the large groups are worth less than when they were purchased and are having trouble competing in a complex media landscape that offers many alternatives to traditional radio. Several radio groups whose stock prices were in the $20 to $30 per share range in 2003 dropped to less than $1 per share by 2008. Now these groups are faced with the challenge of digging themselves out of their financial holes. For example, in an effort to save itself, Clear Channel has rebranded itself as IHeart Radio and heavily promotes the service for online listening.

Radio's distribution model is evolving to keep up with online delivery and radio satellite services. Local radio competes with satellite radio (SiriusXM) and numerous online audio services. To keep listeners, broadcast radio needs to maintain its localism. Coverage of local news and local promotion of businesses and cultural events are what radio stations do that other audio services either cannot or do not provide. Local stations must maintain an online presence to help stations stay close to their audience and provide the opportunity for audience members to stream the station's signal on computers, tablets, and smartphones.

In the near future, it does not appear that radio stations will be changing their operations. Audiences are still listening to radio, and revenue to stations has only decreased slightly. With that in mind, the organization of radio stations will most likely remain stable, at least through 2020. The radio industry will continue to employ general managers, program directors, engineers, IT personnel, announcers, and perhaps most importantly, salespeople. An excellent radio station with a terrific format, great personalities, and a loyal audience will only stay in business if its sales operation is effective.

TELEVISION STATIONS

Television stations have been facing financial pressures for years now that have forced consolidation, contraction in the number of employees, and the need to keep expenses as low as possible. The old wisdom was that expanding the number of stations in a group would lead to higher profitability. Although the economies of scale that come with consolidation makes sense conceptually, it has not worked out that way practically. Advertising sales have been tenuous, and so stations must raise revenues through retransmission contracts with cable companies and new revenue streams like mobile television and digital media that provide opportunities to sell banner ads on their Web sites.

High-quality audio and video production that in the past could have been produced only by expensive equipment manufactured for that purpose is now being created by less expensive equipment that can do a variety of tasks with remarkably high quality.

For example, digital camcorders that cost less than $2,000 are now available in various high-definition (HD) formats and are capable of producing images better than most professional camcorders did only a few years ago. Some camcorders come with digital storage devices similar to the removable 'jump' or 'flash' card devices. Some video cameras produce images that rival film cameras. Feature films and network programs and commercials that were once shot only on film are now being shot on HD or UHD (4K) video.

Video editing is one area of television production that has changed dramatically in the past 10 years. Almost all computers are now equipped with some basic video editing software. Costing only about $100, new video editing programs permit sophisticated editing that is better than software that cost thousands of dollars a few decades ago.

Camcorders, both consumer and professional models, are vastly improved over older ones. The mechanical (moving) parts that loaded and unloaded videotape sometimes jammed or broke on older models. Current models use digital storage on solid-state memory. Fewer moving parts mean less downtime and repair costs to stations.

Better, smaller, lighter equipment strengthens the trend for television stations to hire fewer personnel. In the past, a news crew consisted of a producer, reporter, videographer, and sound person. Better equipment that is easy to use, lighter, and more compact enables a smaller crew to cover news stories in the field. Instead of a four-person crew, many stations now expect a single multimedia journalist to set up the equipment (tripod, camera, sound, and lights), do a stand-up introduction to the story in front of the camera, conduct an interview, shoot video of the scene or location of the story, edit the story, and send it back to the station for airing. In some cases, the location shot is shown live on air. The video may also be posted on the Web. This changes the job preparation necessary for entry-level video journalism jobs. Television reporters are expected to be able to shoot video, be on camera, write the story, edit the video, and send the video back to the station. This new 'one-man band' or 'multimedia' style of reporting requires a wide range of production, journalistic, and multimedia skills. Having one person do the job of three or four people reduces station costs and simplifies hiring.

In the studio, technological changes have also reduced the need for some television crew members. Studio cameras can now be robotically controlled, obviating the need for a camera operator for each studio camera. Instead of three camera operators, one person can sit in the control room and remotely control all camera functions and camera moves needed for a live news show.

The switchover to FCC–mandated digital broadcasting created the ability for multicasting (having multiple signals on one channel). Digital makes it possible for one station to send several programs at once to the audience. For example, the local NBC affiliate might show *Dateline NBC* from the NBC network, but the audience will also be able to get several other programs from the station by simply tuning in to one of its subchannels. Individual broadcast television stations have the capability to become multichannel video program distributors.

Conceptually, multicasting held the promise of increasing advertising revenue by increasing advertising availabilities on newly created subchannels, but the reality has not met expectations. The dollars spent on television advertising did not necessarily increase just because a station offered more programming options. Advertisers may be lured into trying the various digital subchannels but at the trade-off of removing dollars spent on the main channel.

Although local television stations do offer some localism, primarily in local news, much of what is available on local television stations is available online through a variety of services. In addition, streaming television provided through cell phone service and Wi-Fi may supplant much of what is provided by local television stations.

The biggest challenge to television stations lies in the seismic change that is occurring with viewing habits. People are cutting the cord by cancelling their cable and satellite television subscriptions and going over the top and getting their programs directly from the Internet. Although the broadcast networks are adapting by creating subscription opportunities for streaming directly to the audience, local stations may not be able to do so. When audiences can get their favorite network primetime programming directly from the network, local affiliates will have to find compelling reasons to keep their audience.

Over-the-Top (OTT) Television

One research group, the Diffusion Group, predicts that broadcast network and local television revenue that reached almost $60 billion in 2015 will decline to about $47.5 billion in 2018. This research group believes that by 2020, there will be no distinction between OTT and 'legacy' TV (broadcast/network) advertising buys. OTT advertising revenue was $8.4 billion in 2015 and is predicted to grow to $31.5 billion in 2018. This huge increase follows the audience viewing behaviors. In 2015, audiences watched a little more than 3.5 hours per week of OTT television. Some predict that this amount of viewing "will nearly double to 6.9 hours (in 2016), with steady increases each subsequent year through 2020" (Blattberg, 2015). By 2020, average weekly OTT television viewing is expected to be about 19 hours per week. This seismic shift in the way audiences view television will lead to an enormous shake-up in the television industry in the next 5 to 10 years.

ZOOM IN 11.5

How Many Video Cameras?

The idea of individuals creating television 'programs' has been around since camcorders became available at reasonable prices to the general public in the 1980s. The show *America's Funniest Home Videos* has been on network television since 1989 and has depended upon people sending in their funny home videos. Although not every household had a camcorder, the show never lacked for material. In 1999, Nokia cell phones had a feature that allowed them to send video messages. At the time, there were probably a few million video cameras available, most stored in closets waiting for children's birthday parties and some in the hands of video professionals who shot video regularly. At that time, none of the cameras were connected to the Internet.

Technology has dramatically changed the availability of video cameras among the population. It is estimated that now there are almost 2 billion video cameras in use in the world. Obviously, this is because smartphones can also shoot video. Also, these phones can be easily connected to the Internet. The Internet is a network with a potential of 2 billion sources of video information. How does this change our picture of the world?

What would happen if an app were created that captured whatever video is on the screen of the phone, like the latest edition of *Orange Is the New Black*, and sent it directly to another app, like Twitter? Would everyone following you be able to watch that program for free? How would this affect subscription rates to services like Netflix or Amazon Prime Video?

CABLE TELEVISION

Due to the nature of the equipment, cable systems must have many local employees to provide service for the head end, maintain wiring and servicing amplifiers, and to satisfy customers needs, Unlike group-owned radio stations, which can operate with fewer employees because of automation and voice tracking, cable systems cannot make similar staff reductions. For example, a subscription to a broadband service involves a cable modem installation, a service that requires local personnel who have computer technology knowledge and experience and who can physically install the equipment. MSO-owned systems rely upon upper management to decide about programming packages, special deals to encourage people to subscribe, and other promotions that involve convincing subscribers to add premium channels or Internet and telephone (VoIP) service to their subscriptions.

The biggest challenge for cable systems is the trend for people to drop their cable subscriptions and get their television directly from the Internet. This trend changes the cable business from an emphasis on being a multichannel program provider that also provides Internet service and telephone service to a business that may have Internet service as its main selling point. If the cord-cutting trend continues, which appears highly likely, cable companies will hire fewer personnel involved in programming and fewer engineers and technical people and more people to work in Internet and telephone service.

One strategy to stave off dramatic declines in the number of subscribers to cable television service is a change in how channels are bundled, or grouped in tiers. Some companies like Verizon are heeding complaints of television subscribers who dislike paying for hundreds of channels to get the 10 channels they really want. Cable viewers in the future will want more of an 'a la carte' or 'skinny bundle' menu for selecting channels. The cable companies are taking notice, primarily since cable's reach into homes in the U.S. is shrinking. The future may hold slimming down of the size of the bundles that cable companies offer their subscribers. Unless cable companies can stop or at least slow down the cord-cutting trend, the operation of systems will change, based upon less revenue which will lead to hiring fewer personnel.

SATELLITE RADIO AND TELEVISION

SATELLITE RADIO

The satellite industry—including radio and television delivery—reaps the advantage that, unlike broadcast radio, its signals cover huge geographic areas. In fact, two satellites in geostationary orbit can cover the entire contiguous 48 states.

Satellite radio has an advantage over terrestrial radio of nearly continuous coverage for subscribers while traveling. Terrestrial radio has geographic limitations that satellite radio does not have. A traveler can drive across the country with uninterrupted satellite radio service. FM broadcast stations are limited to a 75- to 100-mile radius around the transmitter. AM stations have a broader reach that increases at night, but AM signals do not have the audio fidelity that FM has. In other words, music played on an AM station does not sound nearly as good as it would on FM. Satellite radio has excellent sound fidelity on all of its channels. Satellite radio requires a special receiver that is more costly than a typical AM/FM radio, but these receivers are common in newer model cars.

Satellite radio has many studios that are used to produce talk and news shows and many music channels that require programming. The studios require personnel to help produce the programming, maintain the technical operation of the facility, and schedule the activities in the studio. If revenue shrinks for SiriusXM, the number of studios and the personnel to operate them will decrease.

SATELLITE TELEVISION

Unlike cable and its ability to provide two-way communication through an infrastructure already in place, satellites are restricted in the two-way arena: that is, consumers cannot send or uplink content to a satellite. Because it cannot provide what consumers want most—a high-speed Internet connection—satellite television delivery will most likely continue to operate as it has for the past 20 years, as a multichannel provider of entertainment and information that competes directly with cable for subscribers, but without the prospect of making huge inroads into the subscription market. The merger of AT&T with DirecTV changes the bundling opportunities for satellite television in the future.

Because most of the revenue for both satellite radio and satellite television comes from subscriptions, most of the promotional and sales activity deals with marketing to individual subscribers. The satellite industry faces the same threat as cable television: cutting the cord. As people find other, more economical ways to obtain television entertainment, satellite television will have to find solutions to the problem of a shrinking subscription base. Although some of the loss in revenue because of decreased subscribers has been offset by higher prices and added revenue from movies on demand, this may be a short-term solution to a long-term problem.

Both satellite radio and satellite television have added an online streaming component to their business model; both businesses must support a sophisticated and expensive satellite uplink and downlink system. The radio and television satellite industries are buoyed by millions of subscribers who pay steep monthly fees. The satellite industries are middlemen in the entertainment business. They deliver programming from content creators to an audience whose viewing habits favor a passive audience accustomed to paying high fees for many channels that are never listened to or watched. Both industries have an "antiquated business model, and suffer from . . . a lack of innovation" (Gerber, 2013). Some investors seem to think that the satellite business is in danger of being overtaken by a company or companies that will create content, simplify the viewing experience, and reduce the number of remote controls the audience must use to access the various programming services. In other words, to be successful in the future, satellite distribution companies need to change their operation to become innovators that produce content (similarly to Netflix and Amazon Video) and innovate technology to simplify the listening and viewing experience.

SUMMARY

Radio became a dominant medium in the 1920s. Programming was live, for the most part, and included mainly musical performances. Programming was mostly live music performances, broadcast either from the studio or a remote location.

Television inherited radio's network system and then it proceeded to steal its programming ideas and its audience. By the early 1950s, it had pushed radio out of the way as the dominant electronic entertainment and news medium, leaving it to fend for itself. Radio hung in there and eventually bounced back into public consciousness when it ceded narrative programs to television and focused on what an audio-only medium does best—music. Rock 'n' roll led the way to radio's recovery. Once management learned that formatting a station by type of

music was the best way to compete with other stations, radio boomed.

Meanwhile, the television industry had its own unique challenges to face. As viewers demanded more sophisticated and live-action shows, stations and networks had to find a way to produce them. Higher quality shows meant larger studios, more rehearsal time, and additional personnel such as directors.

Radio and television production has not changed much over the years. Program formats are similar, and production techniques have changed gradually with technology. The signal flow in both audio production and television production is much the same today as it has been for the past 50 years. The most significant change in production has come about with the introduction of digital equipment. Compared to its analog counterparts, digital equipment produces a better-quality product and gives production personnel more flexibility to experiment.

Distribution patterns have changed over the past 50 years. Although over-the-air broadcast radio transmission is still the dominant form of audio transmission, satellite radio and Internet-based audio services have eroded the listening audience a bit. SiriusXM is a subscription service that provides an array of audio channels, some of which are commercial free. Satellite's biggest advantage is in its diverse program offerings that can be heard in almost any location in the continental U.S. Internet audio services provide user customized listening.

Television program distribution has changed more than radio. The network-to-local station program distribution model that has existed since television's inception is no longer the only way to distribute programs.

Technology will continue to change the radio, television, cable, and satellite businesses and how they operate. As consumers rely more upon the use of in-home broadband connections and the mobile Web, the media will need to reconsider their business models and their method of distribution to keep up with the demands of technology and the audience.

BIBLIOGRAPHY

Blattberg, E. (2015, April 13). The future of digital TV advertising, in 5 charts. DIGIDAY. Retrieved from: digiday.com/platforms/future-digital-tv-advertising-5-charts/

Bond. P. (2009, September 25). DVRs in 36% of households. *The Hollywood Reporter.* Retrieved from: www.hollywoodreporter.com/hr/content_display/technology/news/e3i470b0d4b3627285705f84f75e2ab3902

Cheng. R. (2009, April 20). Telcos, satellite join cable's push to build pay wall on Web. *Wall Street Journal.* Cited in Benton Foundation Newsletter (April 21, 2009). Retrieved from: www.benton.org/node/24497

Dick, S. (2001). Satellites. In E. Thomas & B. Carpenter (Eds.), *Mass media in 2025* (pp. 101–114). Westport, CT: Greenwood Press.

Direct-broadcast satellite television. (2015, April 26). *Wikipedia.* Retrieved from: https://en.wikipedia.org/wiki/Direct-broadcast_satellite_television

Eastman, S., & Klein, R. (1991). *Promotion and marketing for broadcasting and cable* (2nd ed.). Prospect Heights, IL: Waveland Press.

Gerber, R. (2013, September 18). Investing in the future of television. *Forbes.* Retrieved from: www.forbes.com/sites/greatspeculations/2013/09/18/investing-in-the-future-of-television/

Gross, L. (2003). *Telecommunications* (8th ed.). Boston: McGraw-Hill.

Hampp, A., & Learmonth, M. (2009, March 2). TV everywhere—as long as you pay for it. *Advertising Age*. Cited in Benton Foundation Newsletter (March 2, 2009), Retrieved from: www.benton.org/node/22751

Head, S., Sterling, C., & Schofield, L. (1994). *Broadcasting in America*. Boston: Houghton Mifflin.

Hilliard., R., & Keith, M. (2001). *The broadcast century and beyond* (3rd ed.). Boston: Focal Press.

How cable television works. (2004). *Howstuffworks.com*. Retrieved from: *entertain ment.howstuffworks.com/cable-tv1.htm* [February 26, 2004]

Hutchinson, T. (1950). *Here is television*. New York: Hastings House.

Internet-ready TV. (2009, January 13). *Los Angeles Times*. Cited in Benton Foundation Newsletter. www.benton.org/node/20638

Janowiak, G., Sheth, J., & Saghafi, M. (1998). Communications in the next millennium. *Telecommunications, 3*, 47–54.

Jessell, H. (2009, July 1). Gainers rare among TV groups in '08. *TV News day*. Cited in Benton Foundation Newsletter (July 1, 2009). Retrieved from: www.benton.org/node/26238

Jessell, H. (2009, August 13). WPVI's new home lets it do it all. *TV News Check*. Retrieved from: www.tvnewscheck.com/articles/2009/08/13/daily.9/

Johnson, B. (2009, October 5). Media revenue set for historic 2009 decline. *Advertising Age*. Cited in Benton Foundation Newsletter. Retrieved from: www.benton.org/node/28496

Kaye, B., & Medoff, N. (2001). *The World Wide Web: A mass communication perspective*. Mountain View, CA: Mayfield.

Learmonth, M. (2010, January 18). Thinking outside the box: Web TVs skirt cable giants. *Advertising Age*. Retrieved from: adage.com/print?article_id 5141554

Palmer, S. (2015, May 9).Up periscope: I see the future of Video. *Shelly Palmer*. Retrieved from: www.shellypalmer.com/2015/05/up-periscope-i-see-the-future-of-video/

Parsons., R., & Frieden, R. (1998). *The cable and satellite television industries*. Boston: Allyn & Bacon.

Quaal, W., & Brown, J. (1976). *Broadcast management*. New York: Hastings House.

Romano, A. (2002, September). "Dearth of women" in top spots. *Broadcasting & Cable*, p. 9.

Settel, I. (1960). *A pictorial history of radio*. New York: Citadel Press.

Sherman, B. (1995). *Telecommunications management*. New York: McGraw-Hill.

Sterling, C., & Kittross, J. (2002). *Stay tuned: A history of American broadcasting*. Mahwah, NJ: Erlbaum.

Thottam, G. (2001). Cable. In E. Thomas & B. Carpenter (Eds.), *Mass media in 2025* (pp. 15–26). Westport, CT: Greenwood Press.

Top 25 MSOs. (2009). *National Cable & Telecommunications Association*. Retrieved from: www.ncta.com/Stats/TopMSOs.aspx [October 10, 2009]

Tribbey, C. (2009, August 27). *Parks Associates: Network-connected TVs to triple in 2010*. Retrieved from: www.homemediamagazine.com/electronic-delivery/parks-associates-network-connected-tvs-triple-2010–16875

Turner, E., & Briggs, P. (2001). Radio. In E. Thomas & B. Carpenter (Eds.), *Mass media in 2025* (pp. 75–84). Westport, CT: Greenwood Press.

TV basics: Alternate delivery systems-national. (2009). *Television Bureau of Advertising*. Retrieved from: www.tvb.org/rcentral/mediatrendstrack/tvbasics/12_ADS-Natl.asp [October 10, 2009]

Vogt, N. (2015, April 29). State of the news media 2015. *Pew Research Center*. Retrieved from: www.journalism.org/2015/04/29/audio-fact-sheet/

Willey, G. (1961). End of an era: The daytime radio serial. *Journal of Broadcasting, 5*, 97–115.

Feature Films: 'The Movies' 12

Ross Helford

Contents

The instinct to tell a story visually is a fundamental human characteristic that dates back to the cave paintings of our earliest ancestors. In the millennia that followed, language was developed, which allowed stories to be passed orally from generation to generation. But unlike the cave paintings, these stories lasted only as long as the civilizations that passed them on. It was only with the innovation of written language that these stories were given a similar sense of permanence to those ancient cave paintings.

Words and images became intertwined with storytelling. The innovation of theater gave a new sense of dramatic realism to some of these stories. And whether the content of these stories, artworks, and plays pertained to the boundless limits inherent in the human imagination or if they strove to represent reality with as much accuracy as the best technology of the time would allow, it was not until the innovation of photography in the mid-19th century that the dynamic film industry we know today was born.

SEE IT THEN

In film's infancy, during the latter half of the 19th century, audiences were captivated merely by the novelty of moving images. But it was in those days, with exhibitors charging audiences pennies to view these moving pictures, that the first seeds of the film industry were planted. What had begun as a series of photographs spun on a zoetrope (a cylinder with vertical slits along the side that, when spun, made the images appear as if they were moving), and evolved into a montage of moving images interconnected to tell a rudimentary story or anecdote, had, by the second decade of the 20th century, become not only a fully realized storytelling art form but ultimately an industry that has proven as mighty, profitable, and enduring as any the world has ever seen.

Although most people think of Hollywood as the birthplace of the film industry, its place of origin is really Fort Lee, New Jersey. There, in 1893, Thomas Edison was experimenting with motion pictures. He built a studio named the Black Maria. It was a tar-paper–covered dark studio room with a retractable roof where short films of various topics were shot. The motion picture business in the area around Fort Lee prospered for about 20 years, until around 1911, when the first studio was built in Hollywood. California's climate, with moderate temperatures, wide-open spaces, and year-round sunshine proved to be cost effective for the fledgling film industry. Another reason southern California attracted filmmakers in those days had to do with Edison's overly litigious nature. Because he held the patents on most filmmaking equipment, the long distance between the east and west coasts made enforcement by Edison much more difficult. As a result, filmmakers flourished in California, creating the origins of the studio-dominated film industry that endures to this day.

AN ENTERTAINMENT INDUSTRY

In the early decades of the film industry, America was not the center of the film world as it is today. Yes, the United States was a major player in the evolution of the early film industry, with luminaries like Charlie Chaplin and D. W. Griffith creating enduring and controversial works that continue to engage, inspire, infuriate, and educate. But

FIG. 12.1 Charlie Chaplin's films featuring the Little Tramp character were enormously popular in the early days of the U.S. film industry.
Courtesy Tri-Star Pictures/Photofest. © TriStar Pictures

a great many of the artistic and technological leaps and bounds were happening an ocean away in France, Russia, and, most notably, Germany. The unrest and violence that swept through Europe during the First and Second World Wars devastated economies and film industries, thus allowing America to emerge as the worldwide film industry leader.

Although the history of cinema worldwide remains a vibrant story of invention and innovation, this chapter will primarily focus on the American film industry.

As America rose to the top of the film industry, California—and, more specifically, sunny Los Angeles—became the film capital of the world.

The 1920s gave rise to the studio system. In those early days, a silent film would be screened in an auditorium, often with an accompanying live organ that would provide both the film's soundtrack and its sound effects. To tell a story (aside from a sparse usage of title cards to convey essential dialogue and descriptions), silent cinema was entirely reliant on continuity editing and the non-verbal expressiveness of its actors. As such, silent movies spoke a universal language that was immediately accessible to non-English speakers, such as America's growing immigrant population, as well as a global audience.

THE PRODUCTION CODE

In its early years, the film industry was largely unregulated for content, and the variety of cinematic offerings during the silent era was not all that different from what it is today, with the eternal truism that sex and violence sell. But there were also family films, as well as children's films as wholesome as anything Walt Disney would stamp its name on today.

It was not until 1934 that guidelines for cinematic content began being enforced. The Motion Picture Production Code adopted by the Motion Picture Producers and Distributors Association (MPPDA, later known as the Motion Picture Association of America, or MPAA) set forth general standards of 'good taste' and specific do's and don'ts about what could and could not be shown in movies. The justification for the code was the "moral importance of entertainment." These guidelines varied from the prohibition of foul language and overt sexuality (which is why in those days married couples were shown to sleep in separate beds) to forbidding mixed-race romances and requiring that all bad deeds be punished in the end. The Production Code (also known as the Hays Code) lasted for more than thirty years, at last ending in 1968 when the industry adopted the MPAA film rating system that endures to this day.

The Hollywood studios enforced the Production Code, just as they would later become complicit in the anti-Communist hysteria of the 1950s, blacklisting a great many of its creative talents. The blacklist (also mentioned in Chapter 3) was literally a list of names of people who were suspected of being Communists. An accusation was all that was needed to get people fired and blacklisted from the industry. Careers and the lives of innocent people were ruined. Writers who were blacklisted often used pseudonyms or 'fronts' through which their work would be submitted. Screenwriter Dalton Trumbo won Oscars in 1953 (*Roman Holiday*) and 1956 (*The Brave One*) under other names while blacklisted. Trumbo also became the symbol for the end of this shameful chapter in Hollywood's history when producer–star Kirk Douglas proved even more heroic than the title character Trumbo penned for him, insisting that he be hired to write 1960's *Spartacus*.

ZOOM IN 12.1

The Motion Picture Production Code General Principles

1. No picture shall be produced that will lower the moral standards of those who see it. Hence the sympathy of the audience should never be thrown to the side of crime, wrongdoing, evil, or sin.
2. Correct standards of life, subject only to the requirements of drama and entertainment, shall be presented.
3. Law, natural or human, shall not be ridiculed, nor shall sympathy be created for its violation.

To see the entire Production Code, go to www.artsreformation.com/a001/hays-code.html.

BUSINESS

At the beginning of the Great Depression, from 1930 to 1933, weekly movie attendance dropped from 110 million to 60 million. In an attempt to lure back audiences, Hollywood studios introduced the double feature. The first feature would be the A-list production, with its big budget and big stars, and the second would be the low-budget B genre flick—thrillers, westerns, gangster, horror, and science fiction. The double feature proved a marketing masterstroke, and during the Depression years, Hollywood entered its Golden Age. This was the era of the Hollywood dream factory, in which American success stories and happy endings were mass produced for a population in desperate need of temporary relief from their real-life woes.

The year 1939 is often looked at as the epitome of the Golden Age. An abbreviated list of the classic films released this year include *Gone With the Wind*, *The Wizard of Oz*, *Gunga Din*, *Mr. Smith Goes to Washington*, *Stagecoach*, *Dark Victory*, *Love Affair*, *Wuthering Heights*, *Beau Geste*, *Intermezzo*, *The Adventures of Sherlock Holmes*, *Destry Rides Again*, *Another Thin Man*, *The Hunchback of Notre Dame*, *Ninotchka*, and *Only Angels Have Wings*.

In the late 1940s, another innovation would profoundly affect the movie industry: television. People stayed home to watch television, because in their newly built suburbs, there often was not a movie theater nearby. Television, not movies, became the entertainment of choice, and theater audiences declined.

FIG. 12.2 *Gone With the Wind*: set during the Civil War; acknowledged as one of the best films ever made.
Photo courtesy of MGM/Photofest. © MGM

FIG. 12.3 *The Wizard of Oz*. This 1939 film is one of the best-known films of all time. It used extensive special effects, was shot in Technicolor, and was captivating because of its use of fantasy storytelling and unusual characters.
Photo courtesy of MGM/Photofest. © MGM

THE HOLLYWOOD EMPIRE FIGHTS BACK

If there is one constant in the film industry, it's the whirring speed of technological innovation. Silent movies were quickly phased out after the unprecedented success of the first 'talkie,' 1927's *The Jazz Singer*. By the late 1930s, Technicolor movies would slowly supplant black-and-white films, and in the 1960s, movies in full color would become the norm.

One of the ways in which Hollywood responded to declining attendance caused by television was to develop widescreen formats, changing the size of the screen (for the first time in film's half-century history) from the standard, nearly square 1.33:1 aspect ratio to a widescreen ratio that was more than twice as wide as it was high. On this new widescreen, Hollywood gave audiences epics that were too big and colorful for small, black-and-white televisions at home. Westerns, war sagas, biblical and costume films like *The Ten Commandments* and *Ben Hur*, and literary adaptations like *From Here to Eternity* and *Moby Dick* would come to define the early years of the wide-screen era.

By the late 1960s, movie censorship virtually ended, with a new rating system that would finally replace the unpopular Production Code. Rating films at that time from G to X based upon suitability for children, the new system allowed filmmakers to explore themes of violence, sex, and liberation from societal constraints and norms in films like *The Graduate*, *Bonnie and Clyde*, *Easy Rider*, *Midnight Cowboy*, and *Rosemary's Baby*.

The 1970s were a new cinematic Golden Age in which movies like *The Godfather*, *Jaws*, and *Star Wars* were to become the epitome of creative and popular success, leading to the first increase in movie attendance since the 1940s. A movie generation had been born. The first film school–educated movie-makers emerged, and the public was more sophisticated, informed, and interested in film as an art than it had ever been before. The 1970s were the epoch of director-driven films. Francis Ford Coppola, Steven Spielberg, Martin Scorsese, Robert Altman, George Lucas, Brian De Palma, Roman Polanski, Woody Allen, Mel Brooks, and Clint Eastwood were a few of the director **auteurs** (directors so distinctive that they are perceived as a film's primary creative influence) whose films defined the era.

Forty years prior, *Jaws* and *Star Wars* would have been the low-budget B movie at the tail end of a double feature. Filmmakers like Spielberg and Lucas grew up loving these films, and they would redefine how Hollywood made genre films, and in so doing, defining the modern-day blockbuster.

FIG. 12.4 *The Graduate*, a 1967 comedy drama that is listed as one of the top films by the American Film Institute and was selected for preservation by the National Film Registry because of its cultural and aesthetic significance.
Photo courtesy of Embassy Pictures Corporation/Photofest. © Embassy Pictures Corporation

FIG. 12.5 *Jaws*, made in 1976, is an American thriller directed by Steven Spielberg. The film has been regarded as one of the first 'summer blockbuster' movies. It has also been selected as one of America's Top 100 movies by the American Film Institute.
Photo courtesy of Universal Pictures/Photofest. © Universal Pictures

FIG. 12.6 Big movie stars like Cary Grant became icons in our culture. He even got a postage stamp with his picture on it in 2002.

STARS AND HEROES

Whereas in decades past, a megawatt star like Cary Grant or Clark Gable would be called on to headline any big-budget venture, the 1980s saw the rise of the action hero, making superstars of the likes of Arnold Schwarzenegger and his lesser ilk: actors of limited range whose bulging biceps, menacing scowls, and well-placed one-liners sold tickets to a worldwide audience to feed its seemingly limitless appetite for action. It might be said that action movies, in the tradition of the silent movies of old, had became cinema's newest international language.

By the 1980s, special effects in Hollywood had come a long way from the stop-motion wonders of 1933's *King Kong*—though it should be noted that until CGI (*computer-generated imagery*) effects became commonplace in the 1990s after the visual triumphs of movies like *Terminator 2* and *Jurassic Park*, stop motion was used extensively in special effects—as were model miniatures; rudimentary computer graphics; advances in animatronics and puppetry from such effects masters as Jim Henson and Stan Winston; and optical special effects, which were used to extraordinary effect in movies like *Superman* (1978), *Star Wars*, and *Close Encounters of the Third Kind*. In 1984's special effects–laden *Ghostbusters*, the apartment building where the ghosts converge in the climactic sequence is a real 20-story building onto which models and a matte painting were added to make it appear as if it stretches into the stratosphere.

By the mid-1990s, CGI was on its way to becoming a fully integrated cinematic technology, as well as a viable art form and industry. Pixar Studios led the way, with mega-hits and critical darlings like *Toy Story* and *A Bug's Life*. Today, CGI can be integrated virtually seamlessly (and cost effectively) into live-action features, giving audiences a glimpse into a visual world that in the past could have been realized only in animated films, fantasy novels, and comic books. But it should be further noted that films too heavily reliant on CGI often find themselves instantly dated—this is because CGI technology continues its own rapid evolution, thanks in no small part to the exponential growth of the multibillion-dollar videogame industry.

Looking back on the early years of CGI, the filmmakers who used the technology most judiciously—as James Cameron did in *Terminator 2* and Spielberg in *Jurassic Park*—created movies that were both hits in their time while remaining visually impressive for decades to come. This is contrasted with the great many movies that relied all too heavily on computer effects, only to have their films look like outdated videogames after a few short years (have a look at *Terminator 3*). Perhaps there is no better example of this than Peter Jackson's *Lord of the Rings* franchise, which, more than a decade after its release (2001–2003), remains a visual marvel. Like Cameron and Spielberg, Jackson used CGI when necessary, such as for the computer-generated character Gollum or for the spectacular battle sequences, but in other instances, he relied on decades-old tried-and-true cinematic 'cheats' like forced perspective, use of body doubles, and make-up effects to get the desired visual effect. Ironically, nearly a decade after *The Return of the King* earned the Academy Award for best picture, Jackson's second crack at a Tolkien-adapted film franchise, *The Hobbit*, initially received a lackluster reaction when it became the first feature film to be shot in the high frame rate 48fps (frames per second), the resolution of which is supposed to create the cleanest, clearest resolution available to modern technology; however, to many, it seemed to more clearly expose the artifice, making the picture look more like something that would be seen on a low-budget made-for-TV movie from the '80s.

SEE IT NOW AND SEE IT LATER

BLOCKBUSTERS IN THE NEW MILLENNIUM AND TODAY

The first decade of the 2000s typified the blockbuster trend, with studios becoming more and more dependent on the huge rewards that a big movie might pay off handsomely at the box office and, better yet, spawn a high-grossing franchise like *Lord of the Rings*, *Harry Potter*, *X-Men*, or the surprise hit *The Bourne Identity*.

It could be argued that the film industry today has been defined in some part by how the studios realigned themselves leading up to, during, and in the wake of the Writers Guild strike that virtually shut down Hollywood in November 2007. Many had surmised that, much like after the resolution of the 1988 writers strike, the industry would pick up in a furious buying frenzy as it worked to make up for lost time. But instead, studios did some realigning of their own, revamping their business model to be more economically efficient, buying and developing fewer original and speculative scripts, and relying even more on previously existing material such as remakes, sequels, novels, and comic books. In this new landscape, many of the non–A-list members of the Hollywood creative community found earning a living increasingly difficult, if not altogether impossible.

In addition to the Writers Guild strike (resolved in March 2008) and the ensuing labor strife with the Screen Actors Guild, there was the U.S. recession, which had a profound effect on the economics of the film industry more than a year before the 'too big to fail' banking crisis that unfolded in the climactic moments of the 2008 presidential race.

By the end of the first decade of the 2000s, another dimension to the film industry appeared. Actually, a third dimension—3D films—was somewhat of a rerun. This was not, after all, the industry's first try at getting audiences excited about adding a depth dimension to the flat cinema screen. Stereoscopic films first appeared in the United States before the Great Depression, in the early 1930s, and then came back in the 1950s with the horror blockbuster *House of Wax* (1953). Periodically since then, 3D films have appeared with mixed results. The 1969 soft-core porn film *The Stewardesses* made lots of money, and Andy Warhol gave 3D a try with *Andy Warhol's Frankenstein* in 1973. But since 2005, 3D has been shot on digital video, and the results are far more compelling than previous attempts. Instead of using two 65-m 150-lb. film cameras to shoot each scene, filmmakers can use two 13-lb. digital video cameras to get the 3D effect. These days, it is expected that any major popcorn flick will be given the 3D treatment. While a case can certainly be made that 3D can be a bit 'stunty,' not adding much, if anything, to the overall viewing experience, there are some movies, like *Gravity* or *Avatar*, whose revolutionary and very deliberate use of 3D, it could be argued, elevate the technology as a legitimate art form.

3D has also become an essential way of drawing crowds to the theater. Much as the double feature reinvigorated the film industry during the Depression, and widescreen helped counteract the threat of television, the film industry continues to adapt to draw audiences from competing formats. These days, with home entertainment systems often as good as an average movie theater, coupled with content being more easily accessible than at any point in history (to say nothing of a general fatigue from an over-saturation of a stunt now a decade old), 3D movies are becoming less of a draw, and theaters are working other angles to attract crowds, including giant IMAX screens, roomy, comfortable reserved seating, serving food and alcohol, and careful quality control in which talkers and texters are removed from the premises.

THE STUDIO SYSTEM AND BRANDING

An argument could be made that the studio system of today is not all that different from the past. The major studios, after all, continue to finance, market, and distribute feature films that are intended—in their first run—to be viewed in a theater.

Moreover, much like today, in decades past, 'branding' was very much a part of how each studio positioned itself in the marketplace. For example, in the 1930s, at the dawn of the talkie, Warner Brothers was known for producing low-budget, gritty films, Paramount for its sophisticated comedies, MGM for its glossy productions, RKO for its special effects, Fox for its biographies and musicals, Universal for its horror films, Republic for its westerns, Disney for its animation. Columbia had been a straight-up B-movie studio until the legendary Frank Capra directed films like *Mr. Deeds Goes to Town*, *Mr. Smith Goes to Washington*, and, after World War II, *It's a Wonderful Life* (which was produced by Capra's own independent production company), earning the studio multiple Oscars and propelling it into a major studio player.

Fast forward to the present day, and while RKO, Republic, and even the once-great MGM have alternately folded, declared bankruptcy, and had their libraries sold and acquired, Warner Brothers, Fox, Paramount, Universal, and Disney remain major players, as does Columbia—though under another name, as it was acquired by Sony in 1989.

Of course, in other ways, the studio system is virtually indistinguishable from what it once was. Starting in the 1980s, studios (that in the past were owned by movie moguls) had become profitable arms of the multinational corporations that own them today: Sony Corporation, News Corp, Time Warner, NBCUniversal, Viacom, and the Walt Disney Company.

A case could be made that this new influx of corporate capital was one of the primary factors leading to ever-increasing budgets, to the point where today a movie that costs $100 million (a sum that just 30 years ago was unthinkable) is considered on the low end of the blockbuster budgets. And why would studios and their corporate parents be willing to spend so much to make and market a single movie? Perhaps this can best be answered by James Cameron's *Titanic* (1996), a movie whose budget kept going up and up, whose release date kept getting pushed back, and whose pundits speculated disaster for all involved. Instead, *Titanic* became the highest-grossing movie ever made—until Cameron broke his own record with 2009's *Avatar*. And for the first time, the notion that a single film could gross more than a billion dollars in global profits had become a reality.

The concept of branding has also come a long way from the 1930s. Nowadays, it is hard to see much of a difference in what type of movie one studio makes as compared with another. A perusal of the top grossing movies year in and year out, for example, reveals multiple studios making big-budget forays into comic book adaptations, CGI family features, sequels, and remakes.

The branding for the modern era, then, is more about studios being able to generate properties with franchise potential. For example, Fox has *X-Men*, Paramount *Transformers*, and Warner Brothers *Batman*. And while 2014 saw Warner Brothers, Fox, and Disney neck and neck in terms of international box office dominance, looking into the future, perhaps Disney is best positioned to dominate the franchise market. After all, this is the company that has made a strategy of paying exorbitant sums to acquire wildly successful studios, beginning with Pixar in 2006, then Marvel Entertainment in 2009, and, most recently, Lucasfilm in 2012, before immediately beginning production on the next three *Star Wars* films as well as a host of spinoffs.

And, while it is easy (and common) to bemoan the 'dumbing down' of the modern movie-going experience, it should not be forgotten that even if the multiplexes are more jam packed than ever with loud, effects-heavy blockbusters, thanks to the democratization of the Internet and the rise of streaming services, consumers also have access to more edgy, challenging content than ever before.

SPEAKING OF *STAR WARS*

Few were surprised that 2015's *Star Wars: The Force Awakens* was destined to become the highest-grossing domestic movie of all-time (and third internationally behind *Avatar* and *Titanic*). Perhaps the overwhelming critical positivity and fan satisfaction the latest chapter in the *Star Wars* saga generated was not even surprising—particularly after *Episodes I–III* (1999–2005) are considered by so many as huge disappointments (though one can see making a case that much of this vitriol is not entirely fair).

One *Star Wars* fan who was not particularly enamored with the newest chapter happened to be none other than its creator, George Lucas. In an interview with Charlie Rose, a grumpy Lucas compared Disney to "white slavers" (Bryant, 2015, Dec. 30).

While Lucas later apologized for the remark, it does not mean he did not have a point (however hyperbolic and arguably offensive the statement might have been). It could be argued that Disney, with its media and merchandising tendrils seeming to permeate virtually every corner of the globe, does take on certain characteristics of the kind of imperialistic planet-crushing agent of the Dark Side commonly associated with *Star Wars* villains.

The original *Star Wars* (1977) was an homage to the matinee B-movie genre flicks the film school–educated Lucas grew up loving, coupled with inspiration ranging from Akira Kurasawa's 1958 masterpiece *Hidden Fortress* to Joseph Campbell's seminal book on comparative mythology, *The Hero With a Thousand Faces* (1949). By comparison, enjoyable as they are, there is a certain focus-grouped-to-death formula to these billion-dollar franchises that make Lucas's original indeed feel as if it came from a long, long time ago and a galaxy far, far away.

THEATRICAL DISTRIBUTION

Outside of big cities, chances are, options for theatrical diversity are more limited than ever before. For those who prefer seeing movies on the big screen, the problem with streaming, on demand, cable, and so on, is that smaller niche movies are discovering new ways of reaching audiences and recouping expenses—and this does not necessarily need to be theatrical distribution.

Even in major metropolitan areas, those small, personal, independent films are rarely going to be getting wide or lengthy releases and are often in theaters one week only to be gone the next. Film distribution and the race for the almighty box office dollar can be cutthroat and risk averse. Why, for example, use up valuable multiplex space for *Rosewater* when three screens showing *Dumb and Dumber To* and another three for *Big Hero 6* are drawing larger crowds?

Moreover, it should not be forgotten that the overwhelming majority of ticket sales do *not* go to the movie theater but instead back to the movie studios. This means that theaters are highly dependent on their overpriced concessions, as well as increasing luxury amenities, to turn a profit. This also begins to explain why catering more heavily to a youth demographic (and its highly coveted 'disposable income') makes better economic sense than to an older audience that might be less willing to line up for a highly anticipated movie on opening night, to say nothing about dropping—in addition to $15 per ticket—$8 for a tub of popcorn and another $5 for a large soda.

THE INDEPENDENT REVOLUTION

In the 1980s, the formative years of another influential movement in American film became evident. This movement had its roots in the renegade filmmaking of the 1960s and 1970s from the likes of Dennis Hopper and John Cassavetes. Hopper starred in, co-wrote, and directed 1969's *Easy Rider*, which, from drug-dealing protagonists to a shocking burst of violence seconds before the end credits, bucked all Hollywood conventions and become a megahit. Cassavetes wrote and directed a slew of offbeat films in the 1970s, including *Minnie and Moskowitz*, *A Woman Under the Influence*, and *The Killing of a Chinese Bookie*. The artists inspired by Hopper, Cassavetes, and other visionary provocateurs of the 1970s quickly found blockbuster-minded Hollywood a hostile environment to make their small, personal, and often twisted, violent, and highly stylized films. So filmmakers like John Sayles, Jim Jarmusch, Spike Lee, David Cronenberg, Joel and Ethan Coen, Sam Raimi, David Lynch, and Steven Soderbergh would blaze the trail for the independent revolution that would reach its maturity in the mid- to late 1990s.

Independent films are, by definition, movies that raise their money outside of the studio system, though connotatively, the term tends to make people think of movies that are low budget, serious, quirky, experimental, and/or nonmainstream.

In fact, because indies experienced so much success in the 1990s, every major studio now has an 'indie' wing—finding, acquiring, and occasionally developing the kinds of challenging, edgy movies that tend to be released at the end of each year in hopes of garnering an Oscar nomination or two.

Crowdfunding, or the invitation of public investors, has emerged as a revelation in the next evolutionary step of independent filmmaking. Web platforms (most famously Kickstarter) invite stakeholders to contribute funds for all kinds of projects, including financing movies.

And it is not only filmmakers outside the Hollywood mainstream who turn to crowdfunding to make their films. And it is not only fringe talent that signs on to be a part of movies that find funding through such sources. Case in point: In April 2015, *Salty*, the satirical novel by Mark Haskell Smith, about an amiable heavy metal bassist on vacation in Thailand who inadvertently becomes ensnared in a terrorist kidnapping plot, received a UK crowdfunding record £1.9 million (approximately $2.9 million) through a platform called SyndicateRoom (Flint, 2015).

The director backing this venture, Simon West, boasts on his résumé such Hollywood blockbusters as *Con Air* and *Tomb Raider*. *Salty* is a project, moreover, that attracted A-level talent (ultimately casting Antonio Banderas in the lead role), proving once again that even in this era of

FIG. 12.7 The success of *Easy Rider* helped stimulate American filmmaking starting in the late 1960s. It is considered part of the 'American New Wave' of filmmaking that flourished from the late 1960s until the early 1980s.
Photo courtesy of Columbia Pictures/Photofest. © Columbia Pictures; Photographer: Peter Sorel

risk-averse blockbusters, there is an appetite on the talent, consumer, and investor levels for movies that might otherwise never have a prayer of being made (Flint, 2015).

UNIONS AND MAKING A LIVING IN THE 'BIZ'

Careers in the entertainment industry, even at the lowest rung of the ladder, are incredibly difficult to come by. Young people stream to Southern California by the scores, nursing the ambition to make their dreams come true.

Rare, however, is the breakthrough talent who takes Hollywood by storm. The truth of the matter is, despite those who luck into a job as a production assistant or intern, the overwhelming majority of Hollywood dreamers will give up, go home, and do something else with their lives.

But the exceptional few who, through luck, perseverance, talent, connections, and other intangibles, do stay, these are the people who will form the backbone for the entertainment industry's next generation. The production assistant today could, in 10 years, be a studio executive, staff writer, producer, editor, or camera operator, or one of hundreds of different careers that keep the entertainment industry running.

Hollywood is a union town, and virtually every job that goes into the producing a film (from camera operators to grips to sound engineers to transportation) is done through organized labor. Likewise, the 'creative' fields, such as writers, actors, directors, and editors, are also organized into unions. With all the money that flows through the entertainment industry, members of union film crews enjoy high wages. And although it is difficult to join these unions, the truth is, once a person is in, it is possible to enjoy steady work.

This is not always the case with those in the creative unions, where separate guilds representing the interests of writers, actors, and directors certainly do a lot toward guaranteeing fair (and handsome) wages, as well as royalties and protections from exploitation. However, these guilds do not help the individual creative entities find work, and it is not uncommon for a writer, director, or actor to enjoy a brief string of success before washing out and never working again.

While it might seem antithetical for a movie studio to spend $20 million on a movie star and then penny pinch at the margins, this practice not only happens, it is, essentially, business as usual. Money people, after all, are always trying to maximize profits and minimize costs.

In the past, the studio system worked well because everything was done 'in house.' Movie studios were essentially mini-cities. Every aspect, from the initiation of the screenplay to the final cut, from the wardrobes and makeup to the props and sets, was created on the studio lot. Even the biggest stars (as well as those rising hopefuls) would be contractually attached to a single studio.

And while studio lots are still in business churning out entertainment round the clock, the studio system as previously organized is no more. Thanks to the globalization of the entertainment industry and particularly the digitization of video, audio, and special effects, production need not all be done in the same place, and it is often much cheaper when done elsewhere.

The phenomenon of 'runaway production,' for example, has become increasingly commonplace. Ever concerned about the bottom line, studios and production companies will move their productions away from Los Angeles to places in the country and across the globe where the labor laws are more relaxed (or nonexistent), where the dollar is stronger and costs are lower.

While it remains true that one can make a great living in Hollywood, it is important to note that the vast majority of people working in the entertainment industry enjoy little to no job security.

DIVERSITY

In 2014, *Nightly Show* host Larry Wilmore Tweeted, "Blackest Oscars ever! Thanks Academy. #keepinitreal #Oscars." That year, *12 Years a Slave* took home gold statuettes for Best Picture, Supporting Actress (Lupita Nyong'o), and Adapted Screenplay (John Ridley). The subsequent 2 years, however, did not build on that momentum, stoking no small amount of outrage over an entire lack of minority nominations for major roles. The trending hashtag for the 2016 Oscars was #OscarsSoWhite, and several prominent celebrities, including Will Smith, Jada Pinkett Smith, and Spike Lee, boycotted the ceremony.

This lack of diversity, however, hardly begins with the Academy Awards. Every year, the Ralph J. Bunche Center for African American Studies at UCLA publishes a diversity report, and while the 2015 edition shows modest gains for women and minorities, the statistics are alarmingly misrepresentative. Some statistics that pop out include:

- In leading roles, minorities and women are underrepresented 2 to 1.
- As directors, minorities are underrepresented 2 to 1, women 8 to 1.
- As writers, minorities are underrepresented 3 to 1, women 4 to 1.
- Film studio heads are 94% White and 100% male.

So, yes, it is easy (and not at all inaccurate) to blame the Academy Awards for its lack of diversity, but this is merely the canary in the coal mine. Until the entertainment industry is more representative of the larger population, the problem will remain.

FEATURE FILMS TODAY AND TOMORROW

Though the notion that the film industry remains recession proof (and *Depression* proof) continues to hold true, because of all the corporate consolidation that has occurred in recent years, movie studios are no longer strictly in the movie business; thus, business decisions must go through ever-increasing layers of corporate scrutiny.

More and more, the people who know how to *sell* movies are being picked to head the major studios—a departure from the past, when studios were run by those who are remembered more for their abilities to *make* movies and develop talent.

If a single term could be applied to the labor unrest that led to the writers' strike, it would probably be 'new media,' and specifically how profits in this new frontier would be equitably dispersed to the talent unions. 'New media' might be defined as the myriad technological advances that have permeated delivery and viewing of content beyond (or possibly in the future, in conjunction with) movie and television screens. However, it should be noted that today's new media is tomorrow's old technology, so perhaps the term 'digital media' would be a more appropriate term.

Regardless of what it is called, a new business model is here and evolving at exponential speed. With more and more data streaming capabilities, the notion of taking up shelf space with DVDs and Blu-rays is becoming as antiquated as LPs and VHS tapes, leaving the need for physical libraries in the hands of ardent fans and collectors, just as digital music has severely cut into CD consumption.

Furthermore, the fundamental restrictions that used to be placed on making a movie have been greatly reduced, thanks to an abundance of inexpensive digital cameras—or even just your phone—and editing software. Now, almost anybody can make and distribute a movie. As for turning a profit well, that is another story.

Of course, just as with the music industry, there is a dark side to the digital media revolution: the high volume of piracy. Piracy of intellectual copyright is nothing new, but what used to be the domain of the street-corner bootlegger is now a mouse click away, all done from the comfort of one's own home, and sapping the film industry and its creative talent of hundreds of millions of dollars in revenue every year. And it should be further noted that it is not only mega-gazillion-aire movie stars, producers, directors, and writers suffering from piracy but also the legions of working-class 'talent' and crew members.

The demand for content today is as high as it has ever been, and this is only the tip of the iceberg in terms of the potential for huge profits in digital media. In the meantime, this is an exciting era to be a filmmaker. Much as D. W. Griffith revolutionized filmmaking a century ago by making a new movie every week, today's filmmakers are testing these vast, uncharted waters.

A maverick of independent cinema, Steven Soderbergh, while finding his way into the Hollywood mainstream with *Ocean's Eleven* and its sequels, was the first mainstream filmmaker to experiment with simultaneously releasing a film theatrically and streaming it online with 2006's *Bubble*. Nowadays, this release strategy has become a common practice for smaller films, and for some movies, such as 2013's Richard Gere thriller *Arbitrage*, this has proven a highly successful business model.

The film industry's future is being written every day—every time we go to a movie, turn on the TV, buy a Blu-ray, download from iTunes, watch a bootleg, or try our hand at joining the swelling ranks of the creative community. Although there are many unknowns as we gaze into the future, of one thing we can be certain: Just as it always has, the film industry will continue to change, adapt, and evolve.

SUMMARY

The instinct to tell a story is a fundamental characteristic that dates back to humanity's sentient awakening. Visual storytelling has evolved from ancient cave paintings to live theater to—in the latter half of the 19th century—moving photo collages, which would become the starting point for cinema, an enterprise that has endured and evolved to become the trillion-dollar industry it is today.

Although most think of Hollywood as film's birthplace, its story in fact began in Fort Lee, New Jersey, where, in 1893, Thomas Edison began experimenting with motion pictures. It was in part to escape the litigation-happy Edison (who held the patents to most filmmaking equipment) that filmmakers moved west to California, where the first movie studio was built around 1911.

In the early decades of the film industry, America was one of many countries (along with France, Russia, and Germany) leading the way in terms of artistic, technical, and commercial achievement. However, the unrest and violence that swept through Europe during the First and Second World Wars devastated those economies, allowing America to emerge as the worldwide film industry leader.

The 1920s gave rise to the studio system. Silent cinema, with its reliance on continuity editing and the nonverbal expressiveness of its actors, spoke a universal language that was immediately accessible to English speakers and non–English speakers alike, helping to forge a shared American identity that incorporated both the growing immigrant population as well as those who had been living in the country for generations.

In 1934, the Production Code (or Hays Code) was enforced by the studios, setting forth standards of 'good taste' in which, among other things, there could be no foul language or overt sexuality. Mixed-race romances were also forbidden, and all bad deeds had to be punished in the end. The Production Code lasted until 1968, when the industry adopted the MPAA film rating system.

Hollywood studios enforced the Production Code, just as they would later become shamefully complicit in the anti-Communist hysteria of the 1950s, blacklisting a great many of its creative talents and ruining lives and careers.

If there is one constant to the story of cinema, it is the speed with which technology either becomes incorporated into the product or else triggers other kinds of innovation. In the late 1920s, 'talkies' gradually replaced silent cinema. By the late 1930s, color began to supplant black-and-white films. When television began to threaten box office receipts, cinema's then-nearly square aspect ratio was transformed to the widescreen format we know today. In the 21st century, with increasingly impressive home entertainment setups and streaming technology, theaters have

ZOOM IN 12.2

The Amazing Saga of *The Interview*

The Interview, about two bumbling media personalities tasked with assassinating North Korean dictator Kim Jong-un, might have been just another silly Seth Rogan–James Franco comedy that could have overcome critical apathy to be a box office success—or maybe not. No one will ever know, because in a stunning example of life imitating art, the film's content so outraged the hermit republic that Sony—which was distributing the movie—suffered a crippling cyber attack that the FBI linked to North Korea. Whether this attack was primarily motivated by Kim Jong-un's documented antipathy toward *The Interview* or if there were other forces at work might not ever be fully known. However, the fall-out from the hack has been monumental indeed. Fearful of threats of violence, movie theaters across the country refused to screen *The Interview*, which led Sony to pull its December 2014 release, which led to much criticism that the studio was caving to censorship, which ultimately led to *The Interview* being released as a streaming download on iTunes and Google Play before getting a limited release on fewer than 100 screens a month later.

The fallout from the Sony hack, meanwhile, has proven even more devastating, including the theft of the studio's intellectual property, as well as the personal data of thousands Sony employees past and present and the leaking of embarrassing emails from longtime studio head Amy Pascal—which ultimately led to her resignation. The fact that anyone would lose their job in the wake of a patently illegal act of Internet espionage—no matter how offensive this person's private statements might be—has rightfully provoked a considerable amount of concern, considering that all those involved in the hack, including Pascal, were victims of a crime.

FIG. 12.8 *The Interview*
Photo courtesy of Getty Image #46083152

adapted with 3-D, IMAX, luxury seating, and serving food and drinks. Meanwhile, CGI (computer-generated imagery) would redefine what could be possible in a live-action movie, beginning in the 1990s with movies like *Jurassic Park*, *Terminator 2*, and *A Toy Story* and continuing through the first two decades of the 21st century with the likes of *Avatar*, *The Avengers*, and *Star Wars: The Force Awakens*.

In the 1970s, the first generation of film school–educated filmmakers gained prominence, bringing about a new epoch of director-driven films whose works defined the era. Whereas 40 years earlier, movies like *Jaws* and *Star Wars* would have been the low-budget B movie at the tail end of a double feature, in this new era, these would be the precursors to the modern-day blockbuster.

The 1980s saw the formative years of another important movement: independent cinema (or indies). Independent films are, by definition, movies that raise their money outside of the studio system, though connotatively, the term tends to make people think of movies that are low budget, serious, quirky, experimental, and/or nonmainstream. In the 1990s, thanks in no small part to movies like *Pulp Fiction* and *Clerks*, indies experienced so much success that every major studio now has an 'indie' wing—finding, acquiring (and occasionally developing) the kinds of challenging, edgy movies that tend to be released at the end of each year in hopes of garnering an Oscar nomination or two.

In the early years of Hollywood, each studio was characterized by the types of movies it made. Warner Brothers was known for low-budget, gritty films, Paramount for sophisticated comedies, MGM for glossy productions, RKO for special effects, Fox for biopics and musicals, Universal for horror, Republic for westerns, and Disney for animation, while Columbia

FIG. 12.9 James Cameron's 3D science fiction epic film *Avatar* was a film conceived in 1994, but it could not be produced until 2009, when the necessary technology for the film became available.
Photo courtesy of Photofest/Twentieth Century Fox Film Corporation. © Twentieth Century Fox Film Corporation

began as a low-budget studio before Frank Capra helped transform it into a major player with a string of Oscar-winning hits. These days, movie studios are profitable arms of giant, multinational corporations. As such, today's studios tend to brand themselves less by the types of movies they make as much as the types of franchises they bank on (*Star Wars, X-Men, Hunger Games,* etc.) to deliver massive profits for their corporate overlords.

Although multiplexes are more bloated than ever with what some might bemoan as mindless popcorn flicks, it is important to recognize that thanks to the democratization of the Internet and the rise of streaming services, today consumers have access to more edgy, challenging content than ever before.

As for careers in the entertainment industry, there are the exceptional few who become the big stars, directors, producers, and writers. There are also legions of employees who make the movie industry run. Hollywood is a strong union town, and in so being, it guarantees crew members high wages; but it has also contributed to the phenomenon of 'runaway productions,' in which bottom line–conscious producers and studios seek less expensive incentives elsewhere in the country or overseas to save money.

The film industry has evolved by leaps and bounds from those early days of Zoetropes and nickelodeons. However, that initial attraction to experiencing storytelling through moving images remains as captivating today as it was nearly a century and a half ago. The film industry is a global force that continues to permeate more markets, and the fact that today

FIG. 12.10 This device, the zoetrope, was used for visual entertainment in the 19th century.

anyone can make a film on a cell phone means anyone with a story to tell is not only able to do so but also able to make it available for anyone to see.

Glimpsing into the future, while there are certain to be plenty of surprises along the way, we can expect a movie industry that will continue to adapt and evolve while attracting new generations of fans and talent.

BIBLIOGRAPHY

Barsam, R. (2007). *Looking at movies*. New York: Norton.

Biskind, P. (1998). *Easy riders, raging bulls*. New York: Simon & Schuster.

Bryant, J. (2015, Dec. 30). George Lucas says he sold *Star Wars* to "White Slavers." Variety. Retrieved from   http://variety.com/2015/film/news/star-wars-george-lucas-disney-white-slavers-1201669959/

Flint,K. (2015, April 14). Con Air director takes Salty path to a crowd funding record. https://www.theguardian.com/film/2015/apr/14/simon-west-salty-crowd funding-con-air-tomb-raider

Frascella, L., & Weisel, A. (2005). *Live fast, die young, the wild ride of making rebel without a cause*. New York: Touchstone.

Harris, M. (2009). *Pictures at a revolution*. London: Penguin Books.

Mast, G., & Kawin, B. (2007). *A short history of the movies*. New York: Pearson.

Newcomer, R. (2006). *Moments in film*. Dubuque, IA: Kendall/Hunt.

Statistics. (10, October 2014). *YouTube*. Retrieved from: www.youtube.com/yt/press/statistics.html

Wolk, A. (2015, February 19). Networks as an Anachronism. Retrieved from: http://tdgresearch.com/networks-as-an-anachronism

The Personal and Social Influence of Media 13

Contents

It seems that television is rife with violence, profanity, and sexual images. Although some viewers do not mind or even notice on-screen sex and violence, there is great concern about individual, social, and cultural effects. Parents and legislators are especially concerned about the effects of violent content on children and young adults. In fact, television content is blamed for many social ills, such as violent behavior, increased crime rates, a lower literacy level, and the breakdown of the family. But at the same time, television plays a very important and positive role. Television programs are socializing agents that teach us how to behave, show what is right and wrong, expose us to other cultures and ideas, enforce social norms, and increase knowledge about life in general. Contemporary concerns about mediated messages are not limited to television but extend to the Internet and to song lyrics. The Internet is blamed for exposing children and young adults to unsavory material, and many claim that the very act of using the Internet for long periods of time may lead to social isolation.

Radio stations get heat for playing songs with racy, violent, and antisocial lyrics, as parents are afraid that their children will do whatever these songs suggest. Others, however, doubt whether media has such a strong influence.

As Dick Cavett, long-time humorist and talk show host, once commented, "There's so much comedy on television. Does that cause comedy in the streets?"

Although television and other media have positive effects, it seems as though most of the attention is focused on the negative effects. There is much debate concerning the degree to which the media influence behaviors, values, and attitudes. Some critics claim that mediated content is very harmful and has a strong influence, especially on children. Others claim that while television may have a mild influence, existing values and attitudes filter out the negative images and words. For example, if a young woman sees an unethical act on television but has a strong sense of right and wrong, she will not mimic what she sees.

The debate about the effects of media content is complex because humans are complex. There are no simple answers to the many questions concerning how mediated content influences media consumers. This chapter begins with a historical look at concerns about mediated messages and the development of theories that help explain the connection between mediated messages and human behaviors and attitudes. The chapter provides an overview of how media effects are studied. Contemporary issues concerning violence on television and the effects of viewing such material are examined next. The chapter ends with a look at the implications of the effects of mediated content and the implementation of rating systems, V-chips, and other tools that screen offensive content.

SEE IT THEN

STRONG EFFECTS

Concern about the negative influence of the media is not a modern-day phenomenon. Religious and government organizations have attempted to suppress printed works that they deem ideologically contrary and unfit for consumption since the advent of the printing press in the mid- to late 1400s. In addition to religious and government elites, members of the upper social strata have long tried to stifle the written word and keep new ideas from the masses, because knowledge is power, and they wanted to keep the power among themselves. With increased literacy and access to mass-produced writings came an even greater need to keep information out of the hands of the general public to silence opposition.

In the 1800s, the penny press sizzled with sensationalized accounts of criminal activity, sexual exploits, scandals, and domestic problems, which prompted community outcry about the negative social effects of reading such scintillating material. In the 1920s, parents protested violent content in motion pictures, and in the 1930s and 1940s, they focused on comic book violence.

The years between the end of World War I and the onset of the Great Depression were a time of growth in the United States. People were migrating from their rural homes into the industrialized cities, leaving behind networks of friends and family and old ways of life, in which traditions, behaviors, and attitudes were passed along from one generation to the next. Moving to the city disconnected individuals from family and social ties and left them to assimilate into a new culture. Without family and friends to depend on, newcomers to a city turned to newspapers, magazines, and books to learn about new ways of life and to keep up with current events. The mass media became a central and influential part of their lives.

Later, concerns about content extended to the social consequences of exposure to indecent, violent, and sexually explicit materials. Underlying these concerns was the strong belief that the mass media were all powerful and that a gullible public could be easily manipulated. Scientists and others sought to explain the social and cultural changes brought on by migration to the cities, industrialization, and increased dependence on the media. They tied personal behavioral, attitudinal, and cognitive changes to the media and later developed theories to help explain how certain aspects of the media affect our lives.

A **theory** is basically an explanation of observed phenomena. Although researchers offer varying definitions of this term, one simple explanation states, in part, "Theories are stories about how and why events occur" (c.f. Baran &

Davis, 2000, p. 29). Another definition claims, "Theories are sets of statements asserting relationships among classes of variables" (Baran & Davis, 2000, p. 30). Mass communication theories are explanations, or stories, of the relationship between the media and the audience and how this relationship influences or affects audience members' everyday lives.

MAGIC BULLET THEORY

In the 1920s, the United States was still struggling with the effects of World War I and rebuilding its economic and social structures. Newspapers were the main source of information, but radio was beginning to build an audience.

Along with these new information and entertainment providers came the fear that the media could take over people's minds and control the way they thought and behaved. Many thought of the media as a '**magic bullet**' or '**hypodermic needle**' that could penetrate people's bodies and minds and cause them to all react the same way to a mediated message. This concept of an all-powerful media was a widely held and frightening belief. These fears were not unfounded when considering the successful propaganda campaigns waged during World War I, the newness of the mass media, and the move to the cities that left many people without close social networks.

PROPAGANDA AND PERSUASION THEORIES

During World War I, propaganda was used to spread hatred across nations, to concoct lies to justify the war, and to mobilize armies. Although propaganda was used more intensely in Europe, it quickly spread to the United States. Starting in the 1920s, the world followed Adolf Hitler's rise to power, which was aided by his domination over radio and carefully crafted propaganda campaigns. During this time, there was also a broad range of social movements in the United States. Radio was an especially powerful tool for spreading propaganda and persuasive messages.

Definitions of propaganda vary slightly and often overlap with definitions of persuasion, but psychologist Harold Brown has distinguished between the two concepts. According to Brown, propaganda and persuasive techniques are the same, but their outcomes differ. **Propaganda** is "when someone judges that the action which is the goal of the persuasive effort will be advantageous to the persuader but not in the best interests of the persuadee," whereas **persuasion** is when the goal is perceived to have greater benefits to the receiver than to the source of the message (Severin & Tankard, 1992, p. 91).

ZOOM IN 13.1

Examples of World War I and World War II propaganda can be found at these sites:

- German Propaganda Archive (speeches, posters, writings): www.calvin.edu/academic/cas/gpa
- Snapshots of the Past—World War I and II posters: http://bir.brandeis.edu/handle/10192/23520
- Propaganda Postcards: www.ww1-propaganda-cards.com

FIG. 13.1 A Nazi Party election poster, urging German workers to vote for Adolf Hitler
© *Lebrecht Music & Arts/Corbis*

LIMITED EFFECTS

Imagine that it is the night before Halloween in 1938. You live in a rural farmhouse in New Jersey. You do not have a telephone or a television, your nearest neighbor is half a mile away, and you depend on your radio to link you to the outside world. The radio airwaves are filled with news about Hitler's rise to power in Germany, and rumors abound that outsiders are infiltrating the United States to initiate the fall of democracy. It is a scary world.

You settle in after dinner and tune your radio to the *Mercury Star Theater* program. You hear that tonight's program is a recreation of H. G. Wells's book *The War of the Worlds*. But if you were not listening carefully or had tuned in a little late, you would not have heard that announcement. In that case, what would you have done when dance music playing as part of the program was interrupted with a 'news report' that a spaceship had landed in New Jersey and that we were at war with Martians? Would you have tuned to another radio station to find out if the report was true? Many radio receivers in those days could pick up only one or two stations. You would not have had a television to turn on or a telephone to call your friends. Even people with telephones could not call out, because the phone lines were jammed. Would you have just laughed off the report, or would you have panicked?

When radio was still a relatively new medium, many listeners relied on it as their primary news source and believed what they heard. So when a 'news report' broke in to the music, many listeners believed that Martians were indeed invading Earth. Mass panic ensued. People jumped in their cars and headed somewhere, anywhere. They hid in closets and under beds. Some even thought of committing suicide. Other listeners, however, realized the broadcast was merely a rendition of *The War of the*

FIG. 13.2 An alien spacecraft opens fire in a scene from the 1953 movie *The War of the Worlds*.
Photo courtesy of Paramount/the Kobal Collection

Worlds. They listened with bemusement and thought the program was quite clever and entertaining.

Considering what was thought about the magic bullet theory at the time, it would make sense to conclude that everyone who heard *The War of the Worlds* broadcast panicked. According to the theory, audience members should have had the same reaction to the message. But that is not what happened. Researchers wondered why some listeners went out of their minds with fear while others enjoyed the show. Scholars at Princeton University took the lead by conducting many studies about audience reaction to the broadcast.

Basically, scientists found that listeners' reactions to the show were influenced by several factors, such as education, religious beliefs, socioeconomic status, political beliefs, whether listeners tuned to the broadcast at the beginning of the program or sometime during the show, where listeners were during the broadcast (e.g., rural home, city apartment), and whether they were alone or with others when listening.

Research about *The War of the Worlds* panic and other studies show that the media are not as all powerful as once thought. Audience members do not all react in the same way to the same mediated stimulus, because other factors in their lives filter messages such that they interpret them in their own ways. The findings from this line of research led to the *limited-effects perspective*, which states that media have the power to influence beliefs, attitudes, and behaviors, but that influence is not as strong as once thought. Moreover, the media are not just evil political instruments but have positive effects as well. The fact that the media's influence is limited by personal characteristics, group membership, and existing values and attitudes makes us less vulnerable and not easily manipulated by what we see and hear.

ZOOM IN 13.2

To learn more about *The War of the Worlds*, visit these sites:

- www.museumofhoaxes.com/hoax/archive/permalink/the_war_of_the_worlds/
- www.war-ofthe-worlds.co.uk

Also read original newspaper accounts of *The War of the Worlds* panic at this site:

- www.war-of-the-worlds.org/Radio/Original.shtml

Listen to the entire 1938 broadcast, visit:

- www.youtube.com/results?search_query=war+of+the+worlds+radio+broadcast+1938
- www.archive.org/details/OrsonWellesMrBruns

Several other perspectives came out of limited-effects research. After conducting a series of studies, researchers Paul Lazersfeld and Elihu Katz discovered that rather than media directly influencing the audience, messages are filtered through a two-step flow process. Two-step flow theory explains that messages flow from the media to opinion leaders and then to opinion followers. The process starts with gatekeepers, such as news producers, newspaper editors, and others who filter media messages. The messages are passed on to opinion leaders, or influential members of a community, who then pass on the messages to opinion followers, or the people the gatekeepers and opinion leaders are trying to influence. For example, suppose that a neighborhood group opposes building a new road through its community. Rather than just send antiroad messages to the mass public, the group will be more effective if it persuades a smaller number of influential homeowners (opinion leaders) that the road will harm the neighborhood and then have the opinion leaders influence the larger group of neighbors (opinion followers) to rally against the road. The **two-step flow** theory supports limited effects by demonstrating that opinion leaders are often more influential than the media.

Further support for the limited-effects perspective emerged with the identification of selective processes. **Selective processes** are "defense mechanisms that we routinely use to protect ourselves (and our egos) from information that would threaten us. Others argue that they are merely routinized procedures for coping with the enormous quantity of sensory information constantly bombarding us" (Baran & Davis, 2000, p. 139).

There are three basic ways in which selective processes operate:

1. **Selective exposure** is the tendency to expose ourselves to media messages that we already agree with and that are consistent with our own values and beliefs.
2. **Selective perception** is the tendency to change the meaning of a message in our own mind so it is consistent with our existing attitudes and beliefs.
3. **Selective retention** is the tendency to remember those messages that have the most meaning to us.

Selective processes support the limited-effects model by explaining how we filter mediated messages so they do not affect us directly. Rather, we choose what messages to expose ourselves to and then screen and alter the meanings of those messages so they are consistent with our current attitudes and beliefs. Long-lasting effects are further limited because we remember only the messages that had meaning to us in the first place.

MODERATE EFFECTS

The mid- to late 1960s were marked with social unrest, instability, riots, and protests. Concerns arose about the effects of watching news footage of real-life violence. It was commonly believed that the viewing public, and children in particular, could not discern between fictionalized violent content and violent news content and that viewing aggressive behavior caused people to act more aggressively in real life.

On the one hand, the media claimed that there was no relationship between viewing violence and increased aggression; in other words, the media claimed they had a limited effect on the viewing public. Yet on the other hand, the media claimed to their advertisers that they could indeed persuade the public to purchase certain products; in other words, the media had a strong effect on viewers. The limited-effects perspective was questioned in light of this inconsistency, and new research indicated that media effects were not as limited as previously believed.

In the early 1960s, Stanford University psychologist Albert Bandura was studying the effects of filmed violence on children. For example, in one variation of Bandura's 'Bobo doll' experiments, children watched a short film of other children playing with Bobo dolls. (A Bobo doll is an air-filled plastic doll that is weighted at the bottom, so when it is punched, it bounces back and forth.) One group of children was shown a film of kids punching and kicking their Bobo dolls and yelling angry, nonsensical words while doing so. A second group of children was shown a film of children playing nicely with their Bobo dolls. After viewing one of the films, each child was given his or her own Bobo doll to play with. As it turns out, the children who viewed the film of the kids playing nicely with their dolls were also gentle with their own dolls, but the children who watched the Bobo dolls being subjected to violent play also kicked and punched their dolls and spouted nonsensical words while doing so.

Bandura and other researchers demonstrated that children and adults learn from observation and model their own behavior after what they see, whether in real life or in films or on television. Further, the media teach people how to behave in certain situations, how to solve problems and cope in certain situations, and in general present a wide range of options upon which to model their own behavior, thus lending support for media as a strong influence. Violence and aggression are very complex, and despite all of the studies, researchers are still grappling with the role media play in the development of these human characteristics.

ZOOM IN 13.3

To see Dr. Albert Bandura and original clips from the experiments, visit www.youtube.com/results?search_query=albert+bandura+bobo+doll+experiment.

POWERFUL EFFECTS

While the debate about media effects continued, studies were commissioned to specifically test the influence that television content (especially violence) had on viewers, particularly children and young adults. In 1969, the U.S. Surgeon General's Scientific Advisory Committee on Television and Social Behavior was created to conduct research on television's effect on children's behavior. After 2 years of extensive study, the committee concluded that there was enough evidence to suggest a strong link between viewing televised violence and engaging in anti-social behavior and that the link was not limited just to children who were already predisposed to aggressive behavior.

FIG. 13.3 In one study, children imitated film-mediated aggression by beating up Bobo dolls. *Source: LearningSpace at the Open University*

The report stirred up much controversy. On one side, parents, medical associations, social groups, teachers, and mental health specialists called for restrictions on televised violence. On the other side, the television industry pointed to research on limited and moderate effects and lobbied hard against any new Federal Communications Commission (FCC) regulations on television. The television industry finally bowed to pressure and agreed to limit the extent of violence in children's programs and to times when children would be less likely to watch. Violence in television programs in general, however, has not decreased by much, and so the struggle to limit such content continued.

FYI: Violence in the 1960s

Although the decade began peacefully, many factors contributed to violence in American society that sent shock waves throughout the country and the government. The civil rights movement raised awareness about racial inequality, President John F. Kennedy was assassinated in 1963, and civil unrest and even civil disobedience resulted from U.S. involvement in the Vietnam War. Television newscasts delivered stark images of the realities of the war, which caused many Americans to turn against it. Antiwar and civil rights demonstrations turned violent in cities across the country. The Reverend Dr. Martin Luther King Jr., a civil rights advocate and leader, and Senator Robert F. Kennedy, a presidential candidate and brother of the late president, were both assassinated in 1968.

Congress and the public turned their attention to the causes of violence and social unrest. The media, and especially violent television programs, were blamed for contributing to social ills. Near the end of the decade, Congress started gathering scientific information about the causes of violence in society. The U.S. Surgeon General supported numerous studies that investigated the relationship between violence in the media and violence in society.

MILLER V. CALIFORNIA

In addition to mediated violence, attention also focused on the effects of obscene words and images. The U.S. Supreme Court set new standards in 1973 with its *Miller v. California* ruling. The Court ruled that **obscenity** was not allowed on any broadcast medium, and it established the following criteria for determining obscenity:

- Whether the average person, applying contemporary community standards, would find that the work, taken as a whole, appeals to the prurient interest, and
- Whether the work depicts or describes, in a patently offensive way, sexual conduct specifically defined by the applicable state law, and
- Whether the work taken as a whole, lacks serious literary, artistic, political, or scientific value (Gillmor et al., 1996, p. 144).

FCC V. PACIFICA FOUNDATION

Television was not the only medium under fire. In late 1973, comedian George Carlin's expletive-filled monologue "Filthy Words," which repeated seven words too 'dirty' ever to say on the airwaves, was broadcast in the middle of the day on a noncommercial New York radio station owned by the Pacifica Foundation. The ensuing public outcry spurred the FCC to bar the broadcast of words that are indecent, or "language that describes, in terms patently offensive as measured by contemporary community standards for the broadcast medium, sexual or excretory activities or organs" (*FCC v. Pacifica Foundation*, 1978).

After years of challenges, including a ruling by an appeals court that found that the FCC was practicing censorship, the FCC ruling against Pacifica was upheld, and in 1978, the U.S. Supreme Court affirmed the definition of **indecency**. The FCC asserted that children were less likely to be in the audience between 10:00 p.m. and 6:00

FIG. 13.4 Police tangled with demonstrators at the 1968 Democratic National Convention, one of many violent incidents that year.
© JP Laffont/Sygma/CORBIS

FIG. 13.5 Comedian George Carlin became well known for a monologue that featured seven 'dirty' words.
Photo courtesy of Photofest

a.m. Therefore, indecent content not intended for children could be broadcast without penalty during this 'safe harbor' time period, but obscene material could not be broadcast at any time.

OFFENSIVE SONG LYRICS

Ten years after the hoopla surrounding George Carlin's on-air comic dirty-words routine, attention focused on indecent words found within song lyrics. After being offended by the lyrics in the Prince song "Darling Nikki," Tipper Gore, then wife of then-Senator Al Gore, founded the now defunct Parents Music Resource Center (PMRC) to protect children and young adults from the influences of indecent and violent lyrics. This group believed that young people would do what the song lyrics suggested and would be inclined to accept deviant behavior as normal.

Hearings before the Senate Commerce Committee in 1985 led to the record industry voluntarily putting parental advisory stickers on record albums and CDs that contained "profanity, violent or sexually explicit lyrics, including topics of fornication, sado-masochism, incest, homosexuality, bestiality and necrophilia."

The courts felt justified in their rulings as more evidence supporting the *powerful-effects perspective* came to light in the 1980s, when several major violence studies concluded that viewing violence is strongly related to aggressive behavior. Additionally, *The Great American Values Test* demonstrated that television could influence beliefs and values. This half-hour television program special was actually a research project in which viewers first assessed their own values, and then the program pointed out inconsistencies to get viewers to question their values and change their minds. The researchers were amazed that a half-hour-long program was influential enough to change viewers' attitudes and values. Even more important, the researchers found that those viewers who were more dependent on television were more likely to change their attitudes. *The Great American Values Test* and other similar research lent further credence to the idea that the

media do indeed have a powerful influence on the viewing public.

In this chapter, the effect magnitudes are neatly delineated, but this has been done to simplify a very complex issue. Although there has been a dominant perspective at any given time, research findings have not always been consistent. Within each era of thought, numerous studies have demonstrated different outcomes.

The **powerful-effects** perspective dominates today, but it is very different from the magic bullet theory of yesterday. Enough is known about the influence of the mass media to know that not all people respond in the same way to the same message. The powerful-effects model is complex, and the circumstances must be right for certain effects to occur. Most young viewers today have grown up watching an enormous amount of violent television, yet they are not all aggressive and violent, as suggested by the magic bullet theory. Rather, some viewers may be influenced and might become more aggressive than others under some circumstances.

SEE IT NOW

RESEARCH ON THE MASS MEDIA

The discussion up to this point has shown that researching the effects of media content is not a simple matter. To date, about 3,500 research studies have been conducted that look just at the effects of mediated violence. These studies are not conducted casually but use strict procedures and methods. Researchers follow scientific protocol for learning about how people use the mass media and how the mass media influence people socially and culturally. Media effects research informs the public about the benefits and consequences of exposure to mediated content.

SURVEY RESEARCH

Gathering information about the media audience involves a systematic method of observation that results in data that can be measured, quantified (counted), tested, and verified. **Survey research** is one of the oldest research techniques and perhaps the most frequently used method of measuring the electronic media audience. Surveys are used to explain and describe human behaviors, attitudes, beliefs, and opinions. Survey research usually entails some sort of questionnaire or observation. Individuals may be asked questions about what they think about new television programs, what programs they watch, how often they watch, what Web sites they depend on most heavily, and how many hours per day they listen to the radio. Survey research is designed to be as objective and unbiased as possible. No attempt is made to manipulate behaviors, opinions, or attitudes but rather to record them as accurately as possible.

CONTENT ANALYSIS

Content analysis is a research method used to study the content of television programs, song lyrics, Web sites, and other mediated messages. For example, content analysis reveals the number of times indecent language is used on television, the number of times violence is promoted in song lyrics, and the number of times the nation's economy is mentioned as the lead story on network news programs or their Web sites. Although content analysis does not tell about media effects or audience use of media, it does tell about media content.

LABORATORY EXPERIMENTS

Experimental methods allow researchers to isolate certain factors they want to study. Laboratory experiments usually involve a **test group**, which is exposed to a variable or condition under study, and a **control group**, which is not exposed to the variable or condition. For example, researchers may be interested in knowing whether viewers with digital cable service change channels more often than those with satellite television. In a laboratory setting, individuals in the test group would watch digital cable, and those in the control group would watch satellite television. The number of times the viewers in the test group changed channels would be compared to channel changes made by those in the control group. Researchers would then know whether having digital cable service results in more channel switching.

The biggest drawback to laboratory experiments is that people may not behave or react in a lab as they do in real life. For instance, viewers at home might switch channels less often than in a lab, where perhaps factors such as boredom and knowing they are part of an experiment might change their behavior.

FIELD EXPERIMENTS

The purpose of a **field experiment** is to study people in their natural environment instead of in an artificial laboratory setting. Field researchers do not have much control over outside factors, but the tradeoff is that they get to observe real-life behaviors. It may be more valuable to observe viewers switching channels in their own homes, where they are more likely to behave as they normally do, than in a lab, where they may behave differently.

MEDIA CONTENT

TELEVISION VIOLENCE

Concerns about media violence and its effects on viewers, especially young adults and children, generates heated debate among families, parents, educators, activists, the media industry, and legislators. Although everyone is out to protect his or her own interests, there is a general consensus that viewing too much violent or objectionable content might, under some circumstances, have serious social and personal consequences. Thus, the argument centers on what, if anything, should be done to curb violent images.

One of the most comprehensive television violence studies was a joint effort by researchers at the University of California, Santa Barbara, and three other universities in the late 1990s. Here are some of the major findings from the National Television Violence Study:

Finding: Six out of 10 programs contain violence.

Consequence: Viewers are overexposed to mediated aggression and violence.

Finding: Television violence is still glamorized. Seven out of 10 violent acts go unpunished.

Consequence: Unpunished violence is more likely to be imitated by viewers.

Finding: Physically attractive characters initiate 4 out of 10 violent acts.

Consequence: Viewers are more likely to imitate characters they judge as being good looking.

Finding: Four out of 10 violent scenes include humor.

Consequence: Humor trivializes the violence and thus contributes to desensitization.

Finding: About half of all violent scenes show pain or harm to the victim.

Consequence: Showing pain and suffering reduces the chance that viewers will learn aggression from media violence.

Finding: Fewer than 5% of violent programs feature an antiviolence message.

Consequence: Viewers are not exposed to alternatives to violence or shown nonviolent ways to solve problems.

Research shows that the number of violent acts during prime time had increased 75% to 4.41 acts per hour between 1998 and 2005 and then decreased to 3.7 in 2010. Additionally, Parents Television Council found that nearly one-half of prime-time broadcast shows that aired in early 2013 contained some form of violence, and almost one-third depicted gun violence.

FIG. 13.6 Researchers are very interested in how and why viewers watch television

Photo courtesy of iStockphoto. © mikkelwilliam, image #0116545

FYI: Media's Contribution to Violence

When Americans were asked if various media contribute to crime in the United States:

- 92% said television contributes to crime.
- 91% said local television news contributes to crime.
- 82% said videogames contribute to crime.
- 80% worry about Internet predators.
- 73% of parents say they know a lot about what their children do online.
- 65% of parents say they closely monitor their children's media use.
- 59% said the Internet is mostly a positive influence on their children.
- 55% think sexual content contributes a lot to inappropriate sexual behaviors.
- 7% of parents say they know little or nothing about what their children do online.

Sources: Common sense media poll, 2009; Parents, children and media, 2007; Potter, 2003

ZOOM IN 13.4

For more information about media violence, read "The Eleven Myths of Media Violence," by W. James Potter (2003).

TELEVISION PROFANITY

A Parents Television Council study discovered that the amount of **profane language**, which is considered verbal aggression, was up 78% to 2.6 instances per hour from 2000 to 2001, and it also increased 69.3% from 2005 to 2010. Academic studies found that on programs aired by broadcast networks during prime time, profanities were uttered 5.5 times per hour in 1990, 7.2 times in 2001, and by 2005, curse words were said 9.8 times per hour. Moreover, mild words have decreased, while strong curse words have increased, as evidenced by a 2,409% jump in the frequency of the f-word (partially bleeped or partially muted).

Because of FCC oversight, broadcast programs have long contained fewer incidents of cursing than cable programs, which are not under the FCC's purview. For several decades, those numbers held true. For example, in 2005, cable shows contained 15.37 occurrences of crude language per hour compared to 9.8 on broadcast, a 44.2% difference. However, 2013 research by the Parents Television Council indicates that cable programs now contain only 6% more incidents of cursing than broadcast shows. The research further indicates that the gap is narrowing because cursing in broadcast programs is increasing, not because cursing on cable is decreasing.

During prime-time evening hours, children are most likely to watch during the first hour (8–9 p.m. EST, PST) than the last hour (10 p.m.–11 p.m. EST, PST). Thus it would be safe to assume that the earlier hour contains less objectionable language than the later hour, but that is no longer the case, the earliest hour contains almost as many offensive words as the latest hour.

VIDEOGAMES

The earliest **videogames** were developed for playing with a console connected to a television. One of the first home **gaming systems** was the Magnavox Odyssey. The Odyssey, released in 1971, was the forerunner to Sony's PlayStation and Microsoft's Xbox. After using the Odyssey system, Nolan Bushnell, cofounder of Atari, came up with the very popular *Pong* arcade game, which was later released as a console version. When millions of copies of the home version of *Pong* were sold in 1977, it solidified Atari's position as the preeminent gaming company. When Nintendo introduced Game Boy in 1989, it added an 'on-the-go' way to play videogames.

Videogames are played in arcade, console, computer, portable/handheld, and mobile modes. **Arcade gaming** usually refers to a coin-operated gaming machine played in a public place like a bar or casual restaurant. Early arcade videogames such as *Pac Man* and *Space Invaders* were very popular and set the gaming industry on a successful trajectory. Arcade games hit their peak in the late 1990s but have since given way to the other modes of gaming.

Today's **console games** have come a long way from the soundless Odyssey system and feature an array of interactive multimedia features displayed on a connected television screen and manipulated by a joystick or some other type of controller. PlayStation and Xbox are two of the most popular video console gaming devices.

PC or computer games are those that are played directly on a computer. Computer games these days are sold through the Internet and downloaded onto a computer. A big draw to computer gaming is playing with others online. With a fast enough broadband connection, players compete in real time with little lag between moves. *Minecraft*, *World of Warcraft*, *Diablo III*, and *The Sims III* are among the best-selling computer games.

Gaming systems such as Game Boy Advance and PlayStation Portable are dedicated portable/handheld devices. In other words, these devices are used only to play games. Portable/handhelds are different from mobile games that can be played on smartphones or tablets. *Tetris*, introduced in 1994, was the first game that could be played on a mobile phone, and 3 years later, Nokia brought on the game *Snake*. Mobile games are sometimes embedded in the device, but most games these days are downloaded as apps. Games are played either solo within the device or with others through a mobile cloud. *Candy Crush*, *Angry Birds*, *Subway Surfer*, and *Sudoku* were some of the most popular games in 2013, 2014, and 2015.

There are three primary types of online shared game playing: (1) identity and **self-presentation**, (2) **collective identity**, and (3) **phatic communication**. With *self-presentation*, gamers determine how they want to be seen by others. For example, filling out a personality test; grading their parents on the originality of their name; posting their horoscope or tarot readings; sharing movie preferences; or playing games in which they reveal their personality sets a public persona. It is a way to let friends know who you are, or at least who you want them to think you are.

With *collective identity*–formation tools, friends define their friends with questions such as: "What's the best way to make [friend's name] happy?" "How would you describe [friend's name] sense of style?" "How would [friend's name] occupy him or herself if he or she was thrown in jail?" Social organization tools that identify the most loyal Facebook followers and other group memberships reveal a collective identity.

Social network interactions are also a form of *phatic communication*, which is a linguistic term that defines a type of expression that is used only for social reasons instead of for the purpose of sharing information. For example, "How are you?" and "Fine, thanks" is phatic communication, because it is mostly used as a polite recognition and not as a conversation starter or information exchange. The same can be said for some types of games and online interactions, such as sending someone a heart icon or a hug or updating one's latest game accomplishments through Facebook. These actions keep the social network going and maintain contact with friends but do not necessarily start a dialogue or inform them of anything meaningful.

EFFECTS OF MEDIATED WORDS AND IMAGES

Despite the numerous studies that have examined the effects of mediated words and images, a consensus about the power of those effects is still lacking. The question now turns from the amount of controversial content, the strength of media effects, to the media effects themselves. Generally, media content influences the way people behave, the way they think, and the way they react emotionally.

What happens when viewers repeatedly see characters being killed? What happens when young adults see on-screen nudity? What happens when children hear others on television yelling profanities? People react differently to such content, and just how they are affected depends on many factors. Sometimes the effects are long lasting and sometimes they are fleeting. Sometimes they are more intense than at other times, depending on social, psychological, and situational factors. After watching a violent show with a friend, a viewer could feel fine, but the friend might be too hyped up to sleep. We all like to think that we are immune to the influences of the mass media, and we have probably all heard someone say something like, "I watched a lot of TV when I was growing up, and I'm not a murderer, so it didn't hurt me." This **third-person effect**, in which individuals claim that they are not as susceptible to mediated messages as others, often leads to an emphasis on others' viewing habits

rather than on our own. But researchers have shown that there are some commonalities in the ways we react to various mediated content.

BEHAVIORAL EFFECTS

We learn to behave by watching what others do and then following those examples. The same principle applies to media effects. Viewers take behavioral cues from the media and apply them to their own lives. For instance, media content such as song lyrics and television violence influence the way we behave, whether positively or negatively. The following types of behavioral effects have been documented.

Imitation

There is particular concern that people may imitate the same behavior they saw on screen. In a 1993 lawsuit, a parent alleged that her 5-year-old son set fire to a house after witnessing the cartoon characters Beavis and Butthead commit the same act. Although there has been much publicity about cases like this, in which children have engaged in extreme behavior that they claim to have imitated from television, these situations are rare. It is also very difficult to say whether a viewer directly imitated what he or she saw or already had a predisposition to violent behavior and aggressive actions and merely used television as an excuse.

Identification

It could be that viewers, especially children and young adults, identify with media figures in a broader sense. For example, young viewers may want to be like their favorite television characters and so take on similar characteristics without really imitating their behaviors. Children may wear T-shirts, for instance, imprinted with pictures or names of their favorite television personalities or characters or even walk or talk the same as their idols, but they will not usually imitate taboo behaviors.

Inhibition/Disinhibition

Punishments and rewards influence the likelihood of modeling and imitating mediated behavior. When negative or aggressive behavior is punished, the likelihood that viewers will behave in a similar manner decreases because their level of inhibition increases. That is, viewing a character being punished or dealing with negative consequences creates an **inhibitory effect** in viewers. They do not want to experience the same punishment and will therefore refrain from engaging in the same negative behavior. On the other hand, positive reinforcement creates a **disinhibitory effect**. When a negative action is rewarded, inhibition decreases, and so the likelihood of repeating the bad behavior increases. Disinhibition also occurs when an authority figure directs a person to behave badly, therefore displacing responsibility from the perpetrator ("He told me to do it").

Arousal

Viewing televised violence also arouses emotions. If a viewer is feeling somewhat aggressive or stressed and then watches a violent program, she may become more stimulated, and her initial aggressiveness or stress may intensify and lead to aggressive or violent behavior. Some coaches believe that showing aggressive sports films to a team shortly before a game heightens the players' levels of arousal and makes it more likely that they will play more aggressively.

Catharsis

Although most evidence supports the contention that viewing aggression leads to increased arousal, there is some support for the opposite perspective: that viewing violence leads to catharsis or the release of aggressive feelings. Catharsis supporters contend that viewing violence satisfies violent or aggressive urges and that watching others act out feelings of anger relieves aggressive feelings, so that we may behave more passively after watching onscreen violence. Although some empirical evidence points to a catharsis effect, the validity is largely based on tradition rather than scientific observation and testing.

Desensitization

Viewing certain types of content may also lead to desensitization, or the dulling of natural responses due to repeated exposure. In other words, the shock value of seeing violence, uncomfortable situations, or sexual acts is diminished the more times they are repeated. For example, when car alarms were still a novelty, everyone would stop and look around and wonder if a vehicle was being broken into when they heard one go off. Now when an alarm shrieks, most people do not pay any attention because car alarms have become a nuisance. Similarly, repeated exposure to intense television images and offensive online graphics has lessened their effect. And just as reactions to mediated violence have dulled, so, too, have reactions to real-life situations. When we are desensitized to violence in the media, we are less likely in real life to help someone in trouble, or call the police, or feel aroused or upset when witnessing a violent act.

Four-Factor Syndrome

Studies have found a link between viewing pornography and criminal sexual behavior. The major effects of viewing pornography are known as the *four-factor syndrome*, which consists of addiction, escalation, desensitization, and the tendency to act out or imitate violent sexual acts. The syndrome is sequential; that is, the effects occur in this sequence over time. *Addiction* to pornography may develop after repeated exposure to the material. *Escalation* occurs next when the viewers want more and stronger stimuli. Viewers become *desensitized* to pornography (and even antisocial or illegal behavior) and thus tend to believe the pornographic behavior is acceptable. Finally, viewers tend to *imitate* the pornographic behavior.

AFFECTIVE (EMOTIONAL) EFFECTS

When viewers see a character being mutilated, shot, thrown over a cliff, stabbed, beaten up, run over by a train, mangled in a car crash, or harmed or killed in other horrendous ways, they cannot help but react emotionally. Viewers also react emotionally to positive images, such as characters getting married, performing acts of kindness, and showing physical affection.

Everyone experiences some sort of emotional reaction to a mediated image, even if that reaction has become blunted due to overexposure. But it is the negative images that get the most attention. Most Americans were glued to their television sets and the Web as they watched the horrors of September 11, 2001. Viewing the intense violence of the day increased levels of posttraumatic stress symptoms and anxiety, especially among heavy television viewers. About 18% of viewers who watched more than 12 hours of television per day reported increased distress levels after 9/11, compared to 7.5% of those who watched television less than 4 hours per day. These findings are further evidence of how viewing violence and other catastrophes affect emotions—and often for a long period of time.

COGNITIVE EFFECTS

The media also influence how and what viewers think about the world. Mediated images influence perceptions of real life. Portrayals of minorities, women, families, and relationships influence social and cultural attitudes. For example, some studies count the *number* of television programs in which women and minorities appear. But recently, the concern has been more with how these groups are *portrayed* and the mediated *effects* they have on viewers. Through their portrayals of women, minorities, gays, and others, the mass media teach and reinforce societal norms and values. This socialization process is life-long but is especially powerful for young people. Even though not all **stereotypical depictions** are negative, they are nonetheless harmful because they objectify, depersonalize, and even deny individuality. In addition, television and the other mass media may provide the dominant or perhaps the only view of certain groups in society.

FYI: Television Boosts Women's Rights Internationally

Although much attention is focused on the negative influence of television, new evidence points to the positive effects on women in developing nations. Studies show that exposure to television is associated with higher school enrollment for girls and greater autonomy and rights for women, along with a lower birth rate, which frees women from the home.

Programs such as *Baywatch*—which is the most-watched program around the world and has been seen by more than 1 billion people—and soap operas often portray women as having equal rights to men, as being equal partners in marriages, as having an education and a career, and as standing up to men and challenging the traditional roles. Further, girls are often named after strong, independent female television characters, which could further signify a desire to increase women's power in societies dominated by men.

Source: Soap operas boost rights, 2009

Cultivation

Researcher George Gerbner and associates spent many years examining how television cultivates a worldview. In particular, they conducted many content analyses that compared television content to real-life situations to viewers' perceptions of actual life. In content analysis, researchers first count how many times a certain action occurs on television, for example, the number of characters who die as a result of a gunshot. Next they ask viewers how often they think gunshot deaths occur in real life. Then they research public records to find out how many people actually died from gunshots. Following this method of data collection, comparisons can be made between real-life occurrences, viewers' perceptions of real life, and what is shown on television.

Assess your own perceptions of the actual world:

1. What percentage of all working males in the United States is employed in some aspect of law enforcement?

    In the real world: 1%
    On television: 12%

2. What are your chances of being involved in an act of violence in any given week?

    In the real world: 0.41% (less than half of 1%)
    On television: 64% of all characters encounter violence

3. What percentage of all U.S. crimes are violent crimes (murder, rape, robbery, and assault)?

    In the real world: 10%
    On television: 77%

                                         (Baran & Davis, 2000)

Were your answers closer to the "real world" or to the "on television" answers? Many viewers say that because much of what they see on television is fictional, it does not influence their perceptions because they know what they are seeing is not real. However, research does indeed show that our cognitive perceptions are shaped by television. Moreover, the influence is especially strong on viewers who watch television more than the average number of hours. Heavy television viewers' estimations of real-life violence and crime levels are more closely in accord with those portrayed on television. For example, compared to light viewers of violent television, heavy viewers are more likely to fear being a victim of violent crime, to exaggerate their chance of being victimized, and to go out of their way to secure their homes and buy guns for protection. In sum, heavy viewers believe the world is a scarier place than it really is. There are also concerns that cultivation may lead viewers to accept violent acts as a normal part of life and to think of violence as an acceptable way to solve conflicts. Television cultivates a 'mean-world syndrome,' in which viewers believe that the world is a mean and scary place to live and alter their behaviors accordingly, all based on television's warped depiction of real life.

It is especially difficult to dismiss violence and not believe that the world is a scary place when the news media blow crime coverage way out of proportion. For instance, the national homicide rate decreased 33% between 1990 and 1998, yet television coverage of murders increased 473%. Television news is filled with child abduction stories or reports on how to protect children from what seems to be a spike in kidnappings. But in reality, the number of child abductions has actually decreased since the 1980s. In a given year, of 50 million school-age children in the United States, about 200 are abducted by strangers, and of these 50 are murdered. In comparison, about 4,400 children die in car accidents and 2,000 are killed by firearms each year, but these deaths receive much less media coverage than kidnappings. Overall, crime rates in the U.S. have steadily declined from 758.2 violent crimes per 100,000 in population in 1992 to 429 per 100,000 in 2009, yet two-thirds of respondents to a Gallup study thought that crime is on the increase.

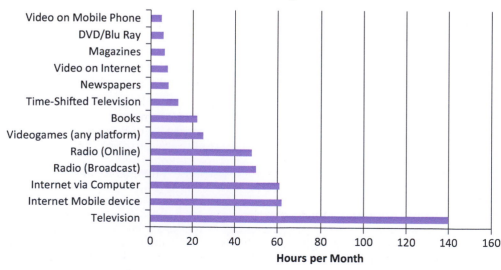

Average Media Use Per Month

FIG. 13.7 Media use per month
Sources: Good, 2014; Internet radio listeners, 2013; Lafayette, 2013; Radio facts and figures, 2013; Richter, 2013

MULTITASKING

There was a time when computers were touted as time savers. It was prognosticated that when computers became integrated into daily life, there would be more time for other activities. But that does not seem to be the case. To have enough time to sleep, work, watch television, use the Internet, listen to the radio, read books, and fit in other activities, we need 38 hours per day. It all gets fit in by **multitasking**.

Humans have always been able to multitask-—singing while showering, walking while chewing gum—without decreasing their concentration. But that is not the case with electronic devices because they command a higher level of cognition than, say, cooking several different dishes at the same time. Media multitasking becomes problematic depending on the number of different actions being taken simultaneously and the media being used—some media are more conducive to multitasking than others. For example, listening to music while surfing the Web or reading a novel go together more readily than watching television while reading a chemistry textbook or using a computer to write a research paper while talking on a cell phone.

When we engage in several activities at the same time, we are more easily distracted, less organized, and generally perform poorly as compared to those attending to one task at a time. It takes too much prefrontal brainpower to focus on various stimuli at the same time. Even though humans can perceive two stimuli at the same time, we cannot process them simultaneously—thus we experience a delay in our responses. Experiments find that when presented with new stimuli, physiologic brain processes that drive concentration are stretched to accommodate attention to several different tasks. A person has finite attention and concentration abilities, and the greater the number of tasks being done at once, the less able a person is to concentrate on any one of them. Moreover, it can take up to 20 minutes for the brain to 'reboot' after an interruption. The more distractions when learning, the less we are able to remember. The more tasks that are performed at the same time, the higher the error rate. Students often study with the television on in the background, thinking that it really does not interfere with learning, but their brains are still processing the images, which leaves less power to concentrate on their schoolwork.

If multitasking reduces productivity and performance, then why do we keep doing it? Because not only do we get social rewards through the misperception that we must be very hardworking and organized to juggle it all, but we also get a kick of dopamine, a neurotransmitter "that some believe is the master molecule of addiction" (Quittner, 1999). Every time we complete a task, our brains give us a shot of the chemical dopamine, which enhances our mood. The more we multitask, the more hits of dopamine, the better we feel.

Talking on a cell phone is particularly disruptive. Psychologists assert that the act of talking on a cell phone decreases cognition and awareness and leads to preoccupation or 'inattention blindness.' Cell phone use decreases auditory as well as visual functions—talkers look right at an object but do not really see it. Using another medium while on the cell phone, such as emailing while talking,

FYI: Multitasking

Percentage of Adults Who Watch Television and . . .

Surf the Internet on a computer	56%
Read a newspaper	44%
Go on a social network site	40%
Text on a mobile phone	37%
Shop online	29%
Surf the Internet on a mobile phone	18%
Read a book on an e-reader	7%
Surf the Internet on a tablet	7%
Do not multitask while watching TV	14%

Source: www.frankwbaker.com/mediause.htm

sends the brain into overtime. Although a little stimulation, such as music while exercising, might boost performance, too much induces stress, which puts the mind in overactive mode, which decreases focus.

In our fast-paced society, multitasking has almost become a badge of honor. Young people grow up thinking that multitasking is the best way to accomplish their work. Just over 8 of 10 young people aged 8 to 18 frequently multitask. Workers crow about how much they can get done by performing many tasks at once. Students brag about how they are so busy that they just have to text and surf online while sitting through a course lecture. But students are fooling themselves if they think their grades are not negatively affected by studying at the same time they are watching television or listening to their iPod. The strangest part of all is that those who think they are good at multitasking actually perform poorly when compared to those who acknowledge that multitasking is difficult.

EFFECTS OF VIDEOGAMES

Videogames have become a part of everyday in-home and mobile entertainment. Some people are hardcore players, and others just play to pass the time like when they are waiting for a bus or in a boring meeting or class. But no matter who plays and how the games are played—arcade, console, computer, portable/handheld, or mobile modes, by 2015, video gaming had become a $70 billion industry worldwide, with the U.S. accounting for $23.5 billion.

In 2000, about 49 million people in the United States played some type of online computer game. By 2009, that number had jumped to 114 million, and by 2013 to about 183 million (about 58% of the U.S. population). Further, 65% of U.S. households own a videogame system. A game console system is the most common way to play, followed by PC and tablets and cell phones.

Gamers play between 13 and 18 hours per week. The average age of a U.S. gamer is 35. Further, one-quarter of people under the age of 18, one-quarter over the age of 50, and one-half of those 18 to 49 play videogames. The ratio of male to female gamers is 60:40, and one in five gamers play every day.

A whopping 97% of all teens have played a videogame. More striking is that about one-third of teens play a videogame daily. College students are players too, with about 65% doing so on a regular basis. Two findings that are disturbing to most college professors indicate that about one-third of college students have played games during class sessions and that the average college freshman has spent twice as much time playing videogames (10,000 hours) as reading (5,000 hours).

As videogames have become more popular, concern has been growing about the social affects, especially increased aggression, that stem from playing such games. Studies suggest that videogames may be more influential than television, for several reasons: (1) they are more interactive, which increases involvement; (2) a large percentage of games involve violence as the main activity, and players are encouraged to 'kill' and 'injure' as many of the 'enemy' as possible to win; and (3) the games' portability makes them somewhat of a companion, as they can be played almost anywhere on a mobile device.

Online games and videogames are played for many reasons, such as for the thrill, to relieve stress, to escape, for something to do, for a sense of social and group belongingness, self-satisfaction, achievement, relaxation, and to make online friends. But is video gaming just an innocent pastime? Some people contend that videogames have become the newest scapegoat for all of society's ills, as evidence shows that playing videogames leads to increased aggression and increased perception of aggressiveness in others. As stated by researchers, "Children identify quite closely with electronic characters of all sorts, and . . . these identifications may have important implications for their emotional well being as well as for the development of their personality" (McDonald & Kim, 2001, p. 241). Research has indeed shown that people act a bit more rudely and aggressively for a short while immediately after playing a violent videogame. Also, elementary and high school students who are heavily into videogames get in more schoolyard scrapes than do students who play less often. But researchers question whether violent videogames lead to aggression or whether aggressive children are drawn to videogames in the first place. Yet other research has drawn a relationship between rising videogame sales and falling crime rates among juveniles from 1993–2011. In 1993, for teens 12 to 17, the offending rate was 52 crimes per 1,000 juveniles, but by 2011, that number plummeted to just 6 crimes per 1,000 juveniles. The inference is that videogames actually decrease juvenile crime, possibly because they have a cathartic effect (games release aggression) and turn attention away from aggressive behavior, or that videogame violence has no affect on juvenile crime and the issue is totally overblown.

Playing videogames may also be addictive. Playing violent videogames stimulates psychoneurological receptors that give the player a 'high,' producing symptoms similar to those induced by drugs and other pleasurable activities. High levels of body and brain involvement also lead to the production of dopamine.

Many organizations list possible warning signs of videogame addiction. Some of these signs include a preoccupation with games, negative physical and emotional symptoms when trying to stop playing, spending less time with friends and family, curtailing other enjoyable activities, lying about the amount of time spent gaming, and falling GPA.

EFFECTS OF THE INTERNET AND MOBILE DEVICES

Studies have shown that people are spending slightly less time with traditional media and more time online. In fact, some people are actually spending *too* much time on the Web. They have become so used to being online that other activities, like socializing with friends and studying for college, have suffered. Internet addiction is a topic of concern for many who find themselves addicted to online games, newsgroups, email, file sharing, and simply surfing. About 61% of online users show signs of Internet addiction. Moreover, about three-quarters of online users would rather give up chocolate, coffee, or alcohol than the Internet, and 20% would give up sex. College students, who enjoy freedom from parental supervision and unlimited access to the Internet through wireless and campus broadband connections, are at particular risk for Internet addiction, with about 71% saying they could not function without the Internet. They sometimes find that the stress of school, work, and social situations leads them to seek a safer, less demanding environment on the Internet. There, they can escape from real-life problems and socialize with new and perhaps anonymous social networked friends.

Overuse of digital technology may seem like an escape, but it is also the cause of many social problems. Digital connections have become more important than face-to-face ones. In some ways, cell phones make us feel less lonely, because we are in constant contact with others. Because of the low cost of making a cell call, we often call just to alleviate loneliness or boredom, not necessarily because we have anything interesting to say.

A *Time* magazine survey found American adults spend an average of 144 minutes per day using a smartphone. Mobile device addiction is called **nomophobia** (for no-mobile-phone phobia), the fear of being disconnected from the virtual world. It is estimated that about three-quarters of 14- to 24-year-olds and about 7 out of 10 25- to 34-year-olds are nomophobic. College students are very susceptible to mobile device addiction. The average college student compulsively checks his or her cell phone 60 times and sends almost 110 texts each day.

For addicts, being without a cell phone or tablet is paralyzing. They cannot leave home without one even if just for a walk around the block or a 15-minute trip to a convenience store. About 84% of Americans said they cannot go a single day without their mobile device in hand, almost three-quarters would feel panicked if they lost their cell phone, and 14% would feel totally desperate. Further, 63% of cell phone users check their phone every hour, 20% every 30 minutes, and 9% look at it every 5 minutes. Moreover, 75% of those ages 25 to 29 sleep with their phones under their pillows.

FYI: Internet Addiction

About 1 out 8 Americans experience signs of Internet addiction.

90% of Millennials check their smartphone before getting out of bed.

65% use the Internet to escape their problems.

33% of university students spend 5 hours a day online.

21% of women wake up in the middle of the night to check Facebook.

When students disconnected for 24 hours, they felt:

"like some sort of withdrawal"

"I felt inferior"

"I really panicked"

"Media is like a drug"

Source: ansonalex.com

FYI: Teenagers and Media Exposure Per Week

- Television content: 30 hours and 3 minutes
- Music/audio: 16 hours and 7 minutes
- Computer use: 9 hours and 3 minutes
- Videogames: 8 hours and 54 minutes
- Print media: 4 hours and 33 minutes

Source: www.frankwbaker.com/mediause.htm

Excessive texting may lead to failing grades, social anxiety, dependence, stress, and other psychological disorders. According to a recent poll conducted by the National Sleep Foundation, more than half of 15- to 17-year-olds sleep about 7 hours a night, 90 minutes less than the minimum recommendation. Tech-addicted teenagers are getting even fewer hours of sleep. Some young people even have a term for it: vamping, a reference to those other legendary creatures of the night. They document their all-nighters by posting selfies on Instagram from bed, with the hashtags #teen and #vamping.

ZOOM IN 13.5

To learn more about online addiction and its treatment, visit www.netaddiction.com

Are you addicted to the Internet? Take this test to find out: http://netaddiction.com/internet-addiction-test/

The popularity of mobile devices has had some unintended and even dangerous consequences. Mobile communications are linked to a significant increase in distracted driving, resulting in injury and loss of life. The National Highway Traffic Safety Administration reported that in 2012, driver distraction was the cause of 18% of all fatal crashes—with 3,328 people killed and 421,000 people injured. Forty percent of all American teens say they have been in a car when the driver used a cell phone in a way that put people in danger, according to a Pew survey. Some cities and states have made driving and using a cell phone illegal. The FCC is working with safety organizations to educate the public about the dangers of distracted driving and is seeking to develop innovative technologies that could reduce the incidence of distracted driving.

EFFECTS OF DIGITAL MEDIA ON CHILDREN

Children and young adults may be particularly susceptible to mediated words and images, both positive and negative. Programs like *Sesame Street*, which are educational and nonviolent, help children learn positive social behaviors, enhance their imaginative powers, and even develop problem-solving skills. Studies conducted over the years have shown that children who watched *Sesame Street* when they were young later demonstrated higher academic achievement and better reading skills than children who did not watch the program. This effect lasted through grade school and even into high school.

Conversely, negative images, especially depictions of violence, can cause short- and long-term harm. Because children watch television on average more than 3 hours per day, there is little doubt that it influences their perceptions of the world, their attitudes toward society, and their behaviors. For instance, children may become anxious around strangers and be afraid to go out, or they may develop long-term, debilitating social problems. Between two-thirds and three-quarters of children ages 6 to 11 feel intensely anxious about violence, guns, and death. Studies that have examined children and their reactions to frightening content have shown that their reactions differ by age. Scary characters and situations frighten very young children, but older children are more affected by threats of either realistic or abstract stimuli, not just scary images by themselves.

Looking at the time children spend watching television versus interacting with their parents, an unsettling trend emerges. In the mid-1960s, American children spent an average of 30 hours a week with their parents, whereas now, they spend an average of 17 hours with them. Not only has parent–child time decreased, but the typical child of today spends an average of 44.5 hours per week in front of the television and using a computer, radio, or other electronic media. Additionally, the average preschooler watches more than 20 hours of television and videos per week. An average of 20 to 25 acts of violence are shown in children's television programs each hour. Most children have witnessed about 8,000 killings before the age of 12 and 200,000 acts of violence before the age of 18. Furthermore, by the time a child graduates from high school, he or she will have spent about 33,000 hours watching television as compared to about 13,000 hours in school. It is hard to make the argument that media does not influence children and young adults when so much of their lives are media-centered.

FIG. 13.8 Among children, the incidence of obesity increases with the amount of time they spend watching television.
Photo courtesy iStockphoto. © stray_cat, image #2397753

Children's eating habits are also influenced by watching television and by watching commercials, in particular. The percentage of body fat and incidence of obesity are highest among children who watch 4 or more hours of television a day and lowest among those who watch an hour or less a day. What is more, the incidence of obesity increases the more time children spend watching television, and children who are heavy television viewers tend to eat more snacks between meals than light television viewers.

Even though parents are the strongest influence on children's eating habits, commercials help set food preferences. With the number of overweight children (6–11 years of age) doubling and adolescents (12–19 years of age) quadrupling from the early 1980s to 2014, there is no denying that the United States is witnessing an epidemic of obese children. Today, about 18% of children and 21% of adolescents are considered severely overweight. Hospital costs for treating obese children have tripled since the early 1980s. If the obesity trend continues, the Center for Disease Control estimates that by 2050, one in three adults will have diabetes.

Blame is being placed on the food industry for making sugar- and fat-laden foods and on the advertising industry for making such foods attractive to children. Food commercials make up 50% of all advertising on children's shows, and most ads are for unhealthy fare (34% for candy and snacks, 28% for cereal, 10% fast food, 4% dairy, 1% fruit juice, 0% for fruits and vegetables). Children see between 4,000 and 7,600 food commercials every year in contrast to only 50 to 150 public service announcements regarding healthy eating. Food advertisements influence caloric intake. Children exposed to television content containing food commercials consume 45% more food than children who watch television without food ads.

The food and beverage industry and the fast food industry together spend about $7 billion per year marketing unhealthy food to children. McDonald's alone spends more than $998 million on U.S. advertising in 2013. In stark contrast, the National Cancer Institute has only $1 million a year to promote the healthy eating of fruits and vegetables. Likewise, the federal government's entire budget for promoting nutritional education is one fifth of what Altoids spends annually on advertising its mints.

Calls have gone out to curb the growing epidemic of overweight and otherwise unhealthy children by limiting or even banning junk food advertising during children's programming and on child-oriented Web sites. Proponents of food advertising regulations claim that junk food advertising is a public health issue, just like the consumption of tobacco products, and that unhealthy foods should be banned from the airwaves, just as are tobacco products.

The abundance of junk food advertising has parents, activists, and other concerned entities urging their legislators to battle it out with the food and advertising industries. Sweden, Norway, Ireland, Great Britain, Finland, and Germany already have some type of ban on junk food advertising, and other countries may soon follow suit.

AGENDA SETTING

Agenda setting refers to the mass media's power to influence the importance of certain news events. In other words, the more airtime and Web space that is devoted to an event, the more important it seems to the audience. Agenda setting is a function of the **gatekeeping** process that news media practice daily. Because of the time and space constraints of radio and television, news producers select (or 'gatekeep') stories and events to cover and then present them on the air and on their Web sites. These stories and events become increasingly important to the audience with repetition and consistency across various media. When an event is the lead story on the broadcast and cable networks' news segments, people believe it to be of utmost importance, when, in fact, other events may be more noteworthy but for various reasons (e.g., political pressure, lack of compelling video) are largely ignored. If the media do not air a story, people do not think it is important.

Agenda setting is of concern—especially because as the number of independently owned news outlets continues to decrease because of aggressive consolidation, there will be fewer individual 'voices' (or media owners). Consumer advocates, civil rights groups, religious groups, independent broadcasters, writers, and concerned citizens fear that instead of getting several different perspectives about a certain issue, they will get one perspective—likely that of the corporate owners. They can give prominence and positive 'spin' to issues that may help them (such as tax breaks) and squelch stories that may hurt them (such as recalls on products with which the media corporation has financial ties).

For the past 20 years, agenda setting has been studied in the context of media effects. A **cumulative effects model** of agenda setting looks at the repetition of certain messages and themes in the media. After viewers repeatedly see and hear about events and topics in the newspaper headlines or on the evening television newscast, they begin to believe that these are the important issues of the day. The cumulative effects model assumes that these effects are observable only after repeated exposure.

Agenda setting can be thought of as the ability of the news media to focus attention and concerns on certain issues. Newscasts are often criticized for focusing attention on, say, a murder trial to move attention away from a politically heated issue, such as the national debt or high unemployment rate. In this sense, the news media do not tell us what to think but rather what to think about.

USES AND GRATIFICATIONS OF THE MASS MEDIA

So far, this chapter has examined the ways in which the media, especially television, influence their audiences—or, as some would say, how the media *use* their audiences. The chapter now focuses on how audiences use the media. The uses and gratifications approach examines how audiences use the media and the gratifications derived from this use. This perspective is based on the premise that people have certain needs and desires that are fulfilled by their media choices, either through use of the medium itself or through exposure to specific content. The uses and gratifications model is used to answer such media use questions as these: Why do some people prefer listening to television news to reading the newspaper? Under what conditions is an individual more likely to watch a sitcom rather than a violent police drama? What satisfactions are derived from watching soap operas or reading blogs? The model is based on these assumptions: (1) the audience actively and freely chooses media and content; (2) individuals select media and content with specific purposes in mind; (3) using the media and exposure to content fulfills many gratifications; and (4) media and content choice are influenced by needs, values, and other personal and social factors.

The more a particular medium or content gratifies a viewer's needs, the more likely the viewer is to continue using that medium or depend on that content. For example, if a viewer's need for feeling smart is fulfilled by a particular game show, he or she will probably continue watching that show. Audiences watch particular shows, listen to particular music, and even select specific Web sites for many reasons, such as to escape, to pass time, to unwind, to relax, to feel less lonely, and to learn about new things.

Sometimes it is not just the content that is gratifying but rather the medium itself. The act of watching television, regardless of what is on, may be relaxing, or the act of surfing the Internet, regardless of the sites being accessed, may gratify the need to feel productive. Think of the number of times that you have flopped down on the sofa, grabbed the remote, turned on the television, and surfed the channels. You did not care too much about what was on, but somehow, just being in front of the television felt good.

Television is watched in two primary ways: instrumentally and ritualistically. **Instrumental viewing** tends to be goal oriented and content based; viewers watch television with a certain type of program in mind. Conversely, **ritualistic viewing** is less goal oriented and more habitual in nature; viewers watch television for the act of watching, without regard to program content. Research suggests that perhaps television-viewing behavior should not be thought of as either instrumental or ritualistic but as falling along a continuum. In other words, sometimes a viewer might watch instrumentally, sometimes ritualistically, or sometimes a bit of both.

The uses and gratifications approach is important in understanding how audiences use the mass media and their reasons for doing so. The theory connects media use to the audience's psychological needs and attitudes toward the media, and it also explains how personal factors influence media use. Additionally, the uses and gratifications model tracks how new communication technologies change media use habits and how social and cultural changes influence content selection.

ZOOM IN 13.6

Next time you watch television, think of why you turned on the set and why you chose the particular program you are watching. Think of the last time you watched television. Did you watch ritualistically, or did you turn on the television to watch a particular program?

CURBING HARMFUL CONTENT AND BEHAVIOR

TELEVISION: OBSCENITY, INDECENCY, PROFANITY

Miller v. California (1973)

Legal definitions of obscenity and indecency are somewhat vague and require subjective judgment. For example, pornography could be considered obscene but only if a court deems it so. What exact content is obscene and what is indecent but allowable is difficult to determine. Therefore, commercial television broadcasters walk a fine line when airing programming fare that could be considered obscene or indecent. According to the *Miller v. California* ruling, obscenity is not allowed on any broadcast medium. The ruling has later been applied to cable, satellite, and the Internet.

Miller v. California restricts obscenity, which has stricter legal standards than indecency. Whereas obscenity cannot be shown, indecency, in some circumstances is allowable. Because the Internet and subscription services—like cable companies and satellite television providers—are not regulated by the FCC, they are free to air indecency. But broadcast networks and stations are regulated by the FCC and thus must follow its restrictions on indecency as imposed by the *Pacifica* ruling of 1978. The broadcast media, then, must be mindful not to offend viewers, but it is difficult to determine what is indecent. Is violence indecent? Is partial nudity indecent? Is a mild curse word indecent?

Telecommunications Act of 1996

Despite concern about the personal and social effects of media violence, profanity, and other objectionable content, televised fare has become coarser, more violent, and edgier. In an attempt to appease concerned parents, legislators, and others who rallied to clean up the airwaves, Congress passed the **Telecommunications Act of 1996**. The act set parameters for protecting the public from objectionable content. For example, it mandated that all television sets with screens larger than 13 inches be equipped with V-chips or other means of blocking objectionable content and required a rating system to alert viewers to violent and sexual content.

The television industry first adopted an age-based rating system similar to the one used by the movie industry, but the system was criticized for inadequate protection from offensive content. The age-based system was enhanced with a system of content warnings. Current television ratings are a blend of both age and content indicators.

Age Ratings

TV-Y: Appropriate for all ages
TV-Y7: Appropriate for age 7 and above
TV-G: Most parents would find content appropriate for all ages
TV-PG: Contains material that most parents would find unsuitable for younger children
TV-14: Contains material that most parents would find unsuitable for children under 14
TV-MA: For mature audiences, unsuitable for children under 17

Content Ratings

V: Violent content
S: Sexual content
L: Coarse language
D: Sexually suggestive dialogue
FV: Fantasy violence (children's programming only)

These age- and content-based ratings are combined into an icon that appears at the beginning of most television programs.

Ratings critics fear that the system might increase the number of violent and sexual images, because producers can now justify such content because viewers are warned and given the opportunity to avoid such programming. Some broadcasters do not comply with the content ratings system and object to its use. One executive producer expressed the fear that content ratings will jeopardize the success of certain programs. He pointed out, as an example, that the now cancelled hospital drama *ER* could have been be considered violent because it showed car accidents, victims of gunshots, and other medical emergencies, but he qualified that the violence is realistic and not gratuitous and therefore does not have the same negative effects on viewers as does violence on other programs.

Although the ratings system is hailed as a way to protect children from age-inappropriate content, many parents do not rely on it, claiming it does not accurately reflect program content. One study showed that 10 years after the implementation, only 8% of adults could correctly identify the content descriptors, and only 30% of parents with 2- to 6-year-old children could do so.

Even the **V-chip** system is rarely used. Although 7 out of 10 parents are aware that televisions come with a built-in V-chip, only 5 to 12% use it. Despite the fact that the ratings system and V-chips are generally ignored, other means of blocking objectionable content are available in the marketplace. TVGuardian and other similar devices mute more than 150 offensive words and phrases from television programs and edit profanities out of closed-captioned scripts.

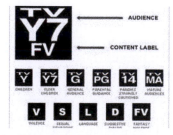

FIG. 13.9 Example of TV content and age rating system

Broadcast Decency Act of 2005

Several high-profile incidents on live television led to examination of the FCC's indecency policy. During the 2002 live Billboard Music Awards program, Cher exclaimed, "f*** 'em." The next year on the same program, Nicole Ritchie asked, "Have you ever tried to get cow s*** out of a Prada purse? It's not so f***ing simple." The singer Bono uttered, ". . . this is really, really f***ing brilliant" during a broadcast of the Golden Globe Awards in 2003. Although the FCC first ruled that Bono used the word as an adjective and not a verb, and thus in that context it did not violate FCC rules, it later changed its position by declaring the f-word indecent in any context and levied fines against the broadcaster.

Interest in indecent content surged after the 2004 Super Bowl halftime show, which featured the infamous 'wardrobe malfunction' that partially exposed Janet Jackson's breast, Kid Rock wearing the American flag as a poncho, questionable lyrics and crotch grabbing by hip-hop artists, and a closing act that included feathers and a Native American theme that was deemed offensive by many viewers. The FCC and CBS were immediately flooded with phone calls, letters, and emails from viewers who were fed up with such indecent and irreverent displays, especially on a program watched by so many young viewers. Network executives were forced to explain their programming standards in special congressional hearings prompted by the post–Super Bowl uproar.

Stemming from the Super Bowl and other incidents and viewer disgust and frustration with raunchy television in general, Congress passed the **Broadcast Decency Enforcement Act** in 2005. The bill allowed the FCC to increase fines for indecency violations from $32,500 to $325,000. Additionally, rather than fining a broadcast television station on a per-program basis, the FCC could issue fines for *each* indecent incident.

Threatened with heavy fines, the broadcast television industry fought back. It claimed that FCC indecency guidelines were unclear and arbitrary and that there was no way of knowing whether programs were in compliance until after they aired. In response, the FCC temporarily deferred enforcement action on most indecency cases while waiting for direction from the courts.

In 2012, the Supreme Court threw out sanctions against broadcasting companies ruling that the "commission failed to give FOX or ABC fair notice prior to the broadcasts in question that fleeting expletives and momentary nudity could be found actionably indecent" (Liptak, 2012, B1). In the wake of the ruling, the FCC is free to modify its indecency policy, and networks, at least for now, are free to self-regulate. But the ruling has left parents and watchdog groups concerned that the amount of cursing on television will increase and the words will become more explicit.

Viewers who are outraged by what they deem to be objectionable content continue to pressure legislators and networks to curb such programming. Broadcast television networks have 'standards and practices' departments that review program scripts to assure that content is in compliance with FCC guidelines and societal norms. These departments have been referred to as 'censors.' Censorship, however, is a government action that prohibits the distribution of certain program content, but networks are nongovernmental entities; therefore, they do not practice censorship, per se, but instead exercise editorial oversight to keep themselves out of trouble with government regulators and the viewing public. The FCC has repeatedly insisted that it is not interested in regulating programming content and has no plans to dictate programming to the industry. Specifically, the FCC does not intend to require more children's or public affairs programming. However, there continues to be government interest in curtailing indecency in radio and violence and sex on television.

VIDEOGAMES

In keeping with the First Amendment, the U.S. government does not restrict videogame content. Videogame retailers, however, abide by a voluntary ratings system. The Entertainment Software Rating Board (ESRB) has established the following age-based ratings system:

- C: Content intended for young children
- E: Suitable for all ages
- E10+: Suitable for viewers older than 10
- T: Suitable for ages 13 and older
- M: Suitable for ages 17 and older
- AO: Adults Only. Not suitable for players under the age of 17. These games may contain high levels of violence, profanity, or sexual content.

INTERNET AND GAMING ADDICTION

Health professionals have been dealing with online addiction for some years and are now also attending to addiction to mobile devices. The consequences of addiction are many, such as depression, antisocial behaviors, lost jobs, dropping out of school, broken relationships, and increased drug and alcohol use. There are many centers and organizations that help with problems related to unhealthy online behaviors.

FIGURE 13.10 Example of videogame rating icon

FIG. 13.11 Some feel that playing videogames may be addictive given the high levels of body and brain involvement.

FIG. 13.12 College students are at particular risk for Internet addiction.
Photo courtesy of iStockphoto. © mdmilliman, image #4528806

TECHNOLOGY-INDUCED WORKPLACE STRESS

Workplace managers are also beginning to recognize that email is one of the main sources of employee stress and dissatisfaction. The typical office employee spends 13.5 hours per workweek reading, responding to, and managing email, and about three-quarters reply to a message within an hour of receiving it. Additionally, 81% of employees check their work email accounts during nonwork hours. All this checking and responding makes employees less efficient, more stressed out, and less productive.

With the goal of increasing productivity and satisfaction, some businesses are turning off their email servers for several hours each day. Germany and France are both calling for antistress regulations to restrict bosses from emailing employees during nonworking hours. Volkswagen in Germany turns on the email servers 30 minutes before the work day and turns them off 30 minutes after work hours. Savvy managers are beginning to recognize a significant reduction in employees' stress levels when they check mail less frequently.

SEE IT LATER

In today's world, television, the Internet, and mobile communication are the central focus of media use.

The remainder of this chapter will therefore discuss the implications of media effects and how television and Web content is shaping our future.

OBJECTIONABLE CONTENT

Concerned viewers, parents, and policymakers have aligned themselves with the more conservative members of the FCC, who have proposed designating 1 hour each evening as family viewing time for wholesome programming. They have also asked television networks to voluntarily cut back on violent, sexual, and offensive programs. The industry has countered that V-chips, language-muting devices, and a content- and age-based ratings system all protect children from objectionable content.

On the one hand, many believe that objectionable content should not be curtailed because of the possibility of its causing negative effects or because some viewers are offended by it. Many strongly believe that it is the parents' responsibility to monitor what their children watch on television, what music they listen to and buy, and what Internet sites they visit. Moreover, if viewers are offended by certain content, they should stop watching or listening to it. In other words, viewers are responsible for their own exposure to offensive content. On the other hand, many believe that the media are responsible for the content

FIG. 13.13 By the time a child graduates from high school, he or she will have spent about 33,000 hours watching television, or about 2.5 times as many hours as spent in school.
Photo courtesy of iStockphoto. © mdmilliman, image #4528806

they air and should curb violent and negative images for the good of society as a whole.

It is difficult to sift through all the arguments, especially when parents are leaning on legislators to clean up television yet taking their kids to see decidedly violent and sexual movies. Given these and other contradictory behaviors, it is understandable why the television industry is reluctant to voluntarily censor its content but instead rallies hard against further regulations on violent, sexual, and verbal content.

MEDIA SUBSTITUTION

New communication technologies have given rise to new uses of media. Activities such as remote channel changing, digitally recording programs, fast-forwarding through commercials, watching videos and DVDs, and watching television programs on the Web all employ new media technologies that have altered existing television-viewing patterns. Satellite radio is also changing existing radio-listening habits, just as satellite television has altered how people select programs when offered hundreds of channels instead of 60 or so.

The Internet has already changed and will continue to change uses of traditionally delivered media, such as radio and television. For example, one study found that in 2012, broadcast television, cable television, newspapers, news magazines, and radio news all took a backseat to the Internet for political information. But other studies indicate that when it comes to hearing about breaking news, such as when Osama bin Laden was killed, U.S. adults still turn to television first. Television is still king among 25- to 54-year-olds, who spend 3.75 hours per day watching television as compared to 1.75 hours on the Internet, 36 minutes reading a newspaper, and 15 minutes reading a magazine. Generally, older individuals watch more television but use the Internet less often than younger people. Baby Boomers (ages 43–61) watch about 19.2 hours per week, Generation X (ages 26–42) spend about 15.1 hours in front of the television set, and Millennials (ages 14–25) watch only 10.5 hours per week.

The Internet is also taking a big bite out of newspaper readership. In 1996, about half of adults read a daily newspaper; now only about one-third do so. In contrast, in 1996, only 2% read news online, but by 2009, that percentage leaped to 37%, and by 2012, about 46% got their news online at least 3 days a week. But what these numbers do tell is that although printed newspaper readership is down, online readership is increasing substantially. In 2012, digital subscriptions accounted for 20% of newspaper readership. Perhaps it is not that newspaper content that is falling out of favor but that digital delivery is preferred to printed delivery.

SWIMMING AGAINST THE TIDE OF TECHNOLOGY

Whether they are called digital dissenters, techno-skeptics, or luddites, there is a growing number of activists who are resisting new communication technologies. Digital critics question the benefits and fear the consequences of technologies that are moving forward at warp speed. In many ways the computer and later the Internet were looked to as saviors that would usher in a utopian society. Although in many ways they have lived up to expectations, in other ways they have disappointed and led to a world of limited privacy, techno-elites, big data collection, and a gig economy of contract workers without benefits and whose offices are their laptops. In a machine-versus-human society, what is good for machines takes precedent over what is good for humans. Personal preferences, purchases, hobbies, everyday movements, and habits are tracked and recorded and processed by algorithms that know more about an individual than she knows about herself. Digital dissenters are attempting to rein in runaway data and help shape a digital world in which man and machine can live side by side but with humans as first priority.

SUMMARY

Various perspectives are offered to explain the effects of media content individually, socially, and culturally. Those perspectives include the strong-effects model and magic bullet theory; the limited-effects model and the research that stemmed from the broadcast of *The War of the Worlds*; the moderate-effects model; and the powerful-effects model, which takes many factors into consideration when examining media influence.

Viewing media violence may affect people behaviorally, emotionally, and cognitively. Viewers, especially children, may imitate aggressive behaviors, identify with unsavory characters, become anxious and fearful, and become desensitized to violence in real life. Repeated exposure to mediated violence breaks down social barriers and may influence antisocial behaviors.

The media, especially television, socializes and shapes people's attitudes, values, and beliefs about the world around them. Program content and commercials both strongly influence the way we think about ourselves and others, as well as sex, food, tobacco, and other life concerns. The Internet also strongly influences our behavior, especially socially. Further, some individuals are addicted to watching television and using the Internet, and as a consequence, they forget about the world and fail to meet their responsibilities.

The television and Internet industries, parents, psychologists, educators, legislators, activists, and other interested parties are struggling with the many social and cultural issues surrounding media use and content. In this struggle, rights to freedom of speech go head-to-head with the desire to protect viewers from objectionable words and images. Arguments include whether viewers need protection and what negative effects mediated content may have. Some worry that our fascination with television and the Internet is turning us into media junkies who live in darkened rooms, transfixed by our screens.

These concerns will not likely be settled in the near future and, in fact, are more likely to grow as television screens get larger and more involving and our dependence on computers and the Internet deepens.

BIBLIOGRAPHY

Achenbach, J. (2006, June 25). Dropping the f-bomb. *The Washington Post*, pp. B1, B4.

Achenbach, J. (2015, December 27). The resistance. *The Washington Post*, pp. A1, A16.

The adult video gamer market in the U.S.: Tapping into the new diversity of video-game players. (2009, January 1). *Market Research.com*. Retrieved from: www.marketresearch.com/search/results.asp?sid538903034–458950752–420123802&query5video1game&submit15Go [September 28, 2009]

Ahrens, F. (2006, June 8). The price for on-air decency goes up. *The Washington Post*, pp. D1, D6.

Andersen, R. E., Crespo, C. J., Bartlett, S. J., Cheskin, L. J., & Pratt, M. (1998). Relationship of physical activity and television watching with body weight and level of fatness among children. *Journal of the American Medical Association, 279*, 938–942.

Associated Press. (2009, August 3). Jury awards $675,000 in music lawsuit. *Arizona Republic*.

Baker, P. (2006, June 16). Bush signs legislation on broadcast decency. *The Washington Post*, p. A6.

Baran, S. J., & Davis, D. K. (2000). *Mass communication theory* (2nd ed.). Belmont, CA: Wadsworth.

Barnes. R. (2009, April 29). Supreme Court rules that government can fine for "fleeting expletives." *The Washington Post*. Retrieved from: www.washingtonpost.com/wp-dyn/content/article/2009/04/28/AR2009042801283.html

Barnes, R. (2012, June 21). Supreme Court overturns FCC sanctions on network, sidesteps larger issue. *The Washington Post*. Retrieved from: www.washingtonpost.com/politics/supreme-court-overturns-fcc-sanctions-on-networks-sidesteps-larger-issue/2012/06/21/gJQAwffxsV_story.html

Bond, P. (2008, December 18). Study: Young people watch less TV. *Reuters*. Retrieved from: ca.retuers.com/article/entertainmentNews/idCATRE4BH10Y20081218 [October 10, 2009]

Canary, D. J., & Spitzberg, B. H. (1993). Loneliness and media gratifications. *Communication Research, 20*(6), 800–821.

Cantor, J. (1998). *"Mommy I'm scared": How TV and movies frighten children and what we can do to protect them.* San Diego, CA: Harcourt Brace.

Carey, B. (2013, February 11). Shooting in the dark. *The New York Times*. Retrieved from: www.nytimes.com/2013/02/12/science/studying-the-effects-of-playing-violent-video-games.html?_r=0 [December 12, 2014]

The changing news landscape. (2009). *Media Literacy Clearinghouse*. Retrieved from: www.frankwbaker.com/mediause.htm [October 1, 2009]

Child obesity facts. (n.d.). *Centers for Disease Control and Prevention*. Retrieved from: www.cdc.gov/HealthyYouth/obesity/facts.htm [December 11, 2014].

Children and media violence (2009). *National Institute on Media 1 Family*. Retrieved from: www.mediafamily.org/facts/facts_vlent.shtml [October 1, 2009]

Clancey, M. (1994). The television audience examined special insert. *Journal of Advertising Research, 34*(4), special insert.

Commission reports to Congress on kids, content and protection. (2009, September 2). *Zogby International*. Retrieved from: www.zogby.com/Soundbites/ReadClips.cfm?ID519055 [September 28, 2009]

Common sense media poll on parents, kids & Internet use (2009, August). *Common Sense Media*. Retrieved from: www.commonsensemedia.org/pressroom?id523 [February 6, 2010]

Consumer choice vs. pointless publicity (2007, March 20). *Zogby International*. Retrieved from: www.zogby.com/templates/printsb.cfm?id514649 [September 28, 2009]

Crespo, C. J., Smit, E., Troiano, R. P., Bartlett, Susan J., Macera, C. A., & Anderson, R. E. (2001). Television watching, energy intake, and obesity in US children. *Archives of Pediatric and Adolescent Medicine, 155*, 360–365.

Curtin, C., Bandini, L. G., Perrin, E. C., Tybor, D. J., & Must, A. (2005). Prevalence of overweight in children and adolescents with attention deficit hyperactivity disorder and autism spectrum disorders: A chart review. *BioMed Center Pediatrics*. Retrieved from: www.pubmedcentral.nih.gov/articlerender.fcgi?artid51352356 [October 1, 2009]

The deadline team. (2014, June 18). New study: TV violence makes people more afraid of crime, but not afraid there is more crime. *Deadline*. Retrieved from: http://deadline.com/2014/06/new-study-tv-violence-makes-people-more-afraid-of-crime-but-not-afraid-there-is-more-crime-792399/

Dennison, B. A., Erb, T. A., & Jenkins, P. L. (2002). Television viewing and television in bedroom associated with overweight risk among low-income preschool children. *Pediatrics, 109*, 1028–1035.

Dietz, W. H. (1990). You are what you eat—what you eat is what you are. *Journal of Adolescent Health Care, 11*, 76–81.

Dietz, W. H., & Gortmaker, S. L. (1985). Do we fatten our children at the television set? Obesity and television viewing in children and adolescents. *Pediatrics, 75*, 807–812.

Dying to entertain. (2007, January). *Parents Television Council*. Retrieved from: http://w2.parentstv.org/main/MediaFiles/PDF/Studies/PTC-DyingtoEnt-Jan07.pdf [December 10, 2014]

ESRB ratings guide. (N.D.). *Entertainment Software Rating Board*. Retrieved from: www.esrb.org/ratings/ratings_guide.jsp#rating_categories [December 12, 2014]

Facts on junk food marketing and kids. (n.d.). *Prevention Institute*. Retrieved from: http://preventioninstitute.org/focus-areas/supporting-healthy-food-a-activity/supporting-healthy-food-and-activity-environments-advocacy/get-involved-were-not-buying-it/735-were-not-buying-it-the-facts-on-junk-food-marketing-and-kids.html [December 11, 2014]

Facts the Parents Television Council doesn't want you to know. (2007, April 19). *Television Watch*. Retrieved from: www.televisionwatch.org/NewsPolls/FactSheets/FS001.html [September 4, 2009]

FCC crackdown on indecency. (2006, March 16). *PBS News Hour with Jim Lehrer*. Retrieved \ from www.pbs.org/newshour/bb/media/jan-june06/fcc_3–16.html

Federal Communications Commission (2003). *Complaints Against Various Broadcast Licensees Regarding their Airing of the "Golden Globe Awards" Program 1*. File No. EB-03-IH-0110.

Federal Communications Commission (2006). *Frequently Asked Questions*. Retrieved from: www.fcc.gov/eb/oip/FAQ.html [September 14, 2006]

Federal Communications Commission (2012a). *Regulation of Obscenity, Indecency, and Profanity*. Retrieved from: http://transition.fcc.gov/eb/oip/Welcome.html

Federal Communications Commission (2012b). *V-Chip—Putting Restrictions on What Your Children Watch*. Retrieved from: www.fcc.gov/guides/v-chip-putting-restrictions-what-your-children-watch

Federal Communications Commission v. Pacifica Foundation. 438 U.S. 726 (1978).

Federman, J. (1998). National television violence study, Vol. 3, Executive summary (Center for Communication and Social Policy—University of California, Santa Barbara), Retrieved from: www.ccsp.ucsb.edu/execsum.pdf

Feshbach, S. (1961). The stimulating versus cathartic effects of a vicarious aggressive activity. *Journal of Abnormal Psychology, 63*, 381–385.

Feshbach, S., & Singer, R. (1971). *Television and aggression: An experimental field study*. San Francisco: Jossey-Bass.

Gartner says video game market to total $93 billion in 2013. (2013, Oct. 29). *Gartner*. Retrieved from: www.gartner.com/newsroom/id/2614915 [December 12, 2014]

Gerbner, G., & Gross, L. (1976). Living with television: The violence profile. *Journal of Communication, 26*, 173–199.

Gerbner, G., Gross, L., Jackson-Beeck, M., Jeffries-Fox, S., & Signorielli, N. (1978). Cultural indicators: Violence profile no. 9. *Journal of Communication, 27*, 171–180.

Gillmor, D., Barron, J., Simon, T., & Terry, H. (1996). *Fundamentals of mass communication law*. Minneapolis: West.

Good, O. S. (2014, June 1). Nielsen survey of gaming time shows consoles overtaking PC again. *Polygon*. Retrieved from: www.polygon.com/2014/6/1/5769368/nielsen-ratings-video-games-pc-console-mobile

Greenya, J. (2005, November). Can they say that on the air? The FCC and indecency. *DC Bar for Lawyers*. Retrieved from: www.dcbar.org/for_lawyers/

Gregoire, C. (2014, December 1). Immediately responding to email is destroying your health. *Huffington Post*. Retrieved from: www.huffingtonpost.com/2014/12/01/work-email-health_n_6226912.html [December 12, 2014]

Gun Violence is Hollywood's favourite type of violence. (2013, March 20). *Parents Television Council*. Retrieved from: http://w2.parentstv.org/Main/News/Detail.aspx?docID=2755 [December 10, 2014]

Haughney, C. (2013, April 30). Newspapers post gains in digital circulation. *The New York Times*. Retrieved from: www.nytimes.com/2013/05/01/business/media/digital-subscribers-buoy-newspaper-circulation.html?_r=0 [December 12, 2014]

Ho, D. (2006, April 15). Networks appeal TV indecency rulings. *The Atlanta Journal-Constitution*, pp. b1, b5.

Hopkinson, N. (2003, October 23). For media-savvy tots, TV and DVD compete with ABCs. *The Washington Post*. Retrieved from: www.washingtonpost.com

How teens use the media. (2009, June). *Nielsen*. Retrieved from: www.scribd.com/doc/16753035/Nielsen-Study-How-Teens-Use-Media-June-2009-Read-in-Full-Screen-Mode [February 6, 2010]

Hume, S. (2014, March 30). McDonald's spent more than $988 million on advertising in 2013. *The Christian Science Monitor*. Retrieved from: www.csmonitor.com/Business/The-Bite/2014/0330/McDonald-s-spent-more-than-988-million-on-advertising-in-2013 [December 12, 2014]

The impact of food advertising on childhood obesity. (n.d.). *American Psychological Association*. Retrieved from: www.apa.org/topics/kids-media/food.aspx [December 11, 2014]

Internet addiction statistics 2012. (2012, April 24). *Ansonalex.* Retrieved from: http://ansonalex.com/infographics/internet-addiction-statistics-2012-infographic/ [December 12, 2104]

Internet radio listeners average 12 hours per week. (2013). *Radio News.* Retrieved from: www.radiostreamingnews.com/2013/04/internet-radio-listeners-average-12.html

The issues. (2009). *Mc Spot Light.* Retrieved from: www.mcspotlight.org/issues/advertising/index.html [October 1, 2009]

Jamieson, P. E., & Romer, D. (2014). Violence in popular US prime time TV dramas and the cultivation of fear: A time series analysis. *Media and Communication, 2*(2), 31.

Jones, S. (2003). Let the games begin: Gaming technology and college students. *Pew Internet & American Life Project.* Retrieved from: www.pewinternet.org/Reports/2003/Let-the-games-begin-Gaming-technology-and-college-students.aspx

Kalb, C. (2002, August 19). How are we doing? *Newsweek,* p. 53.

Kaye, B. K., & Johnson, T. J. (2003). From here to obscurity: The Internet and media substitution theory. *Journal of the American Society for Information Science and Technology, 54*(3), 260–273.

Kaye, B. K., & Johnson, T. J. (2014). The shot heard around the World Wide Web: Who heard what where about Osama bin Laden's death. *Journal of Computer-Mediated Communication, 19*(3), 643–662.

Kaye, B. K., & Sapolsky, B. S. (2009). Taboo or not taboo? That is the question: Offensive language on prime time broadcast and cable programming. *Journal of Broadcasting & Electronic Media, 53*(1), 22–37.

Kellogg, C. (2013, July 2). Hours spent reading books around the world. *Los Angeles Times.* Retrieved from: http://articles.latimes.com/2013/jul/02/entertainment/la-et-jc-hours-reading-books-around-the-world-20130702

Kushlev, K., & Dunn, E. W. (2015, January 11). Stop checking email so often. *The New York Times,* p. 12

Labaton, S. (2006, April 18). U.S. networks turn to the courts in search of clearer rules. *International News Herald,* p. 16.

Lafayette, J. (2013). Nielsen: Time spent watching traditional TV up. *Broadcasting & Cable.* Retrieved from: www.broadcastingcable.com/news/technology/nielsen-time-spent-watching-traditional-tv/50055

Lavers, D. (2002, May 13). The verdict on media violence: It's ugly and getting uglier. *The Washington Post,* pp. 28–29.

Lee, B., & Lee, R. S. (1995). How and why people watch TV: Implications for the future of interactive television. *Journal of Advertising Research, 35*(6), 9–18.

Lenhart, A. (2008, November 2). Teens, video games and civics: What the research is telling us. *Pew Internet & American Life Project.* Retrieved from: www.pewinternet.org/Presentations/2008/Teens-Video-Games-and-Civics.aspx [September 28, 2009]

Lin, C. A. (1993). Adolescent viewing and gratifications in a new media environment. *Mass Comm Review, 20*(1&2), 39–50.

Liptak, A. (2012, June 22). Supreme Court Rejects F.C.C. fines for indecency. *The New York Times,* Retrieved from: www.nytimes.com/2012/06/22/business/media/justices-reject-indecency-fines-on-narrow-grounds.html

Livingston, G. (2002, July 28). Sense and nonsense in the child abduction scares. *San Francisco Chronicle,* p. 6, D.

Macgill, R. (2008, December 7). Video games: Adults are players too. *Pew Internet & American Life Project.* Retrieved from: pewresearch.org/pubs/1048/video-games-adults-are-players-too [September 28, 2009]

Makuch, E. (2016, February 16). US video game industry generated 23.5 billion in 2015, increasing 5%. *GameSpot.* Retrieved from: www.gamespot.com/articles/us-video-game-industry-generated-235-billion-in-20/1100-6434834/

Malamuth, N. M., & Donnerstein, E. (1984). *Pornography and sexual aggression.* Orlando, FL: Academic Press.

Marino-Nachison, D. (2014, December 6). Engineer became the father of video-game industry. *The New York Times,* p. B4.

Marshall, W. L. (1989). Pornography and sex offenders. In D. Zillmann & J. Bryant (Eds.), *Pornography: Research advances and policy considerations* (pp. 185–214). Hillsdale, NJ: Erlbaum.

Massey, K., & Baran, S. J. (2001). *Introduction to mass communication.* Mountain view, CA: Mayfield.

McConnell, B. (2003, January 27). FCC's Martin wants more family-friendly fare. *Broadcasting & Cable,* p. 26.

McDonald, D. G., & Kim, H. (2001). When I die, I feel small: Electronic game characters and the social self. *Journal of Broadcasting & Electronic Media, 45*(2), 241–258.

McGregor, J. (2014, September 14). A scatterbrain? The neuroscience of getting organized. *The New York Times,* p. G7.

McQuail, D. (2000). *Mass communication theory* (4th ed.). London: Sage Publications.

Media Use Statistics (n.d.). *Media Literacy Clearinghouse.* Retrieved from: www.frankwbaker.com/mediause.htm [December 12, 2014]

Media violence study (2013). *Parents Television Council.* Retrieved from: http://w2.parentstv.org/main/Research/Studies/CableViolence/cableviolence2013.aspx [December 10, 2014]

National TV Violence Study executive summary. (1998).

Nelson, S. S. (2014, December 1). German government may say 'nien' to after work emails. *National Public Radio.* Retrieved from: www.npr.org/blogs/parallels/2014/12/01/366806938/german-government-may-say-nein-to-work-emails-after-six [December 12, 2014]

New media study discovers Americans need 38 hours per day to complete their tasks. (n.d.). *Earthtimes.org.* Retrieved from: www.earthtimes.org/articles [September 28, 2009]

The New York Times v. Sullivan. (1964). 376 U.S. 254.

Nielsen: Kids watching TV at eight-year high. (2009, October 26**).** *The Hollywood Reporter.* Retrieved from: www.thrfeed.com/2009/10/nielsen-kids-watching-tv-at-8year-high.html [January 6, 2010]

Nomophobia, the fear of not having a mobile phone, hits record numbers. (2013, June 2). *news.com.au.* Retrieved from: www.news.com.au/technology/nomophobia-the-fear-of-not-having-a-mobile-phone-hits-record-numbers/story-e6frfro0−1226655033189 [December 12]

Online and digital news. (2012, September 27). *Pew Research Center for the People & the Press.* Retrieved from: www.people-press.org/2012/09/27/section-2-online-and-digital-news-2/ [December 12, 2014]

Oppelaar, J. (2000, September 18). Music business rates the ratings. *Variety, 380*(5), 126.

Palmgreen, P., Wenner, L. A., & Rosengren, K. E. (1985). Uses and gratifications research: The past ten years. In K. E. Rosengren, L. A. Wenner, & P. Palmgreen (Eds.), *Media Gratifications Research* (pp. 11–37). Beverly Hills, CA: Sage Publications.

Parents, children and media. (2007). *Kaiser Family Foundation.* Retrieved from: www.frankwbaker.com/mediause.htm [February 6, 2010]

Parents Television Council. (2011). *Habitat for Profanity.* Retrieved from: www.parentstv.org/PTC/news/release/2010/1109.asp

Pember, D. (2001). *Mass media law.* Boston: McGraw-Hill.

Pennebaker. R. (2009, August 30). The mediocre multitasker. *The New York Times,* p. A5.

Perse, E. (2001). *Media effects and society.* Mahwah, NJ: LEA.

Perse, E. M., & Ferguson, D. A. (1992). *Gratifications of newer television technologies.* Paper presented at the annual convention of the Speech Communication Association, Chicago, IL.

Policinski, G. (2009, September 17). Americans still turn to traditional news media first. *First Amendment Center.* Retrieved from: www.firstamendmentcenter.org/commentary.aspx [October 1, 2009]

Poovey, B. (2002, April 15). Device lets viewers bleep at home. *Knoxville News Sentinel,* p. C3.

Potter, W. J. (2003). *The eleven myths of media violence.* Thousand Oaks, CA: Sage.

Publisher reaches out to target automotive market (2004, March 10). *ZiffDavis.* Retrieved from: www.ziffdavis.com/press/releases/040310.0.html [September 28, 2009]

Quittner, J. (1999, May 10). Are video games really so bad? *Time,* pp. 50–59.

Radio facts and figures. (2013). *News Generation.* Retrieved www.newsgeneration.com/broadcast-resources/radio-facts-and-figures/

Richter, F. (2013, August 5). Digital media use set to exceed TV time this year. *Statista.* Retrieved from: www.statista.com/chart/1330/media-use-in-the-us/

Roberts, D. F., & Foehr, U. G. (2008). The future of children. *Children and Electronic Media, 18*(1), 11–37.

Rosengren, K. E., Wenner, L. E., & Palmgreen, P. (1985). *Media gratifications research.* Beverly Hills, CA: Sage Publications.

Rubin, A. M. (1981). An examination of television viewing motives. *Communication Research, 8*(2), 141–165.

Rubin, A. M. (1984). Ritualized versus instrumental viewing. *Journal of Communication, 34,* 67–77.

Rubin, A. M., & Bantz, C. R. (1989). Uses and gratifications of videocassette recorders. In J. L. Salvaggio & J. Bryant (Eds.), *Media use in the information age: Emerging patterns of adoption and consumer use* (pp. 181–195). Hillsdale, NJ: Lawrence Erlbaum Associates.

Ryan, J. (2002). Children now facing adult health issues. *Knoxville-News Sentinel,* p. B4.

Saeler, A. (2011, June). Unreality TV: The media and crime rates. *The Civic Column.* Retrieved from: www.civicinstitute.org/wordpress/wp-content/uploads/2011/07/Civic-Column-Summer-2011.pdf

Severin, W. J., & Tankard, J. W., Jr. (1992). *Communication theories: Origins, methods, and uses in the mass media* (3rd ed.). New York: Longman.

Shute, N. (2013, January 24). If you think you're good at multitasking, you probably aren't. *National Public Radio.* Retrieved from: www.npr.org/blogs/health/2013/01/24/170160105/if-you-think-youre-good-at-multitasking-you-probably-arent [December 10, 2014]

Signorielli, N. (1990). Television's mean and dangerous world: A continuation of the cultural indicators perspective. In N. Signorielli & M. Morgan (Eds.), *Cultivation analysis* (pp. 85–106). Beverly Hills, CA: Sage Publications.

Signorielli, N. (2005). Age-based ratings, content designations, and television content: Is there a problem? *Mass Communication & Society, 8*(4), 277–298.

Smith, L., Wright, J., & Ostroff, D. (1998). *Perspectives on radio and television: Telecommunication in the United States* (4th ed.). Mahwah, NJ: Erlbaum.

Soap operas boost rights, global economist says. (2009). *National Public Radio.* Retrieved from: www.npr.org/templates/story/story.php?storyId5113870313 [October 21, 2009]

Sterling, C., & Kittross, J. (2002). *Stay tuned: A history of American broadcasting.* Mahwah, NJ: Erlbaum.

Steyer, J. P. (2002). *The other parent.* New York: Atria.

Still glued to the tube. (2009). *Media Literacy Clearinghouse.* Retrieved from: www.frankwbaker.com/mediause.htm [October 1, 2009]

Stone, M. (2014, July 31). Smartphone addiction now has a clinical name. *Business Insider.* Retrieved from: www.businessinsider.com/what-is-nomophobia-2014-7

Takahashi, D. (2009, December 14). Gamer population surges—consoles in 60% of households. *Games Beat.* Retrieved from: www.games.venturebeat.com/2009/12/14/video-gamer-population-surges-as-60-percent-of-house holds-now-have-game-consoles

Television and obesity among children (2002). *National Institute on Media and the Family.* Retrieved from: www.mediaandthefamily.org/facts/facts_tvandobchild.shtml

Tsukayama, H. (2015, November 14). Young minds immersed in media. *The Washington Post,* p. A12.

Turkle, S. (2012, April 21). The flight from conversation. *The New York Times, Sunday Review,* p. 1, 8.

TV Watch (2007, June 25). *In the Polls.* Retrieved from: www.televisionwatch.org/NewsPolls/Polls.html [December 12, 2014]

Video games rewire Gen Y brains. (2007, September). *Trends Magazines.* Retrieved from: www.trends-magazine.com/trend.php/Trend/1461/Category/53 [September 28, 2009]

Video game statistics. (n.d.). *Education Database Online.* Retrieved from: www.onlineeducation.net/videogame [December 12, 2014]

Violent video games. (n.d.). *ProCon.* Retrieved from: http://videogames.procon.org/view.resource.php?resourceID=003627 [December 12, 2014]

Walker, J. R., & Bellamy, R. V. (1989). *The gratifications of grazing: Why flippers flip.* A paper presented at the Speech Communication Association conference, San Francisco, CA.

Walker, J. R., Bellamy, R. V., & Truadt, P. J. (1993). Gratifications derived from remote control devices: A survey of adult RCD use. In J. R. Walker & R. V. Bellamy (Eds.), *The remote control in the new age of television* (pp. 103–112). Westport, CT: Praeger Publishers.

Wallis, C. (2009, March 19). The multitasking generation. *Time.* Retrieved from: www.time.com/magazine/ [January 11, 2010]

Wilson, J., & Hudson, W. (2013). Gun violence in PG-13 movies has tripled. *CNN.* Retrieved from: www.cnn.com/2013/11/11/health/gun-violence-movies/index.html

Young, K. (2003). What is Internet addiction? *TechTV.* Retrieved from: www.techtv.com/callforhelp/features/story/0,24330,3322433,00.html

Zill, N. (2001). "Does *Sesame Street* enhance school readiness?" Evidence from a national survey of children. In S. M. Fisch & R. T. Truglio (Eds.), *"G" is for "growing": Thirty years of research on children and Sesame Street* (pp. 115–130). Mahwah, NJ: Erlbaum.

Index

Note: Italicized page numbers indicate a figure on the corresponding page. Page numbers in bold indicate a table on the corresponding page.